VALPARAISO HIGH SCHOOL

Deltoid

$d = \frac{m}{v}$

Conceptual PHYSICS

A High School Physics Program

Written and illustrated by

Paul G. Hewitt

▲▼Addison-Wesley Publishing Company, Inc.
Menlo Park, California · Reading, Massachusetts · Don Mills, Ontario
Wokingham, England · Amsterdam · Sydney · Singapore
Tokyo · Madrid · Bogotá · Santiago · San Juan

Pilot Teachers

Marshall Ellenstein
Ridgewood High School
Norridge, Illinois

Paul Robinson
Edison Computech High School
Fresno, California

Nathan A. Unterman
Glenbrook North High School
Northbrook, Illinois

Nancy T. Watson
Burris Laboratory School
Muncie, Indiana

Consultants

Clarence Bakken
Palo Alto High School
Palo Alto, California

Art Farmer
Gunn High School
Palo Alto, California

Sheron Snyder
Mason High School
Mason, Michigan

Charles A. Spiegel
California State University, Dominguez Hills
Dominguez Hills, California.

Developmental editor: Andrea G. Julian
Text and cover designer: Emily Kissin
Production designer: Rodelinde Albrecht
Photo editor: Barbara B. Hodder
Indexer: Suzanne Aboulfadl

Cover Photograph: © Ben Rose/The Image Bank

ISBN 0-201-20728-1

EFGHIJKL-VH-898

Acknowledgments

I wish to especially thank my two friends Marshall Ellenstein and Paul Robinson, not only for their helpful critiques and for piloting the first draft in their classes, but for urging me to write this book in the first place. Special thanks are also due Charlie Spiegel, whose many suggestions have been incorporated in nearly every chapter. I am thankful to the following teachers for critiquing sample chapters of the pilot manuscript: Hal Eastin, Cortez High School, Phoenix, AZ; Donald G. Gielow, Sequoia High School, Redwood City, CA; Robin Gregg, Menlo-Atherton High School, Atherton, CA; Willa D. Ramsey, Gompers Secondary School Center, San Diego, CA; and Nathan A. Unterman, Glenbrook North High School, Northbrook, IL. I thank my son James (Figure 14-24) for cutting the stencils for the artwork and my nephew Robert Baruffaldi (Figure 20-9) for his suggestions. Finally, I am most grateful to my assistant Helen Yan (Figures 20-8, 23-11, 27-20) for helping through every stage of the book, from planning and typing all manuscript to hand-lettering the final artwork.

San Francisco   Paul G. Hewitt

Contents

To the Student

YOU CAN'T ENJOY A GAME UNLESS YOU FIRST KNOW THE RULES. WHETHER IT'S A BALL GAME, COMPUTER GAME, OR SIMPLY A PARTY GAME --- IF YOU DON'T KNOW THE RULES, THE GAME IS POINTLESS AND OF LITTLE VALUE. YOU MISS OUT. THE SAME APPLIES TO NATURE AND ITS RULES --- WHAT PHYSICS IS ABOUT. NATURE MEANS MORE TO PEOPLE WHO UNDERSTAND ITS RULES THAN TO THOSE WHO DON'T. TO HELP YOU LEARN ABOUT AND COMPREHEND NATURE'S RULES, PHYSICS IS TREATED *CONCEPTUALLY* RATHER THAN MATHEMATICALLY IN THIS BOOK. PHYSICS CONCEPTS ARE IN ENGLISH, WITH EQUATIONS AS "GUIDES TO THINKING" RATHER THAN RECIPES FOR ALGEBRAIC PROBLEM SOLVING. THIS MEANS YOU'LL REALLY BE ABLE TO COMPREHEND THE PHYSICS YOU STUDY. THEN IF YOU TAKE A FOLLOW-UP COURSE IN PHYSICS, YOU'LL GET ALGEBRAIC PROBLEM SOLVING. SO COMPREHENSION NOW, AND IF COMPUTATION FOLLOWS LATER, IT WILL BE WITH UNDERSTANDING!

ENJOY YOUR PHYSICS!

PAUL G. HEWITT

1 About Science

Advances in science have brought many changes to the world. Fifty years ago, people were unfamiliar with television, jet airplanes, and how to prevent polio and simple tooth decay. Five hundred years ago, the earth was considered to be unmoving and the center of the universe. No one knew what makes the stars shine; yet today, we are preparing to travel to them—with the same energy that makes them shine.

Science is not something new. It goes back before recorded history, when people first discovered regularities and relationships in nature. One regularity was the appearance of the star patterns in the night sky. Another was the weather patterns during the year—when the rainy season started or the days grew longer. People learned to make predictions based on these regularities, and to make connections between things that at first seemed to have no relationship. More and more they learned about the workings of nature. That body of knowledge, growing all the time, is part of science. The greater part of science is the methods used to produce that body of knowledge. Science is an activity—a human activity—as well as a body of knowledge.

1.1 The Basic Science—Physics

Science is the present-day equivalent of what used to be called natural philosophy. Natural philosophy was the study of unanswered questions about nature. As the answers were found, they became part of what is now called science.

The study of science today branches into the study of living things and nonliving things: the life sciences and the physical sciences. The life sciences branch into areas such as biology, zoology, and botany. The physical sciences branch into areas such as geology, astronomy, chemistry, and physics.

Physics is more than a part of the physical sciences. It is the most basic of all the sciences. It's about the nature of basic things such as motion, forces, energy, matter, heat, sound, light, and the insides of atoms. Chemistry is about how matter is put together, how atoms combine to form molecules, and how the molecules combine to make up the many kinds of matter around us. Biology is more complex still and involves matter that is alive. So underneath biology is chemistry, and underneath chemistry is physics. The ideas of physics reach up to these more complicated sciences. That's why physics is the most basic science. You can understand science in general much better if you first have some understanding of physics.

1.2   Mathematics—The Language of Science

Science made its greatest headway in the sixteenth century, when it was found that nature can be analyzed and described mathematically. When the ideas of science are expressed in mathematical terms, they are unambiguous. They don't have the "double meanings" that so often confuse the discussion of ideas expressed in common language. When the findings in nature are expressed mathematically, they are easier to verify or disprove by experiment.* The methods of mathematics and experimentation led to the enormous success of science.

1.3   The Scientific Method

The Italian physicist Galileo Galilei (1564–1642) and the English philosopher Francis Bacon (1561–1626) are usually credited as being the principal founders of the **scientific method**—a method that is extremely effective in gaining, organizing, and applying new knowledge. This method is essentially as follows:

---

* Although mathematics is very important to scientific mastery, it will not be the focus of attention in this book. This book focuses instead upon what should come first: the basic ideas and concepts of physics—in English. When you learn physics primarily through word descriptions that help you to visualize ideas and concepts, with only secondary emphasis on mathematical descriptions, and when you postpone to a follow-up course the practice of algebraic problem solving (which often tends to obscure the physics), you gain a better comprehension of the conceptual foundation of physics.

Fig. 1-1 Galileo (left) and Francis Bacon (right) have been credited as the founders of the scientific method.

1. Recognize a problem.
2. Make an educated guess—a **hypothesis**—about the answer.
3. Predict the consequences of the hypothesis.
4. Perform experiments to test predictions.
5. Formulate the simplest general rule that organizes the three main ingredients: hypothesis, prediction, experimental outcome.

Although this cookbook method has a certain appeal, it has not always been the key to the discoveries and advances in science. In many cases, trial and error, experimentation without guessing, or just plain accidental discovery accounts for much of the progress in science. The success of science has more to do with an attitude common to scientists than with a particular method. This attitude is one of inquiry, experimentation, and humility before the facts.

1.4 The Scientific Attitude

In science, a **fact** is generally a close agreement by competent observers of a series of observations of the same phenomena. A scientific hypothesis, on the other hand, is an educated guess that is only presumed to be factual until proven so by experiment. When hypotheses have been tested over and over again and have not been contradicted, they may become known as **laws** or **principles**.

If a scientist believes a certain hypothesis, law, or principle is true, but finds contradicting evidence, then in the scientific spirit, the hypothesis, law, or principle must be changed or abandoned. In the scientific spirit, the idea must be changed or abandoned in spite of the reputation of the person advocating it.

As an example, the greatly respected Greek philosopher Aristotle (384–322 B.C.) claimed that falling objects fall at a speed proportional to their weight. This false idea was held to be true for more than 2000 years because of Aristotle's compelling authority. In the scientific spirit, however, a single verifiable experiment to the contrary outweighs any authority, regardless of reputation or the number of followers or advocates. In modern science, argument by appeal to authority is of no value whatever.

Scientists must accept their findings and other experimental evidence even when they would like them to be different. They must strive to distinguish between what they see and what they wish to see, for scientists, like most people, have a vast capacity for fooling themselves.* People have always tended to adopt general rules, beliefs, creeds, ideas, and hypotheses without thoroughly questioning their validity and to retain them long after they have been shown to be meaningless, false, or at least questionable. The most widespread assumptions are often the least questioned. Most often, when an idea is adopted, particular attention is given to cases that seem to support it, while cases that seem to refute it are distorted, belittled, or ignored.

Scientists use the word *theory* in a different way from its usage in everyday speech. In everyday speech a theory is no different from a hypothesis—a supposition that has not been verified. A scientific **theory**, on the other hand, is a synthesis of a large body of information that encompasses well-tested and verified hypotheses about certain aspects of the natural world. Physicists, for example, speak of the theory of the atom; biologists have the cell theory.

The theories of science are not fixed, but undergo change. Scientific theories evolve as they go through stages of redefinition and refinement. During the last hundred years, the theory of the atom has been refined, as new evidence was gathered. Similarly, biologists have refined the cell theory.

The refinement of theories is a strength of science, not a weakness. Many people feel that it is a sign of weakness to "change your mind." Yet competent scientists must be experts at changing their minds. They change their minds, however, only when confronted with solid experimental evidence to the contrary or when a conceptually simpler hypothesis forces them to a new point of view. More important than defending beliefs is improving them. Better hypotheses are made by those who are honest in the face of fact.

---

* In your education it is not enough to be aware that other people may try to fool you, but mainly to be aware of your own tendency to fool yourself.

1.5 | Scientific Hypotheses Must Be Testable

Before a hypothesis can be classified as scientific, it must conform to a cardinal rule. The rule is that the hypothesis must be testable. It is more important that there be a means of proving it *wrong* than that there be a means of proving it correct. At first thought this may seem strange, for we think of scientific hypotheses in terms of whether they are true or not. For most things we wonder about, we concern ourselves with ways of finding out whether they are true. Scientific hypotheses are different. In fact, if you want to distinguish whether a hypothesis is scientific or not, look to see if there is a test for proving it wrong. If there is no test for its possible wrongness, then it is not scientific.

Consider the hypothesis "Intelligent life exists on other planets somewhere in the universe." This hypothesis is not scientific. Reasonable or not, it is *speculation*. Although it can be proved correct by the verification of a single instance of intelligent life existing elsewhere in the universe, there is no way to prove the hypothesis wrong if no life is ever found. If we searched the far reaches of the universe for eons and found no life, we would not prove that it doesn't exist "around the next corner." A hypothesis that is capable of being proved right but not capable of being proved wrong is not a scientific hypothesis. Many such statements are quite reasonable and useful, but they lie outside the domain of science.

> ▶ **Question**
> Which of these is a scientific hypothesis?
> a. Atoms are the smallest particles of matter that exist.
> b. The universe is surrounded by a second universe, the existence of which cannot be detected by scientists.
> c. Albert Einstein is the greatest physicist of the twentieth century.

▶ **Answer**
Only *a* is scientific, because there is a test for its wrongness. The statement is not only *capable* of being proved wrong, but it in fact *has* been proved wrong. Statement *b* has no test for possible wrongness and is therefore unscientific. Some pseudoscientists and other pretenders of knowledge will not even consider a test for the possible wrongness of their statements. Statement *c* is an assertion, which has no test for possible wrongness. If Einstein was not the greatest physicist, how would we know? It is important to note that because the name Einstein is generally held in high esteem, it is a favorite of pseudoscientists. So we should not be surprised that the name of Einstein, like that of various religious figures, is cited often by charlatans who wish to bring respect to themselves and their points of view.

1.6   Science and Technology

Science and technology are different from each other. Science is a method of answering theoretical questions; technology is a method of solving practical problems. Science has to do with discovering facts and relationships between observable phenomena in nature, and with establishing theories that organize and make sense of these facts and relationships. Technology has to do with tools, techniques, and procedures for putting the findings of science to use.

Another difference between science and technology has to do with its effect on human lives. Science excludes the human factor. Scientists who seek to comprehend the workings of nature cannot be influenced by their own or other people's likes or dislikes, or to popular ideas about what is correct. What scientists discover may shock or anger people—as did Darwin's theory of evolution. If a scientific finding or theory is unpleasant, we have the option of ignoring it. Technology, on the other hand, can hardly be ignored once it is developed. We do not have the option of refusing to breathe polluted air; we do not have the option of refusing to hear the sonic boom of a supersonic jetliner overhead; we do not have the option of living in a nonnuclear age. Unlike science, advances in technology *must* be measured in terms of the human factor.

Fig. 1-2   Science complements technology.

We are all familiar with the abuses of technology. Many people blame technology itself for the widespread pollution and resource depletion and even social decay in general—so much so that the promise of technology is obscured. That promise is a cleaner and healthier world. It is much wiser to combat the dangers of technology with knowledge than ignorance. Wise applications of science and technology *can* lead to a better world.

1.7   In Perspective

More than 2000 years ago enormous human effort went into the construction of great pyramids in Egypt and in other parts of the world. It was only a few centuries ago that the most talented and most skilled artists, architects, and artisans of the world directed their genius and effort to the building of the great stone and marble structures—the cathedrals, synagogues, temples, and mosques. Some of these architectural structures took more than a century to build, which means that nobody witnessed both the beginning and the end of construction. Even the architects and early builders who lived to ripe old ages never saw the finished results of their labors. Entire lifetimes were spent in the shadows of construction that must have seemed without beginning or end. This enormous focus of human energy was inspired by a vision that went beyond world concerns—a vision of the cosmos. To the people of that time, the structures they erected were their "spaceships of faith," firmly anchored but pointing to the cosmos.

Today the efforts of many of our most skilled scientists, engineers, artists, and artisans are directed to building the spaceships that already orbit the earth, and others that will voyage beyond. The time required to build these spaceships is extremely brief compared to the time spent building the stone and marble structures of the past. Many people working on today's spaceships were alive before Charles Lindbergh made the first solo airplane flight across the Atlantic Ocean. Where will younger lives lead in a comparable time?

We seem to be at the dawn of a major change in human growth, not unlike the stage of a chicken embryo before it fully matures. When the chicken embryo exhausts the last of its inner-egg resources and before it pokes its way out of its shell, it may feel it is at its last moments. But what seems like its end is really only its beginning. Are we like the hatching chicks, ready to poke through to a whole new range of possibilities? Are our spacefaring efforts the early signs of a new human era?

The earth is our cradle and has served us well. But cradles, however comfortable, are one day outgrown. With inspiration that in many ways is similar to the inspiration of those who built the early cathedrals, synagogues, temples, and mosques, we aim for the cosmos.

We live at an exciting time!

Fig. 1-3 A NASA conception of a spaceship of the future. New discoveries await the people who will venture beyond our solar system.

1 Chapter Review

Concept Summary

Science is an activity as well as a body of knowledge.
- Physics is the most basic of all the sciences.
- The use of mathematics helps make ideas in science unambiguous.

The scientific method is a procedure for answering questions about the world by testing educated guesses, or hypotheses, and formulating general rules.
- Hypotheses in science must be testable; they are changed or abandoned if they are contradicted by experimental evidence.

A theory is a body of knowledge and well-tested hypotheses about some aspect of the natural world.
- Theories are modified as new evidence is gathered.

Science deals with theoretical questions, while technology deals with practical problems.

Important Terms

fact (1.4)
hypothesis (1.3)
law (1.4)
principle (1.4)
scientific method (1.3)
theory (1.4)

Review Questions

1. Why is physics the most basic of the sciences? (1.1)

2. Why is mathematics important to science? Why is the usage of mathematics minimized in this book? (1.2)

3. What is the scientific method? (1.3)

4. Is a scientific fact something that is absolute and unchanging? Explain. (1.4)

5. Distinguish between a hypothesis and a theory. (1.4)

6. Theories in science undergo change. Is this a strength or a weakness of science? Explain. (1.4)

7. What does it mean to say that if a hypothesis is scientific then there must be a means of proving it wrong? (1.5)

8. Distinguish between science and technology. (1.6)

Think and Explain

1. Why does science tend to be a "self-correcting" way of knowing about things?

2. What is likely the misunderstanding of someone who says, "But that's *only* a scientific theory"?

3. a. Make an argument for bringing to a halt the advances of technology.
   b. Make an argument that advances in technology should continue.
   c. Contrast your two arguments.

I Mechanics

THERE'S A LOT OF PHYSICS IN PLAYING TUG O' WAR. FOR EXAMPLE, ISN'T THE RIGHT END OF THE ROPE BEING PULLED AS HARD AS I PULL THIS LEFT END? ISN'T IT STILL BEING PULLED IF IT'S TIED TO A TREE INSTEAD OF IN MY FRIEND'S HANDS? DON'T I WIN IF I PUSH HARDER ON THE GROUND RATHER THAN PULL HARDER ON THE ROPE? AND DOESN'T IT TAKE *ENERGY* TO EXERT THIS *FORCE* AND IMPART *MOTION*? THESE SAMPLE IDEAS ARE FROM THE FOUNDATION OF PHYSICS--*MECHANICS*-- WHAT THE NEXT 15 CHAPTERS ARE ABOUT. READ, THINK, AND ENJOY!

2 Motion

Motion is all around us. We see it in the everyday activity of people, of cars on the highway, in trees that sway in the wind, and with patience, we see it in the nightime stars. There is motion at the microscopic level that we cannot see directly: jostling atoms make heat or even sound; flowing electrons make electricity; and vibrating electrons produce light. Motion is everywhere.

Motion is easy to recognize but harder to describe. Even the Greek scientists of 2000 years ago, who had a very good understanding of many of the ideas of physics we study today, had great difficulty in describing motion. They failed because they did not understand the idea of *rate*. A **rate** is a quantity divided by *time*. It tells how fast something happens, or how much something changes in a certain amount of time. In this chapter you will learn how motion is described by the rates known as *speed*, *velocity*, and *acceleration*. It would be very nice if this chapter helps you to master these concepts, but it will be enough for you to become familiar with them, and to be able to distinguish among them. The next few chapters will sharpen your understanding of these concepts.

2.1 Motion Is Relative

Everything moves. Even things that appear to be at rest move. They move with respect to, or **relative** to, the sun and stars. A book that is at rest, relative to the table it lies on, is moving at about 30 kilometers per second relative to the sun. And it moves even faster relative to the center of our galaxy. When we discuss the motion of something, we describe its motion relative to something else. When we say that a space shuttle moves at 8 kilometers per second, we mean relative to the earth below. When we say a racing car in the Indy 500 reaches a speed of 300 kilometers per hour, of course we mean relative to the track. Unless stated otherwise, when we discuss the speeds of things in our environment, we mean with respect to the surface of the earth. Motion is relative.

2.2   Speed

A moving object travels a certain distance in a given time. A car, for example, travels so many kilometers in an hour. **Speed** is a measure of how fast something is moving. It is the rate at which distance is covered. Remember, the word *rate* is a clue that something is being *divided by time*. Speed is always measured in terms of a unit of distance divided by a unit of time. Speed is defined as the distance covered per unit of time. The word *per* means "divided by."

Fig. 2-1   A cheetah is the fastest land animal over distances less than 500 meters and can achieve peak speeds of 100 km/h.

Any combination of distance and time units is legitimate for speed—miles per hour (mi/h); kilometers per hour (km/h); centimeters per day (the speed of a sick snail?); lightyears per century—whatever is useful and convenient. The slash symbol (/) is read as "per." Throughout this book we'll primarily use meters per second (m/s). Table 2-1 shows some comparative speeds in different units.

Table 2-1   Approximate Speeds in Different Units
20 km/h = 12 mi/h = 6 m/s
40 km/h = 25 mi/h = 11 m/s
60 km/h = 37 mi/h = 17 m/s
80 km/h = 50 mi/h = 22 m/s
100 km/h = 62 mi/h = 28 m/s
120 km/h = 75 mi/h = 33 m/s

Fig. 2-2 The speedometer for a North American car gives readings of instantaneous speed in both mi/h and km/h. Odometers for the US market give readings in mi; those for the Canadian market give readings in km.

Instantaneous speed

A car does not always move at the same speed. A car may travel down a street at 50 km/h, slow to 0 km/h at a red light, and speed up to only 30 km/h because of traffic. You can tell the speed of the car at any instant by looking at the car's speedometer. The speed at any instant is called the **instantaneous speed**. A car traveling at 50 km/h may go at that speed for only one minute. If it continued at that speed for a full hour, it would cover 50 km in that hour. If it continued at that speed for only half an hour, it would cover only half that distance: 25 km. If it continued for only one minute, it would cover less than 1 km.

Average speed

In planning a trip by car, the driver often wants to know how long it will take to cover a certain distance. The car will certainly not travel at the same speed all during the trip. All the driver cares about is the **average speed** for the trip as a whole. The average speed is defined as follows:

$$\text{average speed} = \frac{\text{total distance covered}}{\text{time interval}}$$

Average speed can be calculated rather easily. For example, if we drive a distance of 60 kilometers in a time of 1 hour, we say our average speed is 60 kilometers per hour (60 km/h). Or if we travel 240 kilometers in 4 hours, we find:

$$\text{average speed} = \frac{\text{total distance covered}}{\text{time interval}} = \frac{240\,\text{km}}{4\,\text{h}} = 60\,\text{km/h}$$

Note that when a distance in kilometers (km) is divided by a time in hours (h), the answer is in kilometers per hour (km/h).

Since average speed is the whole distance covered divided by the total time of travel, it does not indicate the different speeds and variations that may have taken place during shorter time intervals. In practice, we experience a variety of speeds on most trips, so the average speed is often quite different from the speed at any instant, the instantaneous speed. Whether we talk about average speed or instantaneous speed, we are talking about the rates at which distance is traveled.

> ► **Questions**
> 1. a. With the speedometer on the dashboard of every car is an odometer, which records the distance traveled. If the initial reading is set at zero at the beginning of a trip and the reading is 35 km one-half hour later, what has been your average speed?
>   b. Would it be possible to attain this average speed and never exceed a reading of 70 km/h on the speedometer?
>
> 2. If a cheetah can maintain a constant speed of 25 m/s, it will cover 25 meters every second. At this rate, how far will it travel in 10 seconds? In 1 minute?

2.3 Velocity

In everyday language, we can use the words *speed* and *velocity* interchangeably. In physics, we make a distinction between the two. Very simply, the difference is that **velocity** is speed in a given direction. When we say a car travels at 60 km/h, we are specifying its speed. But if we say a car moves at 60 km/h to the north, we are specifying its velocity. Speed is a description of how fast;

► **Answers**

(Are you reading this before you have formulated a reasoned answer in your mind? If so, do you also exercise your body by watching others do push-ups? Exercise your thinking! When you encounter the many questions as above throughout this book, *think* before you read the footnoted answers. You'll not only learn more, you'll enjoy learning more.)

1. a.
$$\text{Average speed} = \frac{\text{total distance covered}}{\text{time interval}} = \frac{35 \text{ km}}{0.5 \text{ h}} = 70 \text{ km/h}$$

   b. No, not if the trip started from rest and ended at rest, for any intervals with an instantaneous speed less than 70 km/h would have to be compensated with instantaneous speeds greater than 70 km/h to yield an average of 70 km/h. In practice, average speeds are usually appreciably less than peak instantaneous speeds.

2. In 10 s the cheetah will cover 250 m, and in 1 minute (or 60 s) it will cover 1500 m, more than fifteen football fields! If we know the average speed and the time of travel, then the distance covered is

$$\text{distance} = \text{average speed} \times \text{time interval}$$
$$\text{distance} = (25 \text{ m/s}) \times (10 \text{ s}) = 250 \text{ m}$$
$$\text{distance} = (25 \text{ m/s}) \times (60 \text{ s}) = 1500 \text{ m}$$

A little thought will show that this relationship is simply a rearrangement of

$$\text{average speed} = \frac{\text{distance}}{\text{time interval}}$$

velocity is how fast and in what direction.* We will see in the next section that there are good reasons for the distinction between speed and velocity.

> ▶ **Question**
> The speedometer of a car moving northward reads 60 km/h. It passes another car that travels southward at 60 km/h. Do both cars have the same speed? Do they have the same velocity?

Fig. 2-3 The car on the circular track may have a constant speed, but not a constant velocity because its direction of motion is changing every instant.

Constant Velocity

From the definition of velocity it follows that to have a constant velocity requires both constant speed *and* constant direction. Constant speed means that the motion remains at the same speed—the object does not move faster or more slowly. Constant direction means that the motion is in a straight line—the object's path does not curve at all. Motion at constant velocity is motion in a straight line at constant speed.

Changing Velocity

If *either* the speed *or* the direction (or both) is changing, then the velocity is considered to be changing. Constant speed and constant velocity are not the same. A body may move at constant speed along a curved path, for example, but it does not move with constant velocity because its direction is changing every instant.

In a car, there are three controls that are used to change the velocity. One is the gas pedal; it is used to increase the speed. The second is the brake; it is used to decrease the speed. The third is the steering wheel; it is used to change the direction.

2.4   Acceleration

We can change the state of motion of an object by changing its speed, by changing its direction of motion, or by changing both. Any of these changes is a change in velocity. Sometimes we are

---

▶ **Answer**
Both cars have the same speed, but they have opposite velocities because they are moving in opposite directions.

* Directional quantities are called *vectors*; velocity is a vector. Nondirectional quantities are called *scalars*; speed is a scalar. Vector and scalar quantities will be covered in Chapter 6.

interested in how fast the velocity is changing. A driver on a two-lane road who wants to pass another car would like to be able to speed up and pass in the shortest possible time. The rate at which the velocity is changing is called the *acceleration*. Because acceleration is a rate, it is a measure of how fast the velocity is changing per unit of time:

$$\text{acceleration} = \frac{\text{change of velocity}}{\text{time interval}}$$

We are all familiar with acceleration in an automobile. The driver depresses the gas pedal, appropriately called the accelerator. The passengers then experience acceleration, or "pickup" as it is sometimes called, as they tend to lurch toward the rear of the car. The key idea that defines acceleration is *change*. Whenever we change our state of motion, we are accelerating. A car that can accelerate well has the ability to change its velocity rapidly. A car that can go from zero to 60 km/h in 5 seconds has a greater acceleration than another that can go from zero to 80 km/h in 10 seconds. So having a good acceleration is being "quick to change" and not necessarily fast.

In physics, the term *acceleration* applies to decreases as well as increases in speed. The brakes of a car can produce large retarding accelerations; that is, they can produce a large decrease per second in the speed. This is often called *deceleration*, or *negative acceleration*. We experience deceleration when the driver of a bus or car slams on the brakes and we tend to lurch forward.

Fig. 2-4   A car is accelerating whenever there is a *change* in its state of motion.

Acceleration applies to changes in *direction* as well as changes in speed. If you ride around a curve at a constant speed of 50 km/h, you feel the effects of acceleration as you tend to lurch toward the outside of the curve. You may round the curve at constant speed, but your velocity is not constant because your direction is changing every instant. Your state of motion is changing; you are accelerating. Now you can see why it is important to distinguish between speed and velocity, and why acceleration is defined as the rate of change of *velocity*, rather than *speed*. Acceleration, like velocity, is directional. If we change either speed or direction, or both, we change velocity and we accelerate.

In most of this book we will be concerned only with motion along a straight line. When straight-line motion is being considered, it is common to use speed and velocity interchangeably. When the direction is not changing, acceleration may be expressed as the rate at which *speed* changes.

$$\text{acceleration (along a straight line)} = \frac{\text{change in speed}}{\text{time interval}}$$

Speed and velocity are measured in units of distance per time. The units of acceleration are a bit more complicated. Since acceleration is the change in velocity or speed per time interval, its units are those of speed per time. If we speed up without change in direction from zero to 10 km/h in 1 second, our change in speed is 10 km/h in a time interval of 1 s. Our acceleration (along a straight line) is

$$\text{acceleration} = \frac{\text{change in speed}}{\text{time interval}} = \frac{10 \text{ km/h}}{1 \text{ s}} = 10 \text{ km/h·s}$$

The acceleration is 10 km/h·s (read as 10 kilometers per hour second). Note that a unit for time enters twice: once for the unit of speed and again for the interval of time in which the speed is changing. If you understand this, you can answer the following questions. If you don't, maybe the answers to the questions will be of help.

▶ **Questions**

1. Suppose a car moving in a straight line steadily increases its speed each second, first from 35 to 40 km/h, then from 40 to 45 km/h, then from 45 to 50 km/h. What is its acceleration?

2. In 5 seconds a car moving in a straight line increases its speed from 50 km/h to 65 km/h while a truck goes from rest to 15 km/h in a straight line. Which undergoes the greater acceleration? What is the acceleration of each vehicle?

▶ **Answers**

1. We see that the speed increases by 5 km/h during each 1-s interval. The acceleration is therefore 5 km/h·s during each interval.

2. The car and truck both increase their speeds by 15 km/hr during the same time interval, so their accelerations are the same. If you realized this without first calculating the accelerations, you're thinking conceptually. The acceleration of each vehicle is:

$$\text{acceleration} = \frac{\text{change in speed}}{\text{time interval}} = \frac{15 \text{ km/h}}{5 \text{ s}} = 3 \text{ km/h·s}$$

Although the speeds involved are quite different, the rates of change of speed are the same. Hence the accelerations are equal.

2.5 | Free Fall: How Fast

Drop a stone and it falls. Does it accelerate while falling? We know it starts from a rest position, and gains speed as it falls. We know this because it would be safe to catch if it fell a meter or two, but not from the top of a tall building. Thus, the stone must gain more speed during the time it drops from a building than during the shorter time it takes to drop a meter. This gain in speed indicates that the stone does accelerate as it falls.

Gravity causes the stone to fall downward once it is dropped. In real life, air resistance affects the acceleration of a falling object. Let's imagine that there is no air resistance, and gravity is the only thing that affects a falling object. Such an object would then be in **free fall**. Table 2-2 shows the instantaneous speed at the end of each second of fall of a freely-falling object dropped from rest. The **elapsed time** is the time that has elapsed, or passed, since the beginning of the fall.

Fig. 2-5 If a falling rock were somehow equipped with a speedometer, in each succeeding second of fall its reading would increase by 10 m/s. Table 2-2 shows the speeds we would read at various seconds of fall.

Table 2-2 Free-Fall Speeds of Object Dropped from Rest	
Elapsed Time (seconds)	Instantaneous Speed (meters/second)
0	0
1	10
2	20
3	30
4	40
5	50
• 6	•
• 7	•
• 8	•
t	$10t$

Note in Table 2-2 the way the speed changes. During each second of fall the instantaneous speed of the object increases by an additional 10 meters per second. This speed gain per second is the acceleration:

$$\text{acceleration} = \frac{\text{change in speed}}{\text{time interval}} = \frac{10 \text{ m/s}}{1 \text{ s}} = 10 \text{ m/s}^2$$

Note that when the change in speed is in m/s and the time interval is in s, the acceleration is in m/s² (read "meters per second squared"). The unit of time, the second, enters twice—once for the unit of speed, and again for the time interval during which the speed changes.

The acceleration of an object falling under conditions where air resistance is negligible is about 10 meters per second squared (10 m/s²). For free fall, it is customary to use the letter g to represent the acceleration (because in free fall, the acceleration is due to gravity). Although g varies slightly in different parts of the world, its average value is nearly 10 m/s². More accurately, it is 9.8 m/s², but it is easier to see the ideas involved when it is rounded off to 10 m/s². Where accuracy is important, the value of 9.8 m/s² should be used for the acceleration during free fall.

Note in Table 2-2 that the instantaneous speed of an object falling from rest is equal to the acceleration multiplied by the amount of time it falls.

$$\text{instantaneous speed} = \text{acceleration} \times \text{elapsed time}$$

The instantaneous speed v of an object falling from rest after an elapsed time t can be expressed in shorthand notation:*

$$v = gt$$

The letter v symbolizes both speed and velocity. Take a moment to check this equation with Table 2-2. You will see that whenever the acceleration $g = 10$ m/s² is multiplied by the elapsed time in seconds, you have the instantaneous speed in meters per second.

> ▶ **Question**
> What would the speedometer reading on the falling rock shown in Figure 2-5 be 4.5 seconds after it drops from rest? How about 8 s after it is dropped? 100 s?

So far, we have been looking at objects moving straight downward under gravity. Now, when an object is thrown upward, it continues to move upward for a while. Then it comes back down. At the highest point, when it is changing its direction of motion from upward to downward, its instantaneous speed is zero. Then it starts downward just as if it had been dropped from rest at that height.

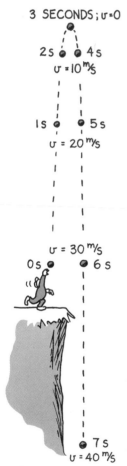

3 SECONDS; $v=0$

2s   4s
$v=10\,^m/s$

1s   5s
$v=20\,^m/s$

$v=30\,^m/s$
0s   6s

$v=40\,^m/s$
7s

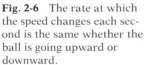

Fig. 2-6 The rate at which the speed changes each second is the same whether the ball is going upward or downward.

▶ **Answer**
The speedometer readings would be 45 m/s, 80 m/s, and 1000 m/s, respectively. You can reason this from Table 2-2 or use the equation $v = gt$, where g is replaced by 10 m/s².

* This relationship follows from the definition of acceleration when the acceleration is g and the initial speed is zero. If the object is initially moving downward at speed v_o, the speed v after any elapsed time t is $v = v_o + gt$. This book will not be concerned with such added complications. You can learn a lot from even the most simple cases!

What about the upward part of the path? During the upward motion, the object slows from its initial upward velocity to zero velocity. We know it is accelerating because its velocity is changing—its speed is decreasing. The acceleration during the upward motion is the same as during the downward motion: g = 10 m/s². The instantaneous *speed* at each point in the path is the same whether the object is moving upward or downward (see Figure 2-6). The *velocities* are different, of course, because they are in different directions. During each second, the speed or the velocity changes by 10 m/s. The acceleration is 10 m/s² the whole time, whether the object is moving upward or downward.

2.6 | Free Fall: How Far

How *fast* a falling object moves is entirely different from how *far* it moves. To understand this, return to Table 2-2. At the end of the first second, the object is moving downward with an instantaneous speed of 10 m/s. Does this mean it falls a distance of 10 m during the first second? No. If it fell 10 m the first second, it would have to have had an *average* speed of 10 m/s. But we know the speed started at zero and increased to 10 m/s only at the end of a full second. How do we find average speed for an object moving in a straight line with constant acceleration, as it is doing here? We find it the same way we find the average of any two numbers: add them and divide by 2. So if we add the initial speed, zero in this case, and the final speed of 10 m/s, and then divide by 2, we get 5 m/s. During the first second, the object has an average speed of 5 m/s. It falls a distance of 5 m. To check your understanding of this, carefully consider the following check question before going further.

▶ **Question**

During the span of the second time interval in Table 2-2, the object begins at 10 m/s and ends at 20 m/s. What is the *average speed* of the object during this 1-second interval? What is its *acceleration*?

▶ **Answer**

The average speed will be

$$\frac{\text{beginning speed} + \text{final speed}}{2} = \frac{(10 \text{ m/s}) + (20 \text{ m/s})}{2} = \frac{30 \text{ m/s}}{2} = 15 \text{ m/s}$$

The acceleration will be

$$\frac{\text{change in speed}}{\text{time interval}} = \frac{(20 \text{ m/s}) - (10 \text{ m/s})}{1 \text{ s}} = \frac{10 \text{ m/s}}{1 \text{ s}} = 10 \text{ m/s}^2.$$

Table 2-3 shows the total distance moved by a freely falling object dropped from rest. At the end of one second it has fallen 5 m. At the end of 2 s it has dropped a total distance of 20 m. At the end of 3 s it has dropped 45 m altogether. These distances form a mathematical pattern: at the end of time t, the object has fallen a distance d of $\frac{1}{2}gt^2$.* Try using $g = 10$ m/s² to calculate the distance fallen for some of the times shown in Table 2-3.

Fig. 2-7 Pretend that a falling rock is somehow equipped with an *odometer*. The readings of distance fallen increase with time and are shown in Table 2-3.

Table 2-3 Free-Fall Distances of Object Dropped from Rest	
Elapsed Time (seconds)	Distance Fallen (meters)
0	0
1	5
2	20
3	45
4	80
5	125
•	• 180
•	• 245
•	• 320
t9 405	$\frac{1}{2}gt^2$
10 500	

▶ **Question**

An apple drops from a tree and hits the ground in one second. What is its speed upon striking the ground? What is its average speed during the one second? How high above ground was it when it first dropped?

▶ **Answer**

Using 10 m/s² for g we find

speed $v = gt = (10$ m/s²$) \times (1$ s$) = 10$ m/s

average speed $\bar{v} = \dfrac{\text{beginning } v + \text{final } v}{2} = \dfrac{(0 \text{ m/s}) + (10 \text{ m/s})}{2} = 5$ m/s

(The bar over the symbol v denotes *average* speed $\bar{v}$.)

distance d = average speed × time interval = (5 m/s) × (1 s) = 5 m

or equivalently,

distance $d = \frac{1}{2}gt^2 = (\frac{1}{2}) \times (10$ m/s²$) \times (1$ s$)^2 = 5$ m

Notice that the distance can be found by either of these equivalent relationships.

* distance = average speed × time interval

$= \dfrac{\text{beginning speed} + \text{final speed}}{2} \times \text{time}$

$= \dfrac{0 + gt}{2} \times t$

$= \frac{1}{2}gt^2$

2.7   Air Resistance and Falling Objects

Drop a feather and a coin and you'll notice that the coin reaches the floor way ahead of the feather. Air resistance is responsible for these different accelerations. This fact can be shown quite nicely with a closed glass tube connected to a vacuum pump. The feather and coin are placed inside. When air is inside the tube and it is inverted, the coin falls much more rapidly than the feather. The feather flutters through the air. But if the air is removed with a vacuum pump and the tube is quickly inverted, the feather and coin fall side by side at acceleration g (Figure 2-8). Air resistance noticeably alters the motion of falling pieces of paper or feathers. But air resistance less noticeably affects the motion of more compact objects like stones and baseballs. Most objects falling in air can be considered to be falling freely. (Air resistance will be covered in more detail in Chapter 4.)

Fig. 2-8   A feather and a coin accelerate equally when there is no air around them.

2.8   How Fast, How Far, How Quick How Fast Changes

Much of the confusion that arrives in analyzing the motion of falling objects comes about from mixing up "how fast" with "how far." When we wish to specify how fast something freely falls from rest after a certain elapsed time, we are talking about speed or velocity. The appropriate equation is $v = gt$. When we wish to specify how far that object has fallen, we are talking about distance. The appropriate equation is $d = \frac{1}{2}gt^2$. Velocity or speed (how fast) and distance (how far) are entirely different from each other.

The most confusing concept, and one of the most difficult encountered in this book, is "how quickly does speed or velocity change": acceleration. What makes acceleration so complex is that it is *a rate of a rate*. It is often confused with velocity, which is itself a rate (the rate at which distance is covered). Acceleration is not velocity, nor is it even a change in velocity; acceleration is the rate at which velocity itself changes.

Please be patient with yourself if you find that you require a few hours to achieve a clear understanding of motion. It took people nearly 2000 years, from the time of Aristotle to Galileo, to achieve as much!

2 | Chapter Review

Concept Summary

Motion is described relative to something.

Speed is a measure of how fast something is moving.
- Speed is the rate at which distance is covered; it is measured in units of distance divided by time.
- Instantaneous speed is the speed at any instant.
- Average speed is the total distance covered divided by the time interval.

Velocity is speed together with the direction of travel.
- The velocity is constant only when the speed and the direction are both constant.

Acceleration is the rate at which the velocity is changing.
- In physics, an object is considered to be accelerating when its speed is increasing, when its speed is decreasing, and/or when the direction is changing.
- Acceleration is measured in units of speed divided by time.

An object in free fall is falling under the influence of gravity alone, where air resistance does not affect its motion.
- An object in free fall has a constant acceleration of about 10 m/s^2.

Important Terms

acceleration (2.4)
average speed (2.2)
elapsed time (2.5)
free fall (2.5)
instantaneous speed (2.2)
rate (2.1)
relative (2.1)
speed (2.2)
velocity (2.3)

Note: Each chapter in this book concludes with a set of Review Questions and Think and Explain exercises. The Review Questions are designed to help you fix ideas and catch the essentials of the chapter material. You'll notice that answers to the questions can be found within the chapter. Think and Explain exercises stress thinking rather than mere recall of information and call for an *understanding* of the definitions, principles, and relationships of the chapter material. In many cases the intention of particular Think and Explain exercises is to help you apply the ideas of physics to familiar situations.

The Activities, which are at the end of most chapters, are pre-lab activities that can be done in or out of class, or simple home projects to do on your own. Their purpose is to encourage you to discover the joy of *doing* physics.

Review Questions

1. What is meant by saying that motion is relative? For everyday motion, what is motion usually relative to? (2.1)

2. Speed is the rate at which what happens? (2.2)

3. You walk across the room at 1 kilometer per second. Express this speed in abbreviated units, or symbols. (2.2)

4. What is the difference between instantaneous speed and average speed? (2.2)

5. Does the speedometer of a car read instantaneous speed or average speed? (2.2)

6. What is the difference between speed and velocity? (2.3)

7. If the speedometer in a car reads a constant speed of 40 km/h, can you say that the car

has a constant velocity? Why or why not? (2.3)

8. What two controls on a car enable a change in *speed*? Name another control that enables a change in *velocity*. (2.3)

9. What quantity describes how quickly you change how fast you're traveling, or how quickly you change your direction? (2.4)

10. Acceleration is the rate at which what happens? (2.4)

11. What is the acceleration of a car that travels in a straight line at a constant speed of 100 km/h? (2.4)

12. What is the acceleration of a car that, while moving in a straight line, increases its speed from zero to 100 km/h in 10 seconds? (2.4)

13. By how much does the speed of a vehicle moving in a straight line change each second when it is accelerating at 2 km/h·s? At 4 km/h·s? At 10 km/h·s? (2.4)

14. Under what conditions can acceleration be defined as the rate at which the *speed* is changing? (2.4)

15. Why does the unit of time enter twice in the unit of acceleration? (2.4)

16. How much gain in speed each second does a freely falling object acquire? (2.5)

17. For a freely-falling object dropped from rest, what is its instantaneous speed at the end of the fifth second of fall? The sixth second? (2.5)

18. For a freely-falling object dropped from rest, what is its *acceleration* at the end of the fifth second of fall? The sixth second? At the end of any elapsed time *t*? (2.5)

19. For an object thrown straight upward, what is its *instantaneous speed* at the top of its path? Its *acceleration*? (Why are your answers different?) (2.5)

20. How far will a freely-falling object fall from a position of rest in five seconds? Six seconds? (2.6)

21. How far will an object move in one second if its average speed during that second is 5 m/s? (2.6)

22. How far will a freely-falling object have fallen from a position of rest when its instantaneous speed is 10 m/s? (2.6)

23. Does air resistance increase or decrease the acceleration of a falling object? (2.7)

24. What is the appropriate equation for how *fast* something freely falls from a position of rest? For how *far* that object falls? (2.8)

25. Why is it said that acceleration is a *rate of a rate*? (2.8)

Activities

1. By any method you choose, determine your average speed of walking.

2. Try this with your friends. Hold a dollar bill so that the midpoint hangs between a friend's fingers (Figure A). Challenge your friend to catch it by snapping his or her fingers shut when you release it. The bill won't be caught! Explanation: It takes at least 1/7 second for the necessary impulses to travel from the eye to the brain to the fingers. But in only 1/8 second, the bill falls 8 cm (from $d = \frac{1}{2}gt^2$), which is half the length of the bill.

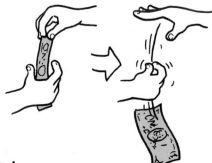

Fig. A

3. You can compare your reaction time with that of a friend by catching a ruler that is dropped between your fingers. Let your friend hold the ruler as shown in Figure B. Snap your fingers shut as soon as you see the ruler released. The number of centimeters that pass through your fingers depends on your reaction time. You can find your reaction time in seconds by solving $d = \frac{1}{2}gt^2$ for time: $t = \sqrt{2d/g} = 0.045 \sqrt{d}$, where d is in centimeters.

Fig. B

Think and Explain

1. Why is it that an object can accelerate while traveling at constant speed, but not at constant velocity?

2. Light travels in a straight line at a constant speed of 300 000 km/s. What is its acceleration?

3. Which has the greater acceleration when moving in a straight line—a car that increases its speed from 50 to 60 km/h, or a bicycle that goes from zero to 10 km/h in the same time? Defend your answer.

4. a. If a freely-falling rock were equipped with a speedometer, by how much would its speed readings increase with each second of fall?
   b. Suppose the freely-falling rock were dropped near the surface of a planet where $g = 20$ m/s². By how much would its speed readings change each second?

5. If a freely-falling rock were equipped with an odometer, would the readings for distance fallen each second be the same, increase with time, or decrease with time?

6. a. When a ball is thrown straight upward in the absence of air resistance, by how much does the speed decrease each second?
   b. After it reaches the top and begins its return downward, by how much does its speed increase each second?
   c. How much time is required in going up compared to coming down?

7. Table 2-2 shows that the instantaneous speed of an object dropped from rest is 10 m/s after 1 s of fall. Table 2-3 shows that the object has fallen only 5 m during this time. Your friend says this is incorrect, because distance traveled equals speed times time, so the object should fall 10 m. What do you say?

8. a. What is the instantaneous speed of a freely-falling object 10 s after it is released from a rest position?
   b. What is its average speed during this time?
   c. How far will it travel in this time?

3 Newton's First Law of Motion— Inertia

If you saw a boulder in the middle of a flat field suddenly begin moving across the ground, you'd look for the reason for its motion. You might look to see if somebody was pulling it with a rope, or pushing it with a stick or something. You'd reason that something was the cause of motion. Nowadays we don't believe that such things happen without cause. In general we would say the cause of the boulder's motion was a force of some kind. We know that something forces the boulder to move.

Fig. 3-1 Do boulders move without cause?

3.1 Aristotle on Motion

The idea that a force causes motion goes back to the fourth century B.C., when the Greeks were developing ideas of science. The foremost Greek scientist was Aristotle, who studied motion and divided it into two kinds: *natural motion* and *violent motion*.

Natural motion on earth was thought to be either straight up or straight down, such as the falling of a boulder toward the ground or the rising of a puff of smoke in air. Objects would seek their natural resting places: boulders on the ground, and smoke high in the air like the clouds. It was natural for heavy things to fall and very light things to rise. Aristotle proclaimed that, for the heavens, circular motion was natural, as it was without beginning or end. So the planets and stars moved in perfect circles about the earth. Since their motions were natural, they were not caused by forces.

Violent motion, on the other hand, was imposed motion. It was the result of forces which pushed or pulled. A cart moved because it was pulled by a horse; a tug-of-war was won by pulling on a rope; a ship was pushed by the force of the wind. The important thing about violent motion was that it had an external cause; violent motion was imparted to objects. Objects in their resting places could not move by themselves, but were pushed or pulled.

It was commonly thought for nearly 2000 years that if an object was moving "against its nature," then a force of some kind was responsible. Such motion was possible only because of an outside force; if there were no force, there would be no motion. So the proper state of objects was one of rest, if they were not pushed or pulled. Since it was evident to most thinkers up to the sixteenth century that the earth must be in its proper place, and that a force large enough to move it was unthinkable, it seemed clear that the earth did not move.

3.2 | Copernicus and the Moving Earth

It was in this climate that the astronomer Nicolaus Copernicus (1473–1543) formulated his theory of the moving earth. Copernicus reasoned from his astronomical observations that the earth traveled around the sun. This idea was extremely controversial in his time, and he worked on his ideas secretly to escape persecution. In the last days of his life, at the urging of close friends, he sent his ideas to the printer. The first copy of his work, *De Revolutionibus*, reached him on the day he died, May 24, 1543.

3.3 | Galileo on Motion

It was Galileo, the foremost scientist of the sixteenth century, who was the first to show that Copernicus's idea of a moving earth was reasonable. Galileo did this by demolishing the notion that a force was necessary to keep an object moving.

A **force** is any push or pull. **Friction** is the name given to the force that acts between materials that are moving past each other. Friction arises from the irregularities in the surfaces of sliding objects. Even very smooth surfaces have microscopic irregularities that act as obstructions to motion. If friction were absent, a moving object would need no force whatever for its continued motion.

Galileo showed that only when friction is present, as it usually is, is a force necessary to keep an object moving. He tested his idea with inclined planes—flat surfaces that are raised at one end. He noted that balls rolling down inclined planes pick up speed (Figure 3-2 left). They rolled to some degree in the direction of the earth's gravity. Balls rolling up inclined planes slowed down (Figure 3-2 center). They rolled in a direction that opposed gravity. What about balls rolling on a level surface, where they would neither roll with nor against gravity (Figure 3-2 right)? He found that for smooth horizontal planes, balls rolled without changing speed. He stated that if friction were entirely absent, a horizontally-moving ball would move forever. No push or pull would be required to keep it moving, once it was set in motion.

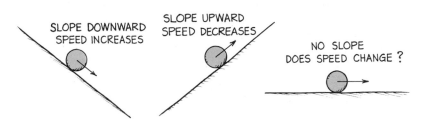

Fig. 3-2 (Left) When the ball rolls down, it moves with the earth's gravity and its speed increases. (Center) When it rolls up, it moves against gravity and loses speed. (Right) When it rolls on a level plane, it moves neither with nor against gravity. Does its speed change?

Galileo's conclusion was supported by another line of reasoning. He placed two of his inclined planes facing each other, as in Figure 3-3. He found that a ball rolling down one plane would roll up the other to nearly the same height. The smoother the planes, the more nearly equal were the initial and final heights. He found that the ball tended to attain the same height even when the second plane was longer and inclined at a smaller angle. In rolling to the same height, the ball had to roll farther. Additional reductions of angle for the upward plane gave the same results. Always the ball went farther as it tended to reach the same height.

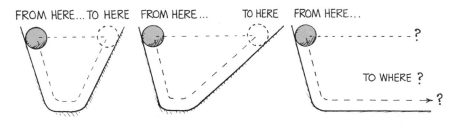

Fig. 3-3 (Left) A ball that rolls down an incline will roll up to its initial height. (Center) As the angle of the upward incline is reduced, the ball must roll a greater distance to reach its initial height. (Right) How far will it roll along the horizontal?

What if the angle of incline of the second plane was reduced to zero, so that the plane was perfectly horizontal? How far would the ball roll? He realized that only friction would keep it from rolling forever. It was not the nature of the ball to come to rest, as Aristotle had claimed. In the absence of friction, the moving ball would naturally keep moving. Galileo said that every material object has a resistance to change in its state of motion. He called this resistance **inertia**.

Galileo's concept of inertia discredited the Aristotelian theory of motion. It would be seen that although a force (gravity) is necessary to hold the earth in orbit around the sun, no force was required to keep the earth in motion. There is no friction in the empty space of the solar system, and the earth therefore coasts around and around the sun without loss in speed. The way was open for Isaac Newton (1642–1727) to synthesize a new vision of the universe.

> ▶ **Question**
>
> A ball is rolled across the top of a pool table and slowly rolls to a stop. How would Aristotle interpret this behavior? How would Galileo interpret it? How would you interpret it?

3.4   Newton's Law of Inertia

Within a year of Galileo's death, Isaac Newton was born. In 1665, at the age of 23, Newton developed his famous laws of motion. They replaced the Aristotelian ideas that had dominated the thinking of the best minds for nearly 2000 years. This chapter covers the first of Newton's three laws of motion. The other two are covered in the next two chapters.

Newton's first law, usually called the **law of inertia**, is a restatement of Galileo's idea.

> Every body continues in its state of rest, or of motion in a straight line at constant speed, unless it is compelled to change that state by forces exerted upon it.

Fig. 3-4 Objects at rest tend to remain at rest.

▶ **Answer**

Aristotle would likely say that the ball comes to a stop because it seeks its proper state, one of rest. Galileo would likely say that once in motion the ball would continue in motion; what prevents continued motion is not its nature or its proper rest state, but the friction between the table and the ball. Only you can answer the last question!

Simply put, things tend to keep on doing what they're already doing. Dishes on a table top, for example, are in a state of rest. They tend to remain at rest, as is evidenced if you snap a table cloth from beneath them. (Try this at first with some unbreakable dishes! If you do it properly, you'll find the brief and small force of friction between the dishes and the fast-moving table-cloth is not significant enough to appreciably move the dishes.) If an object is in a state of rest, it tends to remain at rest. Only a force will change that state.

Now consider an object in motion. If you slide a hockey puck along the surface of a city street, the puck quite soon comes to rest. If you slide it along ice, it slides for a longer distance. This is because the friction force is very small. If you slide it along an air table where friction is practically absent, it slides with no ap-

Fig. 3-5   An air table. Blasts of air from many tiny holes provide a friction-free surface.

parent loss in speed. We see that in the absence of forces, a moving object tends to move in a straight line indefinitely. Toss an object from a space station located in the vacuum of outer space, and the object will move forever. It will move by virtue of its own inertia.

So we see the law of inertia provides a completely different way of viewing motion. Whereas the ancients thought forces were responsible for motion, we now know that objects will continue to move by themselves. Forces are needed to overcome any friction that may be present and to set objects in motion initially. Once an object is moving in a force-free environment, it will move in a straight line indefinitely. The next chapter will show that forces are needed to accelerate objects but not to maintain motion if there is no friction.

Fig. 3-6 The spacecraft launched in the late 1970s on the Pioneer and Voyager missions have gone past the orbit of Saturn (shown) and are still in motion. Except for the gravitational effects of stars and planets in the universe, their motion will continue without change.

> ▶ **Questions**
> 1. If suddenly the force of gravity of the sun stopped acting on the planets, in what kind of path would the planets move?
>
> 2. Would it be correct to say that the *reason* an object resists change and persists in its state of motion is because of inertia?

▶ **Answers**

1. The planets, like any objects, would move in a straight-line path if no forces acted upon them.

2. In a strict sense, no. Scientists don't know the reason that objects exhibit this property. Nevertheless, the property of behaving in this predictable way is called *inertia*. We understand many things and have labels and names for these things. There are also many things we do not understand, and we have labels and names for these things as well. Education consists not so much in acquiring new names and labels but in learning what is understood and what is not.

3.5 | Mass—A Measure of Inertia

Kick an empty tin can and it moves. Kick a can filled with sand, and it doesn't move as much. Kick a tin can filled with solid lead, and you'll hurt your foot. The lead-filled can has more inertia than the sand-filled can, which in turn has more inertia than the empty can. The can with the most matter has the greatest inertia. The amount of inertia an object has depends on its **mass**— that is, on the amount of material present in the object. The more mass an object has, the more force it takes to change its state of motion. Mass is a measure of the inertia of an object.

Fig. 3-7 You can tell how much matter is in the can when you kick it.

Mass Is Not Volume

Many people confuse mass with volume. They think that if an object has a large mass, it must have a large volume. But volume is a measure of space and is measured in units such as cubic centimeters, cubic meters, or liters. Mass is measured in **kilograms**. (A liter of milk, juice, or soda—anything that is mainly water— has a mass of about one kilogram.) How many kilograms of matter are in an object, and how much space is taken up by that object, are two different things. Which has the greater mass—a feather pillow or a common automobile battery? Clearly the more difficult to set in motion is the battery. This is evidence of the battery's greater inertia and hence greater mass. The pillow may be bigger—that is, it may have a larger volume—but it has less mass. Mass is different from volume.

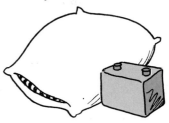

Fig. 3-8 The pillow has a larger size (volume) but a smaller mass than the battery.

Mass Is Not Weight

Mass is most often confused with weight. We say something has a lot of matter if it is heavy. That's because we are used to measuring the quantity of matter in an object by its gravitational attraction to the earth. But mass is more fundamental than weight; mass is a measure of the actual material in a body. It depends only on the number and kind of atoms that compose it. Weight is a measure of the gravitational force that acts on the material, and depends on where the object is located.

The amount of material in a particular stone is the same whether the stone is located on the earth, on the moon, or in outer space. Hence, its mass is the same in any of these locations. This could be shown by shaking the stone back and forth. The same force would be required to shake the stone with the same rhythm whether the stone was on earth, on the moon, or in a force-free region of outer space. That's because the inertia of the stone is solely a property of the stone and not its location.

But the weight of the stone would be very different on the earth

Fig. 3-9 The person in space finds it just as difficult to shake the stone in its weightless state as it is to shake it in its weighted state on earth.

and on the moon, and still different in outer space if the stone were away from strong sources of gravitation. On the surface of the moon the stone would have only one-sixth its weight on earth. This is because gravity is only one-sixth as strong on the moon as compared to on the earth. If the stone were in a gravity-free region of space, its weight would be zero. Its mass, on the other hand, would not be zero. Mass is different from weight.

We can define mass and weight as follows:

Mass: The quantity of matter in a body. More specifically, it is a measure of the inertia or "laziness" that a body exhibits in response to any effort made to start it, stop it, or change in any way its state of motion.

Weight: The force due to gravity upon a body.

Mass and weight are not the same thing, but they are proportional to each other. Objects with great masses have great weights. Objects with small masses have small weights. In the same location, twice as much mass weighs twice as much. Mass and weight are proportional to each other but not equal to each other. Mass has to do with the amount of matter in the object. Weight has to do with how strongly that matter is attracted by the earth's gravity.

▶ **Questions**

1. Does a 2-kilogram iron block have twice as much *inertia* as a 1-kilogram block of iron? Twice as much *mass*? Twice as much *volume*? Twice as much *weight* (when weighed in the same location)?

2. Does a 2-kilogram bunch of bananas have twice as much *inertia* as a 1-kilogram loaf of bread? Twice as much *mass*? Twice as much *volume*? Twice as much *weight* (when weighed in the same location)?

▶ **Answers**

1. The answer is yes to all questions. A 2-kilogram block of iron has twice as many iron atoms, and therefore twice the amount of matter, mass, and weight. The blocks are made of the same material, so the 2-kilogram block also has twice the volume.

2. Two kilograms of *anything* has twice the inertia and twice the mass of one kilogram of anything else. In the same location, two kilograms of anything will weigh twice as much as one kilogram of anything (mass and weight are proportional). So the answer to all questions is yes, except for volume. Volume and mass are proportional only when the materials are the same, or when they are equally compact for their mass—when they have the same *density*. Bananas are denser than bread—enough so that two kilograms of bananas have less volume than one kilogram of ordinary bread.

1 Kilogram Weighs 9.8 Newtons

In the United States it has been common to describe the amount of matter in an object by its gravitational pull to the earth—by its weight. The common, traditional unit of weight is the pound. In most parts of the world, however, the measure of matter is commonly expressed in mass units. The kilogram is the international, metric—SI*—unit of *mass*. The SI symbol for kilogram is kg. At the earth's surface, a 1-kg bag of nails has a weight of 2.2 pounds.

   The SI unit of *force* is the **newton** (named after guess who?). One newton is equal to a little less than a quarter of a pound (like the weight of a quarter-pound burger *after* it is cooked). The SI symbol for newton is N (with a capital letter because it is named after a person). A 1-kg bag of nails has a weight in metric units of 9.8 N. Away from the earth's surface, where the force of gravity is less, it would weigh less.

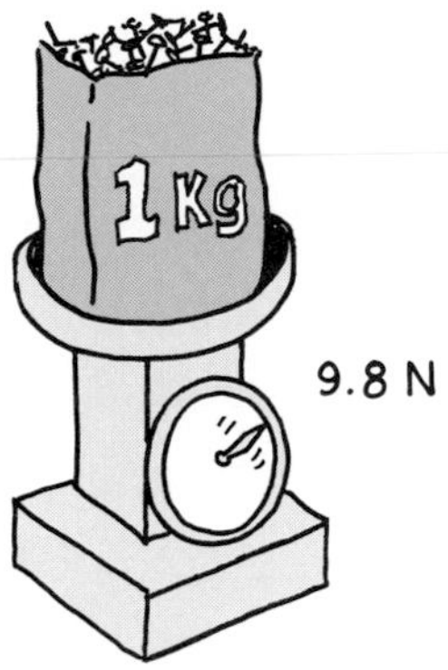

Fig. 3-10   One kilogram of nails weighs 2.2 pounds, which is the same as 9.8 newtons.

> ▶ **Question**
>   The text states that a 1-kg bag of nails weighs 9.8 N at the earth's surface. Does 1 kg of yogurt also weigh 9.8 N?

3.6 | The Moving Earth Again

When Copernicus announced the idea of a moving earth in the sixteenth century, there was much arguing and debating of this controversial idea. One of the arguments against a moving earth was the following: Consider a bird sitting at rest at the top of a tall tree. On the ground below is a fat, juicy worm. The bird sees the worm and drops vertically below and catches it. This would not be possible, it was argued, if the earth moved as Copernicus suggested. If Copernicus were correct, the earth would have to travel at a speed of 107 000 km/h to circle the sun in one year. Convert this speed to kilometers per second and you'll get 30 km/s. Even if the bird could descend from its branch in one

▶ **Answer**
   Yes, at the earth's surface 1 kg of anything weighs 9.8 N. (We used nails in this example because most everybody identifies with nails—at least everybody who likes to build things. But not everybody likes yogurt.)

*   SI stands for the French name, *Le Système International d'Unités*, for the international, metric system of measurement. The short forms of the SI units are called *symbols* rather than *abbreviations*.

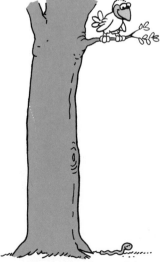

Fig. 3-11   Must the earth be at rest for the bird to catch the worm?

second, the worm would have been swept by the moving earth a distance of 30 kilometers away. For the bird to catch the worm under this circumstance would be an impossible task. But birds in fact do catch worms from high tree branches, which seemed clear evidence that the earth must be at rest.

Can you refute this argument? You can if you invoke the idea of inertia. You see, not only is the earth moving at 30 km/s, but so are the tree, the branch of the tree, the bird that sits on it, the worm below, and even the air in between. All are moving at 30 km/s. A body in motion remains in motion if no unbalanced forces are acting on it. So when the bird drops from the branch, its initial sideways motion of 30 km/s remains unchanged. It catches the worm quite unaffected by the motion of its total environment.

Stand next to a wall. Jump up so that your feet are no longer in contact with the floor. Does the 30-km/s wall slam into you? Why not? Because you are also traveling at 30 km/s—before, during, and after your jump. The 30 km/s is the speed of the earth relative to the sun—not the speed of the wall relative to you.

People three hundred years ago had difficulty with ideas like these not only because they failed to acknowledge the concept of inertia, but because they were not accustomed to moving in high-speed vehicles. Slow, bumpy rides in horse-drawn carriages did not lend themselves to experiments that would reveal inertia. Today we flip a coin in a high-speed car, bus, or plane, and we catch the vertically-moving coin as we would if the vehicle were at rest. We see evidence for the law of inertia when the horizontal motion of the coin before, during, and after the catch is the same. The coin keeps up with us. The vertical force of gravity affects only the vertical motion of the coin.

Our notions of motion today are very different from those of our ancestors. Aristotle did not recognize the idea of inertia because he failed to imagine what motion would be like without friction. In his experience, all motion was subject to resistance, and he made this fact central to his theory of motion. We can only wonder how differently science might have progressed if Aristotle had recognized friction for what it is, namely a force like any other, which may or may not be present.

Fig. 3-12 Flip a coin in a high-speed airplane, and it behaves as if the plane were at rest. The coin keeps up with you: inertia in action.

3 | Chapter Review

Concept Summary

Galileo concluded that if it were not for friction an object in motion would keep moving forever.

According to Newton's first law of motion—the law of inertia—every body continues in its state of rest or of motion in a straight line at constant speed unless forces cause it to change its state.

Inertia is the resistance an object has to a change in its state of motion.
- Mass is a measure of inertia.
- Mass is not the same as volume.
- Mass is not the same as weight.
- The mass of an object depends on the amount and type of matter in it, but does not depend on the location of the object.
- The weight of an object is the gravitational force on it and depends on the location.

Important Terms

force (3.3)
friction (3.3)
inertia (3.3)
kilogram (3.5)
law of inertia (3.4)
mass (3.5)
newton (3.5)
Newton's first law (3.4)

Review Questions

1. What was the distinction that Aristotle made between *natural motion* and *violent motion*? (3.1)

2. Why was Copernicus reluctant to publish his ideas? (3.2)

3. What is the effect of friction on a moving object? How is an object able to maintain a constant speed when friction acts upon it? (3.3)

4. The speed of a ball increases as it rolls down an incline, and the speed decreases as the ball rolls up an incline. What happens to the speed on a smooth horizontal surface? (3.3)

5. Galileo found that a ball rolling down one incline will pick up enough speed to roll up another. How high will it roll compared to its initial height? (3.3)

6. Does the law of inertia pertain to moving objects, objects at rest, or both? Support your answer with examples. (3.4)

7. The law of inertia states that no force is required to maintain motion. Why, then, do you have to keep peddling your bicycle to maintain motion? (3.4)

8. If you were in a spaceship and fired a cannonball into frictionless space, how much force would have to be exerted on the ball to keep it going? (3.4)

9. Does a 2-kilogram rock have twice the mass of a 1-kilogram rock? Twice the inertia? Twice the weight (when weighed in the same location)? (3.5)

10. Does a liter of molten lead have the same volume as a liter of apple juice? Does it have the same mass? (3.5)

11. Why do physics types say that mass is more fundamental than weight? (3.5)

12. An elephant and a mouse would both have the same weight—zero—in gravitation-free space. If they were moving toward you with the same speed, would they bump into you with the same effect? Explain. (3.5)

13. What is the weight of 2 kg of yogurt? (3.5)

14. If you hold a coin above your head while in a bus that is not moving, the coin will land at your feet when you drop it. Where will it land if the bus is moving in a straight line at constant speed? Explain. (3.6)

15. In the cabin of a jetliner that cruises at 600 km/h, a pillow drops from an overhead rack to your lap below. Since the jetliner is moving so fast, why doesn't the pillow slam into the rear of the compartment when it drops? (What is the horizontal speed of the pillow relative to the ground? Relative to you inside the jetliner?) (3.6)

Think and Explain

1. Many automobile passengers have suffered neck injuries when struck by cars from behind. How does Newton's law of inertia apply here? How do headrests help to guard against this type of injury?

2. Suppose you place a ball in the middle of a wagon, and then accelerate the wagon forward. Describe the motion of the ball relative to (a) the ground and (b) the wagon.

3. If an elephant were chasing you, its enormous mass would be most threatening. But if you zigzagged, its mass would be to your advantage. Why?

4. When you compress a sponge, which quantity changes: mass, inertia, volume, or weight?

5. a. A massive ball is suspended by a string from above, and slowly pulled by a string from below (Figure A). Is the string tension greater in the upper or the lower string? Which string is more likely to break? Which property—mass or weight—is important here?
   b. If the string is instead snapped downward, which string is more likely to break? Which property—mass or weight—is important this time?

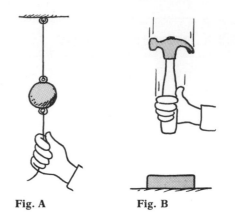

Fig. A   Fig. B

6. If the head of a hammer is loose, and you wish to tighten it by banging it against the top of a work bench, why is it best to hold it with the handle down (Figure B) rather than with the head down? Explain in terms of inertia.

7. Two closed containers look the same, but one is packed with lead and the other with a few feathers. How could you determine which had more mass if you and the containers were orbiting in a weightless condition in outer space?

8. What is your actual (or most desirable) weight in newtons?

9. If you are sitting in a bus that is traveling along a straight, level road at 100 km/h, you are traveling at 100 km/h too.
   a. If you hold an apple over your head, how fast is it moving relative to the road? Relative to you?
   b. If you drop the apple, does it still have the same horizontal motion?

10. As the earth rotates about its axis, it takes three hours for the United States to pass beneath a point above the earth that is stationary relative to the sun. What is wrong with this scheme: To travel from Washington D.C. to San Francisco using very little fuel, simply ascend in a helicopter high over Washington D.C. and wait three hours until San Francisco passes below?

4 Newton's Second Law of Motion— Force and Acceleration

Kick a football and it moves. Its path through the air is not a straight line—it curves due to gravity. Catch the ball and it stops. Most of the motion we see undergoes change. Most things start up, slow down, or curve as they move. The last chapter covered objects at rest or moving at constant velocity. There was no net force acting on these objects. This chapter covers the more common cases, in which there is a change in motion—that is, accelerated motion.

Recall from Chapter 2 that acceleration describes how fast motion is changing. Specifically, it is the change in velocity per certain time interval. In shorthand notation,

$$\text{acceleration} = \frac{\text{change in velocity}}{\text{time interval}}$$

This is the definition of acceleration.* This chapter focuses on the *cause* of acceleration: *force*.

Fig. 4-1 Kick a football and it neither remains at rest nor moves in a straight line.

4.1 Force Causes Acceleration

Consider an object at rest, such as a hockey puck on ice. Apply a force and it moves. Since it was not moving before, it has accelerated—changed its motion. When the stick is no longer in contact with the puck, it moves at constant velocity. Apply another force by striking it with the stick again, and the motion changes. Again, the puck has accelerated. Forces are what produce accelerations.

---

* The Greek letter Δ (delta) is often used as a symbol for "change in" or "difference in." In "delta" notation, $a = \Delta v/\Delta t$, where Δv is the change in velocity, and Δt is the change in time (the time interval).

Fig. 4-2 Puck about to be hit.

Most often, the force we apply is not the only force that acts on an object. Other forces may act as well. The combination of all the forces that act on an object is called the **net force**. It is the net force that accelerates an object.

We see how forces combine to produce net forces in Figure 4-3. If you pull horizontally with a force of 10 N on an object that rests on a friction-free surface, an air track for example, then the net force acting on it is 10 N. If a friend assists you and pulls at the same time on the same object with a force of 5 N in the same direction, then the net force will be the sum of these forces, 15 N (Figure 4-3 top). The object will accelerate as if it were pulled with a single force of 15 N. If, however, your friend pulls with 5 N in the opposite direction, the net force will be the difference of these forces, 5 N (Figure 4-3 center). The acceleration of the object would be the same as if it were instead pulled with a single force of 5 N.

We find that the amount of acceleration depends on the amount of the net force. To increase the acceleration of an object, you must increase the net force. This makes good sense. Double the force on an object, and you will double the acceleration. If you triple the force, you'll triple the acceleration, and so on. We say that the acceleration produced is directly proportional to the net force. We write:

$$acceleration \sim net\ force$$

The symbol $\sim$ stands for "is directly proportional to."

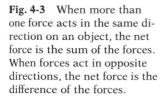

APPLIED FORCES	NET FORCE
5 N / 10 N	15 N
5 N ← / → 10 N	5 N →
5 N ← / → 5 N	0 N

Fig. 4-3 When more than one force acts in the same direction on an object, the net force is the sum of the forces. When forces act in opposite directions, the net force is the difference of the forces.

4.2   Mass Resists Acceleration

Push on an empty shopping cart. Then push equally hard on a heavily loaded shopping cart, and you'll produce much less acceleration. This is because acceleration depends on the mass being pushed upon. For objects of greater mass we find smaller accelerations. Twice as much mass for the same force results in only half the acceleration; three times the mass results in one third the acceleration, and so forth. In other words, for a given force the acceleration produced is inversely proportional to the mass. We write:

$$acceleration \sim \frac{1}{mass}$$

Fig. 4-4 The acceleration produced depends on the mass being pushed.

By **inversely** we mean that the two values change in opposite directions. (Mathematically we see that as the denominator increases, the whole quantity decreases. The quantity $\frac{1}{100}$ is less than the quantity $\frac{1}{10}$, for example.)

4.3 │ Newton's Second Law

Newton was the first to realize that the acceleration we produce when we move something depends not only on how hard we push or pull (the force) but on the mass as well. He came up with one of the most important rules of nature ever proposed, his second law of motion. **Newton's second law** states:

> The acceleration produced by a net force on a body is directly proportional to the magnitude of the net force, in the same direction as the net force, and inversely proportional to the mass of the body.

Or in shorter notation,

$$\text{acceleration} \sim \frac{\text{net force}}{\text{mass}}$$

By using consistent units such as newtons (N) for force, kilograms (kg) for mass, and meters per second squared (m/s²) for acceleration, we get the exact equation.

$$\text{acceleration} = \frac{\text{net force}}{\text{mass}}$$

In briefest form, where a is acceleration, F is net force, and m is mass:

$$a = \frac{F}{m}$$

The acceleration is equal to the net force divided by the mass. From this relationship we can see that if the net force that acts on an object is doubled, the acceleration will be doubled. Suppose instead that the mass is doubled. Then the acceleration will be halved. If both the net force and the mass are doubled, then the acceleration will be unchanged.

Fig. 4-5 The good acceleration of the racing car is due to its ability to produce large forces.

Problem Solving

If the mass of an object is measured in kilograms (kg), and acceleration is expressed in meters per second squared (m/s²), then the force will be expressed in newtons (N). One newton is the force needed to give a mass of one kilogram an acceleration of one meter per second squared. We can arrange Newton's second law to read

$$\text{force} = \text{mass} \times \text{acceleration}$$
$$1 \text{ N} = (1 \text{ kg}) \times (1 \text{ m/s}^2)$$

As you can see,

$$1 \text{ N} = 1 \text{ kg·m/s}^2$$

(The dot between "kg" and "m/s²" means that the units have been multiplied together.)

If you know two of the quantities in Newton's second law, you can easily calculate the third. For example, how much thrust must a 30 000-kg jet plane develop to achieve an acceleration of 1.5 m/s²? Thrust means force, so

$$F = ma$$
$$= (30\ 000 \text{ kg}) \times (1.5 \text{ m/s}^2)$$
$$= 45\ 000 \text{ kg·m/s}^2$$
$$= 45\ 000 \text{ N}$$

Suppose you know the force and the mass and want to find the acceleration. For example, what acceleration will be produced by a force of 2000 N on a 1000-kg car? Using Newton's second law we find

$$a = \frac{F}{m} = \frac{2000 \text{ N}}{1000 \text{ kg}} = \frac{2000 \text{ kg·m/s}^2}{1000 \text{ kg}} = 2 \text{ m/s}^2$$

If the force were 4000 N, what would be the acceleration?

$$a = \frac{F}{m} = \frac{4000 \text{ N}}{1000 \text{ kg}} = \frac{4000 \text{ kg·m/s}^2}{1000 \text{ kg}} = 4 \text{ m/s}^2$$

Doubling the force on the same mass simply doubles the acceleration.

In traditional problem-solving physics courses, the problems are often more complicated than the ones given here. This book will not focus on solving complicated problems, but will instead concentrate on becoming acquainted with the equations as guides to thinking about the basic physics concepts. Mathematical problems can be an objective for follow-up study in physics. Most any other physics textbook will provide you with plenty of mathematical problems if you want them. But please—learn the concepts first!

▶ **Questions**
1. If a car is able to accelerate at 2 m/s², what acceleration can it attain if it is towing another car of equal mass?

2. What kind of motion does an unchanging force produce on an object of fixed mass?

4.4 | Statics

How many forces act on your book as it lies motionless on the table? Don't say one, its weight. If that were the only force acting on it, you'd find it accelerating. The fact that it is at rest, and not accelerating, is evidence that another force is acting. The other force must just balance the weight to make the net force zero. The other force is the **support force** of the table (often called the **normal force***). To see that the table is pushing up on the book, imagine an ant underneath the book. The ant would feel itself being squashed from both sides, the top and the bottom. The table actually pushes up on the book with the same amount of force that the book presses down. If the book is to be at rest, the sum of the forces acting on it must balance to zero.

Hang from a strand of rope. Your weight provides the force which tightens the rope and sets up a tension force in it. How much tension is in the strand? If you are not accelerating, then it must equal your weight. The rope pulls up on you, and the gravitation of the earth pulls down on you. The equal forces cancel because they are in opposite directions. So you hang motionless.

Suppose you hang from a bar supported by two strands of rope, as in Figure 4-7. Then the tension in each rope is half your weight (if the weight of the bar is neglected). The total tension up ($\frac{1}{2}$ your weight + $\frac{1}{2}$ your weight) then balances your downward acting weight. When you do pullups in exercising, you use both arms. Then each arm supports half your weight. Ever try pullups with just one arm? Why is it twice as difficult?

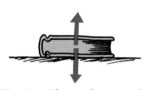

Fig. 4-6 The net force on the book is zero because the table pushes up with a force that equals the downward weight of the book.

Fig. 4-7 The sum of the rope tensions must equal your weight.

▶ **Answers**
1. When the engine produces the same force on twice the mass, the acceleration will be half. It will be 1 m/s².

2. Motion at a constant acceleration, in accord with Newton's second law.

*   This force acts at right angles to the surface; "normal to" means "at right angles to," which is why the force is called a normal force.

> ▶ **Question**
>   When you step on a bathroom scale, the downward pull of gravity and the upward support force of the floor compress a spring which is calibrated to give your weight. In effect, the scale shows the support force. If you stand on two scales with your weight equally divided between each scale, what will each scale read? How about if you stand with more of your weight on one foot than the other?

4.5 | Friction

Even when a single force is applied to an object, it is usually not the only force affecting the motion. This is because of friction. Friction is a force that always acts in a direction to oppose motion. It is due in large part to irregularities in the two surfaces that are in contact. Even very smooth surfaces are bumpy when viewed with a microscope. When an object slides against another, it must either rise over the irregular bumps or else scrape them off. Either way requires a force.

The force of friction between two surfaces depends on the kinds of material and how much they are pressed together. Rubber against concrete, for example, produces more friction than steel against steel. That's why concrete road dividers are replacing steel rails (Figure 4-8). Notice that the concrete divider is wider at the bottom, so that the tire of a sideswiping car rather than the steel body makes contact with it. The greater friction between the rubber tire and the concrete is more effective in slowing the car than the contact between the steel body of the car with the steel rail of the other design.

Friction is not restricted to solids sliding over one another. Friction occurs also in liquids and gases, which are called **fluids** (those which flow). Fluid friction occurs when an object moving through a fluid pushes aside some of the fluid. Have you ever tried running a 100-m dash through waist-deep water? The fric-

Fig. 4-8 Cross-section view of concrete road divider, and steel road divider. Which design is best for slowing an out-of-control, sideswiping car?

▶ **Answer**
  The reading on both scales must add up to your weight. This is because the sum of the scale readings, which equals the support force of the floor, must counteract your weight so the net force on you will be zero. If you stand equally on each scale, each will read half your weight. If you lean more on one scale than the other, more than half your weight will be read on that scale but less on the other, so they will still add up to your weight. For example, if one scale reads two-thirds your weight, the other scale will read one-third your weight. Get it?

tion of liquids is appreciable, even at low speeds. **Air resistance**, the friction that acts on something moving through air, is a very common case of fluid friction. You don't notice it when walking or jogging, but you'll notice it for higher speeds as in skiing downhill or skydiving.

When there is friction, an object may move at constant velocity while a force is applied to it. In this case, the friction force just balances the applied force. The net force is zero, so there is no acceleration. For example, in Figure 4-9 the crate will move at constant velocity when it is pushed just hard enough to balance the friction. The sack will fall at constant velocity when the air resistance balances its weight.

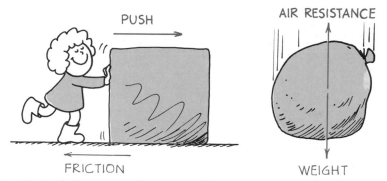

Fig. 4-9   Push the crate to the right, and friction acts toward the left. The sack falls downward, and air friction acts upward. The direction of the force of friction always opposes the direction of motion.

▶ **Questions**

1. Figure 4-6 shows only two forces acting on the book: its weight and the support force from the table. Doesn't the force of friction act as well?

2. Suppose a high-flying jumbo jet cruises at constant velocity when the thrust of its engines is a constant 80 000 N. What is the *acceleration* of the jet? What is the *force* of air resistance acting on the jet?

▶ **Answers**

1. No, not unless the book tends to slide or does slide across the table. For example, if it is pushed toward the left by another force, then friction between the book and table will act toward the right. Friction forces occur only when an object tends to slide or is sliding.

2. The acceleration is zero because the velocity is constant, which means not changing. Since the acceleration is zero, it follows from $a = F/m$ that the net force is zero. This implies that the force of air resistance must just equal the thrusting force of 80 000 N but must be in the opposite direction. So the air resistance is 80 000 N.

4.6 | Applying Force—Pressure

You can put a book on a table, and no matter how you place it—on its back, upright, or even balancing on a single corner—the force of the book on the table is the same. This can be checked by doing this on a bathroom scale, where you'll read the same weight in all cases. Balance the book different ways on the palm of your hand. Although the force will be the same, you'll note differences in the way the book presses against your palm. This is because the area of contact is different in each case. The force *per unit of area* is called **pressure**. More precisely,

$$\text{pressure} = \frac{\text{force}}{\text{area of application}}$$

where the force is perpendicular to the surface area. In equation form,

$$P = \frac{F}{A}$$

where P is the pressure and A is the area against which the force acts. Force, which is measured in newtons, is different from pressure, which is measured in newtons per square meter [called a **pascal** (Pa), a relatively new unit, adopted in 1960].

Many people mistakenly believe that the wider tires of drag-racing vehicles produce more friction. But the larger area reduces only the pressure. The force of friction is independent of the contact area. Wide tires produce less pressure, and narrow tires produce more pressure. The wideness of the tires reduces heating and wear.

You exert more pressure against the ground when you are standing on one foot than when standing on both feet. This is because your area of contact is less. Stand on one toe, like a ballerina, and the pressure is huge. The smaller the area that supports a given force, the greater is the pressure on that surface.

You can calculate the pressure you exert on the ground when you are standing. One way is to moisten the bottom of your foot with water and step on a clean sheet of graph paper marked with squares. Count the number of squares contained in your footprint. Divide your weight by this area and you have the average pressure you exert on the ground while standing still on one foot. How will this pressure compare with the pressure when you stand on two feet?

A dramatic illustration of pressure is shown in Figure 4-12. The author applies appreciable force when he breaks the cement block with the sledge hammer. Yet his friend, who is sandwiched between two beds of sharp nails, is unharmed. This is because

Fig. 4-10 The upright book exerts the same force but a greater pressure against the supporting surface.

Fig. 4-11 Friction between the tire and the ground is the same whether the tire is wide or narrow. The purpose of the greater contact area is to reduce heating and wear.

much of this force is distributed over the more than 200 nails that make contact with his body. The combined surface area of this many nail points results in a tolerable pressure that does not puncture the skin. CAUTION: This demonstration is quite dangerous. Do not attempt it on your own.

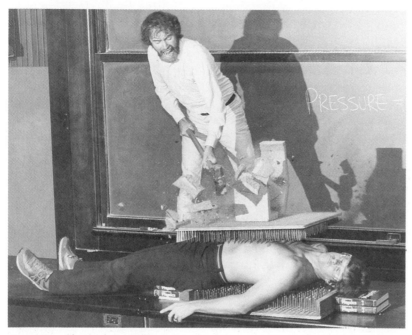

Fig. 4-12   The author applies a force to fellow physics teacher Paul Robinson, who is bravely sandwiched between beds of sharp nails. The driving force per nail is not enough to puncture the skin. CAUTION: Do not attempt this on your own!

▶ **Questions**

1. In attempting to do the demonstration in Figure 4-12, would it be wise to begin with a few nails, and work up-ward to more nails?

2. The massiveness of the cement block plays an important role in this demonstration. Which provides more safety—a less massive or more massive block?

▶ **Answers**

1. No, no, no! There would be one fewer physics teacher if the demonstration were performed with fewer nails, because of the resulting greater pressure.

2. The greater the mass of the block, the smaller is the acceleration of the block and bed of nails toward the friend. Much of the force wielded by the hammer goes into moving this block and *breaking* it. It is important that the block be massive and that it break upon impact.

4.7   Free Fall Explained

Fig. 4-13 Galileo's famous demonstration.

Galileo showed that falling objects will accelerate equally, regardless of their masses. This is *strictly* true if air resistance is negligible, that is, if the objects are falling freely. It is *approximately* true when air resistance is very small compared to the weight of the falling object. For example, a 10-kg cannonball and a 1-kg stone dropped from an elevated position at the same time will fall together and strike the ground at practically the same time. This experiment, said to be done by Galileo from the Leaning Tower of Pisa, demolished the Aristotelian idea that an object that weighs ten times as much as another should fall ten times faster than the lighter object. Galileo's experiment and many others that showed the same result were convincing. But Galileo couldn't say *why* the accelerations were equal. The explanation is a straight-forward application of Newton's second law and is the topic of the cartoon "Backyard Physics" (next page). Let's treat it separately here.

Recall that mass (a quantity of matter) and weight (the force due to gravity) are proportional. A 2-kg bag of nails weighs twice as much as a 1-kg bag of nails. So a 10-kg cannonball has 10 times the force of gravity (weight) of a 1-kg stone. Followers of Aristotle believed that the cannonball should therefore accelerate ten times as much as the stone because they considered only the greater weight. But Newton's second law tells us to consider the mass as well. A little thought will show that ten times as much force acting on ten times as much mass produces the same acceleration as one-tenth the force acting on one-tenth the mass. In symbolic notation,

$$\frac{\boldsymbol{F}}{\boldsymbol{m}} = \frac{F}{m}$$

where $\boldsymbol{F}$ stands for the force (weight) acting on the cannonball, and $\boldsymbol{m}$ stands for the correspondingly large mass of the cannonball. The small F and m stand for the smaller weight and mass of the stone. We see that the *ratio* of weight to mass is the same for these or any objects. All freely falling objects undergo the same acceleration at the same place on earth. This acceleration, which is due to gravity, is represented by the symbol g.

We can show the same result with numerical values. The weight of a 1-kg stone (or 1 kg of *anything*) is 9.8 N at the earth's surface. The weight of 10 kg of matter, such as the cannonball, is 98 N. The force acting on a falling object is the force due to gravity, or weight of the object. The acceleration of the stone is

$$a = \frac{F}{m} = \frac{\text{weight}}{m} = \frac{9.8 \text{ N}}{1 \text{ kg}} = \frac{9.8 \text{ kg·m/s}^2}{1 \text{ kg}} = 9.8 \text{ m/s}^2 = g$$

and for the cannonball,

$$a = \frac{F}{m} = \frac{\text{weight}}{m} = \frac{98 \text{ N}}{10 \text{ kg}} = \frac{98 \text{ kg·m/s}^2}{10 \text{ kg}} = 9.8 \text{ m/s}^2 = g$$

The famous coin-and-feather-in-a-vacuum-tube demonstration was discussed in Chapter 2, but not the *reason* for the equal accelerations. Now we know that the reason both fall with the same acceleration (g) is because the net force on each is only its weight, and the ratio of weight to mass is the same for both.

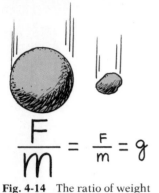

$$\frac{F}{m} = \frac{F}{m} = g$$

Fig. 4-14 The ratio of weight (F) to mass (m) is the same for the 10-kg cannonball and the 1-kg stone.

▶ **Question**

If you were on the moon and dropped a hammer and a feather from the same elevation at the same time, would they strike the surface of the moon together?

4.8   Falling and Air Resistance

The feather and coin fall with equal accelerations in a vacuum, but quite unequally in the presence of air. When air is let into the glass tube and it is again inverted and held upright, the coin falls quickly while the feather flutters to the bottom. Air resistance diminishes the net forces—a tiny bit for the coin and very much for the feather. The downward acceleration for the feather is very brief, for the air resistance very quickly builds up to counteract its tiny weight. The feather does not have to fall very long or very fast for this to happen. When the air resistance of the feather equals the weight of the feather, the net force is zero and no further acceleration occurs. Acceleration terminates. The feather has reached its **terminal speed**. If we are concerned with direction, down for falling objects, we say the feather has reached its **terminal velocity**.

The air resistance on the coin does not have as much effect. At small speeds the force of air resistance is very small compared to the weight of the coin, and its acceleration is only slightly less

▶ **Answer**

Yes. On the moon's surface they would weigh only one-sixth their earth weight, which would be the only force acting on them since there is no atmosphere to provide air resistance. The ratio of moonweight to mass for each would be the same, and they would each accelerate at $\frac{1}{6}g$.

than the acceleration of free fall, g. The coin might have to fall for a few minutes before its speed would be great enough for the air resistance to increase to its weight. Its speed at that point, perhaps 200 km/h or so, would no longer increase. It would have reached its terminal speed.

The terminal speed for a human skydiver varies from about 150 to 200 km/h, depending on weight and position of fall. A heavier person will attain a greater terminal speed than a lighter person. The greater weight is more effective in "plowing through" the air. A heavy and light skydiver can remain in close proximity if the heavy person spreads out like a flying squirrel while the light person falls head or feet first. A parachute greatly increases air resistance, and terminal speed can be cut down to an acceptable 15 to 25 km/h.

Fig. 4-15 Terminal speed is reached for the skydiver when air resistance equals weight.

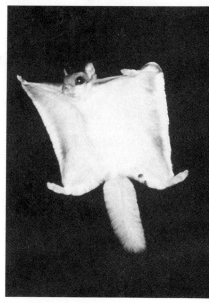

Fig. 4-16 The flying squirrel increases its area by spreading out. This increases air resistance and decreases the speed of fall.

▶ **Question**

If a heavy person and a light person parachute together from the same altitude, and each wears the same size parachute, who should reach the ground first?

▶ **Answer**

The heavy person reaches the ground first. This is because the light person, like the feather, will reach terminal speed sooner while the heavy person continues to accelerate until a greater terminal speed is reached. So the heavy person moves ahead of the light person, who is unable to catch up.

If you hold a baseball and tennis ball at arm's length and release them together, you'll note they strike the floor at the same time. But if you drop them from the top of a high building, you'll notice the heavier baseball strikes the ground first. This is because of the buildup of air resistance with higher speed (like the parachutists in the check question). For low speeds, air resistance may be negligible. For higher speeds, it can make quite a difference. Air resistance is more pronounced on the lighter tennis ball than the heavier baseball, so acceleration of fall is less for the tennis ball. The tennis ball behaves more like a parachute than the baseball does.

▶ **Question**

If the force of air resistance is the *same* for a falling baseball and a falling tennis ball, which will have the greater acceleration?

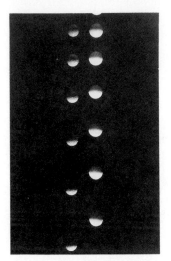

Fig. 4-17 A stroboscopic photo of a golf ball and a Styrofoam ball falling in air. The weight of the heavier golf ball is more effective in overcoming air resistance, so its acceleration is greater. Will both ultimately reach a terminal speed? Which will do so first? Why?

When Galileo reportedly dropped the objects of different weights from the Leaning Tower of Pisa, the heavier one *did* get to the ground first—but only a split second before the other, rather than the pronounced time difference expected by the followers of Aristotle. The behavior of falling objects was never really understood until Newton announced his second law of motion.

Isaac Newton truly changed our way of seeing the world.

▶ **Answer**

Don't say the same! It's true the air resistance is the same for each, but this doesn't mean the net force is the same for each, or that the ratio of net force to mass is the same for each. The heavier baseball will have the greater net force, and greater net force per mass, just like the heavier parachutist in the previous question. (Convince yourself of this by considering the upper limit of air resistance—when it is equal to the weight of the tennis ball. What will be the acceleration of the tennis ball then? Do you see that the baseball has a greater acceleration at that point? And with more thought, do you see that the baseball has the greater acceleration even when the air resistance is less than the weight of the tennis ball?)

4   Chapter Review

Concept Summary

An object accelerates—changes speed and/or direction—when there is a net force that acts on it.

- The acceleration of an object is directly proportional to the net force that acts on it.
- The acceleration of an object is inversely proportional to the mass of the object.
- Acceleration equals net force divided by mass, and is in the same direction as the net force.

An object remains at rest or continues to move with constant velocity when the net force on it equals zero.

- When an object is at rest, its weight is balanced by a support force of equal amount.
- When an object is moving at constant velocity while an applied force acts on it, that force must be balanced by an equal amount of resisting force (usually friction).

The application of a force over a surface produces pressure.

- Pressure equals force divided by area of application, where the force is perpendicular to the surface area.

A falling object is acted on by gravity, which pulls downward with a force equal to the weight of the object.

- In free fall (no air resistance), the acceleration of all objects is the same, regardless of mass.
- When there is air resistance, a falling object will accelerate only until it reaches its terminal speed.
- At terminal speed, the force of air resistance balances the force of gravity.

Important Terms

air resistance (4.5)
fluid (4.5)
inversely (4.2)
net force (4.1)
Newton's second law (4.3)
normal force (4.4)
pascal (4.6)
pressure (4.6)
support force (4.4)
terminal speed (4.8)
terminal velocity (4.8)

Review Questions

1. Distinguish between the relationship that defines acceleration, and the relationship that states how it is produced. (4.1)

2. What is meant by the *net force* that acts on an object? (4.1)

3. Forces of 10 N and 20 N in the same direction act on an object. What is the net force on the object? (4.1)

4. If the forces exerted on an object are 50 N in one direction and 30 N in the opposite direction, what is the net force exerted on the object? (4.1)

5. Suppose a cart is being moved by a certain net force. If the net force is doubled, by how much does its acceleration change? (4.1)

6. Suppose a cart is being moved by a certain net force. If a load is dumped into the cart so its mass is doubled, by how much does the acceleration change? (4.2)

7. Distinguish between the concepts of *directly proportional* and *inversely proportional*.

Support your statement with examples. (4.1–4.2)

8. State Newton's second law in words, and in the form of an equation. (4.3)

9. How much force must a 20 000-kg rocket develop to accelerate 1 m/s²? (4.3)

10. How much support force does a table exert on a book that weighs 15 N when the book is placed on the table? What is the *net* force on the book in this case? (4.4)

11. When a 100-N bag of nails hangs motionless from a single vertical strand of rope, how many newtons of tension are exerted in the strand? How about if the bag is supported by four vertical strands? (4.4)

12. What is the cause of friction, and in what direction does it act with respect to the motion of a sliding object? (4.5)

13. If the force of friction acting on a sliding crate is 100 N, how much force must be applied to maintain a constant velocity? What will be the net force acting on the crate? What will be the acceleration? (4.5)

14. Distinguish between force and pressure. (4.6)

15. Which produces more pressure on the ground—a narrow tire or a wide tire of the same weight? (4.6)

16. The force of gravity is twice as great on a 2-kg rock as on a 1-kg rock. Why then, does the 2-kg rock not fall with twice the acceleration? (4.7)

17. How can a coin and a feather in a vacuum tube fall with the same acceleration? (4.7)

18. Why do a coin and a feather fall with different accelerations in the presence of air? (4.8)

19. How much air resistance acts on a 100-N bag of nails that falls at its terminal speed? (4.8)

20. How do the air resistance and the weight of a falling object compare when terminal speed is reached? (4.8)

21. All things being equal, why does a heavy skydiver have a greater terminal speed than a lighter skydiver? What can be done so that both terminal speeds are equal? (4.8)

22. What is the net force acting on a 25-N freely falling object? What is the net force when it encounters 15 N of air resistance? When it falls fast enough to encounter 25 N of air resistance? (4.7–4.8)

Activities

1. If you drop a sheet of paper and a book side by side, the book will fall faster due to its greater weight compared to the air resistance. If you place the paper against the lower surface of the raised horizontally held book and again drop them at the same time, it will be no surprise that they will hit the surface below at the same time. The book has simply pushed the paper with it as it falls. If you repeat this, only with the paper on *top* of the book, which will fall faster? Try it and see!

2. Drop two balls of different weights from the same height, and for small speeds they practically fall together. Will they roll together down the same inclined plane? If each is suspended from the same size string and each is made into a pendulum, and displaced through the same angle, will they swing to and fro in unison? Try it and see.

3. The net force that acts on an object and the resulting acceleration are always in the same direction. You can demonstrate this with a spool. If the spool is pulled toward you, which way will it roll? Does it make a difference if the string is on the bottom or the top? Try it and maybe you'll be surprised.

Think and Explain

1. What is the difference between saying that one quantity is proportional to another and saying it is equal to another?

2. If a four-engine jet accelerates down the runway at 2 m/s² and one of the jet engines fails, how much acceleration will the other three produce?

3. If a loaded truck can accelerate at 1 m/s² and loses its load so it is only ¾ as massive, what acceleration can it attain for the same driving force?

4. A rocket fired from its launching pad not only picks up speed, but its acceleration increases significantly as firing continues. Why is this so? (*Hint*: About 90% of the mass of a newly launched rocket is fuel.)

5. What is the mass and what is the weight of a 10-kg object on the earth? What is its mass and weight on the moon, where the force of gravity is ⅙ that of the earth?

6. The little girl in Figure A hangs at rest from the ends of the rope. How does the reading on the scale compare to her weight?

Fig. A

7. In football-team blocking, why does a defending lineman often attempt to get his body under that of his opponent and push upward? What effect does this have on the friction force between the opposing lineman's feet and the ground?

8. Why is the force of friction no greater for a wide tire than for a narrow tire?

9. Why does a sharp knife cut better than a dull knife?

10. Harry the painter swings year after year from his bosun's chair. His weight is 500 N and the rope, unknown to him, has a breaking point of 300 N. Why doesn't the rope break when he is supported as shown in Figure B left? One day Harry is painting near a flagpole, and for a change, he ties the free end of the rope to the flagpole instead of to his chair (Figure B right). Why did Harry end up taking his vacation early?

Fig. B

11. What is the advantage of the large flat sole on the foot of an elephant? Why must the small foot (hoof) of an antelope be so hard?

12. When a rock is thrown straight upward, what is its acceleration at the top of its path? (Is your answer consistent with $a = F/m$?)

13. As a skydiver falls faster and faster through the air, does the net force on her increase, decrease, or remain unchanged? Does her acceleration increase, decrease, or remain unchanged? Defend your answers.

14. After she jumps, a skydiver reaches terminal speed after 10 seconds. Does she gain more speed during the first second of fall or the ninth second of fall? Compared to the first second of fall, does she fall a greater or a lesser distance during the ninth second?

15. A regular tennis ball and another filled with heavy sand are dropped at the same time from the top of a high building. Your friend says that even though air resistance is present, they should hit the ground together because they are the same size and "plow through" the same amount of air. What do you say?

5 Newton's Third Law of Motion—Action and Reaction

Fig. 5-1 When you push on the wall, the wall pushes on you.

If you lean over too far, you'll fall over. But if you lean over with your hand outstretched and make contact with a wall, you can do so without falling. When you push against the wall, the wall pushes back on you. That's why you are supported. Ask your friends why you don't topple over. How many will answer, "Because the wall is pushing on you and holding you in place"? Probably not very many people (unless they're physics types) realize that walls can push on us every bit as much as we push on them.*

5.1 Interactions Produce Forces

Fig. 5-2 The interaction that drives the nail is the same as the one that brings the hammer to a halt.

In the simplest sense, a force is a push or a pull. Looking closer, however, we find that a force is not a thing in itself, but is due to an interaction between one thing and another. For example, a hammer hits a nail and drives it into a board. One object interacts with another. Which exerts the force and which receives the force? Newton thought about questions like this, and the more he thought, the more he came to the conclusion that neither object has to be identified as the "exerter" or "receiver." He reasoned that nature is symmetrical and concluded that both objects must be treated equally. The hammer exerts a force against

---

* The terms *push* or *pull* usually invoke the idea of a living thing exerting a force. So strictly speaking, to say "the wall pushes on you" is to say "the wall exerts a force as though it were pushing on you." As far as the balance of forces is concerned, there is no observable difference between the forces exerted by you (alive) and the wall (nonliving).

the nail, but is itself halted in the process. The same interaction that drives the nail also slows the hammer. Such observations led Newton to his third law—the law of action and reaction.

5.2   Newton's Third Law

Newton's third law states:

> Whenever one body exerts a force on a second body, the second body exerts an equal and opposite force on the first.

One force is called the **action force**. The other is called the **reaction force**. It doesn't matter which force we call *action* and which we call *reaction*. They are equal. The important thing is that neither force exists without the other. The action and reaction forces make up a *pair* of forces. Newton's third law is often stated as "to every action there is always opposed an equal reaction."

In every interaction, forces always occur in pairs. For example, in walking across the floor you push against the floor, and the floor in turn pushes against you. Likewise, the tires of a car push against the road, and the road in turn pushes back on the tires. In swimming you push the water backward, and the water pushes you forward. There is a pair of forces acting in each instance. The forces in these examples depend on friction; a person or car on ice, by contrast, may not be able to exert the action force against the ice to produce the needed reaction force.

Fig. 5-3 What happens to the boat when she jumps to shore?

> ▶ **Questions**
> 1. Does a stick of dynamite contain force?
>
> 2. A car accelerates along a road. Strictly speaking, what is the force that moves the car?

▶ **Answers**

1. No, a force is not something a body has, like mass, but is an interaction between one object and another. A body may possess the capability of exerting a force on another object, but it cannot possess force as a thing in itself. Later you will see that something like a stick of dynamite possesses *energy*.

2. It is the road that pushes the car along. Really! Except for air resistance, only the road provides a horizontal force on the car. How does it do this? The rotating tires push back on the road (action). The road, in turn, pushes forward on the tires (reaction). The next time you see a car moving along a road, tell your friends that it is the road that pushes the car along. If at first they don't believe you, convince them that there is more than meets the eye of the casual observer. Turn them on to some physics.

5.3   Identifying Action and Reaction

Often the identification of the pair of action and reaction forces is not immediately obvious. As an example, what are the action and reaction forces in the case of a falling boulder, say, when there is no air resistance? You might say that the earth's gravitational force on the boulder is the action force, but can you identify the reaction force? Is it the weight of the boulder? No, weight is simply another name for the force of gravity. Is it caused by the ground where the boulder strikes? No, the ground does not act on the boulder until the boulder strikes it.

It turns out that there is a simple recipe for treating action and reaction forces. It is this: Make a statement about one of the forces in the pair, say the action force, in this form:

Body A exerts a force on body B.

Then the statement about the reaction force is simply

Body B exerts a force on body A.

This is easy to remember; all you need to do is switch A and B around. So, in the case of the falling boulder, the action is that the earth (body A) exerts a force on the boulder (body B). Then reaction is that the boulder exerts a force on the earth.

ACTION : TIRE PUSHES ROAD   REACTION : ROAD PUSHES TIRE

ACTION : ROCKET PUSHES GAS   REACTION : GAS PUSHES ROCKET

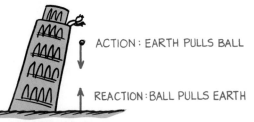

ACTION : EARTH PULLS BALL

REACTION : BALL PULLS EARTH

Fig. 5-4 Force-pair between body A and body B. Note that when action is *A exerts force on B*, the reaction is simply *B exerts force on A*.

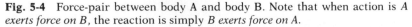

5.4   Action and Reaction on Bodies of Different Masses

Interestingly enough, the boulder pulls up on the earth with every bit as much force as the earth pulls down on it. The forces are equal in strength and opposite in direction. We say that the boulder falls to the earth; could we also say that the earth falls to the boulder? The answer is yes, but not as much. Although the forces on the boulder and the earth are the same, the masses are quite unequal. Recall that Newton's second law states that the respective accelerations will be not only proportional to the net forces, but inversely proportional to the masses as well. Considering the massiveness of the earth, it is no wonder that we don't sense its very small, or infinitesimal, acceleration. Strictly speaking, however, the earth moves up toward the falling boulder. So, when you step off a curb, the street really does come up ever so slightly to meet you!

Fig. 5-5   The earth is pulled up by the boulder with just as much force as the boulder is pulled downward by the earth.

A similar but less exaggerated example occurs in the firing of a rifle. When the rifle is fired, the force the rifle exerts on the bullet is exactly equal and opposite to the force the bullet exerts on the rifle; hence the rifle kicks. At first thought, you might expect the rifle to kick more than it does, or wonder why the bullet moves so fast compared to the rifle. According to Newton's second law, we must also consider the masses involved.

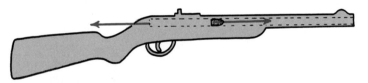

Fig. 5-6   The force that is exerted against the recoiling rifle is just as great as the force that drives the bullet along the barrel. Why then, does the bullet undergo more acceleration than the rifle?

Suppose we let F represent both the action and reaction forces, m the mass of the bullet, and m the mass of the rifle. Then the acceleration of the bullet and rifle are found by taking the ratio of force to mass. The acceleration of the bullet is given by

$$a = \frac{F}{m}$$

while the acceleration of the rifle is

$$a = \frac{F}{m}$$

Do you see why the change in motion of the bullet is so huge compared to the change of motion of the rifle? A given force exerted on a small mass produces a large acceleration, while the same force exerted on a large mass produces a small acceleration. Different sized symbols have been used to indicate the differences in relative masses and resulting accelerations.

> ▶ **Question**
>
> Suppose you're riding in the front seat of a speeding bus and you note that a bug splatters onto the windshield. Without question, a force has been exerted upon the unfortunate bug and it has undergone a sudden deceleration. Is the corresponding force that the bug exerts against the windshield greater, less, or the same? Is the resulting deceleration of the bus greater than, less than, or the same as that of the bug?

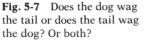

Fig. 5-7 Does the dog wag the tail or does the tail wag the dog? Or both?

Have you ever noticed Newton's third law at work when a dog wags its tail? If the tail is relatively massive compared to the dog, note that the tail also wags the dog! The effect is less noticeable for dogs with tails of relatively small mass.

5.5 | Why Action and Reaction Forces Don't Cancel

Action and reaction forces act on different bodies. If the action is caused by A acting on B, then the reaction is caused by B acting on A. The action force acts on B; the reaction force acts on A. The action and reaction forces never act on the same object. Hence action and reaction forces never cancel one another.

This is often misunderstood. Suppose for example that a friend who hears about Newton's third law says that you can't move a football by kicking it. The reason offered is that the reaction force by the kicked ball is equal and opposite to your kicking force. The net force would be zero. So if the ball was at rest to begin with, it will remain at rest—no matter how hard you kick it! What do you say to your friend?

▶ **Answer**
   The force the bug exerts against the windshield is just as great as the force the windshield exerts against it. The two forces comprise an action-reaction pair. The accelerations, however, are very different. This is because the masses involved are different. The bug undergoes an enormous deceleration, while the bus undergoes a very tiny deceleration. Indeed, a person in the bus doesn't feel the tiny slowing down of the bus as it is struck by the bug. Let the bug be more massive, like another bus for example, and the slowing down is quite evident!

You know that if you kick a football, it will accelerate. Does this acceleration contradict Newton's third law? No! Your kick acts on the ball. No other force has been applied to the ball. The net force on the ball is very real, and the ball accelerates. What about the reaction force? Aha, that doesn't act on the *ball*; it acts on your *foot*. The reaction force decelerates your foot as it makes contact with the ball. Tell your friend that you can't cancel a force on the ball with a force on your foot.

Fig. 5-8   If the action force acts on B, then the reaction force acts on A. Only a single force acts on each, so no cancellation can occur.

Now, if two people kick the same ball with equal and opposite forces at the same time (Figure 5-9), then the net force on the ball would be zero—but not for your single kick.

Fig. 5-9   When A acts on B, and C also acts on B, two opposite forces act on B and cancellation can occur.

5.6   The Horse-Cart Problem

A situation similar to the kicked football is shown in the comic strip "Horse Sense" (next page). The horse believes that its pull on the cart will be cancelled by the opposite and equal pull by the cart on the horse and make acceleration impossible. This is the classical horse-cart problem which is a stumper for many students at the university level. Through careful thinking, you can understand it here.

The horse-cart problem can be looked at from different points of view. First, there is the point of view of the farmer who is concerned only with his cart (the cart system). Then, there is the point of view of the horse (the horse system). Finally, there is the point of the horse and cart together (the horse-cart system).

First look at the farmer's point of view: The farmer is concerned only with the force that is exerted on his cart. A net force on the cart, divided by the mass of the cart, will produce a very real acceleration. The farmer doesn't care about the reaction on the horse.

Now look at the horse's point of view: It's true that the opposite reaction force by the cart on the horse restrains the horse. Without this force the horse could freely gallop to the market.

This force tends to hold the horse back. So how does the horse move forward? By pushing backward on the ground. The ground, in turn, pushes forward on the horse. In order to pull on the cart, the horse pushes backward on the ground. If the horse pushes the ground with a greater force than its pull on the cart, there will be a net force on the horse. Acceleration occurs. When the cart is up to speed, the horse need only push against the ground with enough force to offset the friction between the cart wheels and the ground.

Finally, look at the horse-cart system as a whole. From this viewpoint the pull of the horse on the cart and the reaction of the cart on the horse are internal forces—forces that act and react within the system. They contribute nothing to the acceleration of the horse-cart system. From this viewpoint they can be neglected. The system can be accelerated only by an outside force. For example, if your car is stalled, you can't get it moving by sitting inside and pushing on the dashboard; you must get outside and make the ground push the car. The horse-cart system is similar. It is the outside reaction by the ground that pushes the system.

Fig. 5-10 All the pairs of forces that act on the horse and cart are shown: (1) the pulls P of the horse and cart on each other; (2) the pushes F of the horse and ground on each other; and (3) the friction f between the cart wheels and the ground. Notice that there are two forces each applied to the cart and to the horse. Can you see that the acceleration of the horse-cart system is due to the net force $F - f$?

▶ **Questions**
1. From Figure 5-10, what is the net force that acts on the cart? That acts on the horse? That tends to make the ground recoil?

2. Once the horse gets the cart up to the desired speed, must the horse continue to exert a force on the cart?

5.7 Action Equals Reaction

This chapter began with a discussion of how, when you push against a wall, the wall in turn pushes back on you. Suppose for some reason you punch the wall. Bam, your hand is hurt. Your

▶ **Answers**

1. The net force on the cart is $P - f$; on the horse, $F - P$; on the ground, $F - f$.

2. Yes, but only enough to counteract the wheel friction of the cart and air resistance. Interestingly enough, air resistance would be absent if there were a wind blowing in the same direction and just as fast as the horse and cart. If the wind blows just enough faster to provide a force to counteract friction, the horse could wear rollerskates and simply coast with the cart all the way to the market.

Fig. 5-11 If you hit the wall, it hits you equally hard.

friends see your damaged hand and ask what happened. What can you truthfully say? You can say that the wall hit your hand. How hard did the wall hit your hand? Every bit as hard as you hit the wall. You cannot hit the wall any harder than the wall can hit back on you.

Hold a sheet of paper in midair and tell your friends that the heavyweight champion of the world could not strike the paper with a force of 200 N (nearly 50 pounds). You are correct. And the reason is that the paper is not capable of exerting a reaction force of 200 N. You cannot have an action force unless there can be a reaction force. Now, if you hold the paper against the wall, that is a different story. The wall will easily assist the paper in providing 200 N of reaction force, and then quite a bit more if need be!

For every action, there is an equal and opposite reaction. If you push hard on the world, the world pushes hard on you. If you touch the world gently, the world will touch you gently in return. The way you touch others is the way others touch you.

Fig. 5-12 You cannot touch without being touched—Newton's third law.

5 | Chapter Review

Concept Summary

An interaction between two things produces a pair of forces.
- Each thing exerts a force on the other.
- The two forces are called the action and reaction forces.
- Action and reaction forces are equal in strength and opposite in direction.

Important Terms

action force (5.2)
Newton's third law (5.2)
reaction force (5.2)

Review Questions

1. What evidence can you cite to support the idea that a wall can push on you? (5.1)

2. What is meant by saying that a force is due to an interaction? (5.1)

3. When a hammer interacts with a nail, which exerts a force on which? (5.1)

4. When a hammer exerts a force on a nail, how does the amount of force compare to that of the nail on the hammer? (5.2)

5. Why do we say that forces occur only in pairs? (5.2)

6. When you walk along a floor, what exactly pushes you along? (5.2)

7. When swimming, you push the water backward—call this *action*. Then what exactly is the reaction force? (5.3)

8. If action is a bowstring acting on an arrow, identify the reaction force. (5.3)

9. When you jump up, the world really does recoil downward. Why can't this motion of the world be noticed? (5.4)

10. When a rifle is fired, how does the size of the force of the rifle on the bullet compare to the force of the bullet on the rifle? How do the accelerations of the rifle and bullet compare? Defend your answer. (5.4)

11. Since action and reaction are always equal in size and opposite in direction, why don't they simply cancel one another and make net forces greater than zero impossible? (5.5)

Questions 12–15 refer to Figure 5-10.
12. a. Besides the force of gravity, how many forces are exerted on the cart?
    b. Using the letter symbols shown in the figure, what is the net force on the cart? (5.6)

13. a. How many forces besides gravity are exerted on the horse?
    b. What is the net force on the horse?
    c. How many forces are exerted *by* the horse on other objects? (5.6)

14. a. How many forces are exerted *on* the horse-cart system?
    b. What is the net force on the horse-cart system? (5.6)

15. In order to increase its speed, why must the horse push harder against the ground than it pulls on the wagon? (5.6)

16. If you hit a wall with a force of 200 N, how much force is exerted on you? (5.7)

17. Why cannot you hit a feather in mid-air with a force of 200 N? (5.7)

18. How does the saying "you get what you give" relate to Newton's third law? (5.7)

Think and Explain

1. Newton realized that the sun pulls on the earth with the force of gravity and causes the earth to move in orbit around the sun. Does the earth pull equally on the sun? Defend your answer.

2. Why is it easier to walk on a carpeted floor than on a smooth polished floor?

3. If you walk on a log that is floating in the water, the log moves backward. Why is this so?

4. If you step off a ledge, you noticeably accelerate toward the earth because of the gravitational interaction between the earth and you. Does the earth accelerate toward you as well? Explain.

5. Suppose you're weighing yourself while standing next to the bathroom sink. Using the idea of action and reaction, why will the scale reading be less when you push downward on the top of the sink (Figure A)? Why will the scale reading be more if you pull upward on the bottom of the sink?

Fig. A

6. What is the reaction counterpart to a force of 1000 N exerted by the earth on an orbiting communications satellite?

7. If action equals reaction, why isn't the earth pulled into orbit around the communications satellite in the preceding question?

8. A pair of 50-N weights are attached to a spring scale as shown in Figure B. Does the spring scale read 0, 50, or 100 N? (*Hint:* Would it read any differently if one of the strings were held by your hand instead of being attached to the 50-N weight?)

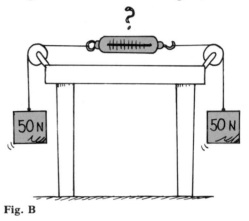

Fig. B

9. If a bicycle and a massive truck have a head-on collision, upon which vehicle is the impact force greater? Which vehicle undergoes the greater change in its motion? Defend your answers.

10. People used to think that a rocket could not be fired to the moon because it would have no air to push against once it left the earth's atmosphere. Now we know that the idea is mistaken, for several rockets have gone to the moon. What exactly is the force that propels a rocket when in a vacuum?

11. Since the force that acts on a bullet when a gun is fired is equal and opposite to the force that acts on the gun, doesn't this imply a zero net force, and therefore the impossibility of an accelerating bullet? Explain.

6 Vectors

Suppose by chance you found yourself sitting beside a physicist when taking a long bus ride. Suppose the physicist was in a talkative mood and told you about some of the things he or she did for a living. To explain things, your companion would likely doodle on a scrap of paper. Physicists love using doodles to explain ideas. Einstein was famous for that. You'd see some arrows in the doodles. The arrows would probably represent the *magnitude* (how much) and the *direction* (which way) of a certain quantity. The quantity might be the electric current that operates a minicomputer, or the orbital velocity of a communications satellite, or the enormous force that lifts an Atlas rocket off the ground. Whenever the length of an arrow represents the magnitude of a quantity, and the direction of the arrow represents the direction of the quantity, the arrow is called a **vector**.

6.1 Vector and Scalar Quantities

Some quantities require both magnitude and direction for a complete description. These are called **vector quantities**. A force, for example, has a direction as well as a magnitude. So does a velocity. Force and velocity are the most familiar vector quantities, but there are a few others treated in later chapters.

Many quantities in physics, such as mass, volume, and time, can be completely specified by their magnitudes. They do not involve any idea of direction. These are called **scalar quantities**. They obey the ordinary laws of addition, subtraction, multiplication, and division. If 3 kg of sand is added to 1 kg of cement, the resulting mixture has a mass of 4 kg. If 5 liters of water is poured from a pail which initially had 8 liters of water in it, the resulting volume is 3 liters. If a scheduled 1-hour trip runs into a 15-minute delay, the trip ends up taking $1\frac{1}{4}$ hours. In each of these cases, no direction is involved. We see that 10 kilograms north, 5 liters east, or 15 minutes south have no meaning. Quantities that involve only magnitude and not direction are scalars.

6.2 Vector Representation of Force

It is easy to draw a vector that represents a force. The length, drawn to some suitable scale, indicates the magnitude of the force. The orientation on the paper and the arrowhead show the direction.

Figure 6-1 left shows a horse pulling a cart and a man pushing the cart from behind. The diagram shows vectors for these two forces acting on the cart. The horse applies twice as much force on the cart as the man. So the vector for the force supplied by the horse is twice as long as the one for the force supplied by the man. The vectors have been drawn to a scale on which 1 cm represents 100 N. The vectors are pointing in the same direction, since the two forces are in the same direction.

Fig. 6-1 The resultant of two forces depends on the directions of the forces as well as on their magnitudes.

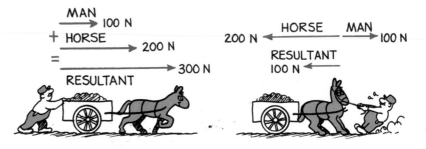

The man pushes with 100 N and the horse pulls with 200 N. Since the two forces act in the same direction, the resulting pull is equal to the sum of the individual pulls and acts in the same direction. The cart moves as if both forces were replaced by a single net force of 300 N. This net force is called the **resultant** of the two forces. We see it is represented by a vector 3 cm long.

Now suppose that the horse is pushing backwards with a force of 200 N while the man is pulling forward with a force of 100 N (Figure 6-1 right). The two forces then act in opposite directions. The resultant (net force) is equal to the difference between them, 200 N − 100 N = 100 N, and acts in the direction of the larger force. It is represented by a vector 1 cm long.

6.3 | Vector Representation of Velocity

Speed is a measure of how fast something is moving; it can be in any direction. When we take into account the direction of motion as well as the speed, we are talking about velocity. Velocity, like force, is a vector quantity.

Consider an airplane flying due north at 100 km/h relative to the surrounding air. There is a tailwind (wind from behind) that also moves due north at a velocity of 20 km/h. This example is represented with vectors in Figure 6-2 left. Here the velocity vectors are scaled so that 1 cm represents 20 km/h. Thus, the 100-km/h velocity of the airplane is shown by the 5-cm-long vector and the 20-km/h tailwind is shown by the 1-cm-long vector. You can see (with or without the vectors) that the resultant velocity is going to be 120 km/h. Without the tailwind, the airplane would travel 100 km in one hour relative to the ground below. With the tailwind, it would travel 120 km in one hour.

Suppose, instead, that the wind is a headwind (wind head-on), so that the airplane flies into the wind rather than with the wind. Now the velocity vectors are in opposite directions (Figure 6-2 right). Their resultant is 100 km/h − 20 km/h = 80 km/h. Flying against a 20-km/h headwind, the airplane would travel only 80 km relative to the ground in one hour.

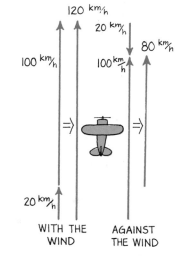

Fig. 6-2 The velocity of an airplane relative to the ground depends on its velocity relative to the air and on the wind velocity.

> ▶ **Questions**
>
> 1. In the cart, horse, and man example shown in Figure 6-1 left, suppose the man's kid brother assists by also pushing forward on the cart, but with a force of 50 N. What would then be the resultant force the men and horse exert on the cart?
>
> 2. Suppose in the previous question that all parties exert forces as stated in a headwind of 10 km/h. What would then be the resultant force the men and horse exert on the cart?

▶ **Answers**

(A reminder: are you reading this before you have thought about the questions and come up with your own answers? Finding, seeing, and remembering the answer is *not* the way to do physics. Learning to *think* about the ideas of physics is more important. Think first, then look! It will make a difference.)

1. The resultant will be the sum of the applied forces: 200 N + 100 N + 50 N = 350 N.

2. The resultant force is the same, 350 N. Be careful here. Adding apples to a pile of oranges doesn't increase the number of oranges. Similarly, we can't add or subtract velocity from force.

6.4 Geometric Addition of Vectors

Consider the forces exerted by the horses towing the barge in Figure 6-3 left. When vectors act at an angle to each other, a simple geometrical technique can be used to find the magnitude and direction of the resultant. The two vectors to be added are drawn with their tails touching (see Figure 6-3 right). A projection of each vector is drawn (dashed lines) starting at the head of the other vector. The four-sided shape that results is known as a *parallelogram* because the opposite sides are parallel and of equal length.* The resultant of the two forces is the diagonal of the parallelogram between the points where the tails meet and the dashed lines meet.**

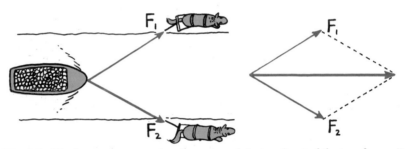

Fig. 6-3   The barge moves under the action of the resultant of the two forces F_1 and F_2. The direction of the resultant is along the diagonal of the parallelogram constructed with sides F_1 and F_2.

We can see that the barge will not move in the direction of either of the forces exerted by the horses, but rather in the direction of their resultant. The resultant is found by using the following rule for the addition of vectors:

> The resultant of two vectors may be represented by the diagonal of a parallelogram constructed with the two vectors as sides.

You can apply this rule to other pairs of forces that act on a common point. Figure 6-4 shows forces of 3 N to the north and

---

*   When the angles in a parallelogram are 90°, it becomes a rectangle; if the four sides of the rectangle are the same length, it is a square.

**   Since vector arrows only *represent* forces, it doesn't matter where you place them on a drawing so long as their directions and lengths are correct. Another way to find their resultant is to rearrange the arrows in any order so they are united tail to tip. A new vector, drawn from the tail of the first vector to the tip of the last vector, represents the resultant or net force.

4 N to the east. Using a scale of 1 N:1 cm, you construct a parallelogram, using the vectors as sides. Since the vectors are at right
angles, your parallelogram is simply a rectangle. If you draw a
diagonal from the tails of the vector pair, you have the resultant.
Measure the length of the diagonal and refer to the scale, and
you have the magnitude of the resultant. The angle can be found
with a protractor.

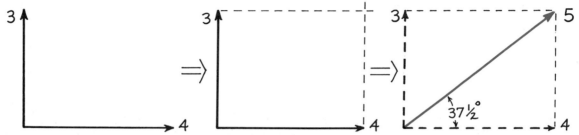

Fig. 6-4   The 3-N and 4-N forces add to produce a force of 5 N.

▶ **Exercises**
1. By the parallelogram method construct the resultants of
   the 3-N and 4-N forces represented by the vectors shown.
   They are drawn to a scale on which 1 cm:1 N. Measure
   your resultants with a ruler and compare them to the
   correct answers given at the bottom of the page.

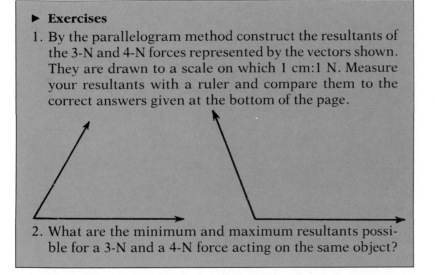

2. What are the minimum and maximum resultants possible for a 3-N and a 4-N force acting on the same object?

   With the technique of vector addition you can correct for the
effect of a crosswind on the velocity of an airplane. Consider a
slow-moving airplane that flies north at 80 km/h and is caught
in a strong crosswind of 60 km/h blowing east. Figure 6-5 shows
vectors for the airplane velocity and wind velocity. The scale

▶ **Answers**
1. Left: 6 N; right: 4 N.

2. The minimum resultant occurs when the forces oppose each other: 4 N − 3 N =
   1 N. The maximum resultant occurs when they are in the same direction: 4 N +
   3 N = 7 N. (At angles to each other, 3 N and 4 N can combine to range between 1 N and 7 N.)

here is 1 cm:20 km/h. The diagonal of the constructed parallelogram (rectangle in this case) measures 5 cm, which represents 100 km/h. So the airplane moves at 100 km/h relative to the ground, in a northeasterly direction.*

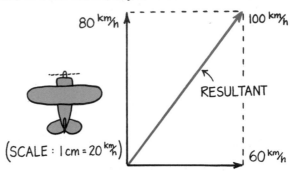

Fig. 6-5 An 80-km/h airplane flying in a 60-km/h crosswind has a resultant speed of 100 km/h relative to the ground.

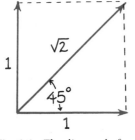

Fig. 6-6 The diagonal of a square is $\sqrt{2}$ the length of one of its sides.

There is a special case of the parallelogram that often occurs. When two vectors that are equal in magnitude and at right angles to one another are to be added, the parallelogram becomes a square. Since for any square the length of a diagonal is $\sqrt{2}$, or 1.414, times one of the sides, the resultant is $\sqrt{2}$ times one of the vectors. For example, the resultant of two equal vectors of magnitude 100 acting at right angles to each other is 141.4.

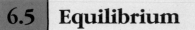

6.5   Equilibrium

The method of combining vectors by the parallelogram rule is an experimental fact. It can be shown to be correct by considering an example that is common and quite surprising the first time— the case of being able to hang safely from a vertical clothesline but not being able to do so when the line is strung horizontally. Invariably, it breaks (Figure 6-7).

---

*   Whenever the vectors are at right angles to each other, their resultant can be found by the Pythagorean Theorem, a well-known tool of geometry. It states that the square of the hypotenuse of a right-angle triangle is equal to the sum of the squares of the other two sides. Note that two right triangles are present in the parallelogram (rectangle in this case) in Figure 6-5. From either one of these triangles we get:

$$\text{resultant}^2 = (60 \text{ km/h})^2 + (80 \text{ km/h})^2$$
$$= 3600 \text{ (km/h)}^2 + 6400 \text{ (km/h)}^2$$
$$= 10\,000 \text{ (km/h)}^2$$

The square root of 10 000 (km/h)² is 100 km/h, as expected.

Fig. 6-7 You can safely hang
from a piece of clothesline
when it hangs vertically, but
you'll break it if you attempt
to make it support your
weight when it is strung
horizontally.

We can understand this with spring scales that are used to
measure weight. Consider a block that weighs 10 N, about the
weight of ten apples. If we suspend it from a single scale, as in
Figure 6-8 left, the reading will be its weight, 10 N. Did you know
that if you stand with your weight evenly divided on two bathroom scales, each scale will record half your weight? This is because the springs in the scales are compressed and, in effect,
push up on you just as hard as gravity pulls down on you. If two
scales support your weight, each will have half the job. The same
is true if you hang by a pair of scales. In this case the springs that
support you are stretched, and each scale has half the job and
reads half your weight. So if we suspend the 10-N block from a
pair of vertical scales (Figure 6-8 right), each scale will read 5 N.
The scales pull up with a resultant force that equals the weight
of the block. The diagram shows a pair of 5-N vectors that have a
10-N resultant that exactly opposes the 10-N weight vector. The
net force on the block is zero, and the block hangs at rest; we say
it is in **equilibrium**. The key idea is this: if a 10-N block is to hang
in equilibrium, the resultant of the forces supplied by the pair of
springs must equal 10 N. For vertical orientation this is easy:
5 N + 5 N = 10 N. This is all Chapter 4 stuff. ☺

Now let's look at a non-vertical arrangement. In Figure 6-9
left, we see that when the supporting spring scales hang at an
angle to support the block, the springs are stretched more, as indicated by the greater reading. At 60° from the vertical, the
readings are 10 N each—double what they were when the scales
were hanging vertically! Can you see the explanation? The resultant of the two scale readings must be 10 N upward to balance
the downward weight of the block. A pair of 5-N vectors will
produce a 10-N resultant only if they are parallel and acting in
the same direction. If the vectors have different directions, then
each vector must be greater than 5 N to produce a resultant of
10 N. For equilibrium, the diagonal of the parallelogram formed
by the vector sides, whatever the angle between them, must remain the same. Why? Because the diagonal must correspond to
10 N, exactly equal and opposite to the 10 N weight of the block.

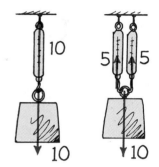

Fig. 6-8 (Left) When a 10-N
block hangs vertically from a
single spring scale, the scale
pulls upward with a force of
10 N. (Right) When it hangs
vertically from two spring
scales, each scale pulls upward with a force of half the
weight, or 5 N.

Fig. 6-9 As the angle between the spring scales increases, the scale readings
increase so that the resultant
(dashed-line vector) remains
at 10 N upward, which is required to support the 10-N
block.

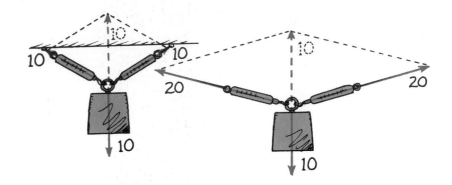

In Figure 6-9 right, where the angle from the vertical has been
increased to 75.5°, each spring must pull with 20 N to produce
the required 10-N resultant. As the angle between the scales is
increased, the scale readings increase. Can you see that as the
angle between the sides of the parallelogram increases, the magnitude of the sides must increase if the diagonal is to remain the
same? If you understand this, you understand why you can't
be supported by a horizontal clothesline without producing a
stretching force that is considerably greater than your weight.
The parallelogram rule turns out to be quite interesting.

▶ **Questions**

1. If the kids on the swings are of equal weight, which swing
   is more likely to break?

2. Two pictures of equal weight are hung in a gallery as
   shown. In which of the two arrangements is the wire
   more likely to break?

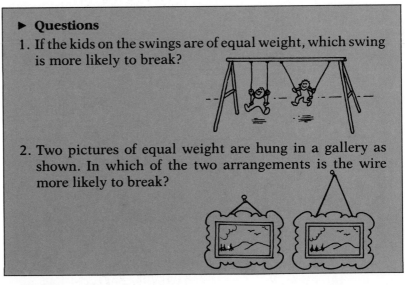

▶ **Answers**

1. The tension is greater in the ropes that hang at an angle, so they are more
   likely to break than the vertical ropes.

2. The tension is greater in the picture to the left, because the supporting rope
   makes a greater angle with respect to the vertical than the picture on the right.
   This is similar to the stretched clothesline.

6.6 | Components of Vectors

Two vectors acting on the same object may be replaced by a single vector (the resultant) that produces the same effect upon the object as the combined effects of the given vectors. The reverse is also true: any single vector may be regarded as the resultant of two vectors, each of which acts on the body in some direction other than that of the given vector. These two vectors are known as the **components** of the given vector that they replace. The process of determining the components of a vector is called **resolution**.

A man pushing a lawnmower applies a force that pushes the machine forward and also against the ground. In Figure 6-10, vector F represents the force applied by the man. We can separate this force into two components. Vector Y is the vertical component, which is the downward push against the ground. Vector X is the horizontal component, which is the forward force that moves the lawnmower.

We can find the magnitude of these components by drawing a rectangle with F as the diagonal. Since X and Y are the sides of a parallelogram, vector F is the resultant of the vectors X and Y. Hence the two components X and Y acting together are equivalent to the force F. That is, the motion of the lawnmower is the same whether we assume that the man exerts two forces, components X and Y, or only one force, F.

The rule for finding the vertical and horizontal components of any vector is relatively simple, and is illustrated in Figure 6-11. A vector V is drawn in the proper direction to represent the force, velocity, or whatever vector is in question (Figure 6-11 left). Then vertical and horizontal lines are drawn at the tail of the vector (Figure 6-11 right). A rectangle is drawn that encloses the vector V in such a way that V is a diagonal and the sides of the rectangle are the desired components. We see that the components of the vector V are then represented in direction and magnitude by the vectors X and Y.

Fig. 6-10 The force F applied to the lawnmower may be resolved into a horizontal component, X, and a vertical component, Y.

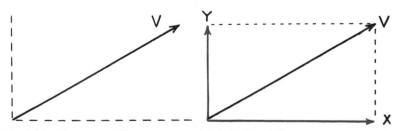

Fig. 6-11 The vector V has component vectors X and Y.

Any vector can be represented by a pair of components that are at right angles to each other. This is neatly illustrated in the explanation of a sailboat sailing against the wind. (See Appendix C, Vector Applications, at the back of this book.)

▶ **Exercise**

With a ruler, draw the horizontal and vertical components of the two vectors shown. Measure the components and compare your findings with the answers given at the bottom of the page.

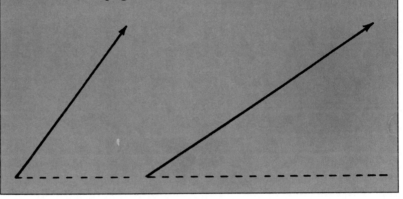

6.7 | Components of Weight

We all know that a ball will roll faster down a steep hill than a hill with a small slope. The steeper the hill is, the greater the acceleration of the ball. We can understand why this is so with vector components. The force of gravity that acts on things gives them weight, which we represent as vector W. The force vector W acts only straight down—toward the center of the earth—but components of W may act in any direction. It is most often useful to consider components that are at right angles to each other.

In Figure 6-12 we see W broken up into components A and B, where A is parallel to the surface and B is perpendicular to the surface. It is component A that makes the ball move. Component B presses the ball against the surface. Pictures are better than words, so study the figure and see how the magnitudes of the components vary for different slopes.

▶ **Answer**

Left vector: the horizontal component is 3 cm; the vertical component is 4 cm.
Right vector: the horizontal component is 6 cm; the vertical component is 4 cm.

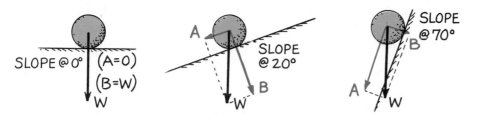

Fig. 6-12   The weight of the ball is represented by vector *W*, which has perpendicular components *A* and *B*. Vector *A* serves to change the speed of the ball, while vector *B* presses it against the surface. Note how the magnitudes of *A* and *B* vary from zero to *W*.

Can you see that only when the slope is zero—when the surface is horizontal—is component *A* equal to zero? That's why the speed of the ball does not change on a horizontal surface. Note another thing when the surface is horizontal. Component *B* is equal to *W*; the ball presses against the surface with the most force. But when the slope is 90°, component *B* becomes zero and *A* equals *W*, so the ball has its maximum acceleration.

> ▶ **Question**
>   At what angle will components *A* and *B* in Figure 6-12 have equal magnitudes? At what angle will *A* equal *W*? At what angle will *A* be greater in magnitude than *W*?

6.8   Projectile Motion

Chapter 2 discussed the vector quantities velocity and acceleration. Since only horizontal and vertical motion was considered, we did not need to know about vector addition or the techniques of vector resolution. But for objects projected at angles other than straight up or straight down, we do.

A **projectile** is any object that is projected by some means and continues in motion by its own inertia. A cannonball shot from a cannon, a stone thrown into the air, or a ball that rolls off the edge of the table are all projectiles. These projectiles follow curved paths that at first thought seem rather complicated. However, these paths are surprisingly simple when we look at the horizontal and vertical components of motion separately.

▶ **Answer**
  Components *A* and *B* have equal magnitudes at 45°; *A* = *W* at 90°; *A* cannot have a greater magnitude than *W* at any angle.

The horizontal component of motion for a projectile is no more complex than the horizontal motion of a bowling ball rolling freely along a level bowling alley. If the retarding effect of friction can be ignored, the bowling ball moves at constant velocity. It covers equal distances in equal intervals of time. It rolls of its own inertia, with no component of force acting in its direction of motion. It rolls without accelerating. The horizontal part of a projectile's motion is just like the bowling ball's motion along the alley (Figure 6-13 left).

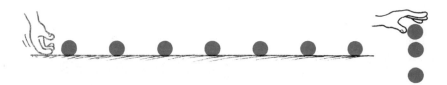

Fig. 6-13   (Left) Roll a ball along a level surface, and its velocity is constant because no component of gravitational force acts horizontally. (Right) Drop it, and it accelerates downward and covers greater vertical distances each second.

The vertical component of motion for a projectile following a curved path is just like the motion described in Chapter 2 for a freely-falling object. Like a ball dropped in mid-air, the projectile moves in the direction of earth gravity and accelerates downward (Figure 6-13 right). The increase in speed in the vertical direction causes successively greater distances to be covered in each successive equal-time interval.

Interestingly enough, the horizontal component of motion for a projectile is completely independent of the vertical component of motion. Each acts independently of the other. Their combined effects produce the variety of curved paths that projectiles follow.

The multiple-flash exposure of Figure 6-14 shows equally-timed successive positions for a ball rolled off a horizontal table. Investigate the photo carefully, for there's a lot of good physics there. The curved path of the ball is best analyzed by considering the horizontal and vertical components of motion separately. There are two important things to notice. The first is that the ball's horizontal component of motion doesn't change as the falling ball moves sideways. The ball travels the same horizontal distance in the equal times between each flash. That's because there is no component of gravitational force acting horizontally. Gravity acts only downward, so the only acceleration of the ball is downward. The second thing to note from the photo is that the vertical positions become farther apart with time. The distances traveled vertically are the same as if the ball were simply dropped. It is interesting to note that the downward motion of the ball is the same as that of free fall.

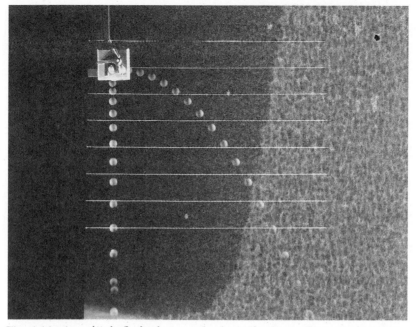

Fig. 6-14   A multiple-flash photograph of a ball rolling off a horizontal table. Notice that in equal times it travels equal horizontal distances but increasingly greater distances vertically. Do you know why?

The path traced by a projectile that accelerates only in the vertical direction while moving at a constant horizontal velocity is called a *parabola*. When air resistance can be neglected—usually for slow-moving projectiles or ones very heavy compared to the forces of air resistance—the curved paths are *parabolic*.

▶ **Question**

At the instant a horizontally held rifle is fired over a level range, a bullet held at the side of the rifle is released and drops to the ground. Which bullet—the one fired down-range or the one dropped from rest—strikes the ground first?

▶ **Answer**

Both bullets fall the same vertical distance with the same acceleration *g* due to gravity and therefore strike the ground at the same time. Can you see that this is consistent with our analysis of Figure 6-14? We can reason this another way by asking which bullet would strike the ground first if the rifle were pointed at an upward angle. In this case, the bullet that is simply dropped would hit the ground first. Now consider the case where the rifle is pointed downward. The fired bullet hits first. So upward, the dropped bullet hits first; downward, the fired bullet hits first. There must be some angle at which there is a dead heat—where both hit at the same time. Can you see it would be when the rifle is neither pointing upward nor downward—when it is horizontal?

6.9 | Upwardly Moving Projectiles

Consider a cannonball shot at an upward angle. Pretend for a moment that there is no gravity; then according to the law of inertia, the cannonball will follow the straight-line path shown by the dashed line in Figure 6-15. But there *is* gravity, so this doesn't happen. What really happens is that the cannonball continually *falls beneath this imaginary line* until it finally strikes the ground. Get this: The vertical distance it falls beneath any point on the dashed line is the same vertical distance it would fall if it were dropped from rest and had been falling for the same amount of time. This distance, as introduced in Chapter 2, is given by $d = \frac{1}{2}gt^2$, where t is the elapsed time.

Fig. 6-15 With no gravity the projectile would follow the straight-line path (dashed line). But because of gravity, it falls beneath this line the same vertical distance it would fall if released from rest. Compare the distances fallen with Table 2-3 in Chapter 2. (With $g = 9.8$ m/s², these distances are more accurately 4.9 m, 19.6 m, and 44.1 m.)

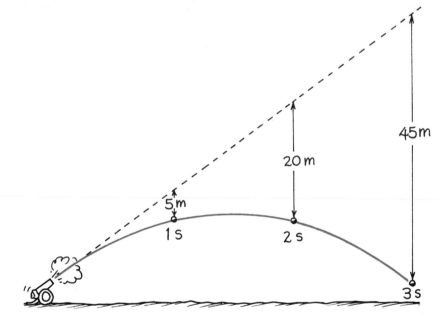

We can put this another way: Shoot a projectile skyward at some angle and pretend there is no gravity. After so many seconds t, it should be at a certain point along a straight-line path. But because of gravity, it isn't. Where is it? The answer is, it's directly below this point. How far below? The answer in meters is $5t^2$ (or more accurately, $4.9t^2$). Isn't that neat?

Note another thing from Figure 6-15. The cannonball moves equal horizontal distances in equal time intervals. That's because no acceleration takes place horizontally. The only acceleration is vertically, in the direction of earth gravity. The vertical distance it falls below the imaginary straight-line path during equal time intervals continually increases with time.

Figure 6-16 shows vectors representing both horizontal and vertical components of velocity for a projectile following a parabolic path. Notice that the horizontal component is everywhere the same, and only the vertical component changes. Note also that the actual velocity is represented by the vector that forms the diagonal of the rectangle formed by the vector components. At the top of the path the vertical component vanishes to zero, so the actual velocity there *is* the horizontal component of velocity at all other points. Everywhere else the magnitude of velocity is greater (just as the diagonal of a rectangle is greater than either of its sides).

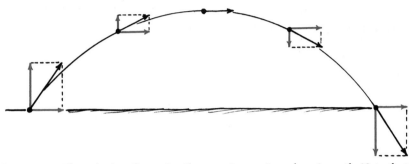

Fig. 6-16 The velocity of a projectile at various points along its path. Note that the vertical component changes and the horizontal component is the same everywhere.

Figure 6-17 shows the path traced by a projectile with the same launching speed at a steeper angle. Notice that the initial velocity vector has a greater vertical component than when the projection angle is less. This greater component results in a higher path. But the horizontal component is less so the range is less.

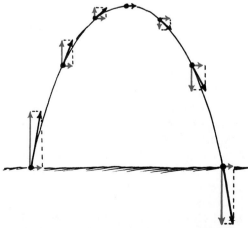

Fig. 6-17 Path for a steeper projection angle.

Figure 6-18 shows the paths of several projectiles all having the same initial speed but different projection angles. The figure neglects the effects of air resistance, so the paths are all parabolas. Notice that these projectiles reach different *altitudes*, or heights above the ground. They also have different *horizontal ranges*, or distances traveled horizontally.

Fig. 6-18 Ranges of a projectile shot at the same speed at different projection angles.

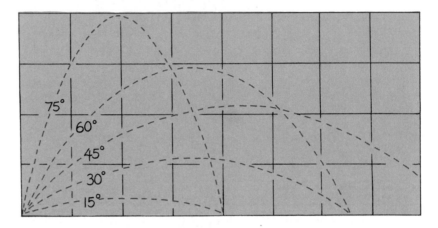

The remarkable thing to note from Figure 6-18 is that the same range is obtained from two different projection angles—angles that add up to 90 degrees! An object thrown into the air at an angle of 60 degrees, for example, will have the same range as if it were thrown at the same speed at an angle of 30 degrees. For the smaller angle, of course, the object remains in the air for a shorter time.

Fig. 6-19 Maximum range is attained when the ball is batted at an angle of nearly 45°. (In cases where the weight of the projectile is comparable to the applied force, as when a heavy javelin is thrown, the applied force does not produce the same speed for different projection angles, and maximum range occurs for angles quite a bit less than 45°.)

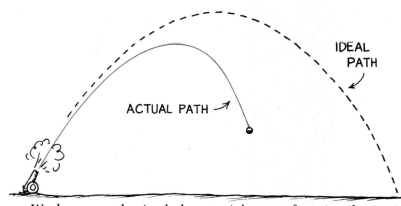

IDEAL
PATH

ACTUAL PATH

Fig. 6-20   In the presence of air resistance, the path of a high-speed projectile falls short of a parabola (dashed curve).

We have emphasized the special case of projectile motion without air resistance. When there is air resistance, the range of a projectile is somewhat shorter and is not a true parabola (Figure 6-20).

▶ **Questions**

1. A projectile is shot at an angle into the air. If air resistance is negligible, what is its downward acceleration? Its horizontal acceleration?

2. At what part of its path does a projectile have minimum speed?

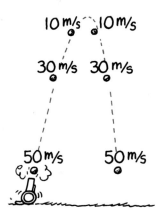

Fig. 6-21   Without air resistance, speed lost while going up equals speed gained while coming down; time up equals time down.

If air resistance is small enough to be negligible, a projectile will rise to its maximum height in the same time it takes to fall from that height to the ground. This is because its deceleration by gravity while going up is the same as its acceleration by gravity while coming down. The speed it loses while going up is therefore the same as the speed it gains while coming down. So the projectile arrives at the ground with the same speed it had when it was projected from the ground.

If an object is projected fast enough so that its curvature matches the curvature of the earth, and it is above the atmosphere so that air resistance does not affect its motion, it will fall all the way around the earth and be an earth satellite. This interesting topic is treated in Chapter 12.

▶ **Answers**

1. Its downward acceleration is g because the force of gravity is downward; its horizontal acceleration is zero because no horizontal forces act on it.

2. The speed of a projectile is minimum at the top of its path. If it is launched vertically, its speed at the top is zero. If it is projected at an angle, the vertical component of velocity is zero at the top, leaving only the horizontal component. So the speed at the top is equal to the horizontal component of the projectile's velocity at any point. Isn't that neat?

6 Chapter Review

Concept Summary

Vector quantities have both magnitude and direction.
- A vector is an arrow whose length represents the magnitude of a vector quantity and whose direction represents the direction of the quantity.

The resultant of several forces or several velocities can be determined from a vector diagram drawn to scale.
- When something is in equilibrium, the resultant of all the forces supporting it must exactly oppose its weight.

Any single vector can be replaced by two components that add to form the original vector.
- It is often convenient to study the horizontal and vertical components of forces or velocities.
- When gravity is the only force acting on a projectile, the horizontal component of its velocity does not change.

Important Terms

component (6.6)   resultant (6.2)
equilibrium (6.5)   scalar quantity (6.1)
projectile (6.8)   vector (6.1)
resolution (6.6)   vector quantity (6.1)

Review Questions

1. How does a vector differ from a scalar? (6.1)

2. If a vector that is 1 cm long represents a force of 5 N, how many newtons does a vector 2 cm long, drawn to the same scale, represent? (6.2)

3. a. What is the resultant of a pair of forces, 100 N upward and 75 N downward?
   b. What is their resultant if they both act downward? (6.2)

4. Why is speed classified as a scalar, and velocity as a vector? (6.3)

5. What is the resultant velocity of an airplane that normally flies at 200 km/h if it experiences a 50-km/h tailwind? A 50-km/h headwind? (6.3)

6. What is a parallelogram? (6.4)

7. When a parallelogram is constructed in order to add forces, what represents the resultant of the forces? (6.4)

8. What is the magnitude of the resultant of two vectors of magnitudes 4 and 3 that are at right angles to each other? (6.4)

9. What is the magnitude of the resultant of a pair of 100-N vectors that are at right angles to each other? (6.4)

10. The tension in a clothesline carrying a load of wash is appreciably greater when the clothesline is strung horizontally than when it is strung vertically. Why? (6.5)

11. What is the net force, or equivalently, the resultant force that acts on an object when it is in equilibrium? (6.5)

12. Compared to your weight, what is the stretching force in your arm when you let yourself hang motionless by one arm? By both arms vertically? Is this force greater or less if you hang with your hands wide apart? Why? (6.5)

13. Distinguish between the method of geometric addition of vectors and vector resolution. (6.4, 6.6)

14. What are the magnitudes of the horizontal and vertical components of a vector that is 100 units long, and oriented at 45°? (6.6)

15. The weight of a ball rolling down an inclined plane can be broken into two vector components: one acting parallel to the plane, and the other acting perpendicular to the plane.
    a. At what slope angle are these two components equal?
    b. At what slope angle is the component parallel to the plane equal to zero?
    c. At what slope angle is the component parallel to the plane equal to the weight? (6.7)

16. Why does a bowling ball move without acceleration when it rolls along a bowling alley? (6.8)

17. In the absence of air resistance, why does the horizontal component of velocity for a projectile remain constant, and why does only the vertical component change? (6.8)

18. How does the downward component of the motion of a projectile compare to the motion of free fall? (6.8)

19. At the instant a ball is thrown horizontally over a level range, a ball held at the side of the first is released and drops to the ground. If air resistance can be neglected, which ball—the one thrown or the one dropped from rest—strikes the ground first? (6.8)

20. a. How far below an initial straight-line path will a projectile fall in one second?
    b. Does your answer depend on the angle of launch or on the initial speed of the projectile? Defend your answer. (6.9)

21. a. A projectile is fired straight upward at 100 m/s. How fast is it moving at the instant it reaches the top of its trajectory?
    b. What is the answer if the projectile is fired upward at 45° instead? (6.9)

22. At what angle should a slingshot be oriented for maximum altitude? For maximum horizontal range? (6.9)

23. Neglecting air resistance, if you throw a ball straight upward with a speed of 20 m/s, how fast will it be moving when you catch it? (6.9)

24. a. Neglecting air resistance, if you throw a baseball at 20 m/s to your friend who is on first base, will the catching speed be greater than, equal to, or less than 20 m/s?
    b. How about if air resistance *is* a factor? (6.9)

Activity

Place a coin at the edge of a smooth table so that it overhangs slightly. Then place a second coin on the table top some distance from the overhanging coin. Set the second coin sliding across the table (such as by snapping it with your finger) so that it strikes the overhanging coin and both coins fall to the floor below. Which—if either—hits the floor first? Does your answer depend on the speed of the sliding coin?

Think and Explain

1. a. What is the maximum possible resultant of a pair of vectors, one of magnitude 5 and the other of magnitude 4?
   b. What is the minimum possible resultant?

2. A boat is rowed at 8 km/h directly across a river that flows at 6 km/h (Figure A).
   a. What is the resultant speed of the boat?
   b. How fast and in what direction can the boat be rowed to reach a destination directly across the river?

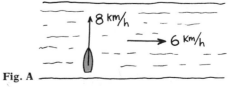

Fig. A

3. By whatever means, find the direction of the airplane in Figure 6-5.

4. In which position is the tension the least in the arms of the weightlifter shown in Figure B? The most?

Fig. B

5. Why do electric power lines sometimes break in winter when a small weight of ice forms on them?

6. Why cannot the strong man in Figure C pull hard enough to make the chain straight?

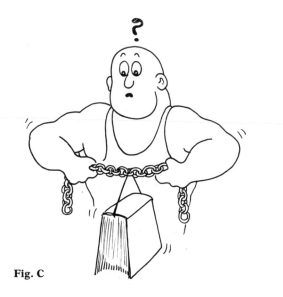

Fig. C

7. Why are the main supporting cables of suspension bridges designed to sag the way they do (Figure D)?

Fig. D

8. The boy on the tower (Figure E) throws a ball 20 m downrange, as shown. What is his pitching speed?

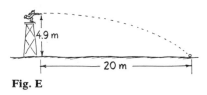

Fig. E

9. Why does a ball rolling down an incline undergo more acceleration the steeper the incline?

10. Why is less force required to push a barrel up a sloping ramp (Figure F) than to lift it vertically?

Fig. F

7 Momentum

Have you ever wondered how a karate expert can sever a stack of cement bricks with the blow of her bare hand? Or why a fall on a wooden floor is not nearly as damaging as a fall on a cement floor? Or why "follow through" is important in golf, baseball, and boxing? To understand these things, you first need to recall the concept of inertia, introduced in Chapter 3 in Newton's first law of motion, and developed further in Chapters 4 and 5 in Newton's second and third laws of motion. Inertia was discussed in terms of objects both at rest and in motion. Now we concern ourselves only with the inertia of moving objects. The idea of inertia in motion is *momentum*, which refers to moving things.

7.1 Momentum

We all know that a massive truck is harder to stop than a small car moving at the same speed. We say the truck has more momentum than the car. By **momentum** we mean inertia in motion, or more specifically, the mass of an object multiplied by its velocity. That is:

$$\text{momentum} = \text{mass} \times \text{velocity}$$

or, in shorthand notation,

$$\text{momentum} = mv$$

When direction is not an important factor, we can say:

$$\text{momentum} = \text{mass} \times \text{speed},$$

which we still abbreviate mv.

We can see from the definition that a moving object can have a large momentum if either its mass is large, or its speed is large,

Fig. 7-1 A truck rolling down a hill has more momentum than a roller skate moving at the same speed, because the truck has more mass. But if the truck is at rest and the roller skate moves, then the skate has more momentum because only it has speed.

or if both its mass and speed are large. A truck has a larger momentum than a car moving at the same speed because its mass is larger. An enormous ship moving at a small speed can have a large momentum, whereas a small bullet moving at a high speed can also have a large momentum. And, of course, a huge object moving at a high speed, such as a massive truck rolling down a steep hill with no brakes, has a huge momentum, whereas the same truck at rest has no momentum at all.

▶ **Question**
Can you think of a case where the roller skate and truck shown in Figure 7-1 both would have the same momentum?

7.2   Impulse Equals Change in Momentum

If the momentum of an object changes, either the mass or the velocity or both changes. If the mass remains unchanged, as is most often the case, then the velocity changes. Acceleration occurs. And what produces an acceleration? The answer is a *force*. The greater the force that acts on an object, the greater will be the change in velocity, and hence, the change in momentum.

But something else is important also: *time*—how long the force acts. Apply a force briefly to a stalled automobile, and you produce a small change in its momentum. Apply the same force over an extended period of time, and a greater change in momentum results. A long sustained force produces more change in momentum than the same force applied briefly. So for changing the momentum of an object, both force and time are important.

Interestingly enough, Newton's second law ($a = F/m$) can be re-expressed to make the *time* factor more evident when the term for acceleration is replaced by its definition, change in velocity per time.* Then the equation becomes "force × time interval" is

▶ **Answer**
The roller skate and truck can have the same momentum if the speed of the roller skate is very much greater than the speed of the truck. How much greater? As many times as the mass of the truck is greater than the mass of the roller skate! Get it?

---

*   We can state Newton's second law as $F = ma$. Since a = (change in velocity) ÷ (time interval), we can say $F = m \times [(\text{change in } v)/t]$. Simple algebraic rearrangement gives Ft = change in mv, or in delta notation, $F\Delta t = \Delta mv$.

equal to the change in "mass × velocity." The quantity "mass × velocity" has already been defined as momentum. So Newton's second law can be re-expressed in the form

$$Ft = \text{change in } mv$$

which reads, "force multiplied by the time-during-which-it-acts equals change in momentum."

The quantity "force × time interval" is called **impulse**. Thus,

$$\text{impulse} = \text{change in momentum}$$

The impulse-momentum relationship helps us to analyze a variety of circumstances where momentum is changed. We will consider familiar examples of impulse for the cases of (1) increasing momentum, (2) decreasing momentum over a long time, and (3) decreasing momentum over a short time.

Case 1: Increasing Momentum

It makes good sense that in order to increase the momentum of an object, we should apply the greatest force we can. Also, we should extend the time of contact as much as possible. A golfer and baseball player do both when they swing as hard as possible and "follow through" when hitting a ball.

The forces involved in impulses are usually not steady, but vary from instant to instant. For example, a golf club that strikes a ball exerts zero force on the ball until it comes in contact; then the force increases rapidly as the club and ball are distorted (Figure 7-2); the force then diminishes as the ball comes up to speed and returns to its original shape. So when we speak of such impact forces in this chapter, we mean the *average* force of impact.

Fig. 7-2 The force of a golf club against a golf ball varies throughout the duration of impact.

Case 2: Decreasing Momentum over a Long Time

If you were in a car that was out of control, and you had your choice of hitting a concrete wall or a haystack, you wouldn't have to call on your knowledge of physics to make up your mind. But knowing some physics helps you to understand *why* hitting something soft is entirely different from hitting something hard. In the case of hitting either the wall or the haystack, your momentum will be decreased by the same impulse. The same impulse means the same *product* of force and time, not the same force or the same time. You have a choice. By hitting the haystack instead of the wall, you extend the time of impact—*you extend the time during which your momentum is brought to zero.* The longer time is compensated by a lesser force. If you extend the time of impact 100 times, you reduce the force of impact by 100. So whenever you wish the force of impact to be small, extend the time of impact.

Fig. 7-3 If the change in momentum occurs over a long time, the force of impact is small.

Everyone knows that a padded dashboard in a car is safer than a bare metal one, and that airbags save lives. Most people also know that if you're going to catch a fast baseball with your bare hand, you extend your hand forward so you'll have plenty of room to let your hand move backward after you make contact with the ball. You wouldn't intentionally hold your hand stationary, as in catching a ball with your bare hand against a hard wall! In these cases you extend the time of impact and thereby reduce the force of impact.

When you jump off something to the ground below, you don't keep your legs straight and stiff (ouch!). Instead, you bend your knees upon making contact. By doing this, you extend the time during which your momentum is decreasing by 10 to 20 times that of a stiff-legged abrupt landing. The forces your bones experience are thus reduced 10 to 20 times by such knee bending.

A wrestler thrown to the floor tries to extend his time of arrival on the mat by relaxing his muscles and spreading the impact into a series of smaller impacts, as his foot, knee, hip, ribs, and shoulder fold onto the mat in turn. Of course, falling on a mat is preferable to falling on a solid floor, for this also increases the time of impact.

Everyone knows that it is less harmful to fall on a wooden floor than on a concrete floor. And most people know that this is because the wooden floor has more "give" than the concrete floor. Ask most people why a floor with more give makes for an easier fall, and you'll get a puzzled response. They may say, "Because it has more give." But your question is, "Why does a floor with more give produce a safer fall?" The answer is, we know, because an impulse is required to bring your momentum to a halt, and the impulse is composed of two variables, impact force and impact time. By extending the impact time as the floor gives, the impact force is correspondingly reduced. The safety net used by acrobats provides an obvious example of small impact force over a long time to provide the required impulse to reduce the momentum of fall.

A boxer confronted with a high-momentum punch wishes to minimize the force of impact. If he cannot avoid being hit, at least he can control the length of time it takes for his body to absorb the incoming momentum of his opponent's fist. So he wisely extends the impact time by "riding or rolling with the punch." This lessens the force of impact.

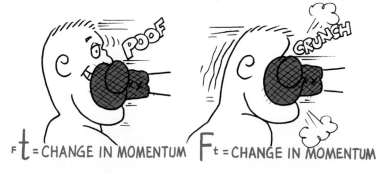

Fig. 7-4 In both cases the impulse provided by the boxer's jaw reduces the momentum of the punch. (Left) When the boxer moves away (rides with the punch), the chief ingredient of impulse is time. (Right) When the boxer moves into the glove, the time is reduced and the chief ingredient of impulse is force.

▶ **Question**
If the boxer in Figure 7-4 is able to make the duration of impact five times as long by riding with the punch, by how much will the force of impact be reduced?

Case 3: Decreasing Momentum over a Short Time
Ride a bicycle into a concrete wall, and you're in trouble. When you catch a high-speed baseball, move your hand toward the ball instead of away upon contact, and your hand is messed up. When boxing, move into a punch instead of away, and your opponent has a better chance of scoring a knockout. In these cases the principal ingredients of impulses required to reduce momentum are impact forces because the times of impact are brief.

Fig. 7-5 If the change in momentum occurs over a short time, the impact force is large.

▶ **Answer**
The force of impact will be 5 times less than if he didn't pull back.

The idea of short time of contact explains how a karate expert can sever a stack of bricks with the blow of her bare hand (Figure 7-6). She brings her arm and hand swiftly against the bricks with considerable momentum. This momentum is quickly reduced when she delivers an impulse to the bricks. The impulse is the force of her hand against the bricks multiplied by the time her hand makes contact with the bricks. By swift execution she makes the time of contact very brief, and correspondingly makes the force of impact huge. If her hand is made to *bounce* upon impact, the force is even greater.

Fig. 7-6 A large impulse to the bricks in a short time produces a considerable force.

▶ **Question**
    A boxer being hit with a punch contrives to extend time for best results, whereas a karate expert delivers a force in a short time for best results. Isn't there a contradiction here?

7.3 | Bouncing

If a flower pot falls from a shelf onto your head, you may be in trouble. If it bounces from your head, you're certainly in trouble. Impulses are greater when bouncing takes place. This is because

▶ **Answer**
    There is no contradiction because the best results for each are quite different. The best result for the boxer is reduced force, accomplished by maximizing time, and the best result for the karate expert is increased force delivered in minimum time.

the impulse required to bring something to a stop and then, in effect, "throw it back again" is greater than the impulse required merely to bring something to a stop. Suppose, for example, that you catch the falling pot with your hands. Then you provide an impulse to catch it and reduce its momentum to zero. If you were to then throw the pot upward, you would have to provide additional impulse. So it would take more impulse to catch it *and* throw it back up than merely to catch it. The same greater impulse is supplied by your head if the pot bounces from it.

The fact that impulses are greater when bouncing takes place was employed with great success in California during the gold rush days. The water wheels used in gold-mining operations were inefficient. A man named Lester A. Pelton saw that the problem had to do with their flat paddles. He designed curved-shape paddles that would cause the incident water to make a U-turn upon impact—to "bounce." In this way the impulse exerted on the water wheels was greatly increased. Pelton patented his idea and made more money from his invention, the Pelton wheel, than any of the gold miners.

IMPULSE

Fig. 7-7 The Pelton wheel. The curved blades cause water to bounce and make a U-turn, which produces a greater impulse to turn the wheel.

7.4 | Conservation of Momentum

Newton's second law tells you that if you wish to accelerate an object, you must apply a force to it. This chapter is saying much the same thing, but in different language. If you wish to change the momentum of an object, exert an impulse on it.

In either case, the force or impulse must be exerted on the object by something outside the object. Internal forces do not count.

For example, the molecular forces within a basketball have no effect upon the momentum of the basketball, just as your push against the dashboard of a car you're sitting in will have no effect in changing the momentum of the car. This is because these forces are internal forces. They act and react within the object. To change the momentum of the basketball or car, an outside push or pull is required. If no outside force is present, then no change in momentum is possible.

Consider a rifle being fired. The force that pushes on the bullet when it is inside the rifle barrel is equal to the force that makes the rifle recoil (Newton's third law, action and reaction). These forces, interestingly enough, are internal to the "system" that comprises the rifle and bullet. So they don't change the momentum of the rifle-and-bullet system. Before the firing, the system is at rest and the momentum is zero. After the firing, the *net*, or total, momentum is *still* zero. No net momentum is gained and no net momentum is lost. Let's look at this carefully.

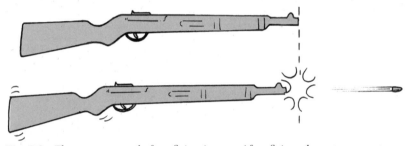

Fig. 7-8   The momentum before firing is zero. After firing, the net momentum is still zero because the momentum of the rifle cancels the momentum of the bullet.

Momentum, like the quantities velocity and force, has a direction as well as a size—it is a *vector quantity*. Hence, like velocity and force, it can be cancelled. So although the bullet in the preceding example has considerable momentum as it accelerates within the rifle barrel and then continues at high speed outside the barrel, and the recoiling rifle has momentum in and of itself, the *system* of both bullet and rifle has none. The momenta (plural form of momentum) of the bullet and the rifle are equal in size but opposite in direction. They cancel each other for the system as a whole. No external force acted on the system before or during firing. When there is no net force, there can be no net acceleration. Or, when there is no net force, there is no net impulse and therefore no net change in momentum. You can see that *if no net force acts on a system, then the momentum of that system cannot change.*

If you extend the idea of a rifle recoiling or "kicking" from the bullet it fires, you can understand rocket propulsion. Consider a machine gun recoiling each time a bullet is fired. The momen-

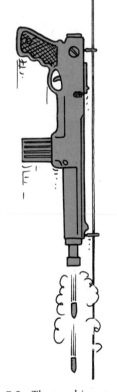

Fig. 7-9   The machine gun recoils from the bullets it fires and climbs upward.

> ▶ **Questions**
> 1. Newton's second law says that if there is no net force exerted on a system, no acceleration is possible. Does it follow from this that no change in momentum can occur?
>
> 2. Newton's third law says that the force a rifle exerts on its bullet is equal and opposite to the force the bullet exerts on the rifle. Does it follow that the *impulse* the rifle exerts on the bullet is equal and opposite to the *impulse* the bullet exerts on the rifle?

tum of recoil increases by an amount equal to the momentum of each bullet fired. If the machine gun is fastened so it is free to slide on a vertical wire (Figure 7-9), it will accelerate upward as bullets are fired downward. A rocket accomplishes acceleration by the same means. It is continually "recoiling" from the ejected exhaust gases. Each molecule of exhaust gas can be thought of as a tiny bullet shot from the rocket (Figure 7-10). It is interesting to note that the momentum of the rocket is equal and opposite to the momentum of the exhaust gases. So if we consider the total system of rocket and exhaust gases, like the rifle and bullet, there is no net change in momentum. As a practical matter, we are usually not concerned with the net momentum of rocket-exhaust, but of the rocket itself.

The momentum of a system cannot change unless prodded by external forces. The momentum possessed by a system before some internal interaction will be the same as the momentum possessed by the system after the interaction. When the momentum (or any quantity in physics) does not change, we say it is **conserved**. The idea that momentum is conserved when no external force acts is elevated to a central law of mechanics, called the **law of conservation of momentum**:

> In the absence of an external force, the momentum of a system remains unchanged.

If a system undergoes changes wherein all forces are internal,

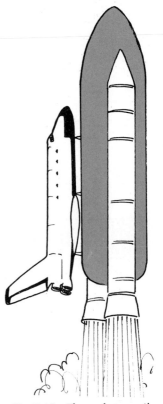

Fig. 7-10 The rocket recoils from the "molecular bullets" it fires and climbs upward.

▶ **Answers**

1. Yes, it follows because the absence of acceleration means there is no change in velocity, which in turn means no change in momentum (mass × velocity). Another line of reasoning is simply that no net force means no net impulse, which means no change in momentum.

2. Yes, because the *time* during which the rifle acts on the bullet and the bullet acts on the rifle is the same. Since time is equal for both, and force is equal and opposite for both, then force × time (impulse) is equal and opposite for both. Impulse, like force, is a vector quantity and can be cancelled.

as for example, in atomic nuclei undergoing radioactive decay, cars colliding, or stars exploding, the net momentum of the system before and after the event is the same.

> **▶ Question**
>   About 50 years ago it was argued that a rocket wouldn't operate in outer space because there is no air for it to push against. But a rocket works even better in outer space precisely because there is no air. How can you explain this?

7.5 | Collisions

The law of conservation of momentum is neatly seen in collisions. Whenever objects collide in the absence of external forces, the total or net momentum never changes:

$$\text{net momentum}_{\text{(before collision)}} = \text{net momentum}_{\text{(after collision)}}$$

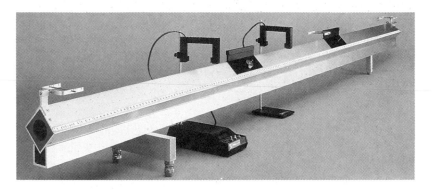

Fig. 7-11 Conservation of momentum is neatly demonstrated with the use of this air track. Jets of air from the many tiny holes in the track provide an air cushion upon which the glider slides nearly friction-free.

Elastic Collisions

When a moving billiard ball makes a head-on collision with another billiard ball at rest, the moving ball comes to rest and the struck ball moves with the initial velocity of the colliding ball. We see that momentum is simply transferred from one ball to

---

▶ **Answer**
  Just as a gun doesn't need air to push against in order to recoil, a rocket doesn't need air to push against in order to accelerate. The presence of air impedes acceleration by offering air resistance, so a rocket actually works better where there is no air. A rocket is not propelled by pushing against air, but by pushing against its own exhaust.

the other. When the colliding objects bound or rebound without lasting deformation or the generation of heat, the collision is said to be an **elastic collision**. Colliding objects bounce perfectly in elastic collisions (Figure 7-12).

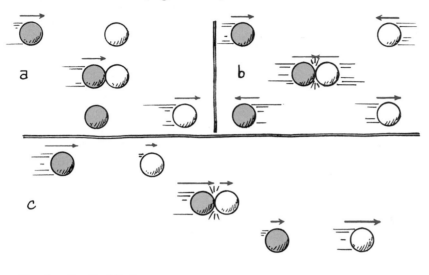

Fig. 7-12 Elastic collisions. (a) The dark ball strikes a ball at rest. (b) A head-on collision between two moving balls. (c) A collision of two balls moving in the same direction. In all cases, momentum is simply transferred or redistributed without loss or gain.

Inelastic Collisions

Momentum conservation holds true even when the colliding objects become distorted and generate heat during the collision. Such collisions are called **inelastic collisions**. Whenever colliding objects become tangled or couple together, we have an inelastic collision. The freight train cars shown in Figure 7-13 provide an illustrative example. Suppose the freight cars are of equal mass m, and one moves at 4 m/s while the other is at rest. Can we predict the velocity of the coupled cars after impact? From the conservation of momentum,

$$\text{net momentum before} = \text{net momentum after}$$
$$(m \times 4 \text{ m/s}) + (m \times 0 \text{ m/s}) = (2m \times ? \text{ m/s})$$

Since twice as much mass is moving after the collision, can you see that the velocity must be half as much as the 4 m/s value before the collision? This is 2 m/s, in the same direction as before.

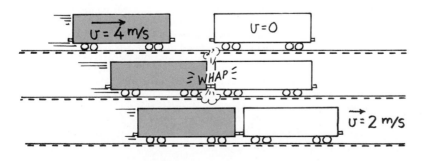

Fig. 7-13 Inelastic collision. The momentum of the freight car on the left is shared with the freight car on the right.

Then both sides of the equation are equal. The initial momentum is shared between both cars without loss or gain. Momentum is conserved.

Fig. 7-14   More inelastic collisions. The net momenta of the vehicles before and after collision are the same.

▶ **Questions**

The following questions refer to the gliders on the air track in Figure 7-11.

1. Suppose both gliders have the same mass. They move toward each other at the same speed and experience an elastic collison. Describe their motion after the collision.

2. Suppose both gliders have the same mass and have Velcro on them so that they stick together when they collide. They move toward each other at equal speed. Describe their motion after the collision.

3. Suppose one of the gliders is at rest and is loaded so that it has 3 times the mass of the moving glider. Again, the gliders have Velcro on them. Describe their motion after the collision.

▶ **Answers**

1. Since the collision is elastic, the gliders will simply reverse directions upon colliding and move away from each other at the same speed as before.

2. Before the collision, the gliders had equal and opposite momenta, since they had equal mass and were moving in opposite directions at the same speed. The net momentum was zero. Since momentum is always conserved, their net momentum after the collision must also be zero. As they are now stuck together, this means they must slam to a dead halt.

3. Before the collision, the net momentum equals the momentum of the unloaded, moving glider. After the collision, the net momentum is the same as before, but now the gliders are stuck together and moving as a single unit. The mass of the stuck-together gliders is four times that of the unloaded glider. Thus, the velocity can be only ¼ that of the unloaded glider before the collision. It is in the same direction as before, since the direction of the momentum is conserved as well as the amount.

In most collisions there are usually some external forces that act on a system. Billiard balls do not continue indefinitely with the momentum imparted to them. The balls encounter some friction with the table and the air they move through. These external forces are usually negligible during the collision itself, so the net momentum does not change during the collision. The net momentum of a couple of trucks that collide is the same before and just after collision. As the combined wreck slides along the pavement, friction provides an impulse to decrease momentum. For a pair of space vehicles docking in outer space, however, the net momentum before and after contact is exactly the same, and persists until the vehicles encounter external forces.

Another thing: perfectly elastic collisions are not common in the everyday world. We find in practice that some heat is generated in collisions. Drop a ball, and after it bounces from the floor, both the ball and the floor are a bit warmer. So even a dropped superball will not bounce to its initial height. At the microscopic level, however, perfectly elastic collisions are commonplace. For example, electrically charged particles bounce off one another without generating heat; they don't even touch in the classic sense of the word. (As later chapters will show, the notion of touching at the atomic level is different from at the everyday level.)

7.6   Momentum Vectors

Momentum is conserved even when colliding objects move at an angle to each other. To analyze momentum for angular directions, we use the vector techniques discussed in Chapter 6. It will be enough in this chapter if you merely become acquainted with momentum conservation for cases that involve angles, so three examples that convey the idea will be considered briefly without going into depth.

In Figure 7-15 you can see that car A has momentum directed due east, and that car B has momentum directed due north. If their individual momenta are equal in magnitude, then after collision their combined momentum will be in a northeast direction. Just as the diagonal of a square is not simply the arithmetic sum of two of its sides, the momentum of the wreck will not be twice the arithmetic sum of the individual momenta before collision.*

---

* It will be $\sqrt{2}$ times the momentum of either vehicle before collision, just as the diagonal of a square is $\sqrt{2}$ the length of a side.

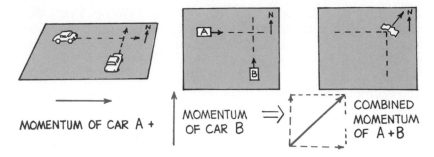

MOMENTUM OF CAR A +   MOMENTUM OF CAR B ⟹ COMBINED MOMENTUM OF A + B

Fig. 7-15 Momentum is a vector quantity. The momentum of the wreck is equal to the vector sum of the momenta of cars A and B before collision.

Figure 7-16 shows a falling firecracker that explodes into two pieces. The momenta of the fragments combine by vector rules to equal the original momentum of the falling firecracker.

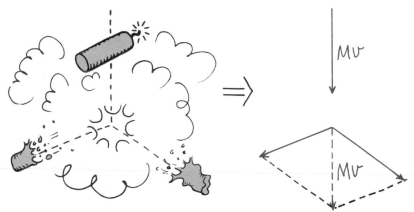

Fig. 7-16 When the firecracker bursts, the momenta of its fragments add up (by the vector parallelogram rule) to the momentum just before bursting.

Fig. 7-17 Momentum is conserved for the high-speed elementary particles, as shown by the tracks they leave in a streamer spark chamber. The relative masses of the particles is determined, among other things, by the paths they take after collision.

Figure 7-17 is a photo of tracks made by subatomic particles in a streamer spark chamber. The masses of these particles can be computed by applying both the conservation of momentum and the conservation of energy (which is covered in the next chapter). The conservation laws are extremely useful to experimenters in the atomic and subatomic realms. A very important feature of their usefulness is the fact that forces don't show up in the equations. The forces of the collision processes, however complicated, need not be of concern.

The law of conservation of momentum and, as the next chapter will discuss, the law of conservation of energy are the two most powerful tools of mechanics. Their application yields detailed information that ranges from understanding the interactions of subatomic particles to measuring the spin rates of entire galaxies.

7 | Chapter Review

Concept Summary

Momentum of an object is the product of mass times velocity.
- The change in momentum depends on the force that acts and on the length of time it acts.
- Impulse is force multiplied by the time during which it acts.
- The change in momentum equals the impulse.

According to the law of conservation of momentum, momentum is conserved when there is no net external force.
- When objects collide in the absence of external forces, momentum is conserved no matter whether the collision is elastic or inelastic.

Momentum is a vector quantity.
- Momenta combine by vector rules.

Important Terms

conserved (7.4)
elastic collision (7.5)
impulse (7.2)
inelastic collision (7.5)
law of conservation of momentum (7.4)
momentum (7.1)

Review Questions

1. a. Which has the greater mass—a heavy truck at rest or a rolling skateboard?
   b. Which has greater momentum? (7.1)

2. When the average force of impact on an object is extended in time, does this increase or decrease the impulse? (7.2)

3. What is the relationship between impulse and momentum? (7.2)

4. a. For a constant force, if the duration of impact upon an object is doubled, by how much is the impulse increased?
   b. By how much is the resulting change in momentum increased? (7.2)

5. a. If both the force that acts on an object *and* the time of impact are doubled, by how much is the impulse increased?
   b. By how much is the resulting change in momentum increased? (7.2)

6. In a car crash, why is it advantageous for an occupant to extend the time during which the collision is taking place? (7.2)

7. If the time of impact in a collision is extended by four times, by how much is the force of impact altered? (7.2)

8. a. Why is it advantageous for a boxer to ride with the punch?
   b. Why is it disadvantageous to move into an oncoming punch? (7.2)

9. When you throw a ball, do you experience an impulse? Do you experience an impulse if you instead catch a ball of the same speed? If you catch it, then throw it out again? Which impulse is greatest? (Visualize yourself on a skateboard.) (7.3)

10. Why is the force of impact greater in a collision that involves bouncing? (7.3)

11. Why is the Pelton wheel design a better one than paddle wheels with flat blades? (7.3)

12. What does it mean to say that momentum is a vector quantity? (7.4)

13. In terms of momentum conservation, why does a gun kick when fired? (7.4)

14. The text states that if no net force acts on a system, then the momentum of that system cannot change. It also states there is no change in momentum when a rifle is fired. Doesn't the fact that a bullet undergoes a considerable change in momentum as it accelerates along the barrel contradict this? Explain. (7.4)

15. What does it mean to say that momentum is conserved? (7.4)

16. How can a rocket be propelled above the atmosphere where there is no air to "push against"? (7.4)

17. Distinguish between an *elastic* and *inelastic* collision. (7.5)

18. What effect does friction have on the momentum of an object? (7.5)

19. Imagine that you are hovering next to a space shuttle in earth orbit and your buddy of equal mass who is moving at 4 km/h with respect to the ship bumps into you. If he holds onto you, how fast do you both move with respect to the ship? (7.5)

20. Is momentum conserved for colliding objects that are moving at angles to one another? Explain. (7.6)

Think and Explain

1. When you ride a bicycle at full speed, which has the greater momentum—you or the bike? (Does this help explain why you go over the handlebars if the bike is brought to an abrupt halt?)

2. For the least harm to your hand, should you catch a fast-moving baseball while your hand is at rest, while it moves toward the ball, or while it is pulling back away from the ball? Explain.

3. You can't throw a raw egg against a wall without breaking it, but you can throw it with the same speed into a sagging sheet without breaking it. Explain.

4. Everybody knows that you will be harmed less if you fall on a floor with "give" than a rigid floor. In terms of impulse and momentum, why is this so?

5. If you throw a heavy rock from your hands while standing on a skateboard, you roll backwards. Would you roll backwards if you didn't actually throw a rock but went through the motions of throwing the rock? Defend your answer.

6. A bug and the windshield of a fast-moving car collide. Tell whether the following statements are true or false.
   a. The forces of impact on the bug and on the car are the same size.
   b. The impulses on the bug and on the car are the same size.
   c. The changes in speed of the bug and of the car are the same.
   d. The changes in momentum of the bug and of the car are the same size.

7. When a space shuttle dips back into the atmosphere from orbit, it turns to the left and right in giant S-curves. How does this maneuver increase the impulse delivered to the shuttle so that it will slow before landing?

8. Who is in greater trouble—a person who comes to an abrupt halt when he falls to the pavement below or a person who bounces upon impact? Explain.

9. A railroad diesel engine weighs 4 times as much as a flatcar. If a diesel coasts at 5 km/h into a flatcar that is initially at rest, how fast do the two coast after they couple together?

10. An alpha particle is a nuclear particle of known mass. Suppose one is accelerated to a high speed in an atomic accelerator and directed into an observation chamber. There it collides and sticks to a target particle that is initially at rest. As a result of the impact, the combined target particle and alpha particle is observed to move at half the initial speed of the alpha particle. Why do observers conclude that the target particle is itself an alpha particle?

8 Energy

Energy is the most central concept that underlies all of science. Surprisingly, the idea of energy was unknown to Isaac Newton, and its existence was still being debated in the 1850s. The concept of energy is relatively new, and today we find it ingrained not only in all branches of science, but in nearly every aspect of human society. We are all quite familiar with it—energy comes to us from the sun in the form of sunlight, it is in the food we eat, and it sustains life. Energy may be the most familiar concept in science; yet it is one of the most difficult to define. Persons, places, and things have energy, but we observe energy only when something is happening—only when energy is being transformed. We will begin our study of energy by observing a related concept: work.

8.1 Work

The last chapter showed that changes in an object's motion are related both to force and to how long the force acts. "How long" meant time. The quantity "force × time" was called *impulse*. But "how long" need not always mean time. It can mean distance as well. When we consider the quantity "force × distance," we are talking about a wholly different quantity—a quantity called **work**.

We do work when we lift a load against the earth's gravity. The heavier the load or the higher we lift the load, the more work is done. Two things enter into every case where work is done:

Fig. 8-1 Work is done in lifting the barbell. If the barbell could be lifted twice as high, the weightlifter would have to expend twice as much energy.

(1) the *exertion of a force* and (2) the *movement of something* by that force.

Let's look at the simplest case, in which the force is constant and the motion takes place in a straight line in the direction of the force. Then the work done on an object by an applied force is defined as the product of the force and the distance through which the object is moved.* In shorter form:

$$\text{work} = \text{force} \times \text{distance}$$
$$W = Fd$$

If you lift two loads one story up, you do twice as much work as you do in lifting one load, because the *force* needed to lift twice the weight is twice as great. Similarly, if you lift a load two stories instead of one story, you do twice as much work because the *distance* is twice as much.

Note that the definition of work involves both a force *and* a distance. A weightlifter who holds a barbell that weighs 1000 N overhead does no work on the barbell. He may get really tired doing so, but if the barbell is not moved by the force he exerts, he does no work on the barbell. Work may be done on the muscles by stretching and contracting, which is force times distance on a biological scale, but this work is not done on the barbell. Lifting the barbell, however, is a different story. When the weightlifter raises the barbell from the floor, he is doing work.

Work generally falls into two categories. One of these is work done to change the speed of something. This kind of work is done in bringing an automobile up to speed or in slowing it down.

The other category of work is the work done against another force. When an archer stretches her bowstring, she is doing work against the elastic forces of the bow. When the ram of a pile driver is raised, it is exerted against the force of gravity. When you do pushups, you do work against your own weight. You do work on something when you force it to move against the influence of an opposing force.

The unit of measurement for work combines a unit of force (N) with a unit of distance (m). The unit of work is the newton-meter (N·m), also called the **joule** (rhymes with pool). One joule (symbol J) of work is done when a force of 1 N is exerted over a distance of 1 m, as in lifting an apple over your head. For larger values we speak of kilojoules (kJ)—thousands of joules—or megajoules (MJ)—millions of joules. The weightlifter in Figure 8-1 does work on the order of kilojoules. The energy released by one kilogram of fuel is on the order of megajoules.

---

*   For more general cases, work is the product of only the component of force that acts in the direction of motion, and the distance moved.

8.2 | Power

The definition of work says nothing about how long it takes to do the work. When carrying a load up some stairs, you do the same amount of work whether you walk or run up the stairs. So why are you more tired after running upstairs in a few seconds than after walking upstairs in a few minutes? To understand this difference, we need to talk about how fast the work is done, or **power**. Power is the rate at which work is done. It equals the amount of work done divided by the amount of time during which the work is done:

$$\text{power} = \frac{\text{work done}}{\text{time interval}}$$

An engine of great power can do work rapidly. An automobile engine with twice the power of another does not necessarily produce twice as much work or go twice as fast as the less powerful engine. Twice the power means it will do the same amount of work in half the time. The main advantage of a powerful automobile engine is its acceleration. It can get the automobile up to a given speed in less time than less powerful engines.

We can look at power this way: a liter of gasoline can do a specified amount of work, but the power produced when we burn it can be any amount, depending on how *fast* it is burned. The liter may produce 50 units of power for a half hour in an automobile or 90 000 units of power for one second in a supersonic jet aircraft.

The unit of power is the joule per second, also known as the **watt** (in honor of James Watt, the eighteenth-century developer of the steam engine). One watt (W) of power is expended when one joule of work is done in one second. One kilowatt (kW) equals 1000 watts. One megawatt (MW) equals one million watts. In the United States we customarily rate engines in units of horsepower and electricity in kilowatts, but either may be used. In the metric system of units, automobiles are rated in kilowatts. (One horsepower is the same as 0.75 kilowatt, so an engine rated at 134 horsepower is a 100-kW engine.)

Fig. 8-2   The three main engines of a space shuttle can develop 33 000 MW of power when fuel is burned at the enormous rate of 3400 kg/s. This is like emptying an average-size swimming pool in 20 seconds!

8.3 | **Mechanical Energy**

When work is done by an archer in drawing a bow, the bent bow has the ability of being able to do work on the arrow. When work is done to raise the heavy ram of a pile driver, the ram acquires the property of being able to do work on an object beneath it when it falls. When work is done to wind a spring mechanism, the spring acquires the ability to do work on various gears to run a clock, ring a bell, or sound an alarm.

In each case, something has been acquired. This "something" which is given to the object enables the object to do work. This "something" may be a compression of atoms in the material of an object; it may be a physical separation of attracting bodies; it may be a rearrangement of electric charges in the molecules of a substance. This "something" that enables an object to do work is called **energy**.* Like work, energy is measured in joules. It appears in many forms, which will be discussed in the following chapters. For now we will focus on **mechanical energy**—the energy due to the position or the movement of something. Mechanical energy may be in the form of either potential energy or kinetic energy.

8.4   Potential Energy

An object may store energy by virtue of its position. The energy that is stored and held in readiness is called **potential energy** (PE), because in the stored state it has the potential for doing work. A stretched or compressed spring, for example, has the potential for doing work. When a bow is drawn, energy is stored in the bow. A stretched rubber band has potential energy because of its position, for if it is part of a slingshot, it is capable of doing work.

The chemical energy in fuels is potential energy, for it is actually energy of position when looked at from a microscopic point of view. This energy is available when the positions of electric charges within and between molecules are altered, that is, when a chemical change takes place. Any substance that can do work through chemical action possesses potential energy. Potential energy is found in fossil fuels, electric batteries, and the food we eat.

Work is required to elevate objects against earth's gravity. The potential energy due to elevated positions is called *gravitational potential energy*. Water in an elevated reservoir and the ram of a pile driver have gravitational potential energy.

The amount of gravitational potential energy possessed by an elevated object is equal to the work done against gravity in lift-

---

* Strictly speaking, that which enables an object to do work is called its *available energy*, for not all the energy of an object can be transformed to work.

ing it. The work done equals the force required to move it up-ward times the vertical distance it is moved ($W = Fd$). The upward force required is equal to the weight mg of the object, so the work done in lifting it through a height h is given by the product mgh:

$$\text{gravitational potential energy} = \text{weight} \times \text{height}$$
$$PE = mgh$$

Note that the height h is the distance above some reference level, such as the ground or the floor of a building. The potential energy mgh is relative to that level and depends only on mg and the height h. You can see in Figure 8-3 that the potential energy of the boulder at the top of the ledge does not depend on the path taken to get it there.

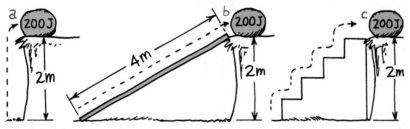

Fig. 8-3   The potential energy of the 100-N boulder with respect to the ground below is the same (200 J) in each case because the work done in elevating it 2 m is the same whether it is (a) lifted with 100 N of force, (b) pushed up the 4-m incline with 50 N of force, or (c) lifted with 100 N of force up each 0.5-m stair. No work is done in moving it horizontally (neglecting friction).

▶ **Questions**

1. How much work is done on a 100-N boulder that you carry horizontally across a 10-m room?

2. a. How much work is done on a 100-N boulder when you lift it 1 m?
   b. What power is expended if you lift it this distance in 1 s?
   c. What is its gravitational potential energy in the lifted position?

▶ **Answers**

1. You do no work on the boulder moved horizontally, for you apply no force (except for the tiny bit to start it) in its direction of motion. It has no more PE across the room than it had initially.

2. a. You do 100 J of work when you lift it 1 m (since $Fd = 100$ N·m = 100 J).
   b. Power = (100 J)/(1 s) = 100 W.
   c. It depends; with respect to its starting position, its PE is 100 J; with respect to some other reference level, it would be some other value.

8.5   Kinetic Energy

Push on an object and you can set it in motion. If an object moves, then by virtue of that motion it is capable of doing work. It has energy of motion, or **kinetic energy** (KE). The kinetic energy of an object depends on the mass of the object as well as its speed. It is equal to half the mass multiplied by the square of the speed.

$$\text{kinetic energy} = \tfrac{1}{2}\text{mass} \times \text{speed}^2$$
$$\text{KE} = \tfrac{1}{2}mv^2$$

When you throw a ball, you do work on it to give it the speed it has when it leaves your hand. The moving ball can then hit something and push against it, doing work on what it hits. The kinetic energy of a moving object is equal to the work required to bring it to that speed from rest, or the work the object can do in being brought to rest:*

$$\text{net force} \times \text{distance} = \text{kinetic energy}$$

or in shorthand notation,

$$Fd = \tfrac{1}{2}mv^2$$

It is important to notice that the speed is squared, so that if the speed of an object is doubled, its kinetic energy is quadrupled ($2^2 = 4$). This means that it takes four times as much work to double the speed of an object, and also that an object moving twice as fast as another takes four times as much work to stop. Accident investigators are well aware that an automobile traveling at 100 km/h has four times as much kinetic energy it would have traveling at 50 km/h. This means that a car traveling at 100 km/h will skid four times as far when its brakes are locked as it would if traveling 50 km/h. This is because speed is squared for kinetic energy.

Kinetic energy underlies other seemingly different forms of energy such as heat, sound, and light.

Fig. 8-4 The potential energy of the drawn bow equals the work (average force × distance) done in drawing the arrow into position. When released, it will become the kinetic energy of the arrow.

Fig. 8-5 Typical stopping distances for cars traveling at various speeds. Notice how the work done to stop the car (friction force × distance of slide) depends on the square of the speed. (The distances would be even greater if reaction time were taken into account.)

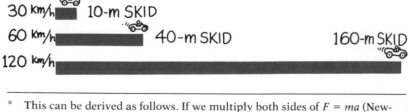

30 km/h 10-m SKID
60 km/h 40-m SKID
120 km/h 160-m SKID

---

*   This can be derived as follows. If we multiply both sides of $F = ma$ (Newton's second law) by d, we get $Fd = mad$. Recall from Chapter 2 that for motion in a straight line at constant acceleration $d = \tfrac{1}{2}at^2$, so we can say $Fd = ma(\tfrac{1}{2}at^2) = \tfrac{1}{2}maat^2 = \tfrac{1}{2}m(at)^2$. Substituting $v = at$, we get $Fd = \tfrac{1}{2}mv^2$.

> ▶ **Question**
>   When the brakes of a motorcycle traveling at 60 km/h
> become locked, how much farther will it skid than if the
> brakes locked at 20 km/h?

8.6 | Conservation of Energy

More important than being able to state *what energy is* is understanding how it behaves—*how it transforms*. You can understand
nearly every process or change that occurs in nature better if
you analyze it in terms of a transformation of energy from one
form to another.

As you draw back the stone in a slingshot, you do work in
stretching the rubber band; the rubber band then has potential
energy. When released, the stone has kinetic energy equal to this
potential energy. It delivers this energy to its target, perhaps a
wooden fence post. The slight distance the post is moved multiplied by the average force of impact doesn't quite match the kinetic energy of the stone. The energy score doesn't balance. But
if you investigate further, you'll find that both the stone and
fence post are a bit warmer. By how much? By the energy difference. Energy changes from one form to another. It transforms
without net loss or net gain.

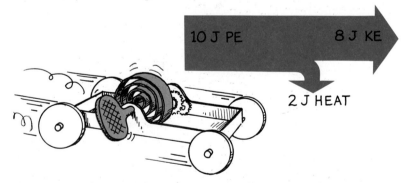

10 J PE   8 J KE

2 J HEAT

Fig. 8-6 Part of the PE of the
wound spring changes into
KE. The rest turns into heating the machinery and the
surroundings due to friction.
No energy is lost.

The study of the various forms of energy and their transformations from one form into another has led to one of the great-

▶ **Answer**
    Nine times farther: the motorcycle has nine times as much energy when it
travels three times as fast: $\frac{1}{2}m(3v)^2 = \frac{1}{2}m9v^2 = 9(\frac{1}{2}mv^2)$. The friction force will ordinarily be the same in either case; therefore, to do nine times the work requires
nine times as much sliding distance.

est generalizations in physics, known as the **law of conservation of energy**:

> Energy cannot be created or destroyed; it may be transformed from one form into another, but the total amount of energy never changes.

When you consider any system in its entirety, whether it be as simple as a swinging pendulum or as complex as an exploding galaxy, there is one quantity that does not change: energy. It may change form, or it may simply be transferred from one place to another, but the total energy score stays the same.

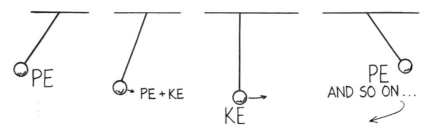

Fig. 8-8   Energy transformations in a pendulum. The PE of the pendulum bob at its highest point is equal to the KE of the bob at its lowest point. Everywhere along its path, the sum of PE and KE is the same. (Because of the work done against friction, this energy will eventually be transformed into heat.)

This energy score takes into account the fact that the atoms that make up matter are themselves concentrated bundles of energy. When the nuclei (cores) of atoms rearrange themselves, enormous amounts of energy can be released. The sun shines because some of this energy is transformed into radiant energy. In nuclear reactors much of this energy is transformed into heat.

Powerful gravitational forces in the deep hot interior of the sun crush the cores of hydrogen atoms together to form helium atoms. This welding together of atomic cores is called *thermonuclear fusion.* This process releases radiant energy, some of which reaches the earth. Part of this energy falls on plants, and part of this later becomes coal. Another part supports life in the food chain that begins with plants, and part of this energy later becomes oil. Part of the energy from the sun goes into the evaporation of water from the ocean, and part of this returns to the earth as rain that may be trapped behind a dam. By virtue of its position, the water in a dam has energy that may be used to power a generating plant below, where it will be transformed to electric energy. The energy travels through wires to homes, where it is used for lighting, heating, cooking, and to operate electric toothbrushes. How nice that energy is transformed from one form to another!

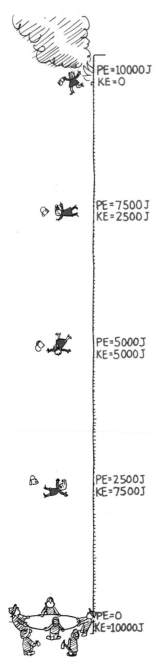

Fig. 8-7   When the lady in distress leaps from the burning building, note that the sum of her PE and KE remains constant at successive positions $\frac{1}{4}$, $\frac{1}{2}$, $\frac{3}{4}$, and all the way down.

> ▶ **Question**
> Suppose a car with a miracle engine is able to convert 100% of the energy released when gasoline burns (40 million joules per liter) to mechanical energy. If the air drag and overall frictional forces on the car traveling at highway speed is 2000 N, what is the upper limit in distance per liter the car could cover at highway speed?

8.7 Machines

A **machine** is a device for multiplying forces or simply changing the direction of forces. Underlying every machine is the conservation of energy concept. Consider one of the simplest machines, the **lever** (Figure 8-9). At the same time we do work on one end of the lever, the other end does work on the load. We see that the direction of force is changed, for if we push *down*, the load is lifted *up*. If the heat from friction forces is small enough to neglect, the work input will be equal to the work output.

$$\text{work input} = \text{work output}$$

Since work equals force times distance, then input force × input distance = output force × output distance.

$$(\text{force} \times \text{distance})_{\text{input}} = (\text{force} \times \text{distance})_{\text{output}}$$

A little thought will show that the pivot point, or **fulcrum**, of the lever can be relatively close to the load. Then a small input force exerted through a large distance will produce a large output force over a correspondingly short distance. In this way, a lever can multiply forces. But no machine can multiply work nor multipy energy. That's a conservation of energy no-no!

Consider the idealized case of the weightless lever shown in Figure 8-10. The child pushes down with a force of only 10 N and is able to lift a load of 80 N. The ratio of output force to input

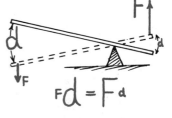

Fig. 8-9 The lever. The work (force times distance) you do at one end equals the work done to the load at the other end.

▶ **Answer**
From the definition of work as force × distance, simple rearrangement gives distance = work ÷ force. If all 40 million joules of energy in one liter were used to do the work of overcoming the air drag and frictional forces, the distance would be:

$$\text{distance} = \frac{\text{work}}{\text{force}} = \frac{40\,000\,000\text{ J}}{2000\text{ N}} = 20\,000\text{ m} = 20\text{ km}$$

The important point here is that even with a perfect engine, there is an upper limit of fuel economy dictated by the conservation of energy.

force for a machine is called the **mechanical advantage**. Here the mechanical advantage is (80 N)/(10 N), or 8. Note that the load is lifted only one eighth the distance that the input force moves. In the absence of friction, the mechanical advantage can also be determined by considering the relative distances through which the forces are exerted.

Fig. 8-10   The output force (80 N) is eight times the input force (10 N), while the output distance ($\frac{1}{8}$ m) is one eighth the input distance (1 m).

There are three different ways to set up a lever (Figure 8-11). Type 1 is the kind of lever commonly seen in a playground with children on each end—the seesaw. The fulcrum is between the input and output ends of the lever. Push down on one end, and you lift a load at the other. You can increase force at the expense of distance.

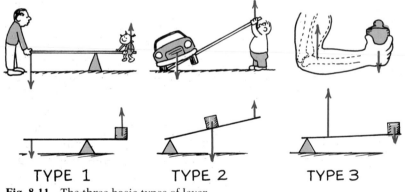

TYPE 1   TYPE 2   TYPE 3

Fig. 8-11   The three basic types of lever.

With the type 2 arrangement, the load is between the fulcrum and the input end. To lift a load, you lift the end of the lever. An example is the placing of one end of a long steel bar under an automobile frame and lifting on the free end to raise the automobile. Again, force is increased at the expense of distance.

In the type 3 arrangement, the fulcrum is at one end and the load is at the other. The input force is applied between them. Your bicep muscles are connected to your forearm in this way. (The fulcrum is your elbow and the load is your hand.) The type

3 lever increases distance at the expense of force. When you move
your bicep muscles a short distance, your hand moves a much
greater distance.

A **pulley** is basically a kind of type 1 lever that is used to change
the direction of a force. Properly used, a pulley or system of pulleys can multiply forces as well.

The single pulley in Figure 8-12 left changes the direction of
the applied force. Here the load and applied force have equal
magnitudes, since the applied force acts on the same single
strand of rope that supports the load. Thus, the mechanical advantage equals one. Also, the distance moved at the input equals
the distance the load moves.

Fig. 8-12   A pulley can (a) change the direction of a force as effort is exerted
downward and load moves upward; (b) multiply force as effort is now half the
load, and (c) when combined with another pulley both change the direction
and multiply force.

In Figure 8-12 center, the load is now supported by two strands.
The force the man applies to support the load is therefore only
half the weight of the load. The mechanical advantage equals
two. The fact that there is only a single rope does not change
this. Careful thought will show that to raise the load 1 m, the
man will have to pull the rope up 2 m (or 1 m for each side).

The pulley system in Figure 8-12 right combines the arrangements of the first two pulleys. The load is supported by two
strands of rope; the upper pulley serves only to change the direction of the force. Try to figure out the mechanical advantage of
the pulley system. Actually experimenting with a variety of pulley systems is much more beneficial than reading about them in
a textbook, so do try to get your hands on some pulleys, in or out
of class. They're fun.

The pulley system shown in Figure 8-13 is a bit more complex,
but the principles of energy conservation are the same. When
the rope is pulled 10 m with a force of 50 N, a 500-N load will be
lifted 1 m. The mechanical advantage is (500 N)/(50 N), or 10.
Force is multiplied at the expense of distance. The mechanical

Fig. 8-13 In an idealized pulley system, applied force × input distance = output force × output distance.

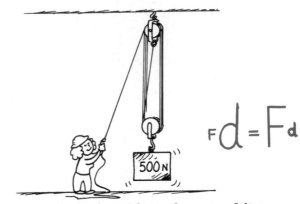

advantage can also be found from the ratio of distances—that is, (input distance) ÷ (output distance)—which is also 10.

No machine can put out more energy than is put into it. No machine can create energy; a machine can only transfer it from one place to another, or transform it from one form to another.

8.8   Efficiency

The previous examples of machines were considered to be ideal; 100% of the work input was transfered to work output. An ideal machine would operate at 100% efficiency. In practice, this does not happen, and we can never expect it to happen. In any machine, some energy is dissipated to atomic or molecular kinetic energy—which makes the machine warmer.

Even a lever rocks about its fulcrum and converts a small fraction of the input energy into heat. We may do 100 J of work and get out 98 J of work. The lever is then 98% efficient, and we waste only 2 J of work input on heat. In a pulley system, a larger fraction of input energy goes into heat. If we do 100 J of work, the forces of friction acting through the distances through which the pulleys turn and rub about their axles may dissipate 40 J of energy as heat. So the work output is only 60 J and the pulley system has an efficiency of 60%. The lower the efficiency of a machine, the greater is the amount of energy wasted as heat.

Efficiency can be expressed as the ratio of useful work output to total work input:

$$\text{efficiency} = \frac{\text{useful work output}}{\text{total work input}}$$

An inclined plane serves as a machine. Sliding a load up an incline requires less force than lifting it vertically. Figure 8-14

Fig. 8-14   Pushing the block of ice 5 times farther up the incline than the vertical distance lifted requires a force of only $\frac{1}{5}$ its weight. Whether pushed up the plane or simply lifted, it gains the same amount of PE.

shows a 5-m inclined plane with its high end elevated 1 m. If we use the plane to elevate a heavy load, we will have to push it five times farther than if we simply lifted it vertically. A little thought will show that if the plane is free of friction, we need apply only one fifth the force required to lift the load vertically. The inclined plane provides a *theoretical* mechanical advantage of 5. A block of ice sliding on an icy plane may have a near theoretical mechanical advantage and approach 100% efficiency, but when the load is a wooden crate and it slides on a wooden plank, both the actual mechanical advantage and the efficiency will be considerably less. Efficiency can also be expressed as the ratio of actual mechanical advantage to theoretical mechanical advantage:

$$\text{efficiency} = \frac{\text{actual mechanical advantage}}{\text{theoretical mechanical advantage}}$$

Efficiency will always be a fraction less than 1. To convert it to percent we simply express it as a decimal and multiply by 100%. For example, an efficiency of 0.25 expressed in percent is 0.25 × 100%, or 25%.

The jack shown in Figure 8-15 is actually an inclined plane wrapped around a cylinder. You can see that a single turn of the handle raises the load a relatively small distance. If the circular distance the handle is moved is 500 times greater than the pitch (the distance between ridges), then the theoretical mechanical advantage of the jack is 500.* No wonder a child can raise a loaded moving van with one of these devices! In practice there is

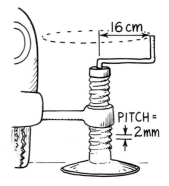

Fig. 8-15   The auto jack is like an inclined plane wrapped around a cylinder. Every time the handle is turned one revolution, the load is raised a distance of one pitch.

---

*   To raise a load by 2 mm the handle has to be moved once around, through a distance equal to the circumference of the circular path of radius 16 cm. This distance is 100 cm (since the circumference is $2\pi r = 2 \times 3.14 \times 16$ cm = 100 cm). A simple calculation will show that the 100-cm work-input distance is 500 times greater than the work-output distance of 2 mm. If the jack were 100% efficient, then the input force would be multiplied by 500 times. The theoretical mechanical advantage of the jack is 500.

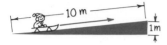

▶ **Questions**

A child on a sled (total weight 500 N) is pulled up a 10-m slope that elevates her a vertical distance of 1 m.

a. What is the theoretical mechanical advantage of the slope?

b. If the slope is without friction, and she is pulled up the slope at constant speed, what will be the tension in the rope?

c. Considering the practical case where friction *is* present, suppose the tension in the rope were in fact 100 N. Then what would be the actual mechanical advantage of the slope? The efficiency?

a great deal of friction in this type of jack, so the efficiency may be around 20%. Thus the jack actually multiplies force by about 100 times. The actual mechanical advantage therefore approximates an impressive 100. Imagine the value of one of these devices during the days when the great pyramids were being built!

An automobile engine is a machine that transforms chemical energy stored in fuel into mechanical energy. The molecules of the petroleum fuel break up when the fuel burns. Burning is a chemical reaction in which atoms combine with the oxygen in the air. Carbon atoms from the petroleum combine with oxygen atoms to form carbon monoxide, and energy is released.

The converted energy is used to run the engine. It would be nice if all this energy were converted to mechanical energy, but a 100% efficient machine is not possible. Some of the energy goes into heat. Even the best designed engines are unlikely to be more than 35% efficient. Some of the energy converted to heat goes into the cooling system and is wasted through the radiator to the air. Some goes out the exhaust, and nearly half is wasted in the friction of the moving engine parts. In addition to these inefficiencies, the fuel does not even burn completely—a certain amount goes unused. We can look at inefficiency in this way: in

---

▶ **Answers**

a. The ideal or theoretical mechanical advantage is (input distance) ÷ (output distance) = (10 m)/(1 m) = 10.

b. 50 N. With no friction, ideal mechanical advantage and actual mechanical advantage would be the same—10. So the input force, the rope tension, will be 1/10 the output force, her 500-N weight.

c. The actual mechanical advantage would be (output force) ÷ (input force) = (500 N)/(100 N) = 5. The efficiency would then be 0.5 or 50%, since (actual mechanical advantage) ÷ (theoretical mechanical advantage) = 5/10 = 0.5. The value for efficiency is also obtained from the ratio (useful work output) ÷ (useful work input).

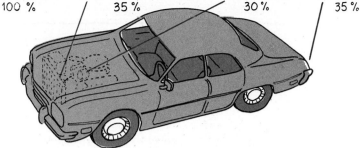

FUEL ENERGY IN = COOLING WATER LOSSES + ENGINE OUTPUT + EXHAUST HEAT
    100 %   35 %   30 %   35 %

Fig. 8-16 Only 30% of the energy produced by burning gasoline in a typically efficient automobile becomes useful mechanical energy.

any transformation there is a dilution of the amount of *useful energy*. Useful energy ultimately becomes heat. Energy is not destroyed—it is simply degraded. Heat is the graveyard of useful energy.

8.9 | Energy for Life

Every living cell in every organism is a living machine. Like any machine, it needs an energy supply. Most living organisms on this planet feed on various hydrocarbon compounds that release energy when they react with oxygen. Like the petroleum fuel previously discussed, there is more energy stored in the molecules in the fuel state than in the reaction products after combustion. The energy difference is what sustains life.

The principle of combustion in the digestion of food in the body and the combustion of fossil fuels in mechanical engines is the same. The main difference is the rate at which the reactions take place. In digestion the rate is much slower, and energy is released as it is required. Like the burning of fossil fuels, the reaction is self-sustaining once started. Carbon combines with oxygen to form carbon dioxide.

The reverse process is more difficult. Only green plants and certain one-celled organisms can make carbon dioxide combine with water to produce hydrocarbon compounds such as sugar. This process is *photosynthesis* and requires an energy input, which normally comes from sunlight. Sugar is the simplest food; all others—carbohydrates, proteins, and fats—are also synthesized compounds of carbon, hydrogen, and oxygen. How fortunate we are that green plants are able to use the energy of sunlight to make food that gives us and all other food-eating organisms our energy. Because of this there is life.

8   Chapter Review

Concept Summary

When a constant force moves an object in the direction of the force, the work done equals the product of the force and the distance through which the object is moved.
• Power is the rate at which work is done.

The energy of an object enables it to do work.
• Mechanical energy is due to the position of something (potential energy) or the movement of something (kinetic energy).

According to the law of conservation of energy, energy cannot be created nor destroyed.
• Energy can be transformed from one form to another.

A machine is a device for multiplying forces or changing the direction of forces.
• Levers, pulleys, and inclined planes are simple machines.
• Normally, the useful work output of a machine is less than the total work input.

Important Terms

efficiency (8.8)
energy (8.3)
fulcrum (8.7)
joule (8.1)
kinetic energy (8.5)
law of conservation of energy (8.6)
lever (8.7)
machine (8.7)
mechanical advantage (8.7)
mechanical energy (8.3)
potential energy (8.4)
power (8.2)
pulley (8.7)
watt (8.2)
work (8.1)

Review Questions

1. A force sets an object in motion. When the force is multiplied by the time of its application, we call the quantity *impulse*, which changes the *momentum* of that object. What do we call the quantity *force × distance*, and what quantity does this change? (8.1)

2. Work is required to lift a barbell. How many times as much work is required to lift the barbell three times as high? (8.1)

3. Which requires more work—lifting a 10-kg sack a vertical distance of 2 m, or lifting a 5-kg sack a vertical distance of 4 m? (8.1)

4. How many joules of work are done on an object when a force of 10 N pushes it a distance of 10 m? (8.1)

5. How much power is required to do 100 J of work on an object in a time of 0.5 s? How much power is required if the same amount of work is done in 1 s? (8.2)

6. What is mechanical energy? (8.3)

7. a. If you do 100 J of work to elevate a bucket of water, what is its gravitational potential energy relative to its starting position?
    b. What would it be if it were raised twice as high? (8.4)

8. If a boulder is raised above the ground so that its potential energy with respect to the ground is 200 J, and then it is dropped, what will its kinetic energy be just before it hits the ground? (8.5)

9. Suppose an automobile has a kinetic energy of 2000 J. If it moves with twice the speed, what will be its kinetic energy? Three times the speed? (8.5)

10. What will be the kinetic energy of an arrow shot from a bow having a potential energy of 50 J? (8.6)

11. What does it mean to say that in any system the total energy score stays the same? (8.6)

12. In what sense is energy from coal actually solar energy? (8.6)

13. Why is there an upper limit on how far an automobile can be driven on a tank of gasoline? (8.6)

14. In what two ways can a machine alter an input force? (8.7)

15. In what way is a machine subject to the law of energy conservation? Is it possible for a machine to multiply energy or work input? (8.7)

16. What does it mean to say that a certain machine has a certain mechanical advantage? (8.7)

17. What are the three basic types of levers? (8.7)

18. What is the efficiency of a machine that requires 100 J of input energy to do 35 J of useful work? (8.8)

19. Distinguish between theoretical mechanical advantage and actual mechanical advantage. How will these compare if a machine is 100% efficient? (8.8)

20. What is the efficiency of the body when a cyclist expends 1000 W of power to deliver mechanical energy to her bicycle at the rate of 100 W? (8.8)

Think and Explain

1. When a rock is projected with a slingshot, there are two reasons why the rock will go faster if the rubber is stretched an extra distance. What are the two reasons? Defend your answer.

2. A certain engine is capable of bringing a car from 0 to 100 km/h in 10 s. All other things being equal, if the engine has twice the power output, how many seconds would be required to accelerate it to this speed?

3. If a car that travels at 60 km/h will skid 20 m when its brakes are locked, how far will it skid if it is traveling at 120 km/h when its brakes are locked?

4. A hammer falls off a rooftop and strikes the ground with a certain KE. If it fell from a roof twice as tall, how would its KE of impact compare?

5. Most earth satellites follow an oval shaped (elliptical) path rather than a circular path around the earth. The PE increases when the satellite moves farther from the earth. According to energy conservation, does a satellite have its greatest speed when it is closest or farthest from the earth?

6. Why will a lighter car generally have better fuel economy than a larger car? Will a streamlined design improve fuel economy?

7. Does an automobile consume more fuel when its air conditioner is turned on? When its lights are on? When its radio is on while it is sitting at rest in the parking lot? Explain in terms of the conservation of energy.

8. How many kilometers per liter will a car obtain if its engine is 25% efficient and it encounters an average retarding force of 1000 N at highway speed? Assume the energy content of gasoline is 40 MJ per liter.

9. The pulley shown on the left in Figure A has a mechanical advantage of 1. What is the mechanical advantage when it is used as shown on the right? Defend your answer.

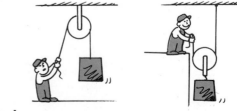

Fig. A

Fig. B

10. What is the theoretical mechanical advantage of the three lever systems shown in Figure B?

11. You tell your friend that no machine can possibly put out more energy than is put into it, and your friend states that a nuclear reactor puts out more energy than is put into it. What do you say?

12. The energy we require for existence comes from the chemically stored potential energy in food, which is transformed into other forms by the process of digestion. What happens to a person whose work output is less than the energy consumed? When the person's work output is greater than the energy consumed? Can an undernourished person perform extra work without extra food? Defend your answers.

9 Center of Gravity

Why doesn't the famous Leaning Tower of Pisa topple over? How far can it lean before it does topple over? Why is it impossible to stand with your back and your heels against a wall and bend over and touch your toes without toppling forward? To answer these questions, you first need to know about center of gravity. Then you need to know how this concept can be applied to balancing and stability. Let's start with center of gravity.

9.1 Center of Gravity

Throw a baseball into the air, and it follows a smooth parabolic path. Throw a baseball bat into the air with a spin, and its path is not smooth. The bat seems to wobble all over the place. But it wobbles about a special point. This point follows a parabolic path, even though the rest of the bat does not (Figure 9-1). The motion of the bat is the sum of two motions: (1) a spin with this point at the center and (2) a movement through the air as if all the weight were concentrated at this point. This point is the **center of gravity** of the bat.

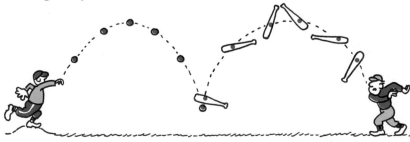

Fig. 9-1 The centers of gravity of the baseball and of the spinning baseball bat each follow parabolic paths.

119

The center of gravity of an object is the point located at the center of an object's weight distribution. For a symmetrical object, such as the baseball, this point is at the geometrical center. But an irregularly shaped object, such as the baseball bat, has more weight at one end, so the center of gravity is toward the heavier end. The center of gravity of a piece of tile cut into the shape of a triangle is one third of the way up from its base. A solid cone has its center of gravity one fourth of the way up from its base.

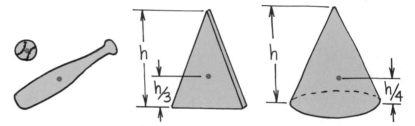

Fig. 9-2   The center of gravity for each object is shown by the colored dot.

Fig. 9-3   The center of gravity of the toy is below its geometric center.

Objects not made of the same material throughout (that is, objects of varying densities) may have the center of gravity quite far from the geometrical center. Consider a hollow ball half filled with lead. The center of gravity would be not at the geometrical center but within the lead. The ball would always roll to a stop in the same position. Make it the body of a lightweight toy clown, and whenever it is pushed over, it will come back right-side-up (Figure 9-3).

The multiple-flash photograph in Figure 9-4 shows a top view of a wrench sliding across a smooth horizontal surface. Notice that its center of gravity, which is marked by the dark X, follows a straight-line path. Other parts of the wrench rotate about this point as the wrench moves across the surface. Notice also that the center of gravity moves equal distances in the equal time intervals of the flashes (because no net force acts in the direction of motion). The motion of the wrench is a combination of straight-line motion of its center of gravity and rotational motion about its center of gravity.

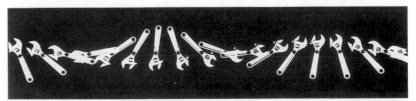

Fig. 9-4   The center of gravity of the rotating wrench follows a straight-line path.

If the wrench were instead tossed into the air, no matter how it rotates its center of gravity would follow a smooth parabola. The same is true even for an exploding projectile, such as a fireworks rocket (Figure 9-5). The internal forces that take place in the explosion do not change the center of gravity of the projectile. Interestingly enough, if air resistance can be neglected, the center of gravity of the dispersed fragments upon reaching the ground will be where the unexploded rocket would have hit.

Fig. 9-5 The center of gravity of the fireworks rocket and its fragments move along the same path before and after the explosion.

9.2 Center of Mass

Center of gravity is also called **center of mass**, which is the point of average *mass* distribution for an object. For almost all objects on and near the earth, the terms are interchangeable. An object in outer space, where gravity forces are practically zero, has a center of mass but no center of gravity. There is also a small difference between center of gravity and center of mass when an object is large enough for gravity to vary from one part to another. The center of gravity of the moon, for example, is a bit closer to the earth than the center of mass is. This is because the nearer parts of the moon are pulled by earth gravity more than the far parts. For everyday objects we can use the terms *center of gravity* and *center of mass* interchangeably.

If you threw a wrench so that it rotated as it moved through the air, you would see it wobble about its center of gravity. The center of gravity itself would follow a parabolic path. Now suppose you threw a lopsided ball—one with its center of gravity off-center. You would see it wobble as it rotated. The sun itself wobbles for a similar reason. The center of mass of the solar system can lie outside the massive sun, not at its geometrical center. This is because the masses of the planets contribute to the overall mass of the solar system. As they orbit their respective distances, the sun actually wobbles off center. Astronomers are seeking similar wobbles in nearby stars, which may indicate that our sun is not the only star with a planetary system.

Fig. 9-6   The center of mass of the solar system is not at the geometrical center of the sun. If all the planets were lined up on one side of the sun, the center of mass would be about 2 solar radii from the sun's center.

9.3 | Locating the Center of Gravity

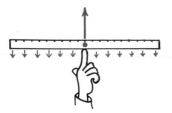

Fig. 9-7   The weight of the entire stick behaves as if it were concentrated at its center.

Fig. 9-8   Finding the CG for an irregularly shaped object with a plumb bob.

The center of gravity (called the CG from here on) of a uniform object (such as a meter stick) is at the midpoint, its geometrical center. The CG is the balance point. Supporting that single point supports the whole object. In Figure 9-7 the many small vectors represent the force of gravity all along the meter stick. All of these can be combined into a resultant force that acts at the CG. The effect is as if the weight of the stick were concentrated at this point. That's why you can balance the stick by a single upward force directed at this point.

If you suspend any object (a pendulum, for example) at a single point, the CG of the object will hang directly below (or at) the point of suspension. To locate the CG, construct a vertical line beneath the point of suspension. The CG lies somewhere along that line. Figure 9-8 shows how a plumb line and bob can be used to construct a line that is exactly vertical. You can locate the CG by suspending the object from some other point and construct a second vertical line. The CG is where the two lines intersect.

The CG of an object may be where no actual material exists. The CG of a ring lies at the geometrical center where no matter exists. The same holds true for a hollow sphere such as a basketball. The CG of even half a ring or half a hollow ball is still outside the physical structure. There is no material at the CG of an empty cup, bowl, or boomerang.

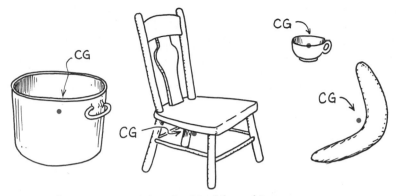

Fig. 9-9 There is no material at the CG of these objects.

> ▶ **Questions**
> 1. Where is the CG of a donut?
>
> 2. Can an object have more than one CG?

9.4 | Toppling

Pin a plumb line to the center of a heavy wooden block and tilt the block until it topples over (Figure 9-10). You can see that the block will begin to topple when the plumb line extends beyond the supporting base of the block.

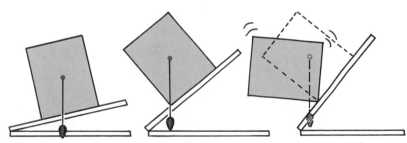

Fig. 9-10 The block topples when the CG extends beyond its support base.

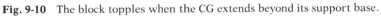

▶ **Answers**

1. In the center of the hole!

2. A rigid object has one CG. If it is nonrigid, such as a piece of clay or putty, and is distorted into different shapes, then its CG may change as its shape is changed. Even then, it has one CG for any given shape.

Fig. 9-11 A "Londoner" double-decker bus undergoing a tilt test. The bus must not topple when the chassis is tilted 28° with the top deck fully loaded and only the driver and conductor on the lower deck. Because so much of the weight of the vehicle is in the lower part, the load of the passengers on the upper deck raises the CG only a little, so the bus can be tilted well beyond this 28° limit without toppling.

The rule for toppling is: If the CG of an object is above the area of support, the object will remain upright. If the CG extends outside the area of support, the object will topple. This principle is dramatically employed in Figure 9-11.

This is why the Leaning Tower of Pisa does not topple. Its CG does not extend beyond its base. As shown in Figure 9-12, a vertical line below the CG falls inside the base, and so the Leaning Tower has stood for centuries. If the tower leaned far enough so that the CG extended beyond the base, the tower would topple.

A support base of an object does not have to be solid. The four legs of a chair bound a rectangular area that is the support base for the chair (Figure 9-13). Practically speaking, supporting props could be erected to hold the Leaning Tower up if it leaned too far. Such props would create a new support base. An object will remain upright if the CG is above its base of support.

Try balancing the pole end of a broom or mop on the palm of your hand. The support base is very small and relatively far

Fig. 9-12 The Leaning Tower of Pisa does not topple over because its CG lies above its base.

Fig. 9-13 The shaded area bounded by the bottom of the chair legs defines the support base of the chair.

> ▶ **Questions**
> 1. When you carry a heavy load—such as a pail of water—with one arm, why do you tend to hold your free arm out horizontally?
>
> 2. To resist being toppled, why does a wrestler stand with (a) his feet wide apart and (b) his knees bent?
>
> 3. How will the support base of the chair in Figure 9-13 change if one of the front legs is removed? Will the chair topple?

beneath the CG, so it is difficult. But after some practice you can do it if you learn to make slight movements of your hand to exactly respond to variations in balance. You learn to avoid under-responding or over-responding to the slightest variations in balance. Similarly, high-speed computers help massive rockets remain upright when they are launched. Variations in balance are quickly sensed. The computers regulate the firings at multiple nozzles to make corrective adjustments, quite similar to the way your brain coordinates your adjustive action when balancing a long pole on the palm of your hand. Both feats are truly amazing.

9.5 | Stability

It is nearly impossible to balance a pen upright on its point but easy to stand it upright on its flat end. This is because the base of support is inadequate for the point and adequate for the flat end. But there is a second reason. Consider a solid wooden cone on a level table. You cannot stand it on its tip (Figure 9-14). Even if you position it so that its center of gravity is exactly above its tip, the slightest vibration or air current will cause the cone to topple.

▶ **Answers**

1. You tend to hold your free arm outstretched to shift the CG of your body away from the load so your combined CG will more easily be above the base of support. To really help matters, divide the load in two if possible, and carry half in each hand. Or, carry the load on your head!

2. (a) Wide-apart feet increase the support base. (b) Bent knees lower the CG.

3. The support base for four legs is a rectangle. With three legs it's a triangle of half the area. The CG will be toward the rear of the chair because of the weight of the back and loss of weight of the front leg, and be within and above the triangular support base. So it will not topple—until somebody sits on it!

When it does, will the center of gravity be raised, be lowered, or not change at all? The answer to this question provides the second reason for stability. A little thought will show that the CG will always be lowered by the movement. We say that an object balanced so that any displacement *lowers* its center of gravity is in **unstable equilibrium**.

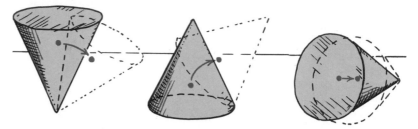

Fig. 9-14 Equilibrium is (a) *unstable* when the CG is lowered with displacement, (b) *stable* when work must be done to raise the CG, and (c) *neutral* when displacement neither raises nor lowers the CG.

A cone balances easily on its base. To make it topple, its CG must be raised. This means its potential energy must be increased, which requires work. We say an object that is balanced so that any displacement *raises* its center of gravity is in **stable equilibrium**.

Place the cone on its side and its CG is neither raised nor lowered with displacement. An object in this configuration is in **neutral equilibrium**.

Like the cone, the pen is in unstable equilibrium when it is on its point. When it is on its flat end, it is in stable equilibrium because the CG must be raised slightly to topple it over (Figure 9-15).

Consider the upright book and the book lying flat in Figure 9-16. Both are in stable equilibrium. But you know the flat book is more stable. Why? Because it would take considerably more work to raise its CG to the point of toppling than to do the same for the upright book. An object with a low CG is usually more stable than an object with a relatively high CG.

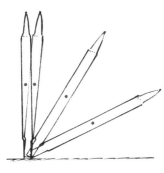

Fig. 9-15 For the pen to topple when it is on its flat end, it must rotate over one edge. During the rotation, the CG rises slightly and then falls.

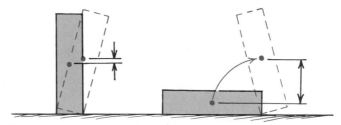

Fig. 9-16 Toppling the upright book requires only a slight raising of its CG; toppling the flat book requires a relatively large raising of its CG. Which requires more work?

The pencil shown balancing horizontally in Figure 9-17 left is barely in stable equilibrium. Its CG must be raised only a slight distance to topple it. But suspend a couple of potatoes from its ends and it is much more stable (Figure 9-17 right). Why? Because the CG is now *below* the point of support. Displacement raises the CG, even if the pencil is tipped almost vertical.

POTATOES

Fig. 9-17 (Left) The balancing pencil is barely in stable equilibrium. (Right) When its ends are stuck into long potatoes that hang below, it is very stable because its new CG remains below the point of support even when it is tipped.

▶ **Question**
Explain why the toy clown shown in Figure 9-3 cannot remain on its side.

Some well-known balancing toys depend on this principle. Their secret is that they have been weighted so that the CG lies vertically underneath the point of support while most of the remainder of the toy is above it (Figure 9-18). Any object that hangs with its CG below its point of support is in stable equilibrium.

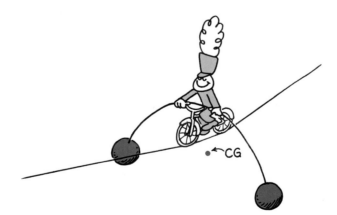

·←CG

Fig. 9-18 The CG of the toy is below the point of support, so the toy is in stable equilibrium.

▶ **Answer**
The CG is lowest when the toy is upright. If it is pushed over, it will easily roll back with the aid of gravity until the CG is the lowest it can be.

Fig. 9-19 The Space Needle in Seattle can no more fall over than can a floating iceberg. In both, the CG is below the surface.

The CG of a building is lowered if much of the structure is below ground level. This is important for tall, narrow structures. An extreme example is the state of Washington's tallest freestanding structure, the Space Needle in Seattle. This structure is so "deeply rooted" that its center of gravity is actually below ground level. It cannot fall over intact. Why? Because falling would not lower its CG at all. If the structure were to tilt intact onto the ground, its CG would be *raised*!

The tendency for the CG to take the lowest position available can be seen by placing a very light object, such as a table-tennis ball, at the bottom of a box of dried beans or small stones. Shake the box, and the beans or stones tend to go to the bottom and force the ball to the top. By this process the CG of the whole system takes a lower position.

Fig. 9-20 (Left) A table-tennis ball is placed at the bottom of a container of dried beans. (Right) When the container is shaken from side to side, the ball is nudged to the top. The result is a lower center of gravity.

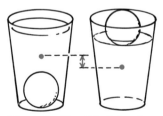

Fig. 9-21 The CG of the glass of water is higher when the table-tennis ball is anchored to the bottom (left) and lower when the ball floats (right).

The same thing happens in water when a light object rises to the surface and floats. If the object weighs less than an equal volume of water, the CG of the whole system will be lowered when the light object is forced to the surface. This is because the heavier (more dense) water can then occupy the available lower space. If the object is heavier than an equal volume of water, it will be more dense than water and sink. In either case, the CG of the whole system is lowered. In the case where the object weighs the same as an equal volume of water (same density), the CG of the system is unchanged whether the object rises or sinks. The object can be at any level beneath the surface without affecting the CG. You can see that a fish must weigh the same as an equal volume of water (have the same density); otherwise it would be unable to remain at different levels in the water. We will return to these ideas in Chapter 19, where liquids are treated in more detail.

Shake a box of stones of different sizes. The shaking enables the small stones to slip down into the spaces between the larger stones and in effect lower the CG. The larger stones therefore tend to rise to the top. The same thing happens when a tray of berries is gently shaken—the larger berries tend to come to the top.

9.6 | Center of Gravity of People

When you stand erect with your arms hanging at your sides, your CG is within your body. It is typically 2 to 3 cm below your navel, and midway between your front and back. The CG is slightly lower in women than in men because women tend to be proportionally larger in the pelvis and smaller in the shoulders. In children, the CG is approximately 5% higher because of their proportionally larger heads and shorter legs.

Raise your arms vertically overhead. Your CG rises 5 to 8 cm. Bend your body into a U or C shape and your CG may be located outside your body altogether. This fact is neatly employed by a high jumper who clears the bar while his CG passes beneath the bar (Figure 9-22).

Fig. 9-22 Dwight Stones executes a "Fosbury flop" to clear the bar while his CG passes beneath the bar.

When you stand, your CG is somewhere above your support base, the area bounded by your feet (Figure 9-23). In unstable situations, as in standing in the aisle of a bumpy-riding bus, you place your feet farther apart to increase this area. Standing on one foot greatly decreases this area. In learning to walk, a baby must learn to coordinate and position the CG above a supporting foot. A pigeon does this by jerking its head back and forth with each step.

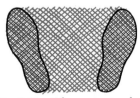

Fig. 9-23 When you stand, your CG is somewhere above the area bounded by your feet.

If you are in reasonable shape, you can probably bend over and touch your toes without bending your knees. In doing so, you unconsciously extend the lower part of your body, as shown in Figure 9-24 left. In this way your CG, which is now outside your body, is nevertheless above your supporting feet. If you try it while standing with your back and heels to a wall, you may be in for a surprise. You cannot do it! This is because you are unable to adjust your body, and your CG protrudes beyond your feet. You are off balance and you topple over.

Fig. 9-24 You can lean over and touch your toes without toppling only if your CG is above the area bounded by your feet.

You don't need to take a course in physics to know where to balance a baseball bat, or how to stand a pencil upright on its flat end. With or without physics, everybody knows that it is easier to hang by your hands below a supporting rope than it is to stand on your hands above a supporting floor. And you don't need a formal study of physics to balance like a gymnast. But maybe it's nice to know that physics is at the root of many things you already know about.

Knowing about things is not always the same as *understanding* things. Understanding begins with knowledge. So we begin by knowing about things, and then progress deeper to an understanding of things. That's where a knowledge of physics is very helpful.

9   Chapter Review

Concept Summary

The center of gravity (CG) of an object is the point at the center of its weight distribution.
- When an object is thrown through the air, its CG follows a smooth parabolic path, even if the object spins or wobbles.
- For everyday objects, the center of gravity is the same as the center of mass.

An object will remain upright if its CG is above the area of support.
- An object is in stable equilibrium when any displacement raises its CG.

Important Terms

center of gravity (9.1)
center of mass (9.2)
neutral equilibrium (9.5)
stable equilibrium (9.5)
unstable equilibrium (9.5)

Review Questions

1. Why is the CG of a baseball bat not at its midpoint? (9.1)

2. What part of an object follows a smooth path when the object is made to spin through the air or across a flat smooth surface? (9.1)

3. Describe the motion of the CG of a projectile, before and after it explodes in mid-air. (9.1)

4. When is the CG and center of mass of an object the same? When are they different? (9.2)

5. What is suggested by a star that wobbles? (9.2)

6. How can the CG of an irregularly shaped object be determined? (9.3)

7. Cite an example of an object that has a CG where no physical material exists. (9.3)

8. Why does the Leaning Tower of Pisa not topple? (9.4)

9. How far can an object be tipped before it topples over? (9.4)

10. How is balancing the pole end of a broom in an upright position on the palm of your hand similar to the launching of a space rocket? (9.4)

11. Distinguish between *unstable*, *stable*, and *neutral* equilibrium. (9.5)

12. Is the gravitational potential energy more, less, or unchanged when the CG of an object is raised? (9.5)

13. Why is it easier to hang by your arms below a supporting cable than to do handstands on a supporting floor? (9.5)

14. What is the "secret" of balancing toys that exhibit stable equilibrium while appearing to be unstable? (9.5)

15. What accounts for the stability of the Space Needle in Seattle? (9.5)

16. If a container of dried beans with a table tennis ball at the bottom is shaken, what happens to the CG of the container? (9.5)

17. What happens to the CG of a glass of water when a table-tennis ball is poked beneath the surface? (9.5)

18. Why do some high-jumpers arch their bodies into a U-shape when passing over the high bar? (9.6)

19. Why do you spread your feet farther apart when standing in a bumpy-riding bus? (9.6)

20. Why can you not successfully bend over and touch your toes when you stand with your back and heels against a wall? (9.6)

Activities

1. Suspend a belt from a piece of stiff wire that is bent as shown in Figure A. Why does the belt balance as it does?

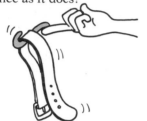

Fig. A

2. Hang a hammer on a loose ruler as shown in Figure B. Then explain why it doesn't fall.

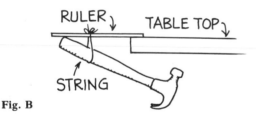

Fig. B

3. Rest a meter stick on two fingers as shown in Figure C. Slowly bring your fingers together. At what part of the stick do your fingers meet? Can you explain why this always happens, no matter where you start your fingers?

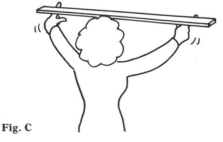

Fig. C

Think and Explain

1. To balance automobile wheels, particularly when tires have worn unevenly, lead weights are fastened to their edges. Where should the CG of the balanced wheel be located?

2. Why does a washing machine vibrate violently if the clothes are not evenly distributed in the tub?

3. Why do we speak of the center of mass of the sun, rather than the center of gravity of the sun?

4. Which glass in Figure D is unstable and will topple?

Fig. D

5. Which balancing act in Figure E is in stable equilibrium? In unstable equilibrium? Nearly at neutral equilibrium?

Fig. E

6. How can the three bricks in Figure F be stacked so that the top brick has maximum horizontal overhang above the bottom brick? For example, stacking them as the dotted lines suggest would be unstable and the bricks would topple. (*Hint*: Start with the top brick and think your way down. At every interface the CG of the bricks above must not extend beyond the end of the supporting brick.)

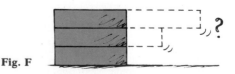

Fig. F

∠1

7. Why don't tall floating icebergs simply tip over?

8. In terms of CG, why is work required to push a table-tennis ball beneath the surface of a glass of water?

9. Why does a pregnant woman during the late stages of pregnancy or a man with a large paunch tend to lean backward when walking?

10. Try the following with a group of males and females. Stand exactly two footlengths away from a wall. Bend over with a straight back and let your head lean against the wall, as shown in Figure G. Then lift a chair that is placed beneath you while your head is still leaning against the wall. With the chair in lifted position, attempt to straighten up.

Give two reasons why females can generally do this and males cannot.

Fig. G

2 FOOTLENGTHS

10 | Universal Gravitation

Things such as leaves, rain, and satellites fall because of gravity. Gravity is what holds tea in a cup and what makes bubbles rise. It made the earth round, and it builds up the pressures which kindle every star that shines. These are things that gravity does. But what *is* gravity? We do not know what gravity is—at least not in the sense that we know what sound, heat, and light are. We have names for things we understand, and we also have names for things we do not understand. Gravity is the name we give to the force of attraction between objects, even though we do not thoroughly understand it. Nevertheless, we understand how gravity affects things such as projectiles, satellites, and the solar system of planets. We also understand that gravity extends throughout the universe; it accounts for such things as the shapes of galaxies. This chapter covers the basic behavior of gravity. In the next chapter we shall investigate more of its consequences.

10.1 | The Falling Apple

The idea that gravity extends throughout the universe is credited to Isaac Newton. According to popular legend, the idea occurred to Newton while he was sitting underneath an apple tree on his mother's farm pondering the forces of nature. Newton understood the concept of inertia developed earlier by Galileo; he knew that without an outside force, moving objects continue to move at constant speed in a straight line. He knew that if an object undergoes a change in speed or direction, then a force is responsible.

A falling apple triggered what was to become one of the most far-reaching generalizations of the human mind. Newton saw the apple fall, or maybe even felt it fall on his head—the story about this is not clear. Looking up through the apple tree branches toward the origin of the apple, did Newton notice the moon? Newton had been giving a lot of thought to the fact that the moon does not follow a straight line path, but instead circles about the earth. Now, circular motion is accelerated motion, which requires a force. But what was this force? Newton had the *insight* to see that the force that pulls between the earth and moon is the same force that pulls between apples and everything else in our universe. This force is the force of gravity.

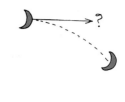

Fig. 10-1 If the moon did not fall, it would follow the straight-line path. Because of its attraction to the earth, it falls along a curved path.

10.2   The Falling Moon

Newton developed this idea further. He compared the falling apple with the falling moon. Does the moon fall? Yes, it does. Newton realized that if the moon did not fall, it would move off in a straight line and leave the earth. His idea was that the moon must be falling *around* the earth. Thus the moon falls in the sense that it *falls beneath the straight line it would follow if no force acted on it*. He hypothesized that the moon was simply a projectile circling the earth under the attraction of gravity.

This idea is illustrated in an original drawing by Newton, shown in Figure 10-2. He compared motion of the moon with a cannonball fired from the top of a high mountain. He imagined that the mountaintop was above the earth's atmosphere, so that air resistance would not impede the motion of the cannonball. If a cannonball were fired with a small horizontal speed, it would follow a parabolic path and soon hit the earth below. If it were fired faster, its path would be less curved and it would hit the earth farther away. If the cannonball were fired fast enough, Newton reasoned, the parabolic path would become a circle and the cannonball would circle indefinitely. It would be in orbit.

Both the orbiting cannonball and the moon have a component of velocity parallel to the earth's surface. This sideways velocity, or **tangential velocity**, is sufficient to insure motion *around* the earth rather than *into* it. If there is no resistance to reduce its speed, the moon "falls" around and around the earth indefinitely.

Newton's idea seemed correct. But for the idea to advance from hypothesis to the status of a scientific theory, it would have to be tested. Newton's test was to see if the moon's "fall" beneath its otherwise straight-line path was in correct proportion to the

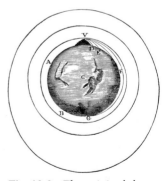

Fig. 10-2 The original drawing by Isaac Newton showing how a projectile fired fast enough would fall around the earth and become an earth satellite. In the same way, the moon falls around the earth and is an earth satellite.

Fig. 10-3   Tangential velocity is the "sideways" velocity. That is, it is the component of velocity that is perpendicular to the pull of gravity.

fall of an apple or any object at the earth's surface. He reasoned that the mass of the moon should not affect how it falls, just as mass has no effect on the acceleration of freely-falling objects on earth. How far the moon falls, and how far an apple at the earth's surface falls, should relate only to their respective *distances* from the earth's center. If the distance of fall for the moon and the apple are in correct proportion, then the hypothesis that earth gravity reaches to the moon must be taken seriously.

The moon was already known to be 60 times farther from the center of the earth than an apple at the earth's surface. The apple will fall nearly 5 m in its first second of fall—or more accurately, 4.9 m. Newton reasoned that gravitational attraction to the earth must be "diluted" by distance. Does this mean the force of earth gravity would reduce to $\frac{1}{60}$ at the moon's distance? No, it is much less than this. As we shall soon see, the influence of gravity should be diluted $\frac{1}{60}$ of $\frac{1}{60}$, or $1/(60)^2$. So in one second the moon should fall $1/(60)^2$ of 4.9 m, which is 1.4 millimeters.*

Using geometry, Newton calculated how far the circle of the moon's orbit lies below the straight-line distance the moon other-

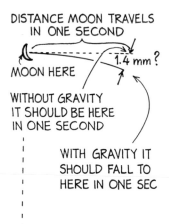

Fig. 10-4   If the force that pulls apples off trees also pulls the moon into orbit, the circle of the moon's orbit should fall 1.4 mm below a point along the straight line where the moon would otherwise be one second later.

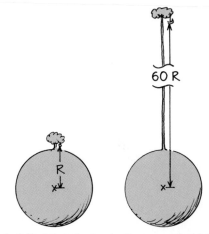

Fig. 10-5   An apple falls 4.9 m during its first second of fall when it is near the earth's surface. Newton asked how far the moon would fall in the same time if it were 60 times farther from the center of the earth. His answer was (in today's units) 1.4 mm.

---

*   Or working backwards, $(0.0014 \text{ m}) \times (60)^2 = 4.9$ m.

wise would travel in one second (Figure 10-4). The distance should have been 1.4 mm. But he was disappointed to end up with a large discrepancy. Recognizing that a hypothesis, however elegant, is not valid if it cannot be backed up with tests, he placed his papers in a drawer, where they remained for nearly 20 years. During this period he laid the foundation and developed the field of geometrical optics for which he first became famous.

It turns out that Newton used an incorrect figure in his calculations. When he finally returned to the moon problem at the prodding of his astronomer friend Edmund Halley (of Halley Comet fame) and used a corrected figure, he obtained excellent agreement. Only then did he publish what is one of the greatest achievements of the human mind—the law of universal gravitation.* Newton generalized his moon finding to all objects, and stated that all objects in the universe attract each other.

10.3 | The Falling Earth

Newton's theory of gravitation confirmed the Copernican theory of the solar system. It was now clear that the earth and planets orbit the sun in the same way that the moon orbits the earth. The mighty sun and planets simply pull each other so that the planets continually "fall" around the sun in closed paths. Why don't the planets crash into the sun? They don't because of their tangential velocities. What would happen if their tangential velocities were reduced to zero? The answer is simple enough: their motion would be straight toward the sun and they would indeed crash into it. Any objects in the solar system with insufficient tangential velocities have long ago crashed into the sun. What remains is the harmony we observe.

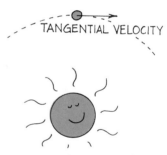

Fig. 10-6 The tangential velocity of the earth about the sun allows it to fall around the sun rather than directly into it. If this tangential velocity were reduced to zero, what would be the fate of the earth?

> ▶ **Question**
>
> Since the moon is gravitationally attracted to the earth, why does it not simply crash into the earth?

▶ **Answer**
The moon would crash into the earth if its tangential velocity were reduced to zero, but because of its tangential velocity, the moon falls around the earth rather than into it. We will return to this idea in more detail in the next chapter.

---

* This is a dramatic example of the painstaking effort and crosschecking that go into the formulation of a scientific theory. Compare this with the lack of "doing one's homework," the hasty judgements, and the absence of cross-check examinations that so often characterize the pronouncement of less-than-scientific theories.

10.4 | The Law of Universal Gravitation

Newton did not discover gravity. What Newton discovered was that gravity was universal. Everything pulls on everything else in a beautifully simple way that involves only mass and distance. Newton's **law of universal gravitation** states that every object attracts every other object with a force that for any two objects is directly proportional to the mass of each object. The greater the masses, the greater the force of attraction between them. Newton deduced that the force decreases as the square of the distance between the centers of mass of the objects. The farther away the objects are from each other, the less the force of attraction between them.

The law can be expressed symbolically as

$$F \sim \frac{m_1 m_2}{d^2}$$

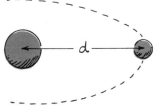

Fig. 10-7 The force of gravity between objects depends on the distance between their centers of mass.

where m_1 is the mass of one object, m_2 is the mass of the other, and d is the distance between their centers of mass. The greater the masses m_1 and m_2, the greater the force of attraction between them. The greater the distance d between the objects, the weaker the force of attraction.

We can express the symbolic statement as an exact equation if we introduce the quantity G, the **universal constant of gravitation**. G is a conversion factor needed to change the units of mass and distance on the right side of the equation to the units of force on the left (just as we use conversion factors to change from miles to kilometers, for example). Then we can express the law of universal gravitation as the exact equation

$$F = G \frac{m_1 m_2}{d^2}$$

In words, the force of gravity between two objects is found by multiplying their masses, dividing by the square of the distance between them, then multiplying this result by the constant G. The value of G was determined by actual measurement long after Newton's time. It is

$$G = 0.0000000000667 \ \frac{\text{N·m}^2}{\text{kg}^2}$$

or, in scientific notation,*

$$G = 6.67 \times 10^{-11} \ \text{N·m}^2/\text{kg}^2$$

Do you understand this????

---

*   Scientific notation is discussed in Appendix A at the end of this book.

The value of *G* tells us that the force of gravity is a very weak force. It is the weakest of the four fundamental forces found in nature. (The other three are the electromagnetic force and two kinds of nuclear forces.) We sense gravitation only when masses like that of the earth are involved. The force of attraction between a pair of 1-kg masses with their centers of gravity 1 m apart is only 6.67 × 10⁻¹¹ N, too tiny for ordinary measurement. The force of attraction between you and the earth, however, can be measured. It is called your *weight*.

Your weight is the gravitational attraction between your mass and the mass of the earth. If you gain mass, you gain weight. Or if the earth somehow gained mass, you'd also gain weight. Your weight also depends on your distance from the center of the earth. At the top of a mountain your mass is no different than it is anywhere else, but your weight is slightly less than at ground level. This is because your distance from the center of the earth is greater.

Fig. 10-8   Your weight is less at the top of a mountain because you are farther from the center of the earth.

> ▶ **Question**
>
> If there is an attractive force between all objects, why do we not feel ourselves gravitating toward massive buildings in our vicinity?

10.5   Gravity and Distance: The Inverse-Square Law

We can understand how gravity is reduced with distance by considering an imaginary "butter gun" used in a busy restaurant for buttering toast (Figure 10-9). Imagine melted butter sprayed through a square opening in a wall. The opening is exactly the size of one piece of square toast. And imagine that a spurt from the gun deposits an even layer of butter 1 mm thick. Consider the consequences of holding the toast twice as far from the butter gun. You can see in Figure 10-9 that the butter would spread out

---

▶ **Answer**

We *are* gravitationally attracted to massive buildings and everything else in the universe. The 1933 Nobel prize-winning physicist Paul A.M. Dirac put it this way: "Pick a flower on earth and you move the farthest star!" How *much* you are influenced by buildings or how much interaction there is between flowers, is another story. The forces between you and buildings are relatively small because the masses are small compared to the mass of the earth. The forces due to the stars are small because of their great distances. These tiny forces escape our notice when they are overwhelmed by the overpowering attraction to the earth.

for twice the distance and would cover twice as much toast vertically and twice as much toast horizontally. A little thought will show that the butter would now spread out to cover four pieces of toast. How thick will the butter be on each piece of toast? Since it has been diluted to cover four times as much area, its thickness will be one-quarter as much, or 0.25 mm.

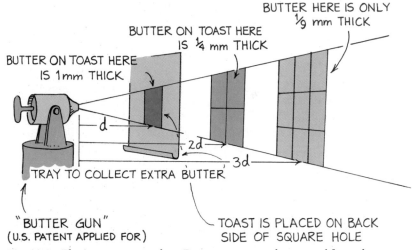

Fig. 10-9   The inverse-square law. Butter spray travels outward from the nozzle of the butter gun in straight lines. Like gravity, the "strength" of the spray obeys the inverse-square law.

Note what has happened. When the butter gets twice as far from the gun, it is only $\frac{1}{4}$ as thick. More thought will show that if it gets 3 times as far, it will spread out to cover 3×3, or 9, pieces of toast. How thick will the butter be then? Can you see it will be $\frac{1}{9}$ as thick? And can you see that $\frac{1}{9}$ is the inverse square of 3? (The inverse of 3 is simply $\frac{1}{3}$; the inverse *square* of 3 is $\left(\frac{1}{3}\right)^2$, or $\frac{1}{9}$.) This law applies not only to the spreading of butter from a butter gun, and the weakening of gravity with distance, but to all cases where the effect from a localized source spreads evenly throughout the surrounding space. More examples are light, radiation, and sound.

The greater the distance from the earth's center, the less an object will weigh (Figure 10-10). If your little sister weighs 300 N at sea level, she will weigh only 299 N atop Mt. Everest. But no matter how great the distance, the earth's gravity does not drop to zero. Even if you were transported to the far reaches of the universe, the gravitational influence of the earth would be with you. It may be overwhelmed by the gravitational influences of nearer and/or more massive objects, but it is there. The gravitational influence of every object, however small or far, is exerted through all space. Isn't that amazing?

> ▶ **Question**
>
>   Suppose that an apple at the top of a tree is pulled by earth gravity with a force of 1 N. If the tree were twice as tall, would the force of gravity on the apple be only $\frac{1}{4}$ as strong? Explain your answer.

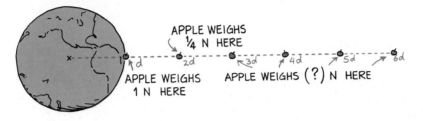

Fig. 10-10   An apple that weighs 1 N at the earth's surface weighs only 0.25 N when located twice as far from the earth's center because the pull of gravity is only $\frac{1}{4}$ as strong. When it is 3 times as far, it weighs only $\frac{1}{9}$ as much, or 0.11 N. What would it weigh at 4 times the distance? Five times?

10.6 | Universal Gravitation

We all know that the earth is round. But why is the earth round? It is round because of gravitation. Since everything attracts everything else, the earth has attracted itself together before it became solid. Any "corners" of the earth have been pulled in so that the earth is a giant sphere. The sun, the moon, and the earth are all fairly spherical because they have to be (rotational effects make them somewhat wider at their equators).

If everything pulls on everything else, then the planets must pull on each other. The net force that controls Jupiter, for example, is not just from its interaction with the sun, but with the planets also. Their effect is small compared to the pull of the more massive sun, but it still shows. When the planet Saturn is near Jupiter, for example, its pull disturbs the otherwise smooth path of Jupiter. Both planets deviate from their normal orbits. This deviation is called a **perturbation**.

▶ **Answer**
   No, because the twice-as-tall apple tree is not twice as far from the *earth's center*. The taller tree would have to have a height equal to the radius of the earth (6370 km) before the weight of the apple reduces to $\frac{1}{4}$ N. Before its weight decreases by one percent, an apple or any object must be raised 32 km—nearly 4 times the height of Mt. Everest, the tallest mountain in the world. So as a practical matter we disregard the effects of everyday changes in elevation.

Up until the middle of the last century astronomers were puzzled by unexplained perturbations of the planet Uranus. Even when the influences of the other planets were taken into account, Uranus was behaving stangely. Either the law of gravitation was failing at this great distance from the sun, or some unknown influence such as another planet was perturbing Uranus.

The source of Uranus's perturbation was uncovered in 1845 and 1846 by two astronomers, John Adams in England, and Urbain Leverrier in France. With only pencil and paper and the application of Newton's law of gravitation, both astronomers independently arrived at the same conclusion: a disturbing body beyond the orbit of Uranus was the culprit. They sent letters to their local observatories with instructions to search a certain part of the sky. The request by Adams was delayed by misunderstandings at Greenwich, England, but Leverrier's request to the director of the Berlin observatory was heeded right away. The planet Neptune was discovered within a half hour.

Other perturbations of Uranus led to the prediction and discovery of the ninth planet, Pluto. It was discovered in 1930 at the Lowell observatory in Arizona. Pluto takes 248 years to make a single revolution about the sun, so it will not be seen in its discovered position again until the year 2178.

The perturbations of double stars and the shapes of distant galaxies are evidence that the law of gravitation extends beyond the solar system. Over still larger distances, gravitation dictates the fate of the entire universe.

Fig. 10-11   Gravitation dictates the shape of the spiral arms in a galaxy.

Current scientific speculation is that the universe originated in the explosion of a primordial fireball some 15 to 20 billion years ago. This is the "Big Bang" theory of the origin of the universe. All the matter of the universe was hurled outward from this event and continues in an outward expansion. We find ourselves in an expanding universe.

This expansion may go on indefinitely, or it may be overcome by the combined gravitation of all the galaxies and come to a stop. Like a stone thrown upward, whose departure from the ground comes to an end when it reaches the top of its trajectory, and which then begins its descent to the place of its origin, the universe may contract and fall back into a single unity. This would be the "Big Crunch." After that, the universe would presumably re-explode to produce a new universe. The same course of action might repeat itself and the process may well be cyclic. If this is true, we live in an oscillating universe.

We do not know whether the expansion of the universe is cyclic or indefinite, because we are uncertain about whether enough mass exists to halt the expansion. The period of oscillation is estimated to be somewhat less than 100 billion years. If the universe does oscillate, who can say how many times it has expanded and collapsed? We know of no way a civilization could leave a trace of ever having existed during a previous cycle, for all the matter in the universe would be reduced to bare subatomic particles during the collapse. All the laws of nature, such as the law of gravitation, would then have to be rediscovered by the higher evolving life forms. And then students of these laws would read about them, as you are doing now. Think about that.

Few theories have affected science and civilization as much as Newton's theory of gravity. The successes of Newton's ideas ushered in the Age of Reason or Century of Enlightenment. Newton had demonstrated that by observation and reason, people could uncover the workings of the physical universe. How profound that all the moons and planets and stars and galaxies have such a beautifully simple rule to govern them, namely

$$F = G \frac{m_1 m_2}{d^2}$$

The formulation of this simple rule is one of the major reasons for the success in science that followed, for it provided hope that other phenomena of the world might also be described by equally simple laws.

This hope nurtured the thinking of many scientists, artists, writers, and philosophers of the 1700s. One of these was the English philosopher John Locke, who argued that observation and reason, as demonstrated by Newton, should be our best judge and guide in all things. Locke urged that all of nature and even society should be searched to discover any "natural laws" that might exist. Using Newtonian physics as a model of reason, Locke and his followers modeled a system of government that found adherents in the 13 British colonies across the Atlantic. These ideas culminated in the Declaration of Independence and the Constitution of the United States of America.

10 | Chapter Review

Concept Summary

The moon and other objects in orbit around the earth are actually falling toward the earth but have great enough tangential velocity to avoid hitting the earth.

According to Newton's law of universal gravitation, everything pulls on everything else with a force that depends upon the masses of the objects and the distances between their centers of mass.

- The greater the masses, the greater is the force.
- The greater the distance, the smaller is the force.

Important Terms

law of universal gravitation (10.4)
perturbation (10.6)
tangential velocity (10.2)
universal constant of gravitation (10.4)

Review Questions

1. Why did Newton think that a force must act on the moon? (10.1)

2. What was it that Newton discovered about the force that pulls apples to the ground and the force that holds the moon in orbit? (10.1)

3. If the moon falls, why doesn't it get closer to the earth? (10.2)

4. What is meant by *tangential* velocity? (10.2)

5. How did Newton check his hypothesis that there is an attractive force between the earth and moon? (10.2)

6. What is required before a hypothesis (an educated guess) advances to the status of a scientific theory (organized knowledge)?

7. Since the planets are pulled to the sun by gravitational attraction, why don't they simply crash into the sun? (10.3)

8. Newton did not *discover* gravity but, rather, something about gravity. What was his discovery? (10.4)

9. What does the very small value of the gravitational constant G tell us about the strength of gravitational forces? (10.4)

10. Exactly what are the two specific masses and the one specific distance that determine your weight? (10.4)

11. In what way is gravity reduced with distance from the earth? (10.5)

12. What would be the difference in your weight if you were five times farther from the center of the earth as you are now? Ten times? (10.5)

13. What makes the earth round? (10.6)

14. What causes planetary perturbations? (10.6)

15. Distinguish between the "Big Bang" and the "Big Crunch." (10.6)

Think and Explain

1. The moon "falls" 1.4 mm each second. Does this mean that it gets 1.4 mm closer to the earth each second? Would it get closer if its tangential velocity were reduced? Defend your answer.

2. Comparing a large boulder with a small stone, which is attracted to the earth with the greater force? Which undergoes the greater acceleration when falling without air resistance? Defend both answers.

3. If the gravitational forces of the sun on the planets suddenly disappeared, in what kind of paths would they move?

4. If the moon were twice as massive, would the attractive force between the earth and moon be twice as large? Between the moon and earth? Defend your answer.

5. a. If the gravitational force between two massive bodies were measured and divided by the product of their masses, and then multiplied by the square of the distance between their centers of mass, what number would result?
   b. Would this number differ for different masses at different distances? Defend your answer.

6. If the moon were twice as far away and remained in circular orbit about the earth, by what distance would it "fall" each second?

7. If you stood atop a ladder that was so tall that you were twice as far from the earth's center, how would your weight compare to its present value?

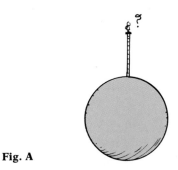

Fig. A

8. By what factor would your weight change if the earth were twice as big and had twice the mass?

9. Evidence indicates that the present expansion of the universe is slowing down. Is this consistent with, or contrary to, the law of gravity? Explain.

10. Some people dismiss the validity of scientific theories by saying they are "only" theories. The law of universal gravitation is a theory. Does this mean that scientists still doubt its validity? Explain.

11 Gravitational Interactions

Everyone knows that objects fall because of gravity. Even people before the time of Isaac Newton knew this. Contrary to popular belief, Newton did not discover gravity. What Newton discovered was that gravity is *universal*—that the same force that pulls an apple off a tree holds the moon in orbit, and that the earth and moon are similarly held in orbit about the sun. And the sun revolves as part of a cluster of other stars about the center of the galaxy, the Milky Way. Newton discovered that all objects in the universe attract each other. This was discussed in the last chapter. In this chapter we shall investigate the role of gravity at, below, and above the earth's surface; in the earth's oceans and its atmosphere; and in stellar objects called black holes. We begin with the simple case of free fall.

11.1 Acceleration Due to Gravity

Recall from earlier chapters that an object in free fall (that is, falling with only the force of gravity acting on it) accelerates downward at the rate of 9.8 m/s². This acceleration is known as g, the acceleration due to gravity. Some people get mixed up between little g and big G. Big G is not acceleration, but is the universal gravitational constant in the equation for the gravitational force between any two objects:

$$F = G \frac{m_1 m_2}{d^2}$$

Thus g and G are quite different quantities that represent quite different things. Nevertheless, there is an interesting relationship between the two. The value of little g was found from measurements of falling objects. But why should it be 9.8m/s² and not some other value? We can find the answer from the equation for gravity.

The weight of any object is the gravitational force of attraction between that object and the earth. This force depends on the mass of the object, m, and the mass of the earth, which we will call M. At the surface of the earth the distance between their centers is simply the radius of the earth, which we will call R. If we make these substitutions ($m_1 = m$, $m_2 = M$, $d = R$) in the law of gravity, we get an equation for the weight of an object at the earth's surface:

Fig. 11-1   The weight of the load is the gravitational force of attraction between the load and the earth.

$$\text{weight} = \frac{GmM}{R^2}$$

From Newton's second law, $F = ma$, the weight of an object is mg. Substituting mg for weight in the preceding equation, we get

$$mg = \frac{GmM}{R^2}$$

Notice that m, the mass of the object, appears on both sides. We can divide it out, which leaves

$$g = \frac{GM}{R^2}$$

Suppose you use a calculator to evaluate g, using the following values for G, M, and R:

$$G = 6.67 \times 10^{-11} \text{ N·m}^2/\text{kg}^2$$
$$M = 5.98 \times 10^{24} \text{ kg}$$
$$R = 6.37 \times 10^{6} \text{ m}$$

Multiply the first two numbers and divide two times by the third. Rounded off to two digits, your answer will be 9.8 N/kg. And, since 1 N equals 1 kg·m/s², the combination unit N/kg is the same as m/s².

Thus, you can see that the numerical value of g depends on the mass of the earth and its radius. If the earth had a different mass or radius, g would have a different value. If you know the mass and radius of any planet, you can calculate the value of g for that planet.

Interestingly enough, the mass of the earth was itself found from this equation after the value of G was experimentally determined in 1798 by an English scientist named Henry Cavendish. In fact, his procedure for determining G with heavy lead weights was called the "weighing-the-earth experiment." You can see that if g, G, and R are known, then the mass of the earth M is easily found. So the combined mass of all the rock, oceans,

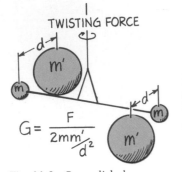

TWISTING FORCE

$$G = \frac{F}{2mm'/d^2}$$

Fig. 11-2 Cavendish determined G by measuring the force of gravity between lead masses. Once he had the value of the force, the masses, and their distances apart, can you see how he determined G?

mountains, trees, sticks, and twigs that make up the planet earth can be easily calculated. The power of a little algebra is enormous!

▶ **Questions**

1. The acceleration of objects on the surface of the moon due to universal gravity is only $\frac{1}{6}$ of 9.8 m/s². From this fact, is it correct to say that the mass of the moon is therefore $\frac{1}{6}$ the mass of the earth?

2. Estimate "g" on the surface of the planet Jupiter by comparing its mass and size with that of the earth. First, it is about 300 times more massive than the earth. (This means that if it had the same radius, "g" for Jupiter would be 300 earth g.) But the radius of Jupiter is about 10 times greater than the radius of earth, so an object at its surface is 10 times farther from Jupiter's center than it would be on earth. So about how much greater is "g" on Jupiter compared to earth g?

11.2   Gravitational Fields

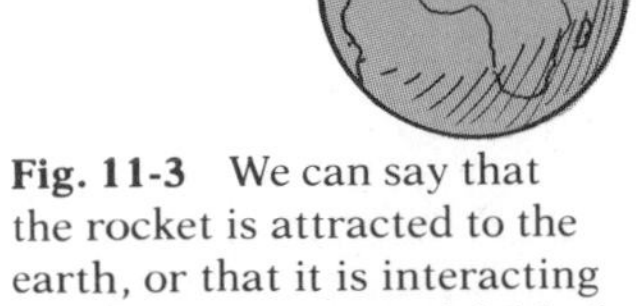

The earth pulls on the moon. We regard this as *action at a distance*, because the earth and moon interact with each other even though they are not in contact. Or we can look at this in a different way: we can regard the moon as interacting with the **gravitational field** of the earth. The properties of the space surrounding any mass can be considered to be altered in such a way that another mass introduced to this region will experience a force. This alteration of space is the gravitational field. It is common to think of distant rockets and space probes as interacting with gravitational fields rather than with the masses of the earth and other planets or stars that are the sources of these fields. The field concept plays an in-between role in our thinking about the forces between different masses.

Fig. 11-3 We can say that the rocket is attracted to the earth, or that it is interacting with the gravitational field of the earth. Both points of view are equivalent.

▶ **Answers**

1. No. We could assume the mass of the moon to be $\frac{1}{6}$ that of earth only if both moon and earth had the same radius. The radius of the moon (1.74×10^6 m) is in fact less than one third the earth's radius, and its mass (7.36×10^{22} kg) is about $\frac{1}{60}$ the mass of the earth.

2. The value of "g" on Jupiter's surface is $G(300M)/(10R)^2 = 300GM/(100R^2) = 3GM/R^2 = 3g$.

A gravitational field is an example of a **force field**, for something in the space experiences a force due to the field. Another force field you may be familiar with is a magnetic field. You have probably seen iron filings lined up in patterns around a magnet. (Look ahead to Figure 36-4 on page 541, for example.) The pattern of the filings shows the strength and direction of the magnetic field at different points in the space around the magnet. Where the filings are closest together, the field is strongest. The direction of the filings shows the direction of the field at each point.

The pattern of the earth's gravitational field can be represented by field lines (Figure 11-4). Like the iron filings around a magnet, the field lines are closer together where the gravitational field is stronger. At each point on a field line, the direction of the field at that point is along the line. Arrows show the field direction. A particle, astronaut, spaceship, or any mass in the vicinity of the earth will be accelerated in the direction of the field line at that location.

The strength of the earth's gravitational field, like the strength of its force on objects, follows the inverse-square law.* It is strongest near the earth's surface, and weakens with increasing distance from the earth.

The gravitational field of the earth exists inside the earth as well as outside. To investigate the field beneath the surface, imagine a hole drilled completely through the earth, say from the North Pole to the South Pole. Forget about impracticalities such as lava and high temperatures, and consider the kind of motion you would undergo if you fell into such a hole. If you started at the North Pole end, you'd fall and gain speed all the way down to the center, then overshoot and lose speed all the way "up" to the South Pole. You'd gain speed moving toward the center, and lose speed moving away from the center. Without air drag, the one-way trip would take nearly 45 minutes. If you failed to grab the edge, you'd fall back toward the center, overshoot, and return to the North Pole in the same time.

Repeat this trip with an accelerometer of some kind. Your acceleration at the beginning of fall is g, but you'd find it progressively less as you continue toward the center of the earth. Why? Because as you are being pulled "downward" toward the earth's center, you are also being pulled "upward" by the part of the earth that is "above" you. In fact, when you get to the center of the earth, the pull "down" is balanced by the pull "up." You

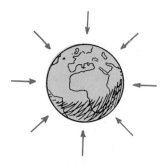

Fig. 11-4   Field lines represent the gravitational field about the earth. Where the field lines are closer together, the field is stronger. Farther away, where the field lines are farther apart, the field is weaker.

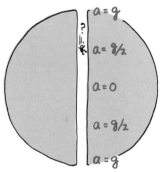

Fig. 11-5   As you fall faster and faster in a hole bored completely through the earth, your acceleration diminishes because the mass you leave behind retards it. At the earth's center your acceleration is zero. Momentum carries you against a growing acceleration past the center to the opposite end of the tunnel where it is again g.

---

*   The strength of the gravitational field at any point is measured by the force on a unit mass placed there. If a force F is exerted on a mass m, the field strength is F/m, and its units are newtons per kilogram (N/kg).

are pulled in every direction equally, so the net force on you is zero. There is no acceleration as you whiz with maximum speed past the center of the earth. The gravitational field of the earth at its center is zero!

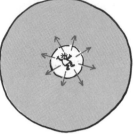

Fig. 11-6 In a cavity at the center of the earth your weight would be zero because you are pulled equally by gravity in all directions. The gravitational field at the earth's center is zero.

Interestingly enough, you'd gain acceleration during the first few kilometers beneath the earth's surface, because the density of surface material is much less than the condensed center. The mass of earth "above you" exerts a small backward pull compared to the disproportionally greater pull of the denser core material. This means you'd weigh slightly more for the first few kilometers beneath the earth's surface. Further in, your weight would decrease and would diminish to zero at the earth's center.

▶ **Questions**

1. If you stepped into a hole through the center of the earth and made no attempt to grab the edges at either end, what kind of motion would you experience?

2. Halfway to the center of the earth, would you weigh more or weigh less than you weigh at the surface of the earth?

▶ **Answers**

1. You would oscillate back and forth, in what is called *simple harmonic motion*. Each round trip would take nearly 90 minutes. Interestingly enough, we will see in the next chapter that an earth satellite in close orbit about the earth also takes the same 90 minutes to make a complete round trip. (This is no co-incidence, for if you study physics further, you'll learn about an interesting relationship between simple harmonic motion and circular motion at constant speed.)

2. You would weigh less, because the part of the earth's mass that pulls you "down" is counteracted by mass above you that pulls you "up." If the earth were of uniform density, halfway to the center your weight would be exactly half your surface weight. But since the earth's core is so dense (about 7 times the density of surface rock) your weight would be somewhat more than half surface weight. Exactly how much depends on how the earth's density varies with depth, information that is not known today.

11.3 | Weight and Weightlessness

The force of gravity, like any force, causes acceleration. Objects under the influence of gravity accelerate toward each other. We are almost always in contact with the earth. For this reason, we think of gravity primarily as something that presses us against the earth rather than as something that accelerates us. The pressing against the earth is the sensation we interpret as weight.

Stand on a bathroom scale that is supported on a stationary floor. The gravitational force between you and the earth pulls you against the supporting floor and scale. By Newton's third law, the floor and scale in turn push upward on you. Located in between you and the supporting floor are springs inside the bathroom scale. The springs are compressed by this pair of forces. The weight reading on the scale is linked to the amount of compression of the springs.

If you repeat this weighing procedure in a moving elevator, you would find your weight reading would vary—not during steady motion, but during accelerated motion. If the elevator accelerates upward, the bathroom scale and floor push harder against your feet, and the springs inside the scale are compressed even more. The scale shows an increase in your weight.

Fig. 11-7   The sensation of weight is equal to the force that you exert against the supporting floor. If the floor accelerates up or down, your weight seems to vary. You feel weightless when you lose your support in free fall.

If the elevator accelerates downward, the scale shows a decrease in your weight. The support force of the floor is now less. If the elevator cable breaks and the elevator falls freely, the scale reading would register zero. According to the scale, you are weightless. Would you really be weightless? The answer to this question depends on your definition of weight.

If you define weight as gravitational force acting on an object, then you still have weight whether or not you are freely falling. Astronauts in earth orbit who float freely inside their capsules are still being pulled by gravity and therefore have weight. But you in the falling elevator and the astronauts in the "falling" satellites don't feel the usual effects of weight. You feel weightless. That is, your insides are no longer supported by your legs and pelvic region. Your organs respond as though gravity were absent. You are in a state of **apparent weightlessness**.

Fig. 11-8   The astronaut is in a state of apparent weightlessness all the time in orbit.

To be truly weightless, you would have to be far out in space—well away from the earth and other attracting stars and planets—where gravitational forces are negligible. In this truly weightless condition, you would drift in a straight-line path rather than in the curved path of an orbit.

In a more practical sense, we define the weight of an object as the force it exerts against a supporting floor (or weighing scales). According to this definition, you are as heavy as you feel. The condition of weightlessness is then not the absence of gravity, but the absence of a support force. That queasy feeling you get when you are in a car that seems to leave the road momentarily when it goes over a hump, or worse, off a cliff, is not the absence of gravity. It is the absence of a support force. Astronauts in orbit are without a support force and are in a sustained state of weightlessness. That causes "spacesickness" until they get used to it.

Fig. 11-9   Both people experience weightlessness.

11.4   Ocean Tides

Seafaring people have always known there was a connection between the ocean tides and the moon, but no one could offer a satisfactory theory to explain the two high tides per day. Newton showed that the ocean tides are caused by *differences* in the gravitational pull between the moon and the earth on opposite sides of the earth. The gravitational force of the moon on the earth is stronger on the side of the earth nearer to the moon, and weaker on the side of the earth farther from the moon. This is simply because the gravitational force is weaker with increased distance.

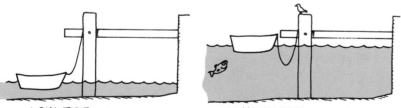

LOW TIDE   HIGH TIDE

Fig. 11-10   Twice a day, every point along the ocean shore has a high tide. In between the high tides is a low tide.

Consider a big spherical ball of gooey taffy. If you exert the same force on every part of the ball, it would remain spherical as it accelerates. But if you pulled harder on one side than the other, its shape would become elongated. Imagine that you attach a rope to the ball of taffy and swing it in a circle around you (Figure 11-11). The ball will bulge outward in two places. One bulge will be toward the center of the circular path. The other bulge will be on the opposite side, on the faster-moving outer part that is "thrown" outward. For an initially spherical ball of uniform taffy, these oppositely facing bulges would be of equal size.

Fig. 11-11 An initially spherical ball of gooey taffy will be elongated when it is spun in a circular path.

This is what happens to this big ball we're living on. You might tend to think of the earth at rest and the moon circling around us. But if you lived on the moon, you'd likely say that the moon is at rest and the earth circles about the moon. It turns out that both the earth and the moon orbit about a common point—their earth-moon center of mass. So the water covering the circling earth is distorted and will bulge like the circling taffy in Figure 11-11.

The earth makes one complete spin per day beneath these ocean bulges. This produces two sets of ocean tides per day. Any part of the earth that passes beneath one of the bulges has a high tide. On a world average, a high tide is about 1 m above the average surface level of the ocean. When the earth makes a quarter turn, 6 hours later, the water level at the same part of the ocean is about 1 m below the average sea level. This is low tide. The water that "isn't there" is under the bulges that make up the high tides. A second high tidal bulge is experienced when the earth makes another quarter turn. So we have two high tides and two low tides daily. It turns out that while the earth spins, the moon moves in its orbit and appears at the same position in our sky every 24 hours and 50 minutes, so the two-high-tide cycle is actually at 24-hour-and-50-minute intervals. That is why tides do not occur at the same time every day.

The sun also contributes to ocean tides, although it is less than half as effective as the moon in raising tides. This may seem puzzling when it is realized that the pull between the sun and earth is about 180 times stronger than the pull between the moon and earth. Why, then, does the sun not cause tides 180 times greater

Fig. 11-12 Two tidal bulges remain relatively fixed with respect to the moon while the earth spins daily beneath them.

than lunar tides? The answer has to do with a key word: *differ-ence*. Because of the sun's great distance from the earth, there is not much difference in the distances from the sun to the near part of the earth and the far part. This means there is not much difference in the gravitational pull of the sun on the part of the ocean nearest it and on the part farthest from it. The relatively small difference in pulls on opposite sides of the earth only slightly elongates the earth's shape and produces tidal bulges less than half those produced by the moon.*

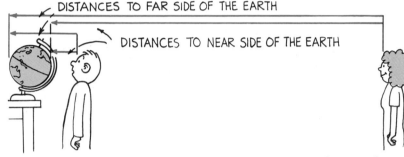

DISTANCES TO FAR SIDE OF THE EARTH

DISTANCES TO NEAR SIDE OF THE EARTH

Fig. 11-13 If you stand close to the globe (as the moon is in relation to the earth), the closest part of the globe is noticeably nearer to you than the farthest part. If you stand far away (as the sun is in relation to the earth), the difference in distance between the closest and farthest part of the globe is less significant.

When the sun, earth, and moon are all lined up, the tides due to the sun and the moon coincide. Then we have higher-than-average high tides and lower-than-average low tides. These are called **spring tides** (Figure 11-14). (Spring tides have nothing to do with the spring season.)

Fig. 11-14 When the attractions of the sun and the moon are lined up with each other, spring tides occur.

---

*   The relative differences in distance is only hinted at in Figure 11-13, which is way off scale. The actual distance between the earth and the sun is nearly 12 000 earth diameters. So for an ordinary globe of $\frac{1}{3}$ meter in diameter, to judge the relative difference between closest and farthest parts of the globe from the moon, you'd have to stand 10 meters away; and from the sun's position, you'd have to stand "across town," 4 kilometers away!

If the alignment is *perfect*, we have an eclipse. A **lunar eclipse** is produced when the earth is directly between the sun and moon (Figure 11-15a). A **solar eclipse** is produced if the moon is directly between the sun and earth (Figure 11-15c). The alignment is usually not perfect. Instead, we have a full moon when the earth is in the middle, and a new moon when the moon is in the middle. So spring tides occur at the times of a new or full moon.

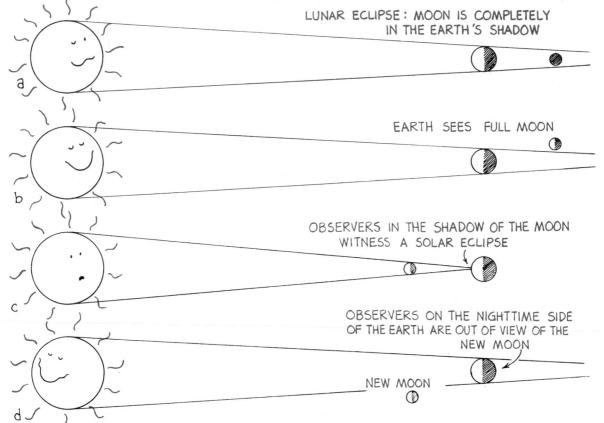

Fig. 11-15 Detail of sun-earth-moon alignment. (a) Perfect alignment produces a lunar eclipse. (b) Non-perfect alignment produces a full moon. (c) Perfect alignment produces a solar eclipse. (d) Non-perfect alignment produces a new moon. Can you see that from the daytime side of the world, the new moon cannot be seen because the dark side faces the earth, and from the nighttime side of the world, the moon is out of view altogether?

Fig. 11-16 When the attractions of the sun and the moon are at right angles to one another (at the time of a half moon), neap tides occur.

When the moon is half way between a new moon and a full moon, in either direction (Figure 11-16), the tides due to the sun and the moon partly cancel each other. Then, the high tides are lower than average and the low tides are not as low as average low tides. These are called **neap tides**.

Another factor that affects the tides is the tilt of the earth's axis (Figure 11-17). Even though the opposite tidal bulges are equal, the earth's tilt causes the two daily high tides experienced in most parts of the ocean to be unequal most of the time.

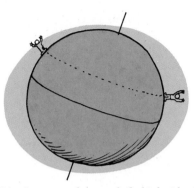

Fig. 11-17   The earth's tilt causes the two daily high tides in the same place to be unequal.

11.5 | Tides in the Earth and Atmosphere

The earth is not a rigid solid but, for the most part, is molten liquid covered by a thin solid and pliable crust. As a result, the moon-sun tidal forces produce earth tides as well as ocean tides. Twice each day the solid surface of the earth rises and falls by as much as 25 cm! As a result, earthquakes and volcanic eruptions have a slightly higher probability of occurring when the earth is experiencing an earth spring tide—that is, near a full or new moon.

We live at the bottom of an ocean of air that also experiences tides. Because of the low mass of the atmosphere, the atmospheric tides are very small. In the upper part of the atmosphere is the ionosphere, so named because it is made up of ions, electrically charged atoms that are the result of intense cosmic ray bombardment. Tidal effects in the ionosphere produce electric currents that alter the magnetic field that surrounds the earth. These are magnetic tides. They in turn regulate the degree to which cosmic rays penetrate into the lower atmosphere. The cosmic ray penetration affects the ionic composition of our atmosphere, which in turn, is evident in subtle changes in the behaviors of living things. The highs and lows of magnetic tides are greatest when the atmosphere is having its spring tides— again, near the full and new moon.*

---

*   This may be why some of your friends seem a bit weird at the time of a full moon!

| 11.6 | **Black Holes** |

There are two main processes that are going on all the time in stars such as our sun. One is gravitation, which tends to crunch all solar material toward the center. The other process is nuclear fusion. The core of the sun is continuously undergoing hydrogen-bomb-like explosions that tend to blow its material out from the center. The two processes balance each other, and the result is the sun of a given size.

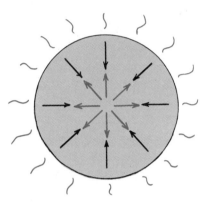

Fig. 11-18   The size of the sun is the result of a "tug-of-war" between two opposing processes: nuclear fusion, which tends to blow it up (colored arrows), and gravitational contraction, which tends to crunch it together (black arrows).

If the fusion rate increases, the sun gets bigger; if the fusion rate decreases, the sun gets smaller. What happens when the sun runs out of fusion fuel (hydrogen)? The answer is, gravitation dominates and the sun collapses. For our sun, this collapse will ignite the nuclear ashes of fusion (helium) and fuse them into carbon. During this fusion process, the sun will expand to become the type of star known as a *red giant*. It will be so big that it will extend beyond the earth's orbit and swallow the earth. Fortunately, this won't take place until 5 billion years from now. When the helium is all fused, the red giant will collapse and die out. It will no longer give off heat and light. It will then be the type of star called a *black dwarf*—a cool cinder among billions of others.

The story is a bit different for stars more massive than the sun. For stars of more than four solar masses, once gravitational collapse takes place—fusion or no fusion—it doesn't stop! The stars not only cave in on themselves, but the atoms that compose the stellar material also cave in on themselves until there are no empty spaces. What is left is compressed to unimaginable densities. Gravitation near the surfaces of these shrunken configurations is so enormous that nothing can get back out. Even light cannot escape. They have crushed themselves out of visible existence. They are called **black holes**.

Interestingly enough, a black hole is no more massive than the star from which it collapsed. The gravitational field near the black hole may be enormous, but the field beyond the original radius of the star is no different after collapse than before (Figure 11-19). The amount of mass has not changed, so there is no change in the field at any point beyond this distance. Black holes will be formidable only to future astronauts who venture too close.

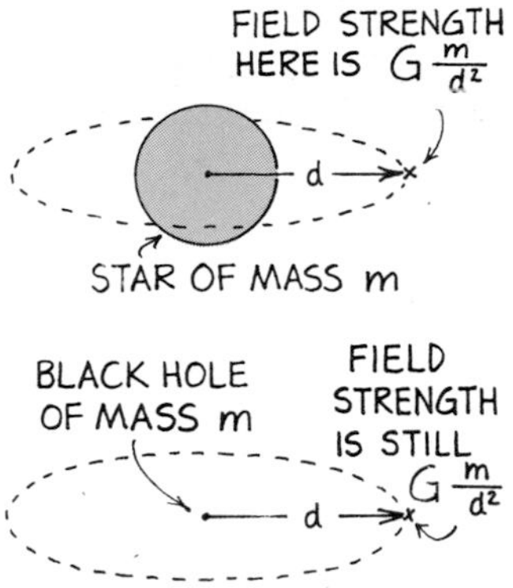

> ▶ **Question**
>
>   If the sun were to collapse from its present size and become a black hole, would the earth be drawn into it?

Fig. 11-19   The gravitational field strength near a giant star that collapses to become a black hole is the same (top) before collapse and (bottom) after collapse.

The configuration of the gravitational field about a black hole represents the collapse of space itself. The field is usually represented as a warped two-dimensional surface, as shown in Figure 11-20. Astronauts could enter the fringes of this warp and, with a powerful spaceship, still escape. After a certain distance, however, they could not, and they would disappear from the observable universe. Don't go too close to a black hole!

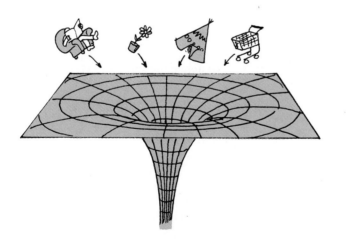

Fig. 11-20   A two-dimensional representation of the gravitational field around a black hole. Anything that falls into the central warp disappears from the observable universe.

▶ **Answer**
    No, the gravitational force between the solar black hole and the earth would not change. The earth and other planets would continue in their orbits. Observers outside the solar system would see the planets orbiting about "nothing" and likely deduce the presence of a black hole. Observations of astronomical bodies that orbit unseen partners indicate the presence of black holes.

11 | Chapter Review

Concept Summary

The acceleration due to gravity at the earth's surface can be derived from the law of universal gravitation.

The earth can be thought of as being surrounded by a gravitational field that interacts with objects and causes them to experience gravitational forces.

Objects in orbit around the earth have a gravitational force acting on them even though they may appear to be weightless.

Ocean tides (and even tides in the solid earth and in the atmosphere) are caused by differences in the gravitational pull of the moon on the earth on opposite sides of the earth.

When a star runs out of fuel for fusion, it collapses under gravitational forces.

Important Terms

apparent weightlessness (11.3)
black hole (11.6)
force field (11.2)
gravitational field (11.2)
lunar eclipse (11.4)
neap tide (11.4)
solar eclipse (11.4)
spring tide (11.4)

Review Questions

1. Distinguish between g and G. Which is a variable (that is, something that can vary in value)? (11.1)

2. Upon what quantities does the acceleration of gravity on the surfaces of various planets depend? (11.1)

3. Why was Henry Cavendish's experiment to determine the value of G called the "weighing-the-earth experiment"? (11.1)

4. Which is more correct—to say that a distant rocket interacts with the mass of the earth or to say that it interacts with the gravitational field of the earth? Explain. (11.2)

5. How does the gravitational field strength about the planet earth vary with increasing distance? (11.2)

6. If you fell into a hole that was bored completely through the earth, would you accelerate all the way through and shoot like a projectile out the other side? Why or why not? (11.2)

7. What is the value of the gravitational field of the earth at its center? (11.2)

8. Where is your weight greatest—at the surface of the earth, slightly below the surface, or above the surface? (11.2)

9. Does your apparent weight change when you ride an elevator at constant speed, or only while it is accelerating? Explain. (11.2)

10. Distinguish between apparent weightlessness and true weightlessness. (11.3)

11. Why is *difference* a key word in explaining tides? (11.4)

12. If the gravitational pull between the moon and the earth were the same over all parts of the earth, would there be any tides? Defend your answer. (11.4)

13. Which force-pair is greater—that between the moon and earth, or that between the sun and earth? (11.4)

14. Which is more effective in raising ocean tides—the moon or the sun? (11.4)

15. Why are tides greater at the times of the full and new moons and at the times of lunar and solar eclipses? (11.4)

16. Distinguish between *spring* tides and *neap* tides. (11.4)

17. Why are ocean tides greater than atmospheric tides? (11.5)

18. What two major processes determine the size of a star? (11.6)

19. Distinguish between a stellar black *dwarf* and a black *hole*. (11.6)

20. Why would the earth not be sucked into the sun if it evolved to become a black hole? (11.6)

Think and Explain

1. If the earth were the same size but twice as massive, what would be the value of G? What would be the value of g? (Why are your answers different? In this and the following question, let the equation $g = GM/R^2$ guide your thinking.)

2. If the earth somehow shrunk to half size without any change in its mass, what would be the value of g at the new surface? What would be the value of g above the new surface at a distance equal to the present radius?

3. A man named Philipp von Jolly suspended a spherical vessel of mercury of known mass on one arm of a very sensitive balance that he put in equilibrium with counterweights, as shown in Figure A. He rolled a 6-ton lead sphere beneath the mercury and had to re-adjust the balance. Why? And how did this enable him to calculate the value of G?

4. We can think of a force field as a kind of extended aura that surrounds a body, spreading its influence to affect things. As later chapters will show, an electric field affects electric charges, and a magnetic field affects magnetic poles. What does a gravitational field affect?

5. The weight of an apple near the surface of the earth is 1 N. What is the weight of the earth in the gravitational field of the apple?

6. If you stand on a shrinking planet, so that in effect you get closer to its center, your weight increases. But if you instead burrow into the planet and get closer to its center, your weight decreases. Explain.

7. If you were unfortunate enough to be in a freely-falling elevator, you might notice the bag of groceries you were carrying hovering in front of you, apparently weightless. Are the groceries falling? Defend your answer.

8. When would the moon be its fullest—just before a solar eclipse or just before a lunar eclipse? Defend your answer.

9. The human body is about 80% water. Is it likely that the moon's gravitational pull causes biological tides—cyclic changes in water flow among the fluid compartments of the body? (*Hint:* Is any part of your body appreciably closer to the moon than any other part? Is there a *difference* in lunar pulls?)

10. A black hole is no more massive than the star from which it collapsed. Why then, is gravitation so intense near a black hole?

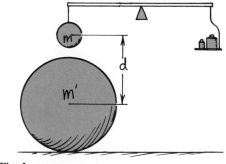

Fig. A

12 Satellite Motion

If you drop a stone from a rest position to the ground below, it will fall in a straight-line path. If you move your hand horizontally as you drop the stone, it will follow a curved path to the ground. If you move your hand faster, the stone lands farther away and the curvature of the path is less pronounced. What would happen if the curvature of the path matched the curvature of the earth? The answer is simple enough: if air resistance can be neglected, you would have an earth satellite.

Fig. 12-1 The greater the stone's horizontal motion when released, the wider the arc of its curved path.

12.1 | Earth Satellites

Fig. 12-2 If you toss the stone horizontally with the proper speed, its path will match the surface curvature of the asteroid.

An earth satellite is simply a projectile that falls *around* the earth rather than *into* it. This is easier to see if you imagine yourself on a "smaller earth," perhaps an asteroid (Figure 12-2). Because of the small size and low mass, you will not have to throw the stone very fast to make its curved path match the surface curvature of the asteroid. If you toss the stone just right, it will follow a circular orbit.

How fast would the stone have to be thrown horizontally in order to orbit the earth? The answer depends on the rate at which it falls and the degree to which the earth curves. Recall from Chapter 2 that a stone dropped from rest will accelerate 9.8 m/s² and fall a vertical distance of 4.9 m during the first second of fall. Also recall from Chapter 6 that the same is true of anything already moving as it starts to fall, that is, a projectile. Remember that in the first second a projectile will fall a vertical distance of 4.9 m from where it would have gone without gravity. (It may be helpful to refresh your memory and review Figure 6-15 on page 78.) So in the first second after a stone is thrown, it will be 4.9 m below the straight-line path it would have taken if there were no gravity.

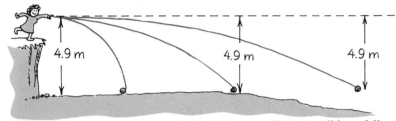

Fig. 12-3   Throw a stone at any speed and one second later it will have fallen 4.9 m below where it would have been without gravity.

A geometrical fact about the curvature of our earth is that its surface drops a vertical distance of 4.9 m for every 8000 m tangent to the surface (Figure 12-4). This means that if you were swimming in a calm ocean, you would be able to see only the top of a 4.9-m mast on a ship 8 km away.

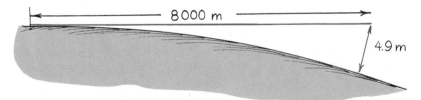

Fig. 12-4   The earth's curvature (not to scale).

If a stone could be thrown fast enough to travel a horizontal distance of 8 km during the time (1 s) it takes to fall 4.9 m, then can you see it will follow the curvature of the earth? Isn't this speed simply 8 km/s? So we see that the orbital speed for close orbit about the earth is 8 km/s. If this doesn't seem to be very fast, convert it to kilometers per hour; you'll see it is an impressive 29 000 km/h (or 18 000 mi/h). At this speed, atmospheric friction would burn a stone to a crisp. That is why a satellite must stay about 150 km or so above the earth's surface to keep from burning up like a "falling star" against the friction of the atmosphere.

12.2 | **Circular Orbits**

Interestingly enough, in circular orbit the speed of a satellite is not changed by gravity. We can understand this by comparing a satellite in circular orbit with a bowling ball rolling along a bowling alley. Why doesn't the gravity that acts on the bowling ball change its speed? The answer is that gravity is pulling neither a bit forward nor a bit backward: gravity pulls straight downward. The bowling ball has no component of gravitational force along the direction of the alley.

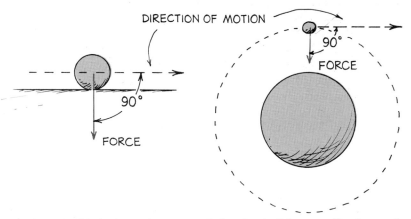

Fig. 12-5   (Left) The force of gravity on the bowling ball does not affect its speed because there is no component of gravitational force horizontally.(Right) The same is true for the satellite in circular orbit. In both cases, the force of gravity is at right angles to the direction of motion.

The same is true for a satellite in circular orbit. In circular orbit a satellite is always moving perpendicularly to the force of gravity. It does not move in the direction of gravity, which would increase its speed, nor does it move in a direction against gravity, which would decrease its speed. Instead, the satellite exactly "criss-crosses" gravity, with the result that no change in speed occurs—only a change in direction. A satellite in circular orbit around the earth moves parallel to the surface of the earth at constant speed.

For a satellite close to the earth, the time for a complete orbit about the earth, called the **period**, is about 90 minutes. For higher altitudes, the orbital speed is less and the period is longer. Communications satellites located in orbit 5.5 earth radii above the surface of the earth, for example, have a period of 24 hours. Their period of satellite rotation matches the period of daily earth rotation. Their orbit is around the equator, so they stay above the

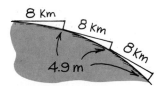

Fig. 12-6   A satellite in circular orbit close to the earth moves tangentially at 8 km/s. During each second, it falls 4.9 m beneath each successive 8-km tangent.

same point on the equator. The moon is even farther away and has a period of 27.3 days. The higher the orbit of a satellite, the less its speed and the longer its period.*

> ▶ **Questions**
> 1. There are usually alternate explanations for things. Is the following explanation valid? Satellites remain in orbit instead of falling to the earth because they are beyond the main pull of earth's gravity.
>
> 2. Satellites in close circular orbit fall about 4.9 m during each second of orbit. How can this be if the satellite does not get closer to the earth?

Recall from the last chapter that satellite motion was understood by Isaac Newton. He stated that at a certain speed a cannonball would circle the earth and coast indefinitely, provided air resistance could be neglected. Newton calculated the required speed to be equivalent to 8 km/s. Since such a cannonball speed was clearly impossible, he did not foresee people launching satellites. Newton did not consider multi-stage rockets.

12.3 | Elliptical Orbits

If a projectile just above the drag of the atmosphere is given a horizontal speed somewhat greater than 8 km/s, it will overshoot a circular path and trace an oval-shaped path, an **ellipse**.

▶ **Answers**

1. No, no, a thousand times no! If any moving object were beyond the pull of gravity, it would move in a straight line and would not curve around the earth. Satellites remain in orbit because they *are* being pulled by gravity, not because they are beyond it.

2. In each second, the satellite falls about 4.9 m below the straight-line tangent it would have taken if there were no gravity. The earth's surface curves 4.9 m beneath a straight-line tangent that is 8 km long. Since the satellite moves at 8 km/s, it "falls" at the same rate that the earth "curves."

---

* If you continue with your study of physics and take a follow-up course, you'll learn that the speed v of a satellite in circular orbit is given by $v = \sqrt{GM/d}$ and the period T of satellite motion is given by $T = 2\pi \sqrt{d^3/GM}$, where G is the universal gravitational constant (see Chapter 10), M is the mass of the earth (or whatever body the satellite orbits), and d is the altitude of the satellite measured from the center of the earth or parent body.

An ellipse is a specific curve: the closed path taken by a point that moves in such a way that the sum of its distances from two fixed points (called **foci**) is constant. In the case of a satellite orbiting a planet, the center of the planet is at one focus (singular of *foci*); the other focus is empty. An ellipse can be easily constructed by using a pair of tacks, one at each focus, a loop of string, and a pencil, as shown in Figure 12-7. The closer the foci are to each other, the closer the ellipse is to a circle. When both foci are together, the ellipse *is* a circle. A circle is actually a special case of an ellipse with both foci at the center.

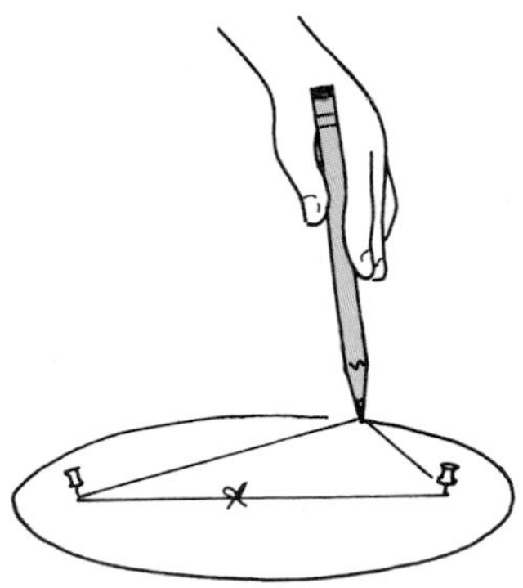

Fig. 12-7   A simple method of constructing an ellipse.

Fig. 12-8   The shadows of the ball are all ellipses with one focus where the ball touches the table.

Whereas the speed of a satellite is constant in a circular orbit, speed varies in an elliptical orbit. When the initial speed is greater than 8 km/s, the satellite overshoots a circular path and moves away from the earth, against the force of gravity. It therefore loses speed. Like a rock thrown into the air, it slows to a point where it no longer recedes, and it begins to fall back toward the earth. The speed it lost in receding is regained as it falls back toward the earth, and it finally crosses its original path with the same speed it had initially (Figure 12-9). The procedure repeats over and over, and an ellipse is traced each cycle.

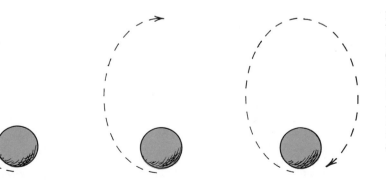

Fig. 12-9   Elliptical orbit. When the satellite exceeds 8 km/s, it overshoots a circle (left) and travels away from the earth against gravity. At its maximum separation (center) it starts to come back toward the earth. The speed it lost in going away is gained in returning, and (right) the cycle repeats itself.

► **Question**

The orbital path of a satellite is shown in the sketch. In which of the marked positions A through D does the satellite have the greatest speed? Lowest speed?

12.4   Energy Conservation and Satellite Motion

Recall from Chapter 8 that an object in motion possesses kinetic energy (KE) by virtue of its motion. An object above the earth's surface possesses potential energy (PE) by virtue of its position. Everywhere in its orbit, a satellite has both KE and PE with respect to the body it orbits. The sum of the KE and PE will be a constant all through the orbit. The simplest case occurs for a satellite in circular orbit.

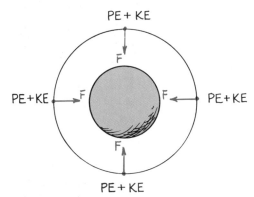

Fig. 12-10 The force of gravity on the satellite is always toward the center of the body it orbits. For a satellite in circular orbit, no component of force acts along the direction of motion. The speed, and thus the KE, cannot change.

► **Answer**

The satellite has its greatest speed as it whips around A. It has its lowest speed at position C. Beyond C it gains speed as it falls back to A to repeat its cycle.

15 /

In circular orbit the distance between the body's center and the satellite does not change. This means that the PE of the satellite is the same everywhere in orbit. Then, by the conservation of energy, the KE must also be constant. In other words, the speed is constant in any circular orbit.

In elliptical orbit the situation is different. Both speed and distance vary. The PE is greatest when the satellite is farthest away (at the **apogee**) and least when the satellite is closest (at the **perigee**). Correspondingly, the KE will be least when the PE is most; and the KE will be most when the PE is least. At every point in the orbit, the sum of the KE and PE is the same.

At all points along the orbit there is a component of gravitational force in the direction of motion of the satellite (zero at the apogee and perigee). This component changes the speed of the satellite. Or we can say: (this component of force) × (distance moved) = change in KE. Either way we look at it, when the satellite gains altitude and moves against this component, its speed and KE decrease. The decrease continues to the apogee. Once past the apogee, the satellite moves in the same direction as the component, and the speed and KE increase. The increase continues until the satellite whips past the perigee and repeats the cycle.

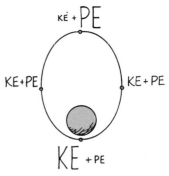

Fig. 12-11 The sum of KE and PE for a satellite is a constant at all points along an elliptical orbit.

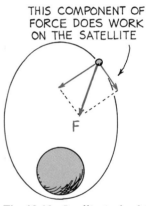

Fig. 12-12 In elliptical orbit, a component of force exists along the direction of the satellite's motion. This component changes the speed and, thus, the KE. (The perpendicular component changes only the direction.)

> ▶ **Questions**
> 1. The orbital path of a satellite is shown in the sketch in the previous question box (opposite page). In which marked positions *A* through *D* does the satellite have the greatest KE? Greatest PE? Greatest total energy?
>
> 2. How can the force of gravity change the speed of a satellite when it is in an elliptical orbit, but not when it is in a circular orbit?

▶ **Answers**

1. The KE is maximum at the perigee *A*; the PE is maximum at the apogee *C*; the total energy is the same everywhere in the orbit.

2. At any point on its path, the direction of motion of a satellite is always tangent to its path. If a component of force exists along this tangent, then the acceleration of the satellite will involve a change in speed as well as direction. In circular orbit the gravitational force is always perpendicular to the direction of motion of the satellite, just as every part of the circumference of a circle is perpendicular to the radius. So there is no component of gravitational force along the tangent, and only the direction of motion changes, not the speed. But when the satellite moves in directions that are not perpendicular to the force of gravity, as in an elliptical path, there is a component of force along the direction of motion which changes the speed of the satellite. From a work-energy point of view, a component of force along the direction the satellite moves does work to change its KE.

12.5 | **Escape Speed**

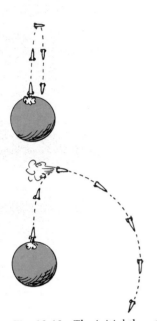

Fig. 12-13 The initial thrust of the rocket lifts it vertically. Another thrust tips it from its vertical course. When it is moving horizontally, it is boosted to the required speed for orbit.

When a payload is put into earth orbit by a rocket, the speed and direction of the rocket are important. For example, what would happen if the rocket were launched vertically and quickly achieved a speed of 8 km/s? Everyone had better get out of the way, because it would soon come crashing back at 8 km/s. To achieve orbit, the payload must be launched *horizontally* at 8 km/s once above the drag of the atmosphere. Launched only vertically, the old saying "What goes up must come down" becomes a sad fact of life.

But isn't there some vertical speed that is sufficient to insure that what goes up will escape and not come down? The answer is yes. Fire it at any speed greater than 11.2 km/s, and it will continually leave the earth, traveling more and more slowly, but will never be brought to a stop by earth gravity.* Let's look at this from an energy point of view.

How much work is required to move a payload against the force of earth gravity to a distance very, very far ("infinitely far") away? We might think that the PE would be infinite because the distance is infinite. But gravity diminishes with distance via the inverse-square law. The force of gravity is strong only close to the earth. Most of the work done in launching a rocket, for example, occurs near the earth's surface. It turns out that the value of PE for a 1-kilogram mass infinitely far away is 60 million joules (MJ). So to put a payload infinitely far from the earth's surface requires at least 60 MJ of energy per kilogram of load. We won't go through the calculation here, but a KE of 60 MJ corresponds to a speed of 11.2 km/s whatever the mass involved. This is the value of the **escape speed** from the surface of the earth.**

If we give a payload any more energy than 60 MJ/kg at the surface of the earth or, equivalently, any more speed than 11.2 km/s, then, neglecting air resistance, the payload will escape from the earth never to return. As it continues outward, its PE increases and its KE decreases. Its speed becomes less and less, though it

---

*   In a more-advanced physics course you would learn how the value of escape speed v, from any planet or any body, is given by $v = \sqrt{2GM/d}$, where G is the universal gravitational constant, M is the mass of the attracting body, and d is the distance from its center. (At the surface of the body d would simply be the radius of the body.)

**   Interestingly enough, this might well be called the *maximum falling speed*. Any object, however far from earth, released from rest and allowed to fall to earth only under the influence of the earth's gravity would not exceed 11.2 km/s.

is never reduced to zero. The payload outruns the gravity of the earth. It escapes.

The escape speeds of various bodies in the solar system are shown in Table 12-1. Note that the escape speed from the sun is 620 km/s at the surface of the sun. Even at a distance equaling that of the earth's orbit, the escape speed from the sun is 42.5 km/s. The escape speed values in the table ignore the forces exerted by other bodies. A projectile fired from the earth at 11.2 km/s, for example, escapes the earth but not necessarily the moon, and certainly not the sun. Rather than recede forever, it will take up an orbit around the sun.

The first probe to escape the solar system, *Pioneer 10*, was launched from earth in 1972 with a speed of only 15 km/s. The escape was accomplished by directing the probe into the path of oncoming Jupiter. It was whipped about by Jupiter's great gravitational field, picking up speed in the process—just as the speed of a ball encountering an oncoming bat is increased when it departs from the bat. Its speed of departure from Jupiter was increased enough to exceed the sun's escape speed at the distance of Jupiter. *Pioneer 10* passed the orbit of Pluto in 1984. Unless it collides with another body, it will wander indefinitely through interstellar space. Like a note in a bottle cast into the sea, *Pioneer 10* contains information about the earth that might be of interest to extra-terrestrials, in hopes that it will one day wash up and be found on some distant "seashore."

Fig. 12-14 *Pioneer 10*, launched from earth in 1972, escaped from the solar system in 1984 and is wandering in interstellar space.

Table 12-1 Escape Speeds at the Surface of Bodies in the Solar System			
Astronomical Body	Mass (earth masses)	Radius (earth radii)	Escape speed (km/s)
Sun	330 000	109	620
Sun (at a distance of the earth's orbit)		23 000	42.5
Jupiter	318	11	61.0
Saturn	95.2	9	37.0
Neptune	17.3	3.4	25.4
Uranus	14.5	3.7	22.4
Earth	1.00	1.00	11.2
Venus	0.82	0.96	10.4
Mars	0.11	0.53	5.2
Mercury	0.05	0.38	4.3
Moon	0.01	0.27	2.4

It is important to point out that the escape speeds for different bodies refer to the initial speed given by a brief thrust, after which there is no force to assist motion. One could escape the earth at any sustained speed more than zero, given enough time. Suppose a rocket is going to a destination such as the moon. If the rocket engines burn out when still close to the earth, the rocket needs a minimum speed of 11.2 km/s. But if the rocket engines can be sustained for long periods of time, the rocket could go to the moon without ever attaining 11.2 km/s.

It is interesting to note that the accuracy with which an unpiloted rocket reaches its destination is not accomplished by staying on a preplanned path or by getting back on that path if it strays off course. No attempt is made to return the rocket to its original path. Instead, the control center in effect asks, "Where is it now with respect to where it ought to go? What is the best way to get there from here, given its present situation?" With the aid of high-speed computers, the answers to these questions are used in finding a *new* path. Corrective thrusters put the rocket on this new path. This process is repeated over and over again all the way to the goal.

Is there a lesson to be learned here? Suppose you find that you are "off course." You may, like the rocket, find it more fruitful to take a course that leads to your goal as best plotted from your present position and circumstances, rather than try to get back on the course you plotted from a previous position and under, perhaps, different circumstances.

12 | Chapter Review

Concept Summary

An earth satellite is a projectile that moves fast enough horizontally so that it falls around the earth rather than into it.
- The speed of a satellite in a circular orbit is not changed by gravity.
- The speed of a satellite in an elliptical orbit increases as the distance from the earth decreases, and vice versa.
- The sum of the kinetic and potential energies is constant all through the orbit.
- If something is launched from earth with a great enough energy per kilogram of mass, it will move so fast that it will never return to earth.

Important Terms

apogee (12.4)
ellipse (12.3)
escape speed (12.5)
focus (12.3)
perigee (12.4)
period (12.2)

Review Questions

1. If you simply drop a stone from a position of rest, how far will it fall vertically in the first second? If you instead move your hand sideways and drop it (throw it), how far will it fall vertically in the first second? (12.1)

2. What do the distances 8000 m and 4.9 m have to do with a line tangent to the earth's surface? (12.1)

3. How does a satellite in circular orbit move in relation to the surface of the earth? (12.2)

4. Why doesn't gravitational force change the speed of a satellite in circular orbit? (12.2)

5. a. What is the period of a satellite in close circular orbit about the earth?
   b. Is the period greater or less for greater distances from the earth? (12.2)

6. What is an ellipse? (12.3)

7. Why does gravitational force change the speed of a satellite in elliptical orbit? (12.4)

8. a. At what part of an elliptical orbit is the speed of a satellite maximum?
   b. Where is it minimum? (12.4)

9. The sum of PE and KE for a satellite in a circular orbit is constant. Is this sum also constant for a satellite in an elliptical orbit? (12.4)

10. Why does the force of gravity do no work on a satellite in circular orbit, but does do work on a satellite in an elliptical orbit? (12.4)

11. What will be the fate of a projectile that is fired vertically at 8 km/s? At 12 km/s? (12.5)

12. Why does most of the work done in launching a rocket take place when the rocket is still close to the earth's surface? (12.5)

13. a. How fast would a particle have to be ejected from the sun to leave the solar system?
    b. What speed would be needed if it started at a distance from the sun equal to the earth's distance from the sun? (12.5)

14. What is the escape speed from the moon? (12.5)

15. Although the escape speed from the surface of the earth is said to be 11.2 km/s, couldn't a rocket with enough fuel escape at any speed? Defend your answer. (12.5)

Activity

Draw an ellipse with a loop of string, two tacks, and a pen or pencil. Try different tack spacings for a variety of ellipses.

Think and Explain

1. A satellite can orbit at a 5-km altitude above the moon, but not at 5 km above the earth. Why?

2. Does the speed of a satellite around the earth depend on its mass? Its distance from the earth? The mass of the earth?

3. If a cannonball is fired from a tall mountain, gravity changes its speed all along its trajectory. But if it is fired fast enough to go into circular orbit, gravity does not change its speed at all. Why is this so?

4. If you stopped an earth satellite dead in its tracks, it would simply crash into the earth.

Why, then, don't the communications satellites that "hover motionless" above the same spot on earth simply crash into the earth?

5. Would you expect the speed of a satellite in close circular orbit about the moon to be less than, equal to, or greater than 8 km/s? Defend your answer.

6. Why do you suppose that a space shuttle is sent into orbit by firing it in an easterly direction (the direction in which the earth spins)?

7. If an astronaut in an orbiting space shuttle wished to drop something to earth, how could this be accomplished?

8. What is the maximum possible speed of impact upon the surface of the earth for a faraway object initially at rest that falls to earth by virtue of only the earth's gravity?

9. Is it possible for a rocket to escape from earth without ever traveling as fast as 11.2 km/s? Explain.

10. If the earth somehow became more massive, would the escape speed be less than, equal to, or more than 11.2 km/s? Defend your answer.

13 Circular Motion

Which moves faster on a merry-go-round—a horse near the outside rail or a horse near the inside rail? Why do the riders of a rotating carnival ride not fall out when the rotating platform is raised as shown in Figure 13-1? If you swing a tin can at the end of a string overhead in a circle, and the string breaks, does the can fly directly outward or does it move tangent to the circle? Why is it that astronauts orbiting in a space shuttle float in a weightless condition, while those who orbit in a rotating facility will experience normal earth gravity? These questions indicate the flavor of this chapter. We begin by distinguishing between rotations and revolutions.

Fig. 13-1 Why do the occupants of this carnival ride not fall out when it is tipped almost vertically?

13.1 | Rotations and Revolutions

The carnival ride shown in Figure 13-1 and an ice skater doing a pirouette each turn around an **axis**, the straight line about which rotation takes place. When this axis is located within the body, the motion is called a **rotation**, or a *spin*. The motion of both the carnival ride and the skater is a rotation.

When an object turns about an *external* axis, the rotational motion is called a **revolution**. Although the carnival ride rotates, the riders along its outer edge revolve about the axis of the device.

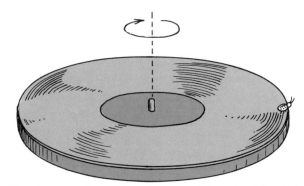

Fig. 13-2   The phonograph record *rotates* around its axis while a ladybug sitting at its edge *revolves* around the same axis.

The earth undergoes both types of rotational motion: it revolves around the sun once every 364 days, and it rotates around an axis through its geographical poles once every 24 hours.

13.2 | Rotational Speed

We began this chapter by asking which moves with the greater speed on a merry-go-round—a horse near the outside rail or a horse near the inside rail? Similarly, which part of a phonograph record passes under the stylus faster—a musical piece at the beginning of the record or one near the end? You will get different answers if you ask people these questions. That's because some people will think about linear speed and some will think about rotational speed.

Linear speed is what we have been calling simply "speed"—the distance in meters or kilometers moved per unit of time. A point on the outside of a merry-go-round or record moves a greater distance in one complete rotation than a point on the inside. The linear speed is greater on the outside of a rotating object than closer to the axis.

Rotational speed (sometimes called angular speed) refers to the number of rotations or revolutions per unit of time. All parts of the rigid merry-go-round and phonograph record circle the axis of rotation *in the same amount of time*. All parts share the same rate of rotation, or *number of rotations or revolutions per unit of time*. It is common to express rotational rates in revolutions per minute (RPM).* A common phonograph record, for example, rotates at $33\frac{1}{3}$ RPM. A ladybug sitting on the surface of the record revolves at $33\frac{1}{3}$ RPM.

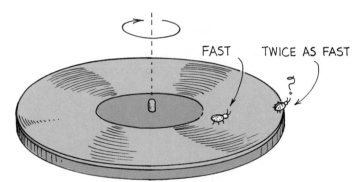

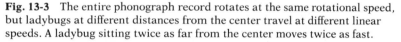

Fig. 13-3   The entire phonograph record rotates at the same rotational speed, but ladybugs at different distances from the center travel at different linear speeds. A ladybug sitting twice as far from the center moves twice as fast.

Linear speed and rotational speed are related. Have you ever ridden on a giant, rotating, round platform in an amusement park? The faster it turns, the faster is your linear speed. Linear speed is directly proportional to rotational speed.

Linear speed, unlike rotational speed, depends on the distance from the center. At the very center of the rotating platform, you have no speed at all; you merely rotate. But as you approach the edge of the platform, you find yourself moving faster and faster. Linear speed is directly proportional to distance from the center. Move out twice as far from the center, and you move twice as fast. Move out three times as far, and you have three times as much linear speed. If you find yourself in any rotating system whatever, your linear speed depends on how far you are from the axis of rotation. When a row of people locked arm in arm at the skating rink makes a turn, the motion of "tail-end Charlie" is evidence of this greater speed.

---

*   Physicists usually describe the rate of rotation in terms of the angle turned in a unit of time. Then the symbol for rotational speed is ω (the Greek letter *omega*).

▶ **Questions**

1. Which is correct—to say that a child on a merry-go-round *rotates* about the rotational axis of the merry-go-round, or to say that the child *revolves* about this axis?

2. On a particular merry-go-round, horses along the outer rail are located three times as far from the axis of rotation as horses along the inner rail. If a boy sitting on an inner horse has a rotational speed of 4 RPM and a linear speed of 2 m/s, what will be the rotational and linear speeds of his sister who sits on an outer horse?

13.3 | Centripetal Force

Fig. 13-4 The only force that is exerted on the whirling can (neglecting gravity) is directed *toward* the center of circular motion. It is called a centripetal force. No outward force is exerted on the can.

If you whirl a tin can on the end of a string, you find that you must keep pulling on the string (Figure 13-4). You pull inward on the string to keep the can revolving over your head in a circular path. A force of some kind is required for any kind of circular motion. Any force that causes an object to follow a circular path is called a **centripetal force**. Centripetal means "center-seeking," or "toward the center." The force that holds the occupants safely in the rotating carnival ride (Figure 13-1) is a center-directed force. Without it, the motion of the occupants would be along a straight line—they would not revolve.

Centripetal force is not a new kind of force. It is simply the name given to *any* force that is directed at right angles to the path of a moving object and that tends to produce circular motion. Gravitational and electrical forces act across empty space as centripetal forces. Gravitational force directed toward the center of the earth holds the moon in an almost circular orbit about the earth. Electrons that revolve about the nucleus of the atom are held by an electrical force that is directed toward the central nucleus.

▶ **Answers**

1. The child *revolves* around the axis, because it is external to the child. This same axis is within the merry-go-round itself, so the merry-go-round rotates about this axis.

2. The rotational speed of the sister is also 4 RPM; her linear speed is 6 m/s. Since the merry-go-round is rigid, all horses have the same rotational speed, but the outer horse at three times the distance from the center has three times the linear speed.

Centripetal force plays the main role in the operation of a centrifuge. A familiar example is the spinning tub in an automatic washing machine. In its spin cycle, the tub rotates at high speed. The tub's inner wall exerts a centripetal force on the wet clothes, which are forced into a circular path. The tub exerts great force on the clothes, but the holes in the tub prevent the tub from exerting the same force on the water in the clothes. The water therefore escapes. It is important to note that a force acts on the clothes, not the water. No force causes the water to fly out. The water will do that anyway, as it tends to move by inertia in a straight-line path (Newton's first law) *unless* acted on by a centripetal or any other force. So interestingly enough, the clothes are forced away from the water, and not the other way around.

When an automobile rounds a corner, the sideways friction between the tires and the road provides the centripetal force that holds the car on a curved path (Figure 13-6). If the force of friction is not great enough, the car fails to make the curve and the tires slide sideways. The car skids.

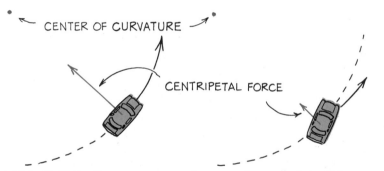

Fig. 13-6 (Left) For the car to go around a curve, there must be sufficient friction to provide the required centripetal force. (Right) If the force of friction is not great enough, skidding occurs.

13.4 | Centripetal and Centrifugal Forces

In the preceding examples, circular motion was described as caused by a center-directed force. Sometimes an outward force is attributed to circular motion. This outward force is called **centrifugal force**.* *Centrifugal* means "center-fleeing," or "away from the center." In the case of the whirling can, it is a common mis-

---

* Both centrifugal and centripetal forces depend on the mass m, tangential speed v, and radius of curvature r of the circularly moving object. If you take a follow-up physics course, you'll learn how the exact relationship is $F = mv^2/r$.

conception to state that a centrifugal force pulls outward on the can. If the string holding the whirling can breaks (Figure 13-7), it is commonly stated that a centrifugal force pulls the can from its circular path. But the fact is that when the string breaks, the can goes off in a tangent straight-line path because *no* force acts on it. We illustrate this further with another example.

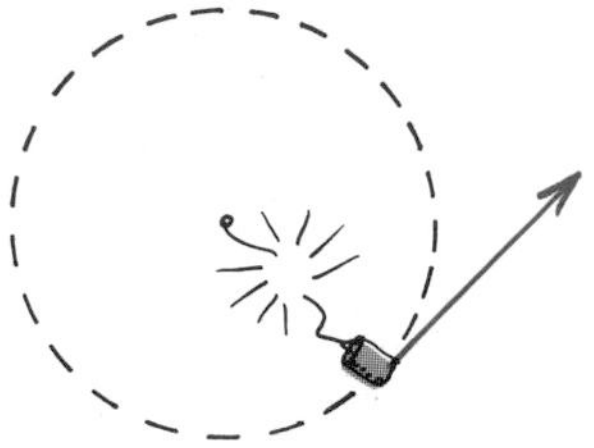

Suppose you are the passenger of a car that suddenly stops short. You pitch forward against the dashboard. When this happens, you don't say that something forced you forward. You know that you pitched forward because of the *absence* of a force, which a seat belt would have provided. Similarly, if you are in a car that rounds a sharp corner to the left, you tend to pitch outward to the right—not because of some outward or centrifugal force, but because there is no centripetal force holding you in circular motion (as a seat belt provides). The idea that a centrifugal force bangs you against the car door is a misconception.

So when you swing a tin can in a circular path, there is no force pulling the can outward. Only the pull by the string acts on the can and pulls the can inward. The outward force is *on the string*, not on the can.

Fig. 13-7 When the string breaks, the whirling can moves in a straight line, tangent to—not outward from the center of—its circular path.

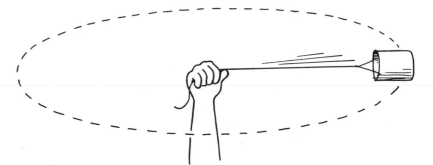

Fig. 13-8 The only force that is exerted *on* the whirling can (neglecting gravity) is directed *toward* the center of circular motion. This is a *centripetal* force. No outward force acts on the can.

Now suppose there is a ladybug inside the whirling can (Figure 13-9). The can presses against the bug's feet and provides the centripetal force that holds it in a circular path. The ladybug in turn presses against the floor of the can, but (neglecting gravity) the *only* force exerted *on the ladybug* is the force of the can on its feet. From our outside stationary frame of reference, we see there is no centrifugal force exerted on the ladybug, just as there was no centrifugal force exerted on the person who lurched against the car door. The "centrifugal-force effect" is attributed not to any real force but to inertia—the tendency of the moving body to follow a straight-line path. But try telling that to the ladybug!

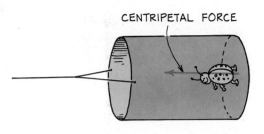

CENTRIPETAL FORCE

Fig. 13-9 The can provides the centripetal force necessary to hold the ladybug in a circular path.

13.5 Centrifugal Force in a Rotating Reference Frame

Our view of nature is very much influenced by the frame of reference from which we view it. When we sit on the seat of a fast-moving train, we have no speed at all relative to the train but an appreciable speed relative to the reference frame of the stationary ground outside. We have just seen that in a nonrotating reference frame, the force that holds an object in circular motion is a centripetal force. For the ladybug, the bottom of the can exerts a force on its feet. No other force was acting on the ladybug.

But nature seen from the reference frame of the rotating system is different. In the rotating frame, in addition to the force of the can on the ladybug's feet, there is a centrifugal force that is exerted *on the ladybug*. Centrifugal force in a rotating reference frame is a force in its own right, as real as the pull of gravity. However, there is a fundamental difference: Gravitational force is an interaction between one mass and another. The gravity we experience is our interaction with the earth. But centrifugal force in the rotating frame is not part of an interaction. It is like gravity, but with nothing pulling. Nothing produces it; it is a result of rotation. For this reason, physicists rank it as a *fictitious force* and not a real force like gravity, electromagnetism, and the nuclear forces. Nevertheless, to observers who are in a rotating system, centrifugal force seems to be, and is interpreted to be, a very real force. Just as from the surface of the earth we find gravity ever present, so within a rotating system centrifugal force is ever present.

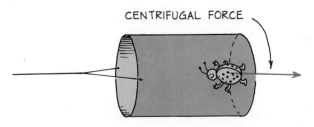

CENTRIFUGAL FORCE

Fig. 13-10 From the reference frame of the ladybug inside the whirling can, it is being held to the bottom of the can by a force that is directed away from the center of circular motion. The ladybug calls this outward force a *centrifugal* force, which is as real to it as gravity.

▶ **Questions**

1. A heavy iron ball is attached by a spring to a rotating platform, as shown in the sketch. Two observers, one in the rotating frame and one on the ground at rest, observe its motion. Which observer sees the ball being pulled outward, stretching the spring? Which sees the spring pulling the ball in a circle?

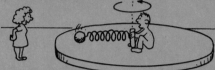

2. The spring in the sketch is observed to stretch 10 cm when the ball is midway between the rotational axis and the outer edge of the circular platform. If the spring support is moved so that the ball is directly over the edge, will the spring be stretched more than, less than, or the same 10 cm?

13.6   Simulated Gravity

Consider a colony of ladybugs living inside a bicycle tire—the old-fashioned balloon kind that has plenty of room inside. If we toss the bicycle wheel through the air or drop it from an airplane high in the sky, the ladybugs will be in a weightless condition. They will float freely while the wheel is in free fall. Now spin the wheel. The ladybugs will feel themselves pressed to the outer part of the inner surface. If the wheel is spun not too fast and not too slowly but just right, the ladybugs will be provided

▶ **Answer**

1. The observer in the reference frame of the rotating platform states that a centrifugal force pulls radially outward on the ball, which stretches the spring. The observer in the rest frame states that a centripetal force supplied by the stretched spring pulls the ball into a circle along with the rotating platform. (The rest-frame observer can state in addition that the reaction to this centripetal force is the ball pulling outward on the spring. The rotating observer, interestingly enough, can state no reaction counterpart to the centrifugal force.)

2. The ball will have twice the linear speed at twice the distance from the rotational axis. The greater the speed, the greater will be the centripetal/centrifugal force (which will also be twice, and stretch the spring 20 cm). (We will see in Chapter 18 that the stretch of a spring is directly proportional to the applied force.)

with *simulated gravity* that will feel like the gravity to which they are accustomed. Gravity is simulated by centrifugal force. The "down" direction to the ladybugs will be what we would call "radially outward," away from the center of the wheel.

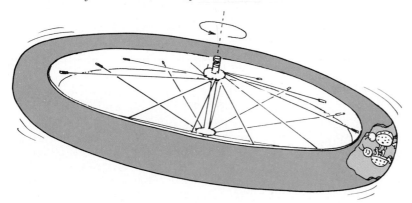

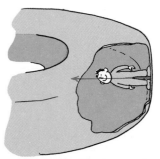

Fig. 13-11 If the spinning wheel freely falls, the ladybugs inside will experience a centrifugal force that feels like gravity when the wheel spins at the appropriate rate. To the occupants, the direction "up" is toward the center of the wheel and "down" is radially outward.

Fig. 13-12 The interaction between the man and the floor as seen at rest outside the rotating system. The floor presses against the man (action) and the man presses back on the floor (reaction). The only force exerted on the man is by the floor. It is directed toward the center and is a centripetal force.

Today we all live on the outer surface of this spherical planet and are held here by gravity. The planet has been the cradle of humankind. But we will not stay in the cradle forever. We are becoming a spacefaring people. Many people in the years ahead will likely live in huge lazily-rotating habitats in space and will be held to the inner surfaces by centrifugal force. The rotating habitats will provide a simulated gravity so that the human body can function normally.

Occupants of a space shuttle are weightless because they lack a support force. Future space travelers need not be subject to weightlessness, however. A rotating space habitat for humans, like the rotating bicycle wheel for the ladybugs, can effectively supply a support force and neatly simulate gravity.

Structures of small diameter would have to rotate at high rates to provide a simulated gravitational acceleration of $1g$. Sensitive and delicate organs in our middle ears sense rotation. Although there appears to be no difficulty at a single revolution per minute (RPM) or so, many people find difficulty in adjusting to rates greater than 2 or 3 RPM (although some easily adapt to 10 or so RPM). To simulate normal earth gravity at 1 RPM requires a large structure—one 2 km in diameter. This is an immense structure, compared to today's space shuttle vehicles. Economics will probably dictate the size of the first inhabited structures. They are likely to be small and not rotate at all. The inhabitants will adjust to living in a weightless environment. Larger, rotating habitats will likely follow later.

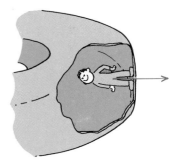

Fig. 13-13 As seen from inside the rotating system, in addition to the man-floor interaction there is a centrifugal force exerted on the man at his center of mass. It seems as real as gravity. Yet, unlike gravity, it has no reaction counterpart—there is nothing out there that he can pull back on. Centrifugal force is not part of an interaction, but results from rotation. It is therefore called a fictitious force.

Fig. 13-14   A NASA depiction of a rotational space colony.

If the structure rotates so that inhabitants on the inside of the outer edge experience 1*g*, then half-way to the axis they would experience 0.5*g*. At the axis itself they would experience weightlessness at 0*g*. The variety of fractions of *g* possible from the rim of a rotating space habitat holds promise for a most different and (at this writing) yet unexperienced environment. We would be able to perform ballet at 0.5*g*; diving and acrobatics at 0.2*g* and lower-*g* states; three-dimensional soccer and new sports not yet conceived in very-low-*g* states. People will be exploring possibilities never before available to them. This time of transition from our earthly cradle to new vistas is an exciting time in which to live—especially for those who will be prepared to play a role in these new adventures.*

---

*   At the risk of stating the obvious, this preparation can well begin by taking your study of physics very seriously.

13 | Chapter Review

Concept Summary

An object rotates when it turns around an axis inside itself; it revolves when it turns around an axis outside itself.
- The rotational speed is the number of rotations or revolutions made per unit of time.

A centripetal force pulls objects toward a center.
- An object moving in a circle is acted on by a centripetal force.
- When an object moves in a circle, there is no force pushing the object outward from the circle.
- From within a rotating frame of reference, there seems to be an outwardly directed centrifugal force, which can simulate gravity.

Important Terms

axis (13.1)
centrifugal force (13.4)
centripetal force (13.3)
linear speed (13.2)
revolution (13.1)
rotation (13.1)
rotational speed (13.2)

Review Questions

1. Distinguish between a rotation and a revolution. (13.1)

2. Does a child on a merry-go-round *revolve* or *rotate* around the merry-go-round's axis? (13.1)

3. Distinguish between linear speed and rotational speed. (13.2)

4. What is the relationship of linear speed to rotational speed? (13.2)

5. When you whirl a can at the end of a string in a circular path, what is the direction of the force that acts on the can? (13.3)

6. Does the force that holds the riders on the carnival ride in Figure 13-1 act toward or away from the center? (13.3)

7. Is it an inward force or an outward force that acts on the clothes during the spin cycle of an automatic washer? (13.3)

8. When a car makes a turn, do seat belts provide you with a centripetal force or a centrifugal force? (13.4)

9. If the string breaks that holds a whirling can in its circular path, what kind of force causes it to move in a straight-line path—centripetal, centrifugal, or no force? What law of physics supports your answer? (13.4)

10. Identify the action and reaction forces in the interaction between the ladybug and the whirling can of Figure 13-9. (13.5)

11. Can you name the action and reaction forces in the interaction in which one force is the centrifugal force acting on an empty tin can swung into a circle? Why or why not? (13.5)

12. Why is centrifugal force in a rotating frame called a "fictitious force?" (13.5)

13. How can gravity be simulated in an orbiting space station? (13.6)

14. Why will orbiting space stations that simulate gravity likely be large structures? (13.6)

15. How will the values of *g* vary at different distances from the hub of a rotating space station? (13.6)

Activity

Half fill a bucket of water and swing it in a circular path overhead. Why doesn't the water spill?

Think and Explain

1. A ladybug sits halfway between the axis and the edge of a phonograph record. What will happen to its linear speed if (a) the RPM rate is doubled? (b) It sits at the edge? (c) Both *a* and *b*?

2. What path would you follow if you fell off the edge of a rotating merry-go-round?

3. A motorcyclist is able to ride on the vertical wall of a bowl-shaped track, as shown in Figure A. Is it centripetal or centrifugal force that acts on the motorcycle?

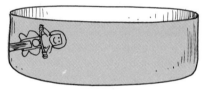

Fig. A

4. When a gliding seagull suddenly turns in its flight, what is the source of the centripetal force that acts on it?

5. Does centripetal force do work on a rotating object? Defend your answer.

6. When a car makes a turn on a level road, friction between the tires and the road supply the centripetal force. On a banked road, a car pushes into the road as it turns. The road in turn, pushes back on the car to produce a "normal force." Make a sketch and show how the horizontal component of the normal force decreases (or eliminates) the friction force. (Can you see that if a curve is banked at an angle to match a certain speed, no friction occurs?)

7. A space habitat rotates so as to provide an acceleration of $1g$ for its inhabitants. If it is rotated at a greater angular speed, how will the acceleration for the inhabitants differ?

8. Explain why the faster the earth spins, the less a person weighs, whereas the faster a space colony spins, the more a person weighs.

9. Consider a too-small space habitat that consists of a rotating sphere of radius 4 m. If a man standing inside is 2 m tall and his feet are at $1g$, what is the acceleration at the elevation of his head? (Do you see why rotating space structures are designed to be large?)

10. Occupants in a single space shuttle in orbit feel weightless. Describe a scheme whereby occupants in a pair of shuttles (or even one shuttle and a large massive object such as an asteroid) would be able to use a long cable to continually experience a comfortable normal earth gravity.

14 Rotational Mechanics

Push on an object that is free to move, and you set it in motion. Some objects will move without rotating, some rotate without moving, and others do both. For example, a kicked football often tumbles end over end. What determines whether an object will rotate when a force acts on it? This chapter is about the factors that affect rotation. We will see that they underlie most of the techniques used by gymnasts, ice skaters, and divers.

14.1 Torque

Every time you open a door, turn on a water faucet, or tighten a nut with a wrench, you exert a turning force. This turning force produces a *torque* (rhymes with fork). Torque is different from force. If you want to make an object move, apply a force. Forces tend to make things accelerate. If you want to make an object turn or rotate, apply a torque. Torques produce rotation.

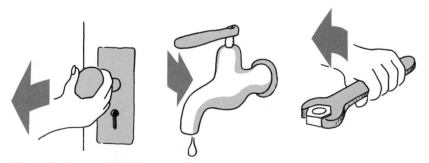

Fig. 14-1 A torque produces rotation.

FORCE

Fig. 14-2 When a perpendicular force is applied, the lever arm is the distance between the doorknob and the edge with the hinges.

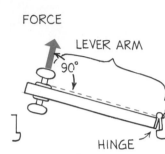

Fig. 14-3 The doorknobs on these doors are placed at the center. What do physics types say about the practicality of this?

A torque is produced when a force is applied with "leverage." A doorknob is placed far away from the turning axis at its hinges to provide more leverage when you push or pull on the doorknob. The direction of your applied force is important. In opening a door, you'd never push or pull the doorknob sideways to make the door turn. You push *perpendicular* to the plane of the door. Experience has taught you that a perpendicular push or pull gives the greatest amount of rotation for the least amount of effort.

If you have ever used both a short wrench and a long wrench, you also know that less effort and more leverage results with a long handle. When the force is perpendicular, the distance from the turning axis to the point of contact is called the **lever arm**.

If the force is not at a right angle to the lever arm, then only the perpendicular component of the force, $F_\perp$, will contribute to the torque. **Torque** is defined as:*

$$\text{torque} = \text{force}_\perp \times \text{lever arm}$$

So the same torque can be produced by using a large force with a short lever arm, or a small force with a long lever arm. Greater torques are produced when both the force and lever arm are large.

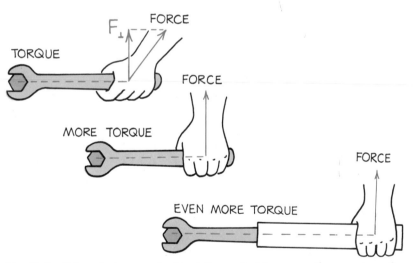

Fig. 14-4 Although the magnitudes of the applied forces are the same in each case, the torques are different. Only the component of forces perpendicular to the lever arm contributes to torque.

---

*   The units of torque are newton-meters. Work is also measured in newton-meters (the same as joules), but work and torque are not the same. With work, the force moves along (parallel to) the distance, while with torque the force moves perpendicular to the distance.

▶ **Questions**
1. If a doorknob were placed in the center of a door rather than at the edge, how much more force would be needed to produce the same torque for opening the door?

2. If you cannot exert enough torque to turn a stubborn bolt, would more torque be produced if you fastened a length of rope to the wrench handle as shown?

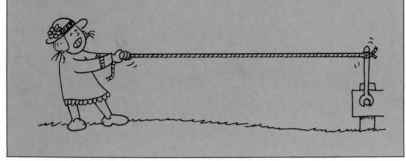

14.2 | Balanced Torques

Torques are intuitively familiar to youngsters playing on a seesaw. Children can balance a seesaw even when their weights are unequal. Weight alone does not produce rotation. Torque does, and children soon learn that the distance they sit from the pivot point is every bit as important as weight (Figure 14-5). The heavier boy sits a short distance from the fulcrum (turning axis) while the lighter girl sits farther away. Balance is achieved if the torque that tends to produce clockwise rotation by the boy equals the torque that tends to produce counterclockwise rotation by the girl.

Scale balances that work with sliding weights are based on balanced torques, not balanced masses. The sliding weights are adjusted until the counterclockwise torque just balances the clockwise torque. Then the arm remains horizontal.

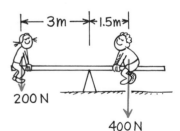

Fig. 14-5 A pair of torques can balance each other.

Fig. 14-6 This scale relies on balanced torques.

▶ **Answers**
1. Twice as much, because the lever arm would be half as long in the center of the door. Mathematically, $2F \times (d/2) = F \times d$, where F is the applied force, d is the distance from the hinges to the edge of the door, and $d/2$ is the distance to the center.

2. No, because the lever arm is the same. To increase the lever arm, a better idea would be to use a pipe that extends upward.

Computational Example

Suppose that a meter stick is supported at the center, and a 20-N block is hung at the 20-cm mark. Another block of unknown weight just balances the system when it is hung at the 90-cm mark. What is the weight of the second block?

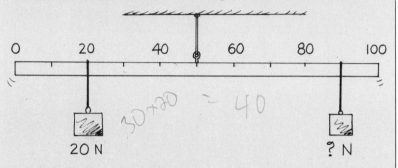

You can compute the unknown weight by applying the principle of balanced torques. The block of unknown weight tends to rotate the system of blocks and stick *clockwise*, and the 20-N block tends to rotate the system *counterclockwise*. The two torques are equal when the system is in balance:

clockwise torque = counterclockwise torque

$$(F_\perp d)_{\text{clockwise}} = (F_\perp d)_{\text{counterclockwise}}$$

It is important to note that the lever arm for the unknown weight is 40 cm, because the distance between the 90-cm mark and the pivot point at the 50-cm is 40 cm. The lever arm for the 20-N block is 30 cm, because its distance from the midpoint of the stick is 30 cm. When known values are substituted into the last equation, it becomes:

$$F_\perp \times (40 \text{ cm}) = (20 \text{ N}) \times (30 \text{ cm})$$

The unknown weight is therefore

$$F_\perp = \frac{(20 \text{ N}) \times (30 \text{ cm})}{(40 \text{ cm})} = 15 \text{ N}$$

The answer makes sense. You can tell that the weight must be less than 20 N because its lever arm is greater than that of the block of known weight. In fact, the lever arm is (40 cm) ÷ (30 cm), or 4/3, that of the first block, so the weight is 3/4 that of the first block. Anytime you use physics to compute something, be sure to consider whether or not your answer makes sense. Computation without comprehension is not conceptual physics!

14.3   Torque and Center of Gravity

If you attempt to lean over and touch your toes while standing with your back and heels to the wall, you soon find yourself rotating. This is because of a torque. Recall from Chapter 9 that if there is no base of support beneath the center of gravity (CG), an object will topple. When the area bounded by your feet is not beneath your CG, there is a torque. Now you can see that the cause of toppling is the presence of a torque.

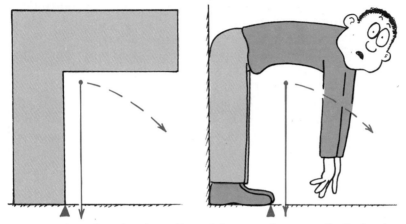

Fig. 14-7 The L-shape bracket will topple because of a torque. Similarly, when you stand with your back and heels to the wall and try to touch your toes, a torque is produced when your CG extends beyond your feet.

This chapter began by asking what determines whether or not a football will tumble end over end when kicked. The answer involves CG, force, and torque. You know that a force is required to launch any projectile, whether it be a football or a Frisbee. If the direction of the force is through the CG of the projectile, all the force can do is move the object as a whole. There will be no torque to turn the projectile around its own CG. If the force, instead, is directed "off center," then in addition to motion of the CG, the projectile will rotate about its CG.

For example, when you throw a ball and impart spin to it, or when you launch a Frisbee, a force must be applied to the edge of the object. This produces a torque which adds rotation to the projectile. If you wish to kick a football so that it sails through the air without tumbling, kick it in the middle (Figure 14-8 top). If you want it to tumble end over end in its trajectory, kick it below the middle (Figure 14-8 bottom) to impart torque as well as force to the ball.

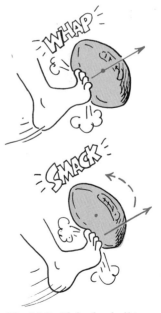

Fig. 14-8 If the football is kicked in line with the center of gravity, it will move without rotating. If it is kicked above or below its center of gravity, it will also rotate.

14.4   Rotational Inertia

Fig. 14-9 Most of the mass of this industrial flywheel is concentrated along its edge so that it has a greater rotational inertia. Once it is set into rotation, it is harder to stop than if its mass were nearer the axis.

In Chapter 3 you learned about the law of inertia: an object at rest tends to stay at rest, and an object in motion tends to remain moving in a straight line. There is a similar law for rotation:

> An object rotating about an axis tends to keep on rotating about that axis.

The resistance of an object to changes in its rotational state of motion is called **rotational inertia**.* Rotating objects tend to keep rotating, while nonrotating objects tend to stay nonrotating.

Just as it takes a force to change the linear state of motion of an object, a torque is required to change the rotational state of motion of an object. In the absence of a net torque, a rotating top keeps rotating, while a nonrotating top stays nonrotating.

Like inertia in the linear sense, rotational inertia depends on the mass of the object. But unlike inertia, rotational inertia depends on the distribution of the mass. The greater the distance between the bulk of the mass of an object and the axis about which rotation takes place, the greater the rotational inertia.

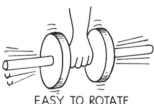

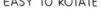

EASY TO ROTATE

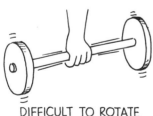

DIFFICULT TO ROTATE

Fig. 14-10 Rotational inertia depends on the distance of mass from the axis.

Fig. 14-11 By holding a long pole, the tightrope walker increases his rotational inertia. This allows him to resist rotation and gives him plenty of time to re-adjust his CG.

---

*   Rotational inertia is sometimes called *moment of inertia*.

A long pendulum has a greater rotational inertia than a short pendulum. It's "more lazy," so it swings to and fro more slowly than a short pendulum. Hang a weight from a string. When the string is short, the pendulum swings to and fro more frequently than when it is long (Figure 14-12). Likewise with the legs of people and animals when their legs are allowed to swing freely. Long-legged animals such as giraffes, horses, and ostriches normally run with a slower gait than hippos, dachshunds and mice.

It is important to note that the rotational inertia of an object is not necessarily a fixed quantity. It is greater when the mass within the object is extended from the axis of rotation. You can try this with your outstretched legs. Allow your outstretched leg to swing to and fro from the hip like a pendulum. Now do the same with your leg bent. In the bent position it swings to and fro more frequently. To reduce the rotational inertia of your legs, simply bend them. That is an important reason for running with your legs bent. They are easier to swing to and fro.

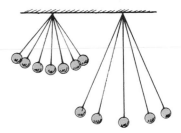

Fig. 14-12   The short pendulum will swing to and fro more frequently than the long pendulum.

Fig. 14-13   For similar mass distributions, short legs have less rotational inertia than long legs. Animals with short legs more easily run with quicker strides than animals with long legs.

Fig. 14-14   You bend your legs when you run to reduce their rotational inertia.

Formulas for Rotational Inertia

When all the mass m of an object is concentrated at the same distance r from a rotational axis (as in a simple pendulum bob swinging on a string about its pivot point, or a thin wheel turning about its center), then the rotational inertia $I = mr^2$. When the mass is more spread out, as in your leg, the rotational inertia is less and the formula is different. Figure 14-15 compares rotational inertias for various shapes and axes. (It is not important for you to learn these values, but you can see how they vary with the shape and axis.)

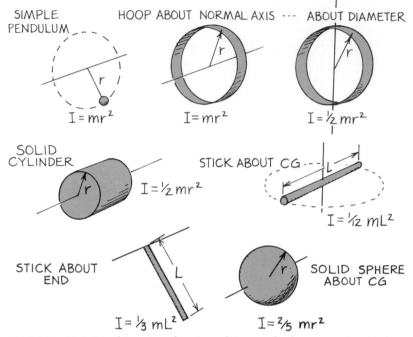

SIMPLE
PENDULUM

$I = mr^2$

HOOP ABOUT NORMAL AXIS --- ABOUT DIAMETER

$I = mr^2$

$I = \frac{1}{2} mr^2$

SOLID
CYLINDER

$I = \frac{1}{2} mr^2$

STICK ABOUT CG

$I = \frac{1}{12} mL^2$

STICK ABOUT
END

$I = \frac{1}{3} mL^2$

SOLID SPHERE
ABOUT CG

$I = \frac{2}{5} mr^2$

Fig. 14-15 Rotational inertias of various objects each of mass m, about indicated axes.

Rolling

Which will roll down an incline with greater acceleration—a hollow or a solid cylinder of the same mass and radius? The answer is, the object with the smaller rotational inertia. Why? Because the object with the greater rotational inertia will require more time to get rolling. Remember that inertia of any kind is a measure of "laziness." Which has the greater rotational inertia—the hollow or the solid cylinder? The answer is, the one with its mass concentrated farthest from the axis of rotation— the hollow cylinder. So a hollow cylinder has a greater rotational inertia than a solid cylinder of the same radius and mass and will be more "lazy" in gaining speed. The solid cylinder will roll with greater acceleration.

Interestingly enough, any solid cylinder will roll down an incline with greater acceleration than any hollow cylinder, regardless of mass or diameter. A hollow cylinder has more "laziness per mass" than a solid cylinder.

Objects of the same shape but different sizes will accelerate equally when rolled down an incline. You should see this for yourself and experiment. If started together, the smaller shape, whether it be ball, disk, or hoop, will rotate more times than a larger one, but both will reach the bottom of the incline in the same time.

Fig. 14-16 A solid cylinder rolls down an incline faster than a hollow one, whether or not they have the same mass or diameter.

> ▶ **Questions**
>
> 1. Why is the rotational inertia of your leg swinging from your hip less when your leg is bent?
>
> 2. A heavy iron cylinder and a light wooden cylinder, similar in shape, roll down an incline. Which will have the greater acceleration?

14.5 | Rotational Inertia and Gymnastics

Consider the whole human body. You can rotate freely and stably about three principal axes of rotation (Figure 14-17). These axes are each at right angles to the others (mutually perpendicular). Each axis coincides with a line of symmetry of the body and passes through the center of gravity. The rotational inertia of the body differs about each axis.

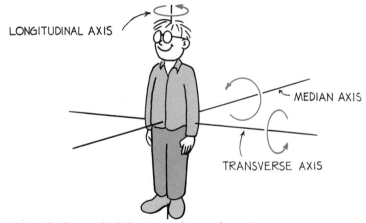

Fig. 14-17 The human body's principal axes of rotation.

▶ **Answers**

1. The rotational inertia of any object is less when its mass is concentrated closer to the axis of rotation. Can you see that a bent leg satisfies this requirement?

2. The cylinders have different masses, but the *same rotational inertia per mass*, so both will accelerate equally down the incline. Their different masses make no difference, just as the different masses of objects in free fall do not affect acceleration. Similarly, different masses suspended from the same length of string (pendulum bobs) will accelerate equally. Sliding different masses down friction-free inclines will result in equal accelerations. Similar round shapes of different mass will also roll down inclines with equal accelerations.

Rotational inertia is least about the longitudinal axis, which is the vertical head-to-toe axis. This is because most of the mass is concentrated along this axis. A rotation of your body about your longitudinal axis is therefore the easiest rotation to perform. An ice skater executes this type rotation when going into a spin. Rotational inertia is increased by simply extending a leg or the arms (Figure 14-18). The rotational inertia when both arms are extended is found to be about three times greater than with arms tucked in, so if you go into a spin with outstretched arms, you will triple your spin rate when you draw your arms in. With your leg extended as well, you can vary your spin rate by as much as 6 times.

Fig. 14-18 Rotations about the longitudinal axis. Rotational inertia in position *d* is about 5 or 6 times as great as in position *a*, so spinning in position *d* and then changing to position *a* will increase the spin rate about 5 or 6 times.

You rotate about your transverse axis when you perform a somersault or a flip. Figure 14-19 shows positions of different rotational inertia, starting with the least (when your arms and legs are drawn inward in the tuck position) to the greatest (when your arms and legs are fully extended in a line). The relative magnitudes of rotational inertia stated in the caption are with respect to the body's center of gravity.

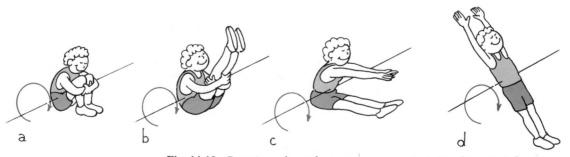

Fig. 14-19 Rotations about the transverse axis. Rotational inertia is least in *a*, the tuck position. It is about 1.5 times as great in *b*, 3 times as great in *c*, and 5 times as great in *d*.

Rotational inertia is greater when the axis is through the hands, such as when doing a somersault on the floor or swinging from a horizontal bar with your body fully extended. The rotational inertia of a gymnast is up to 20 times as great when swinging in fully extended position from a horizontal bar as after dismount when she somersaults in the tuck position. Rotation transfers from one axis to another, from the bar to her center of gravity, and she automatically increases her rate of rotation by up to 20 times. This is how she is able to complete two or three somersaults before contact with the ground.

Fig. 14-20 The rotational inertia of a body is with respect to the rotational axis. When the gymnast pivots about the bar (a), she has a greater rotational inertia than when she spins freely about her center of gravity (b).

The third axis of rotation for the human body is the front-to-back axis, or medial axis. This is a less common axis of rotation and is used in executing a cartwheel. Like rotations about the other axes, rotational inertia can be varied with different body configurations.

▶ **Question**

Why is the rotational inertia greater for a gymnast when she swings by her hands from a horizontal bar with her body fully extended than when she executes a somersault with her body fully extended?

▶ **Answer**

When the gymnast pivots about the bar, her mass is concentrated far from the axis of rotation. Her rotational inertia is therefore greater. When she somersaults, her mass is concentrated close to the axis because she is rotating about her center of gravity.

Fig. 14-21 When Julianne McNamara pivots around the horizontal bar with her body fully extended, her rotational inertia is greatest.

14.6 | Angular Momentum

Anything that rotates, whether it be a colony in space, a cylinder rolling down an incline, or an acrobat doing a somersault, keeps on rotating until something stops it. A rotating object has an "inertia of rotation." Recall from Chapter 7 that all moving objects have "inertia of motion," or momentum, which is the product of mass and velocity. To be clear, let us call this kind of momentum **linear momentum**. Similarly, the "inertia of rotation" of rotating objects is called **angular momentum**.

Like linear momentum, angular momentum is a vector quantity and has direction as well as a magnitude. When a direction is assigned to rotational speed, we call it **rotational velocity**. Rotational velocity is a vector whose magnitude is the rotational speed. (By convention, the rotational velocity vector, as well as the angular momentum vector, have the same direction and lie along the axis of rotation.)

Angular momentum is defined as the product of rotational inertia and rotational velocity.

angular momentum = rotational inertia × rotational velocity

It is the counterpart of linear momentum:

linear momentum = mass × velocity

In this book, we won't treat the vector nature of momentum (or even of torque, which also is a vector) except to acknowledge the remarkable action of the gyroscope. The rotating bicycle wheel in Figure 14-24 shows what happens when a torque by

Fig. 14-22 A phonograph turntable has more angular momentum when it is turning at 45 RPM than at $33\frac{1}{3}$ RPM. It has even more angular momentum if a load is placed on it so its rotational inertia is greater.

Fig. 14-23 A gyroscope. Low-friction swivels can be turned in any direction without exerting a torque on the whirling gyroscope. As a result, it stays pointed in the same direction.

Fig. 14-24 Angular momentum keeps the wheel axle horizontal when a torque supplied by earth's gravity acts on it. Instead of toppling, the torque causes the wheel to slowly precess about a vertical axis.

earth's gravity acts to change the direction of its angular momentum (which is along the wheel's axle). The pull of gravity that acts to topple the wheel over and change its rotational axis causes it instead to *precess* in a circular path about a vertical axis. You must do this yourself to fully *believe* it. Full *understanding* will likely not come until a later time.

For the case of an object that is small compared to the radial distance to its axis of rotation, such as a tin can swinging from a long string or a planet orbiting around the sun, the angular momentum is simply equal to the magnitude of its linear momentum, *mv*, multiplied by the radial distance, *r*. In shorthand notation,

$$\text{angular momentum} = mvr$$

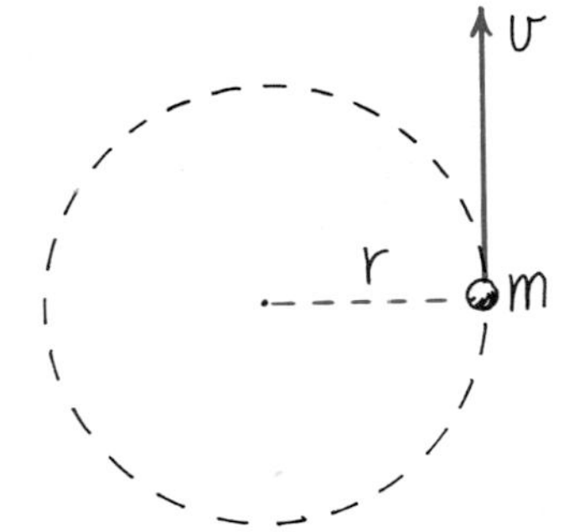

Fig. 14-25 An object of concentrated mass *m* whirling in a circular path of radius *r* with a speed *v* has angular momentum *mvr*.

14.7 Conservation of Angular Momentum

We know that an external net force is required to change the linear momentum of an object. The rotational counterpart is: an external net torque is required to change the angular momentum of an object. We restate Newton's first law of inertia for rotating systems in terms of angular momentum:

> An object or system of objects will maintain its angular momentum unless acted upon by an unbalanced external torque.

We all know that it is not easy to balance on a bicycle that is at rest. The wheels have no angular momentum. If our center of gravity is not above a point of support, a slight torque is produced and we fall over. When the bicycle is moving, however, the wheels have angular momentum. To tip the wheels requires a greater torque than before because the direction of the angular momentum would change. This makes the bicycle easy to balance when it is moving.

Fig. 14-26 Bicycles are easier to balance when their wheels have angular momentum.

Just as the linear momentum of any system is conserved if no net forces are acting on the system, angular momentum is conserved for systems in rotation. The **law of conservation of angular momentum** states:

If no unbalanced external torque acts on a rotating system, the angular momentum of that system is constant.

This means that the product of rotational inertia and rotational velocity at one time will be the same as at any other time.

An interesting example illustrating angular momentum conservation is shown in Figure 14-27. The man stands on a low-friction turntable with weights extended. His rotational inertia, with the help of the extended weights, is relatively large in this position. As he slowly turns, his angular momentum is the product of his rotational inertia and rotational velocity. When he pulls the weights inward, the rotational inertia of himself and the weights is considerably reduced. What is the result? His rotational speed increases! This example is best appreciated by the turning person who feels changes in rotational speed that seem to be mysterious. But it's straight physics! This procedure is used by a figure skater who starts to whirl with his arms and perhaps a leg extended, and then draws his arms and leg in to obtain a greater rotational speed. Whenever a rotating body contracts, its rotational speed increases.

Fig. 14-27 Conservation of angular momentum. When the man pulls his arms and the whirling weights inward, he decreases his rotational inertia, and his rotational speed correspondingly increases.

Similarly, when a gymnast is spinning freely in the absence of unbalanced torques on his body, his angular momentum does not change. However, as we have already seen, he can change his rotational speed by simply making variations in his rotational inertia. He does this by moving some part of his body toward or away from his axis of rotation.

Fig. 14-28 Rotational speed is controlled by variations in the body's rotational inertia as angular momentum is conserved during a forward somersault.

If a cat is held upside down and dropped, it is able to execute a twist and land upright even if it has no initial angular momentum. Zero-angular-momentum twists and turns are performed by turning one part of the body against the other. While falling, the cat rearranges its limbs and tail to change its rotational inertia. Repeated reorientations of the body configuration result in the head and tail rotating one way and the feet the other, so that the feet are downward when the cat strikes the ground.

During this maneuver the total angular momentum remains zero. When it is over, the cat is at rest. This manuever rotates the body through an angle, but it does not create continuing rotation. To do so would violate angular momentum conservation.

Humans can perform similar twists without difficulty, though not as fast as a cat can. Astronauts have learned to make zero-angular-momentum rotations about any principal axis. They do these to orient their bodies in any preferred direction when floating freely in space.

The law of momentum conservation is seen in the planetary motions and in the shape of the galaxies. It will be a fact of everyday life to inhabitants of rotating space habitats who will head for these distant places.

Fig. 14-29 Time-lapse photo of a falling cat.

14 | Chapter Review

Concept Summary

Torque for an object being turned is the product of the lever arm and the component of the force perpendicular to the lever arm.
- When balanced torques act on an object, there is no rotation.
- When the base of support is not under the center of gravity, the gravitational force produces a torque that causes toppling.

The resistance of an object to changes in its rotational state of motion is called rotational inertia.
- The greater the rotational inertia, the harder it is to change the rotational speed of an object.

Angular momentum, the "inertia of rotation," is the product of rotational inertia and rotational velocity.
- Angular momentum is conserved when no external torque acts on a body.

Important Terms

angular momentum (14.6)
law of conservation of angular momentum (14.7)
lever arm (14.1)
linear momentum (14.6)
rotational inertia (14.4)
rotational velocity (14.6)
torque (14.1)

Review Questions

1. Compare the effects of a force exerted on an object and a torque exerted on an object. (14.1)

2. What is meant by the lever arm of a force? (14.1)

3. In what direction should a force be applied to produce maximum torque? (14.1)

4. How do clockwise and counterclockwise torques compare when a system is balanced? (14.2)

5. In terms of *center of gravity*, *support base*, and *torque*, why can you not stand with heels and back to a wall and then bend over to touch your toes and return to your stand-up position? (14.3)

6. Where must a football be kicked so that it won't topple end-over-end as it sails through the air? (14.3)

7. What is the law of inertia for rotation? (14.4)

8. What two quantities make up rotational inertia? (14.4)

9. Why does a short pendulum swing to and fro at a greater rate than a long pendulum? (14.4)

10. Why do your legs swing to and fro more rapidly when they are bent? (14.4)

11. Which will have the greater acceleration in rolling down an incline—a large ball or a small ball? (14.4)

12. Which will have the greater acceleration rolling down an incline—a hoop or a solid disk? (14.4)

13. How can a person vary his or her rotational inertia? (14.5)

14. Distinguish between rotational velocity and rotational speed. (14.6)

15. Distinguish between linear momentum and angular momentum. (14.6)

16. What motion does the torque produced by earth's gravity impart to a vertically spinning bicycle wheel supported at only one end of its axle? (14.6)

17. What is the law of inertia for angular momentum? (14.7)

18. What does it mean to say that angular momentum is conserved? (14.7)

19. If a skater who is spinning pulls her arms in so as to reduce her rotational inertia to half, by how much will her rate of spin increase? (14.7)

20. What happens to a gymnast's angular momentum when he changes his body configuration during a somersault? His rotational speed? (14.7)

Activities

1. Ask a friend to stand facing a wall with toes against the wall. Then ask your friend to stand on the balls of the feet without toppling backward. Your friend won't be able to do it. Now you explain why it can't be done.

2. Place the ends of a uniform horizontal board on a pair of weighing scales. You'll note that each scale reads half the weight of the board. If a toy truck with a heavy brick in it is placed on the center of the board, the reading on each scale will increase by half the weight of the loaded truck. Why? How would you go about figuring how much weight is supported by each scale when the load is *not* in the middle? Make predictions for various places *before* you check by experiment. (Try to do this before you measure torques in the lab part of your course.)

Think and Explain

1. Which is easier for prying open a stuck cover from a can of paint—a screwdriver with a thick handle or one with a long handle? Which is easier for turning stubborn screws? Provide an explanation.

2. What is the mass of the rock shown in Figure A?

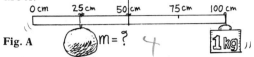

Fig. A

3. What is the mass of the meter stick shown in Figure B?

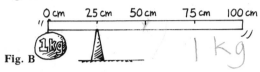

Fig. B

4. You cannot stand with heels and back to the wall and successfully lean over and touch your toes without toppling. Would either stronger legs or longer feet help you to do this? Defend your answer.

5. If you walk along the top of a fence, why does holding your arms out help you to balance?

6. Which will have the greater acceleration rolling down an incline—a bowling ball or a volleyball? Defend your answer.

7. The most popular gyroscope around is the Frisbee. What function, besides being a place for gripping and catching, does its somewhat thicker curved rim serve?

8. Why and how do you throw a football so that it spins about its long axis when traveling through the air?

9. If a trapeze artist rotates twice each second while sailing through the air, and contracts to reduce her rotational inertia to one third, how many rotations per second will result?

10. You sit in the middle of a large, freely rotating turntable at an amusement park. If you crawl toward the outer rim, does the rotation speed increase, decrease, or remain unchanged? What law of physics supports your answer?

15 Special Relativity— Space and Time

Everyone knows that we move in time, at the rate of 24 hours per day. And everyone knows that we can move through space, at rates ranging from a snail's pace to those of supersonic aircraft and the space shuttles. But relatively few people know that motion through space is related to motion in time.

The first person to understand the relationship between space and time was Albert Einstein. Einstein went beyond common sense when he stated in 1905 that in moving through space we also change our rate of proceeding into the future—time itself is altered. This view was introduced to the world in his **special theory of relativity**. This theory describes how time is affected by motion in space at constant velocity, and how mass and energy are related. Ten years later Einstein announced a similar theory, called the *general theory of relativity*, which encompasses accelerated motion as well. These theories have enormously changed the way scientists view the workings of the universe. This book discusses only the special theory and leaves the general theory for followup study later in your education.

This chapter will serve merely to acquaint you with the basic ideas of special relativity as they relate to space and time. Chapter 16 will continue with the relationship between mass and energy. These ideas, for the most part, are not common to your everyday experience. As a result, they don't agree with common sense. So please be patient with yourself if you find that you do not understand them. Perhaps your children or grandchildren will find them very much a part of their everyday experience. If so, they should find an understanding of relativity considerably less difficult.

15.1 | Spacetime

Newton and others before Einstein thought of space as an infinite expanse in which all things exist. We are in space, and we move about in space. It was never clear whether the universe exists in space, or space exists within the universe. Is there space outside the universe? Or is space only within the universe? The same question could be raised for time. Does the universe exist in time, or does time exist only within the universe? Was there time before the universe came to be? Will there be time if the universe ceases to exist? Einstein's answer to these questions is that both space and time exist only within the universe. There is no time or space "outside."

Fig. 15-1 The universe does not exist in a certain part of infinite space, nor does it exist during a certain era in time. It is the other way around: space and time exist within the universe.

Einstein reasoned that space and time are two parts of one whole called **spacetime**. To begin to understand this, consider your present knowledge that you are moving through time at the rate of 24 hours per day. This is only half the story. To get the other half, convert your thinking from "moving through time" to "moving through spacetime." From the viewpoint of special relativity, you travel through a combination of space and time—spacetime—at a constant speed. If you stand still, then all your traveling is through time. If you move a bit, then some of your travel is through space and most is still through time. What hap-

pens to time if you were to travel through space at the speed of light? The answer is that all your traveling would be through space, with no travel through time! You would be as ageless as light, for light travels through space only (not time) and is timeless.

Motion in space affects motion in time. Whenever we move through space, we to some degree alter our rate of moving into the future. This is **time dilation**, a stretching of time that occurs ever so slightly for everyday speeds, but significantly for speeds approaching the speed of light. In the high-speed spacecrafts of tomorrow, people will be able to travel noticeably in time. They will be able to jump centuries ahead, just as today people can jump from the earth to the moon. To understand time dilation and how this can be, you first need to understand several ideas: the relativity of motion and the postulates of special relativity.

Fig. 15-2 When you stand still, you are traveling at the maximum rate in time: 24 hours per day. If you travel at the maximum rate through space (the speed of light), time stands still.

15.2   Motion Is Relative

Fig. 15-3 The bag of groceries has an appreciable speed in the frame of reference of the building, but in the frame of reference of the freely-falling elevator it has no speed at all.

Fig. 15-4 Your speed is 1 km/h relative to your seat, and 100 km/h relative to the road.

Recall from Chapter 2 that whenever we discuss motion, we must specify the position from which the motion is being observed and measured. For example, you may walk along the aisle of a moving bus at a speed of 1 km/h relative to your seat, but at 100 km/h relative to the road outside. Speed is a relative quantity. Its value depends upon the place—the frame of reference—where it is observed and measured. An object may have different speeds relative to different frames of reference.

Suppose your friend always pitches a baseball at the same speed of 60 km/h. Neglecting air resistance and other small effects, the ball is moving at 60 km/h when you catch it. Now suppose your friend pitches the ball to you from the flatbed of a truck that moves toward you at 40 km/h. How fast does the ball meet you? You'll have to be sure to wear a catcher's mitt, because the speed of the ball will be 100 km/h (the 60 km/h relative to the truck plus the 40 km/h relative to the ground). Speed is relative.

Suppose the truck moves away from you at 40 km/h and your friend again pitches the ball to you. This time you need no glove at all, for the ball reaches you at a speed of 20 km/h (since 60 km/h minus 40 km/h is 20 km/h). This is not surprising, for you expect that the ball will be traveling faster when the truck approaches you and that the ball will be traveling more slowly when the truck recedes.

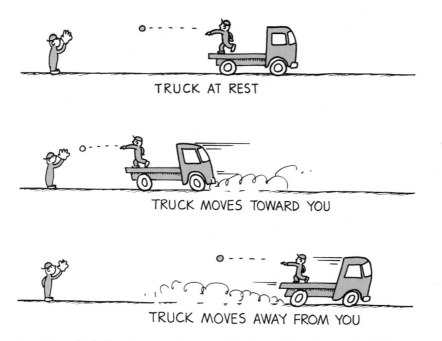

TRUCK AT REST

TRUCK MOVES TOWARD YOU

TRUCK MOVES AWAY FROM YOU

Fig. 15-5 The ball is always pitched at 60 km/h relative to the truck. (a) When both you and the truck are at relative rest the ball is traveling at 60 km/h when you catch it. (b) When the truck moves toward you at 40 km/h, the ball is traveling at 100 km/h when you catch it. (c) When the truck moves away from you at the same speed, the ball is traveling at 20 km/h when you catch it.

The idea that speed is a relative quantity goes back to Galileo and was known long before the time of Einstein. As you will learn in this chapter, Einstein expanded the relativity of speed to include the relativity of things that seem unchangeable.

15.3 | The Speed of Light Is Constant

Suppose you actually caught baseballs thrown off a moving truck out in a parking lot and found that no matter what the speed or direction of the truck, the ball always got to you at only one speed—60 km/h. That is to say, if the truck zooms toward

Fig. 15-6 The speed of light is found to be the same in all frames of reference.

you at 50 km/h and your friend pitches the ball at his speed of 60 km/h, you catch the ball with the same speed as if the truck were not moving at all. Furthermore, if the truck moves away from you at whatever speed, the ball still gets to you at 60 km/h. This all seems quite impossible, for it is contrary to common experience. And if you *did* experience this, you would have to re-evaluate your whole notion of reality. To put it mildly, you would be quite confused.

Baseballs do not behave this way. But it turns out that light does! Every measurement of the speed of light in empty space gives the same value of 300 000 km/s, regardless of the speed of the source or the speed of the receiver.* We do not ordinarily notice this because light travels so incredibly fast.

The fact that light has only one speed in empty space was discovered at the end of the last century.** Light from an approaching source reaches an observer at the same speed as light from a receding source. And the speed of light is the same whether we move toward or away from a light source. How did the physics community regard this finding? They were as perplexed as you would be if you caught baseballs at only one speed no matter how they were thrown. Experiments were done and redone, and always the results were the same. Nothing could vary the speed of light. Various interpretations were proposed, but none were satisfactory. The foundations of physics were on shaky ground.

Albert Einstein looked at the speed of light in terms of the definition of speed. What is speed? It is the amount of *space* traveled compared to the *time* of travel. Einstein recognized that the classical ideas of space and time were suspect. He concluded that space and time were a part of a single entity—spacetime. The constancy of the speed of light, Einstein reasoned, unifies space and time.

The special theory of relativity that Einstein developed rests on two fundamental assumptions, or **postulates**.

---

* The presently accepted value for the speed of light is 299 792 km/s, which we round off to 300 000 km/s. This corresponds to 186 000 mi/s.

** In 1887 two American physicists, A. A. Michelson and E. W. Morley, performed an experiment to determine differences in the speed of light in different directions. They thought that the motion of the earth in its orbit about the sun would cause shifts in the speed of light. The speed should have been faster when light traveled in the same direction as the earth, and slower when it traveled at right angles to the earth. Using a device called an *interferometer*, they found that the speed was the same in all directions. For Michelson's many experiments on the speed of light, he was the first American honored with a Nobel Prize.

15.4 | The First Postulate of Special Relativity

Einstein reasoned that there is no stationary hitching post in the universe relative to which motion should be measured. Instead, all motion is relative and all frames of reference are arbitrary. A spaceship cannot measure its speed relative to empty space, but only relative to other objects. If, for example, spaceship A drifts past spaceship B in empty space, spaceman A and spacewoman B will each observe only the relative motion. From this observation each will be unable to determine who is moving and who is at rest, if either.

Fig. 15-7   Spaceman A considers himself at rest and sees spacewoman B pass by. But spacewoman B considers herself at rest and sees spaceman A pass by. Who is moving and who is at rest?

This is a familiar experience to a passenger in a car at rest waiting for the traffic light to change. If you look out the window and see the car in the neighboring lane start to move backward, you may find to your surprise that the car you observe is really at rest and the car you are in is moving forward. Or vice versa. If you could not see out the windows, there would be no way to determine whether your car was moving with constant velocity or was at rest.

In a high-speed jetliner we flip a coin and catch it just as we would if the plane were at rest. Coffee pours from the flight attendant's coffee pot as it does when the plane is standing on the runway. If we swing a pendulum, it moves no differently when the plane is moving uniformly (constant velocity) than when not moving at all. There is no physical experiment we can perform to determine our state of uniform motion. Of course, we can look outside and see the earth whizzing by, or send a radar signal out. However, no experiment confined within the cabin itself can determine whether or not there is uniform motion. The laws of physics within the uniformly-moving cabin are the same as those in a stationary laboratory.

Fig. 15-8   A person playing pool on a smooth and fast-moving ocean liner does not have to make adjustments to compensate for the speed of the ship. The laws of physics are the same for the ship whether it is moving uniformly or is at rest.

These examples illustrate one of the two building blocks of special relativity. It is Einstein's **first postulate of special relativity**:

> All the laws of nature are the same in all uniformly moving frames of reference.

Any number of experiments can be devised to detect *accelerated* motion, but none can be devised, according to Einstein, to detect the state of uniform motion.

15.5 The Second Postulate of Special Relativity

One of the questions that Einstein as a youth asked his schoolteacher was, "What would a light beam look like if you traveled along beside it?" According to classical physics, the beam would be at rest to such an observer. The more Einstein thought about this, the more convinced he became of its impossibility. He came to the conclusion that *if* an observer could travel close to the speed of light, he would measure the light as moving away from him at 300 000 km/s.

This is the idea that makes up Einstein's **second postulate of special relativity**:

> The speed of light in empty space will always have the same value regardless of the motion of the source or the motion of the observer.

The speed of light in all reference frames is always the same.

Consider, for example, a spaceship departing from the space station shown in Figure 15-9. A flash of light is emitted from the station at 300 000 km/s, which we call simply c. No matter what the speed of the spaceship relative to the space station, an observer on the spaceship will measure the speed of the flash of light passing her as the same speed c. If she sends a flash of her own to the space station, observers on the station will measure the speed of these flashes as c. The speed of the flashes will be no different if the spaceship stops or turns around or approaches. All observers who measure the speed of light find it has the same value c.

Fig. 15-9 The speed of a light flash emitted by either the spaceship or the space station is measured as c by observers on the ship or the space station. Everyone who measures the speed of light will get the same value c.

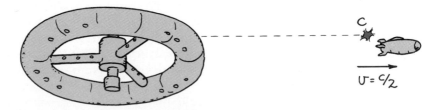

The constancy of the speed of light is what unifies space and time. And for any observation of motion through space, there is a corresponding passage of time. The ratio of space to time for light is the same for all who measure it. The speed of light is a constant.

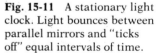

$$\frac{\text{SPACE}}{\text{TIME}} = \frac{\text{SPACE}}{\text{TIME}} = c$$

Fig. 15-10   All space and time measurements of light are unified by c.

15.6 | Time Dilation

Special relativity turns around some of our conceptions about the world. We think that speed is relative, that it depends on the speeds of the source and the observer. Yet, the speed of light is absolute—independent of the speeds of the source or observer. Time, on the other hand, is thought of as absolute. It seems to pass at the same rate regardless of what is happening. Yet, Einstein did not accept this, and proposed that time depends on the motion between the observer and the event being observed.

We measure time with a clock. A clock can be any device that measures periodic intervals, such as the swings of a pendulum, the oscillations of a balance wheel, or the vibrations of a quartz crystal. We are going to consider a "light clock," a rather impractical device, but one that will help to describe time dilation.

Imagine an empty tube with a mirror at each end (Figure 15-11). A flash of light bounces back and forth between the parallel mirrors. The mirrors are perfect reflectors, so the flash bounces indefinitely. If the tube is 300 000 km in length, each bounce will take 1 s in the frame of reference of the light clock. If the tube is 3 km long, each bounce will take 0.00001 s.

Suppose we view the light clock as it whizzes past us in a high-speed spaceship (Figure 15-12). We see the light flash bouncing up and down along a longer diagonal path.

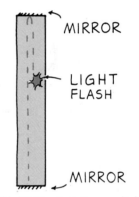

Fig. 15-11   A stationary light clock. Light bounces between parallel mirrors and "ticks off" equal intervals of time.

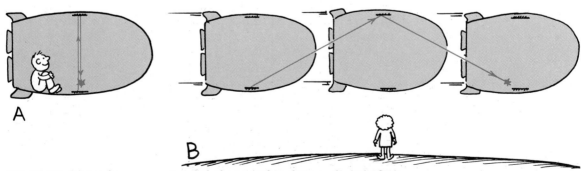

Fig. 15-12   (a) An observer moving with the spaceship observes the light flash moving vertically between the mirrors of the light clock. (b) An observer who is passed by the moving ship observes the flash moving along a diagonal path.

But remember the second postulate of relativity: the speed will be measured by *any* observer as *c*. Since the speed of light will not increase, we must measure more time between bounces! The light clock is a measure of time for the moving spaceship. We would measure more time for bounces in the light clock and for all the seconds and minutes experienced by the inhabitants of the spaceship. We would observe time in the spaceship running more slowly than it does where we are.

Fig. 15-13 The longer distance taken by the light flash in following the diagonal path must be divided by a correspondingly longer time interval to yield an unvarying value for the speed of light.

$$\frac{DISTANCE}{TIME} = \frac{DISTANCE}{TIME} = c$$

The slowing of time is not peculiar to the light clock. It is time itself in the moving frame of reference, as viewed from our frame of reference, that slows. The heartbeats of the spaceship occupants will have a slower rhythm. All events on the moving ship will be observed by us as slower. It is time itself that is dilated.

How do the occupants on the spaceship view their own time? Do they perceive themselves moving in slow motion? Do they experience longer lives as a result of time dilation? As it turns out, they notice none of these things. Time for them is the same as when they do not appear to us to be moving at all. Recall Einstein's first postulate: all laws of nature are the same in all uniformly moving frames of reference. There is no way they can tell uniform motion from rest. They have no clues that events on board are seen to be dilated when viewed from other frames of reference.

Fig. 15-14 Mathematical detail of Figure 15-13.

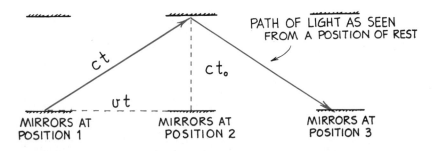

PATH OF LIGHT AS SEEN FROM A POSITION OF REST

ct

ct_0

vt

MIRRORS AT POSITION 1

MIRRORS AT POSITION 2

MIRRORS AT POSITION 3

The Time Dilation Equation*

Figure 15-14 shows three successive positions of the light clock as it moves to the right at constant speed v. The diagonal lines represent the path of the light flash as it starts from the lower mirror at position 1, moves to the upper mirror at position 2, and then back to the lower mirror at position 3.

The symbol t_0 represents the time it takes for the flash to move between the mirrors as measured from a frame of reference fixed to the light clock. This is the time for straight up or down motion. Since the speed of light is always c, the light flash is observed to move a vertical distance ct_0 in the frame of reference of the light clock. This is the distance between mirrors and is at right angles to the horizontal motion of the light clock. This vertical distance is the same in both reference frames.

The symbol t represents the time it takes the flash to move from one mirror to the other as measured from a frame of reference in which the light clock moves to the right with speed v. Since the speed of the flash is c and the time to go from position 1 to position 2 is t, the diagonal distance traveled is ct. During this time t, the clock (which travels horizontally at speed v) moves a horizontal distance vt from position 1 to position 2.

These three distances make up a right triangle in the figure, in which ct is the hypotenuse, and ct_0 and vt are legs. A well-known theorem of geometry (the Pythagorean theorem) states that the square of the hypotenuse is equal to the sum of the squares of the other two sides. If we apply this to the figure, we obtain:

$$c^2t^2 = c^2t_0^2 + v^2t^2$$

$$c^2t^2 - v^2t^2 = c^2t_0^2$$

$$t^2[1 - (v^2/c^2)] = t_0^2$$

$$t^2 = \frac{t_0^2}{1 - (v^2/c^2)}$$

$$t = \frac{t_0}{\sqrt{1 - (v^2/c^2)}}$$

---

* The mathematical derivation of this equation for time dilation is included here mainly to show that it involves only a bit of geometry and elementary algebra. It is not expected that you master it! (If you take a follow-up physics course, you can master it then.)

> ▶ **Questions**
> 1. Does time dilation mean that time *really* passes more slowly in moving systems or that it only *seems* to pass more slowly?
>
> 2. If you are moving in a spaceship at a high speed relative to the earth, would you notice a difference in your pulse rate? In the pulse rate of the people back on earth?
>
> 3. Will observers A and B agree on measurements of time if A moves at half the speed of light relative to B? If both A and B move together at 0.5*c* relative to the earth?

How do occupants on the spaceship view *our* time? Do they see our time as speeded up? The answer is no—motion is relative, and from *their* frame of reference it appears that *we* are the ones who are moving. They see our time running slow, just as we see their time running slow. Is there a contradiction here? Not at all. It is physically impossible for observers in different frames of reference to refer to one and the same realm of spacetime. The measurements in one frame of reference need not agree with the measurements made in another reference frame. There is only one measurement they will always agree on: the speed of light.

15.7   The Twin Trip

A dramatic illustration of time dilation is the classic case of the identical twins, one an astronaut who takes a high-speed round-trip journey while the other stays home on earth. When the trav-

---

▶ **Answers**

1. The slowing of time in moving systems is not merely an illusion resulting from motion. Time really does pass more slowly in a moving system *compared* to one at relative rest, as we shall see in the next section. Read on!

2. There would be no relative speed between you and your own pulse, so no relativistic effects would be noticed. There would be a relativistic effect between you and people back on earth. You would find their pulse rate slower than normal (and *they* would find *your* pulse rate slower than normal). Relativity effects are always attributed to "the other guy."

3. When A and B have different motions relative to each other, each will observe a slowing of time in the frame of reference of the other. So they will not agree on measurements of time. When they are moving in unison, they share the same frame of reference and will agree on measurements of time. They will see each other's time as passing normally, and they will each see events on earth in the same slow motion.

eling twin returns, he is younger than the stay-at-home twin. How much younger depends on the relative speeds involved. If the traveling twin maintains a speed of 50% the speed of light for one year (according to clocks aboard the spaceship), 1.15 years will have elapsed on earth. If the traveling twin maintains a speed of 87% the speed of light for a year, then 2 years will have elapsed on earth. At 99.5% the speed of light, 10 earth years would pass in one spaceship year. At this speed the traveling twin would age a single year while the stay-at-home twin ages 10 years.

Fig. 15-15 The traveling twin does not age as fast as the stay-at-home twin.

The question arises, since motion is relative, why isn't it just as well the other way around—why wouldn't the traveling twin return to find his stay-at-home twin younger than himself? We will show that from the frames of reference of both the earthbound twin and traveling twin, it is the earthbound twin who ages more.

First, consider a spaceship hovering at rest relative to a distant planet. Suppose the spaceship sends regularly-spaced brief flashes of light to the planet (Figure 15-16). Some time will elapse before the flashes get to the planet, just as 8 minutes elapses before sunlight gets to the earth. The light flashes will encounter the receiver on the planet at speed c. Since there is no relative motion between the sender and receiver, successive flashes will be received as frequently as they are sent. For example, if a flash is sent from the ship every 6 minutes, then after some initial delay, the receiver will receive a flash every 6 minutes. With no motion involved, there is nothing unusual about this.

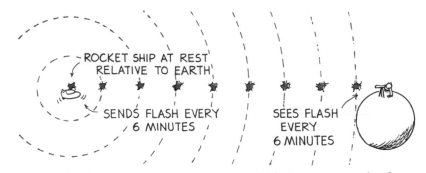

Fig. 15-16   When no motion is involved, the light flashes are received as frequently as the spaceship sends them.

When motion is involved, the situation is quite different. It is important to note that the *speed* of the flashes will still be *c*, no matter how the ship or receiver may move. How *frequently* the flashes are seen, however, very much depends on the relative motion involved. When the ship travels *toward* the receiver, the receiver sees the flashes more frequently. This happens not only because time is altered due to motion, but mainly because each succeeding flash has less distance to travel as the ship gets closer to the receiver. If the spaceship emits a flash every 6 minutes, the flashes will be seen at intervals of less than 6 minutes. Suppose the ship is traveling fast enough for the flashes to be seen twice as frequently. Then they are seen at intervals of 3 minutes (Figure 15-17).

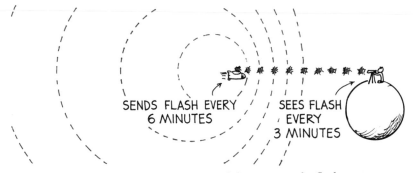

Fig. 15-17   When the sender moves toward the receiver, the flashes are seen more frequently.

If the ship *recedes* from the receiver at the same speed and still emits flashes at 6-min intervals, these flashes will be seen half as frequently by the receiver, that is, at 12-min intervals (Figure 15-18). This is mainly because each succeeding flash has a longer distance to travel as the ship gets farther away from the receiver.

The effect of moving away is just the opposite of moving closer

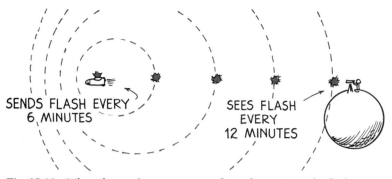

Fig. 15-18   When the sender moves away from the receiver, the flashes are spaced farther apart and are seen less frequently.

to the receiver. So if the flashes are received twice as frequently when the spaceship is approaching (6-min flash intervals are seen every 3 min), they are received half as frequently when it is receding (6-min flash intervals are seen every 12 min).*

The light flashes make up a light clock. In the frame of reference of the receiver, events that take 6 min in the spaceship are seen to take 12 min when the spaceship recedes, and only 3 min when the ship is approaching.

▶ **Questions**

1. Here's a simple arithmetic question: If a spaceship travels for one hour and emits a flash every 6 min, how many flashes will be emitted?

2. A spaceship sends equally spaced flashes while approaching a receiver at constant speed. Will the flashes be equally spaced when they encounter the receiver?

3. A spaceship emits flashes every 6 min for one hour. If the receiver sees these flashes at 3-min intervals, how much time will occur between the first and the last flash (in the frame of reference of the receiver)?

▶ **Answers**

1. Ten flashes, since (60 min)/(6 min/flash) = 10 flashes.

2. Yes, as long as the ship moves at constant speed, the equally-spaced flashes will be seen equally spaced but more frequently. (If the ship accelerated while sending flashes, then they would not be seen at equally spaced intervals.)

3. All 10 flashes will be seen in 30 min [(10 flashes) × (3 min/flash) = 30 min].

---

*   The frequencies for approach and for recession are *reciprocals* of each other. That is, flashes that are seen 2 times as frequently for approach are seen $\frac{1}{2}$ as frequently for recession. For higher speeds, if seen 3 times as frequently for approach, the flashes are seen $\frac{1}{3}$ as frequently for recession; or 4 times for approach, $\frac{1}{4}$ for recession, and so on.

Let's apply this doubling and halving of flash intervals to the twins. Suppose the traveling twin recedes from the earthbound twin at the same high speed for one hour, then quickly turns around and returns in one hour. The traveling twin takes a round trip of two hours, according to all clocks aboard the spaceship. This trip will *not* be seen to take two hours from the earth frame of reference, however. We can see this with the help of the flashes from the ship's light clock.

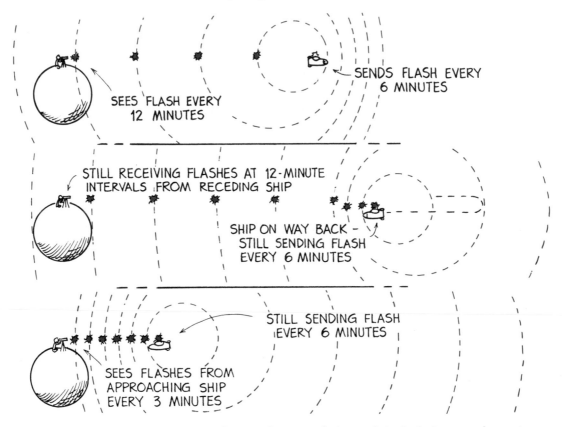

Fig. 15-19 The spaceship emits flashes each 6 min during a two-hour trip. During the first hour, it recedes from the earth. During the second hour, it approaches the earth.

As the ship recedes from the earth, it emits a flash of light every 6 min. These flashes are received on earth every 12 min. During the hour of going away from the earth, a total of 10 flashes are emitted. If the ship departs from the earth at noon, clocks aboard the ship read 1 p.m. when the tenth flash is emitted. What time will it be on earth when this tenth flash reaches the earth? The answer is 2 p.m. Why? Because the time it takes the earth to receive 10 flashes at 12-min intervals is (10 flashes) × (12 min/flash), or 120 min (= 2 h).

Suppose the spaceship is somehow able to suddenly turn around in a negligibly short time and return at the same high speed. During the hour of return it emits 10 more flashes at 6-min intervals. These flashes are received every 3 min on earth, so all 10 come in 30 min. A clock on earth will read 2:30 p.m. when the spaceship completes its 2-hour trip. We see that the earthbound twin has aged a half hour more than the twin aboard the spaceship!

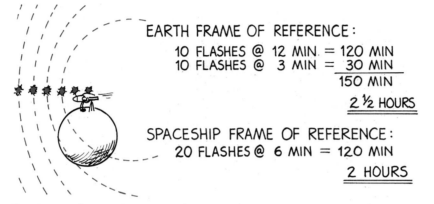

EARTH FRAME OF REFERENCE:
10 FLASHES @ 12 MIN. = 120 MIN
10 FLASHES @ 3 MIN = 30 MIN
                        150 MIN
                       2 ½ HOURS

SPACESHIP FRAME OF REFERENCE:
20 FLASHES @ 6 MIN = 120 MIN
                     2 HOURS

Fig. 15-20 The trip that takes 2 h in the frame of reference of the spaceship takes 2.5 h in the earth's frame of reference.

The result is the same from either frame of reference. Consider the same trip again, only this time with flashes emitted *from the earth* at regularly spaced 6-min intervals in earth time. From the frame of reference of the receding spaceship, these flashes are received at 12-min intervals (Figure 15-21a). This means that 5 flashes are seen by the spaceship during the hour of receding from earth. During the spaceship's hour of approaching, the light flashes are seen at 3-min intervals (Figure 15-21b), so 20 flashes will be seen.

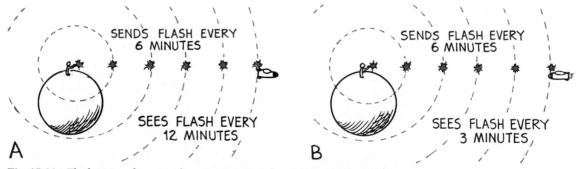

Fig. 15-21 Flashes sent from earth at 6-min intervals are seen at 12-min intervals by the ship when it recedes, and at 3-min intervals when it approaches.

Hence, the spaceship receives a total of 25 flashes during its 2-hour trip. According to clocks on the earth, however, the time it took to emit the 25 flashes at 6-min intervals was (25 flashes) × (6 min/flash), or 150 min (= 2.5 h). This is shown in Figure 15-22.

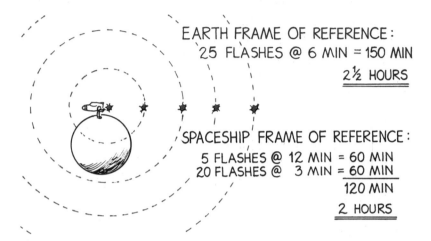

EARTH FRAME OF REFERENCE:
25 FLASHES @ 6 MIN = 150 MIN
2½ HOURS

SPACESHIP FRAME OF REFERENCE:
5 FLASHES @ 12 MIN = 60 MIN
20 FLASHES @ 3 MIN = 60 MIN
120 MIN
2 HOURS

Fig. 15-22 A time interval of 2.5 h on earth is seen to be 2 h in the spaceship's frame of reference.

So both twins agree on the same results, with no dispute as to who ages more than the other. While the stay-at-home twin remains in a single reference frame, the traveling twin has experienced two different frames of reference, separated by the acceleration of the spaceship in turning around. The spaceship has in effect experienced two different realms of time, while the earth has experienced a still different but single realm of time. The twins can meet again at the same place in space only at the expense of time.

15.8   Space and Time Travel

Before the theory of special relativity was introduced, it was argued that humans would never be able to venture to the stars. It was thought that our life span is too short to cover such great distances—at least for the distant stars. Alpha Centauri is the nearest star to earth, after the sun, and it is 4 light years away.* It was therefore thought that a round trip even at the speed of

---

*   A light year is the distance that light travels in one year. One light year equals 9.45×10^{12} km.

light would require 8 years. The center of our galaxy is some 30 000 light years away, so it was reasoned that a person traveling even at the speed of light would require a lifetime of 30 000 years to make such a voyage. But these arguments fail to take into account time dilation. Time for a person on earth and time for a person in a high-speed spaceship are not the same.

A person's heart beats to the rhythm of the realm of time it is in. One realm of time seems the same as any other realm of time to the person, but not to an observer who is located outside the person's frame of reference—for she sees the difference. As an example, astronauts traveling at 99% the speed of light could go to the star Procyon (10.4 light years distant) and back in 21 years in earth time. It would take light itself 20.8 years in earth time to make the same round trip. Because of time dilation, it would seem that only 3 years had gone by for the astronauts. All their clocks would indicate this, and biologically they would be only 3 years older. It would be the space officials greeting them on their return who would be 21 years older.

At higher speeds the results are even more impressive. At a speed of 99.90% the speed of light, travelers could travel slightly more than 70 light years in a single year of their own time. At 99.99% the speed of light, this distance would be pushed appreciably further than 200 years. A 5-year trip for them would take them farther than light travels in 1000 earth-time years.

Such journeys seem impossible to us today. The amounts of energy involved to propel spaceships to such relativistic speeds are billions of times the energy used to put the space shuttles into orbit. The problems of shielding radiation induced by these high speeds seems formidable. The practicality of such space journeys are prohibitive, so far.

If and when these problems are overcome and space travel becomes routine, people will have the option of taking a trip and returning in future centuries of their choosing. For example, one might depart from earth in a high-speed ship in the year 2050, travel for 5 years of so, and return in the year 2500. One might live among earthlings of that period for a while and depart again to try out the year 3000 for style. People could keep jumping into the future with some expense of their own time, but they could not trip into the past. They could never return to the same era on earth that they bid farewell to.

Time, as we know it, travels only one way—forward. Here on earth we constantly move into the future at the steady rate of 24 hours per day. An astronaut leaving on a deep-space voyage must live with the fact that, upon her return, much more time will have elapsed on earth than she has experienced on her voyage. Star travelers will not bid "so long, see you later" to those they leave behind but, rather, a permanent "goodby."

Fig. 15-23   From the earth frame of reference, light takes 30 000 years to travel from the center of the Milky Way galaxy to our solar system. From the frame of reference of a high-speed spaceship, the trip takes less time. From the frame of reference of light itself, the trip takes no time. There is no time in a speed-of-light frame of reference.

15 | Chapter Review

Concept Summary

According to Einstein's special theory of relativity, time is affected by motion in space at constant velocity.

- Time appears to pass more slowly in a frame of reference that is moving relative to the observer.

All the laws of nature are the same in all uniformly moving frames of reference.

- No experiment can be devised to detect whether the observer is moving at constant velocity.

The speed of light in empty space is the same in all frames of reference.

- The speed of light has the same value regardless of the motion of the source or the motion of the observer.

Important Terms

first postulate of special relativity (15.4)
postulate (15.3)
second postulate of special relativity (15.5)
spacetime (15.1)
special theory of relativity (15.1)
time dilation (15.1)

Review Questions

1. What is spacetime? (15.1)

2. Can you travel while remaining in one place in space? Explain. (15.1)

3. Does light travel through space? Through time? Through both space and time? (15.1)

4. What is time dilation? (15.1)

5. What does it mean to say that motion is relative? (15.2)

6. The speed of a ball you catch that is thrown from a moving truck depends on the speed and direction of the truck. Does the speed of light caught from a moving source similarly depend on the speed and direction of the source? Explain. (15.3)

7. What does it mean to say that the speed of light is a constant? (15.3)

8. What is the first postulate of special relativity? (15.4)

9. What is the second postulate of special relativity? (15.5)

10. The ratio of circumference to diameter for any size circle is π. Similarly, what is the ratio of space to time for light waves? (15.5)

11. The path of light in a vertical "light clock" in a high-speed spaceship is seen to be longer when viewed from a stationary frame of reference. Why, then, does the light not appear to be moving faster? (15.6)

12. If we view a passing spaceship and see that the inhabitants' time is running slow, how do they see our time running? (15.6)

13. When a flashing light source approaches you, does the speed of light, or the frequency of arrival of light flashes, or both, increase? (15.7)

14. a. How many frames of reference does the stay-at-home twin experience in the twin trip?
    b. How many frames of reference does the traveling twin experience? (15.7)

15. Is it possible for a person with a 70-year life span to travel farther than light travels in 70 years? Explain. (15.8)

Think and Explain

1. If you were in a smooth-riding train with no windows, could you sense the difference between uniform motion and rest? Between accelerated motion and rest? Explain how you could do this with a bowl filled with water.

2. When you change your position and move through space, what else are you moving through?

3. Suppose you're playing catch with a friend in a moving train. When you toss the ball in the direction of the moving train, how does the speed of the ball appear to an observer standing at rest outside the train? (Does it increase or appear the same as if the observer were riding on the train?)

4. Suppose you're shining a light to a friend on a moving train. When you shine the light in the direction of the moving train, strictly speaking, how would the speed of light appear to an observer standing at rest outside the train? (Does it increase or appear the same as if the observer were riding with the train?)

5. Light travels a certain distance in, say, 10 000 years. Is it possible that an astronaut could travel more slowly than the speed of light and yet travel that distance in a 10-year trip? Explain.

6. Can you get younger by traveling at speeds near the speed of light?

7. Explain why it is that when we look out into the universe, we see into the past.

8. One of the fads of the future might be "century hopping," where occupants of high-speed spaceships would depart from the earth for a few years and return centuries later. What are the present-day obstacles to such a practice?

9. Consider a high-speed spaceship equipped with a flashing light source. If the spaceship approaches you so that the frequency of flashes is doubled, how will your measurements of the time between flashes (the period) differ from similar measurements by occupants of the spaceship? Is this period constant for a constant relative speed? For accelerated motion? Defend your answer.

10. If you were in a high-speed spaceship traveling away from the earth at a speed close to that of light, would you measure your normal pulse to be slower, the same, or faster? How would your measurements be of the pulse of friends back on earth if you could monitor them from your ship? Explain.

Biography: Albert Einstein (1879–1955)

Albert Einstein was born in Ulm, Germany, on March 14, 1879. According to popular legend, he was a slow child and learned to speak at a much later age than average; his parents feared for a while that he might be mentally retarded. Yet his elementary school records show that he was remarkably gifted in mathematics, physics, and playing the violin. He rebelled, however, at the practice of education by regimentation and rote, and was expelled just as he was preparing to drop out at the age of 15.

Largely because of business reasons, his family moved to Italy. Young Einstein renounced his German citizenship and went to live with family friends in Switzerland. There he was allowed to take the entrance examinations for the renowned Swiss Federal Institute of Technology in Zurich, two years younger than normal age. But because of difficulties with the French language he did not pass the examination. He spent a year at a Swiss preparatory school in Aarau, where he was "promoted with protest in French." He tried the entrance exam again at Zurich and passed.

As a student he cut many lectures, preferring to study on his own, and in 1900 he succeeded in passing his examinations by cramming with the help of a friend's meticulous notes. He said later of this, ". . . after I had passed the final examination, I found the consideration of any scientific problem distasteful to me for an entire year."

It was not until two years after graduation that he got a steady job, as a patent examiner at the Swiss Patent Office in Berne. Einstein held this position for over seven years. He found the work rather interesting, sometimes stimulating his scientific imagination, but mainly freeing him of financial worries while providing free time to think about problems in physics.

With no academic connections whatsoever, and with essentially no contact with other physicists, he laid out the main lines along which twentieth-century theoretical physics has developed. In 1905, at the age of 26, he earned his Ph.D. in physics and published three major papers. The first was on the quantum theory of light, including an explanation of the photoelectric effect, for which he won the 1921 Nobel prize in physics. The second paper was on the statistical aspects of molecular theory and Brownian motion, a proof for the existence of atoms. His third and most famous paper was on special relativity. In 1915 he published a paper on the theory of general relativity which presented a new theory of gravitation that included Newton's theory as a special case. These trailblazing papers have greatly affected the course of modern physics.

Einstein's concerns were not limited to physics. He lived in Berlin during World War I and denounced the German militarism of his time. He publicly expressed his deeply felt conviction that warfare should be abolished and an international organization founded to govern disputes between nations. In 1933, while Einstein was visiting the United States, Hitler came to power. Einstein spoke out against Hitler's racial and political policies and resigned his position at the University of Berlin. Hitler responded by putting a price on his head. Einstein then accepted a research position at the Institute for Advanced Study in Princeton, New Jersey.

In 1939, one year before Einstein became an American citizen, and after German scientists fissioned the uranium atom, he was urged by several prominent American scientists to write the famous letter to President Roosevelt pointing out the scientific possibilities of a nuclear bomb. Einstein was a pacifist, but the thought of Hitler developing such a bomb prompted his action. The outcome was the development of the first nuclear bomb, which, ironically, was detonated on Japan after the fall of Germany.

Einstein believed that the universe is indifferent to the human condition, and stated that if humanity were to continue, it must create a moral order. He intensely advocated world peace through nuclear disarmament. Nuclear bombs, Einstein remarked, had changed everything but our way of thinking.

He was a man of unpretentious disposition with a deep concern for the welfare of his fellow beings. Albert Einstein was much more than a great scientist—he was a great human being.

16 Special Relativity—Length, Mass, and Energy

The speed of light is the speed limit for all matter. Suppose that two spaceships are each traveling at nearly the speed of light and are moving directly toward each other. The realms of space-time for each ship are so distorted that the relative speed of approach is still less than the speed of light! For example, if both ships are traveling toward each other at 80% the speed of light with respect to the earth, each would measure their speed of approach as 98% the speed of light. There are no circumstances where the relative speeds of any material objects surpass the speed of light.

Why is the speed of light the universal speed limit? To understand this, we must know how motion through space affects the length and mass of moving objects.

16.1 Length Contraction

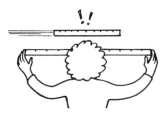

Fig. 16-1 A meter stick traveling at 87% the speed of light relative to an observer would be measured as only half as long as normal.

For moving objects, space as well as time undergoes changes. When viewed by an outside observer, moving objects appear to contract along the direction of motion. The amount of contraction is related to the amount of time dilation. For everyday speeds, the amount of contraction is much too small to be measured. For relativistic speeds, the contraction would be noticeable. A meter stick aboard a spaceship whizzing past you at 87% the speed of light, for example, would appear to you to be only 0.5 meter long. If it whizzed past at 99.5% the speed of light, it would appear to you to be contracted to one-tenth its original length. As relative speed gets closer and closer to the speed of light, the measured lengths of objects contract closer and closer to zero.

226

Do people aboard the spaceship also see their meter sticks—
and everything else in their environment—contracted? The answer is no. People in the spaceship see nothing at all unusual
about the lengths of things in their own reference frame. If they
did, it would violate the first postulate of relativity. Recall that
all the laws of physics are the same in all uniformly moving reference frames. Besides, there is no relative speed between themselves and the events they observe in their own reference frame.
There is a relative speed between themselves and *our* frame of
reference, however, so they will see *our* meter sticks contracted
to half size—and us as well.

Fig. 16-2   In the frame of reference of the meter stick, its length is one meter.
Observers from this frame see *our* meter sticks contracted. The effects of relativity are always attributed to "the other guy."

The contraction of speeding objects is the contraction of space
itself. Space contracts in only one direction, the direction of motion. Lengths along the direction perpendicular to this motion
are the same in the two frames of reference. So if an object is
moving horizontally, no contraction takes place vertically (Figure 16-3).

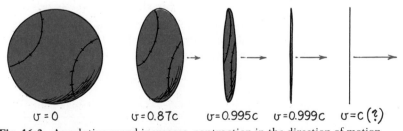

$\upsilon = 0$ $\upsilon = 0.87c$ $\upsilon = 0.995c$ $\upsilon = 0.999c$ $\upsilon = c$ (?)

Fig. 16-3   As relative speed increases, contraction in the direction of motion
increases. Lengths in the perpendicular direction do not change.

Relativistic length contraction can be expressed mathematically as:

$$L = L_0 \sqrt{1 - (v^2/c^2)}$$

In this equation, v is the speed of the object relative to the observer, c is the speed of light, L is the length of the moving object as measured by the observer, and L_0 is the measured length of the object at rest.*

Suppose that an object is at rest, so that $v = 0$. When 0 is substituted for v in the equation, we find $L = L_0$, as we would expect. It was stated earlier that if an object were moving at 87% the speed of light, it would contract to half its length. When $0.87c$ is substituted for v in the equation, we find $L = 0.5L_0$. Or when $0.995c$ is substituted for v, we find $L = 0.1L_0$, as stated earlier. If the object moves at c, its length would contract to zero. This is one of the reasons that the speed of light is the upper limit for the speed of any material object.

> ▶ **Question**
> A spacewoman travels by a spherical planet so fast that it appears to her to be an ellipsoid (egg shaped). If she sees the short diameter as half the long diameter, what is her speed relative to the planet?

16.2 | **The Increase of Mass with Speed**

If we push an object that is free to move, it will accelerate. If we maintain a steady push, it will accelerate to higher and higher speeds. If we push with a greater and greater force, the acceleration in turn will increase. It might seem that the speed should increase without limit, but there is a speed limit in the universe—the speed of light. In fact, we cannot accelerate a material object enough to reach the speed of light, let alone surpass it.

We can understand this from Newton's second law. Recall that the acceleration of an object depends not only on the applied force, but on the mass as well: $a = F/m$. Einstein stated that

---

▶ **Answer**
The spacewoman passes the spherical planet at 87% the speed of light.

---

*   This equation (and those that follow) is simply stated as a "guide to thinking" about the ideas of special relativity. The equations are given here without any explanation as to how they are derived.

when work is done to increase the speed of an object, its mass increases as well. This increase of mass means that a constant applied force produces less and less acceleration as the object's speed increases. The relationship between mass and speed is expressed mathematically by:

$$m = \frac{m_0}{\sqrt{1 - (v^2/c^2)}}$$

Here m is the **relativistic mass** of the moving object—its mass as measured by an observer. Again, v is the speed of the object relative to the observer. The symbol m_0 represents the **rest mass**, the mass the object would have at rest. When an object is given kinetic energy, its mass is greater than its rest mass.

The faster an object is pushed, the more its mass increases, which results in less and less response to the applied force. As v approaches c, the denominator of the equation approaches zero. This means that the mass m approaches infinity! An object pushed to the speed of light would have infinite mass and would require an infinite force, which is clearly impossible. Therefore we find another reason for saying that nothing made of matter can be accelerated to the speed of light.

Subatomic particles have been accelerated to nearly the speed of light, however. The masses of particles accelerated beyond 99% the speed of light increase thousands of times, as evidenced when a beam of electrons is directed into a magnetic field. Charged particles moving in a magnetic field experience a force that deflects them from their normal paths. The amount of deflection is known to depend on the mass—the greater the mass, the less the deflection. The particles are found to deflect less than predicted—unless the relativistic mass increase is taken into account (Figure 16-4). Only when the mass increase is taken into account do the particles strike their predicted targets. Physicists working with high-speed subatomic particles in atomic accelerators find the increase of mass with speed an everyday fact of life.

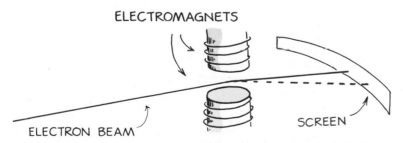

ELECTROMAGNETS

ELECTRON BEAM

SCREEN

Fig. 16-4 If the masses of the electrons did not increase with speed, the beam would follow the dashed line. But because of the increased inertia, the high-speed electrons in the beam are not deflected as much.

16.3   The Mass-Energy Equivalence

The most remarkable insight of Einstein's special theory of relativity is his conclusion that mass and energy are equivalent. As stated earlier, the energy pumped into atomic particles in an accelerator increases their mass. This is simply a consequence of the fact that energy and mass are equivalent to each other.

Consider the energy that goes into running particle accelerators. We all know that the energy that goes into accelerating the particles comes from a power plant someplace. The accelerator facility must pay its electric bill like other consumers of energy.

Let us follow this further. Energy generated at the power plant most likely comes from either the chemical combustion or nuclear reactions of certain fuels. If the process is chemical combustion, the masses of hydrocarbon molecules produced by combustion are reduced by about one part in a billion. If the process is nuclear fission, the masses of the fission fragments after reaction are reduced by about one part in a thousand. That is, the masses of the fission fragments are about a thousandth less than the mass of the initial uranium atoms before reaction. In either the chemical or the nuclear case, mass is given off when energy is given off. Now comes the interesting question: how much mass is gained by the accelerator particles? If we neglect the inefficiencies of power transmission, *the fuel fragments lose just as much mass as the particles at the accelerator gain!*

Fig. 16-5 Saying that a power plant delivers 90 million megajoules of energy to its consumers is equivalent to saying that it delivers one gram of energy to its consumers. This is because mass and energy are equivalent.

So a power company delivers mass every bit as much as it delivers energy. To say it delivers one is the same as saying it delivers the other. This is because mass and energy are in a practical sense one and the same. When something gains energy, it gains mass. When something loses energy, it loses mass.

Einstein realized that anything with mass—even if it is not moving—has energy. Conversely, anything with energy—even if it is not matter (such as light or microwaves, for example)—has mass! The amount of energy E is related to the amount of mass m by the most celebrated equation of the twentieth century:

$$E = mc^2$$

The c^2 is the conversion factor for energy units and mass units.* Because of the large magnitude of c, the speed of light, a small mass corresponds to a huge amount of energy. For example, the energy equivalent of a single gram of matter is greater than the energy used daily by the populations of our largest cities.

The change in mass for energy changes is so slight that it has not been detected until recent times. When we strike a match, for example, a chemical reaction occurs. Phosphorus atoms in the match head rearrange themselves and combine with oxygen in the air to form new molecules. The resulting molecules have very slightly less mass than the separate phosphorus and oxygen molecules. From a mass standpoint, the whole is slightly less than the sum of its parts, but not by very much—by only about one part in 10^9.

According to Einstein, the missing mass has not been destroyed. It has been carried off in the guise of radiant energy and kinetic energy. If E is the amount of energy given off, the "missing" mass is just E/c^2. For all chemical reactions that give off energy, there is a corresponding decrease in mass.

It is important to realize that the *total* amount of mass is the same before and after the reaction *if you remember to include the mass equivalent of the energy*. Similarly, the total energy is the same before and after the reaction if you include the energy equivalent of the rest mass (the mass of matter at rest) involved. Mass *is* conserved and energy *is* conserved. Some people say that mass is converted to energy (or vice versa). It is really more accurate to say that rest mass is converted to "pure" energy (energy that is not due to rest mass). The total amount of mass does not change, nor does the total amount of energy.

---

*   When c is in meters per second and m is in kilograms, then E will be in joules. If the equivalence of mass and energy had been understood long ago when physics concepts were first being formulated, there would likely be no separate units for mass and energy. Furthermore, with a redefinition of space and time units, c could equal 1, and $E = mc^2$ would simply be $E = m$.

Fig. 16-6 In one second, 4.5 billion tons of rest mass are converted to radiant energy in the sun. The sun is so massive, however, that in a million years only one ten-millionth of the sun's rest mass will have been converted to radiant energy.

In nuclear reactions, the decrease in rest mass is considerably more than in chemical reactions—about one part in 10^3. The decrease of rest mass in the sun and other stars by the process of thermonuclear fusion bathes the solar system with radiant energy and nourishes life. The present stage of thermonuclear fusion in the sun has been going on for the past 5 billion years, and there is sufficient hydrogen fuel for fusion to last another 5 billion years. It is nice to have such a big sun!

The equation $E = mc^2$ is not restricted to chemical and nuclear reactions. *Any* change in energy corresponds to a change in mass. The increased kinetic energy of a baseball is accompanied by a slight increase in mass. The filament of a light bulb energized with electricity has more mass than when it is turned off. A hot cup of tea has more mass than the same cup of tea when cold. A wound-up spring clock has more mass than the same clock when unwound. But these examples involve incredibly small changes in mass—too small to be measured by conventional methods. No wonder the fundamental relationship between mass and energy was not discovered until this century.

The equation $E = mc^2$ is more than a formula for the conversion of rest mass into pure energy, or vice versa. It states that energy and mass are the same. Mass is simply congealed energy. If you want to know how much energy is in a system, measure its mass. For an object at rest, its energy *is* its mass. It is energy itself that is hard to shake.

> ▶ **Question**
> Can we look at the equation $E = mc^2$ another way and say that matter transforms into pure energy when it is traveling at the speed of light squared?

16.4 The Correspondence Principle

If a new theory is valid, it must account for the verified results of the old theory. New theory and old must overlap and agree in the region where the results of the old theory have been fully verified.

▶ **Answer**
No, no, no! There are several things wrong with that statement. As matter is propelled faster, its mass increases rather than decreases. In fact, its mass approaches infinity. At the same time, its energy approaches infinity. It has more mass *and* more energy. Matter cannot be made to move at the speed of light, let alone the speed of light squared (which is not a speed!). The equation $E = mc^2$ simply means that energy and mass are "two sides of the same coin."

This requirement is known as the **correspondence principle**. If the equations of special relativity are valid, they must correspond to those of the mechanics of Newton—classical mechanics—when speeds much less than the speed of light are considered.

The relativity equations for time, length and mass are:

$$t = \frac{t_0}{\sqrt{1 - (v/c)^2}}$$

$$L = L_0 \sqrt{1 - (v/c)^2}$$

$$m = \frac{m_0}{\sqrt{1 - (v/c)^2}}$$

We can see that these equations each reduce to Newtonian values for speeds that are very small compared to c. Then, the ratio $(v/c)^2$ is very small, and for everyday speeds may be taken to be zero. The relativity equations become

$$t = \frac{t_0}{\sqrt{1 - 0}} = t_0$$

$$L = L_0 \sqrt{1 - 0} = L_0$$

$$m = \frac{m_0}{\sqrt{1 - 0}} = m_0$$

So for everyday speeds, the length, mass, and time of moving objects are essentially unchanged. The equations of special relativity hold for all speeds, although they are significant only for speeds near the speed of light.

Einstein's theory of relativity has raised many philosophical questions. What, exactly, is time? Can we say that it is nature's way of seeing to it that everything does not all happen at once? And why does time seem to move in one direction? Has it always moved forward? Are there other parts of the universe where time moves backward? Perhaps these unanswered questions will be answered by the physicists of tomorrow. How exciting!

16 Chapter Review

Concept Summary

When an object moves at very high speed relative to an observer, it is measured as contracted in the direction of motion.

When an object moves at very high speed relative to an observer, its mass is measured as greater than the value when it is not moving.

Einstein realized that mass and energy are equivalent—anything with mass also has energy, and anything with energy also has mass, according to the equation $E = mc^2$.
- Only when the release of energy is very great is the release of mass large enough to be detected.
- During any reaction, the total amount of mass remains the same if the mass equivalent of the energy released is taken into account.
- During any reaction, the total amount of energy remains the same if the energy equivalent of the rest mass involved is taken into account.

According to the correspondence principle, the equations for the special effects due to motion at high speed match the old, well tested equations governing motion at everyday speeds when small values of the speed are substituted.

Important Terms

correspondence principle (16.4)
relativistic mass (16.2)
rest mass (16.2)

Review Questions

1. If we witness events in a frame of reference moving past us, time appears to be stretched out (dilated). How do the lengths of objects in that frame appear? (16.1)

2. How long would a meter stick appear if it were traveling like a properly thrown spear at 99.5% the speed of light? (16.1)

3. How long would a meter stick appear if it were traveling at 99.5% the speed of light, but with its length perpendicular to its direction of motion? (Why are your answers to this and the last question different?) (16.1)

4. If you were traveling in a high-speed spaceship, would meter sticks on board appear to you to be contracted? Defend your answer. (16.1)

5. What happens to the mass of an object that is pushed to higher speeds? (16.2)

6. a. What is meant by rest mass?
   b. If an object has a rest mass of 1 kg, will its relativistic mass be greater, less, or the same, if it is accelerated to a high speed? (16.2)

7. What would be the mass of an object if it were pushed to the speed of light? (16.2)

8. a. What is meant by the equivalence of mass and energy?
   b. What does the equation $E = mc^2$ mean? (16.3)

9. Does the equation $E = mc^2$ apply only to reactions that involve the atomic nucleus? (16.3)

10. If an object is supplied with energy of any kind, does it then have more mass? (16.3)

11. The masses of particles in research accelerators gain appreciable mass when they are accelerated to speeds near that of light. Does this mass increase consume energy from the power utility that services the accelerator? Explain. (16.3)

12. In the preceding question, the power utility gets its energy from the mass of fuel, whether coal, oil, or atomic nuclei. If we neglect all inefficiencies at the power plant and in the transmission lines and at the accelerator, how does the mass increase of the accelerated particles compare with the decrease in mass of the fuel at the power plant? Explain. (16.3)

13. Where does solar energy originate? (16.3)

14. What is the correspondence principle? (16.4)

15. How does the relativistic mass of a car moving at ordinary speeds compare to its rest mass? (16.4)

Think and Explain

1. You observe a spaceship moving away from you at speed v_1. A rocket is fired straight ahead from the spaceship, so that it also moves away from you. Suppose that it is fired at speed v_2 relative to the spaceship. It so happens that the speed of the rocket relative to you is *not* $v_1 + v_2$. For motion in a straight line, the relativistic sum V of velocities v_1 and v_2 is given by

$$V = \frac{v_1 + v_2}{1 + \dfrac{v_1 v_2}{c^2}}$$

Suppose v_1 and v_2 are each half the speed of light, or $0.5c$. Show that the speed V of the rocket relative to you is $0.8c$.

2. Substitute small values of v_1 and v_2 into the preceding equation and show that for everyday speeds V is practically equal to $v_1 + v_2$.

3. Pretend that the spaceship of the first question is somehow traveling at speed c with respect to you, and it fires a rocket at speed c with respect to itself. Use the equation to show that the speed of the rocket with respect to you is still c!

4. If a high-speed spaceship appears contracted to half its length, how will its relativistic mass compare to its rest mass?

5. Suppose you travel past the earth at relativistic speeds in a spaceship and earth observers tell you that your ship appears to be contracted. Comment on the idea of checking their observations by putting a meter stick or finer measuring rulers to parts of your spaceship.

6. Give two reasons why we say there is a speed limit for particles in the universe.

7. The two-mile-long linear accelerator at Stanford University in California "appears" to be less than a meter long to the electrons that travel in it. Explain.

8. The masses of electrons that are accelerated in the Stanford accelerator become thousands of times as great as the rest mass by the time the electrons reach the end of their trip. In theory, if you could travel with them, would you notice an increase in their mass? In the mass of the target they are about to hit? Explain.

9. The electrons that illuminate your TV screen travel at about one-fourth the speed of light and have an increased mass of nearly 3 percent. Does this relativistic effect tend to increase or decrease your electric bill?

10. Since there is an upper limit on the speed of a particle, does it follow that there is therefore an upper limit on its kinetic energy or momentum? Defend your answer.

II Properties of Matter

PEOPLE SAY THE BALLOON FLOATS BECAUSE IT'S LIGHTER THAN AIR --- THAT THE HELIUM GAS INSIDE IS COMPOSED OF LIGHTER *ATOMS* THAN THE AIR OUTSIDE --- THAT HELIUM HAS LESS *DENSITY* THAN AIR. BUT DOES THIS EXPLAIN *WHY* THE BALLOON FLOATS? ISN'T THE EXPLANATION THAT THE BOTTOM OF THE BALLOON IS "DEEPER" IN OUR "OCEAN OF AIR" THAN THE TOP, AND THAT THE *ATMOSPHERIC PRESSURE* UP AGAINST THE BOTTOM IS SLIGHTLY GREATER THAN THE PRESSURE DOWN AGAINST THE TOP? AND THIS DIFFERENCE IN UPWARD AND DOWNWARD PRESSURES PRODUCES A *BUOYANT FORCE* THAT EXCEEDS THE WEIGHT OF THE BALLOON? YOU'LL FIND A LOT OF "EVERYDAY PHYSICS" IN UNIT 2!

17 The Atomic Nature of Matter

Suppose you were to break apart a large boulder with a heavy sledgehammer. You break it into rocks. Then you break the rocks into stones, and the stones into gravel. You keep going and break the gravel into sand, and the sand into a powder. The powder is made of fine crystals, which are particles of the minerals that make up the rock. If the crystals are broken down, you have **atoms**, the building blocks of all matter. Everything is made of atoms.

17.1 Elements

All material things that exist—shoes, ships, mice, people, the stars—are made of atoms. Interestingly enough, the incredible number of different things that exist are made of a surprisingly small number of kinds of atoms. Just as only three colors of dots of light combine to form almost every conceivable color on a television screen, only about 100 distinct kinds of atoms combine to form all the materials we know about. Each of these kinds of atoms belongs to a different **element**.

To date (1985), there are 109 known elements. Of these, only 90 are found in nature. The others are made in the laboratory with high-energy atomic accelerators and nuclear reactors. These elements are too unstable (radioactive) to occur naturally in any appreciable amounts.

All matter, however complex, living or nonliving, is some combination of the elements. From a pantry having about 100 bins, each containing a different element, we have all the materials needed to make up any substance occurring in the known universe.

Common materials are composed of not more than about 14 of these elements, for the majority of elements are relatively rare.

Living things, for example, are composed primarily of four elements: carbon (C), hydrogen (H), oxygen (O), and nitrogen (N). The letters in parenthesis represent the chemical symbols for these elements. Table 17-1 lists the 14 most common elements.

Table 17-1 The 14 Most Common Elements	
hydrogen (H)	carbon (C)
nitrogen (N)	oxygen (O)
sodium (Na)	magnesium (Mg)
aluminum (Al)	silicon (Si)
phosphorus (P)	sulfur (S)
chlorine (Cl)	potassium (K)
calcium (Ca)	iron (Fe)

The lightest atoms of all are hydrogen. Hydrogen atoms are also the most abundant. They comprise over 90 percent of the atoms in the known universe. More complex and heavier atoms are simply the result of hydrogen atoms squeezed together in the deep interiors of stars under enormous temperatures and pressures. Nearly all the elements that occur in nature are remnants of stars that exploded long before the solar system came into being.

17.2 Atoms Are Ageless

Atoms are much older than the materials they compose. The age of many atoms goes back to the origin of the universe and the age of most atoms is more than the age of the sun and earth.

Fig. 17-1 Both you and she are made of stardust—in the sense that the carbon, oxygen, nitrogen, and other atoms that make up your body originated in the deep interior of ancient stars that have long since exploded.

Atoms in your body have been around long before the solar system came into existence. They cycle and recycle among innumerable forms, both living and nonliving. Every time you breathe, for example, only part of the atoms that you inhale are exhaled in your next breath. The remaining atoms are taken into your body to become part of you, and most leave your body sooner or later.

Strictly speaking, you don't "own" the atoms that make up your body—you borrow them. We all share from the same atom pool, as atoms migrate around, within, and throughout us. So some of the atoms in the ear you scratch today may have been part of your neighbor's breath yesterday!

Most people know we are all made of the same *kinds* of atoms. But what most people don't know is that we are made of the *same* atoms—atoms that cycle from person to person as we breathe, sweat, and vaporize.

17.3   Atoms Are Small

There are about 10^{23} atoms in a gram of water (a thimbleful). The number 10^{23} is an enormous number. It is roughly the number of drops of water in the Atlantic Ocean. So there are as many atoms in a drop of water as there are drops of water in the Atlantic Ocean! No wonder a cupful of DDT or any material thrown into the ocean spreads around and is later found in every part of the world's oceans. The same is true of materials released into the atmosphere.

Atoms are so small that there are about as many atoms in the air in your lungs at any moment as there are breathfuls of air in the atmosphere of the whole world. It takes about 6 years for one of your exhaled breaths to become evenly mixed in the atmosphere. At that point, every person in the world inhales an average of one of your exhaled atoms in a single breath. And this occurs for *each* breath you exhale! When you take into account the many thousands of breaths that people exhale, there are many atoms in your lungs at any moment that were once in the lungs of every person who ever lived. We are literally breathing each other's breaths.

Atoms are too small to be seen—at least with visible light. You could connect microscope to microscope and never "see" an atom. This is because light travels in waves, and atoms are smaller than the wavelengths of visible light. The size of a particle visible under the highest magnification must be larger than the wavelength of light.

Fig. 17-2 There are as many atoms in a normal breath of air as there are breathfuls of air in the atmosphere of the world.

17.4 | Evidence for Atoms

The first somewhat direct evidence for the existence of atoms was unknowingly discovered in 1827. A Scottish botanist, Robert Brown, was studying the spores of pollen under a microscope. He noticed that the spores were in a constant state of agitation, always jiggling about. At first, Brown thought that the spores were some sort of moving life forms. Later, he found that inanimate dust particles and grains of soot also showed this kind of motion. The perpetual jiggling of particles that are just large enough to be seen is called **Brownian motion**. Brownian motion is now known to result from the motion of neighboring atoms too small to be seen.

More direct evidence for the existence of atoms is available today. A photograph of individual atoms is shown in Figure 17-3. The photograph was made not with visible light but with an electron beam. A familiar example of an electron beam is the one that sprays the picture on your television screen. Although an electron beam is a stream of tiny particles, electrons, it has wave properties. It so happens that a high-energy electron beam has a wavelength more than a thousand times smaller than the wavelength of visible light. With such a beam, atomic detail can be seen. The historic (1970) photograph in Figure 17-3 was taken with a powerful and very thin electron beam in a scanning electron microscope. It is the first photograph of clearly distinguishable atoms.

More recently, IBM researchers have developed an electron microscope small enough to be held in your hand; it is called a scanning tunneling microscope. An image of graphite taken with this remarkable instrument in 1985 is shown in Figure 17-4. The gray "hill top" areas indicate the location of individual carbon atoms in the graphite layer.

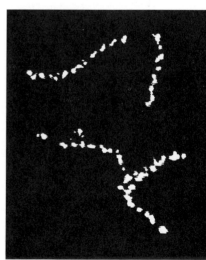

Fig. 17-3 The strings of dots are chains of thorium atoms taken with a scanning electron microscope by researchers at the University of Chicago's Enrico Fermi Institute.

Fig. 17-4 An image of graphite obtained using the small scanning tunneling microscope. The "bumps" indicate the location of individual carbon atoms.

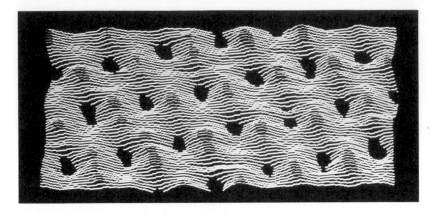

17.5 Molecules

Sometimes, atoms combine to form larger particles called **molecules**. For example, two atoms of hydrogen (H) are combined with a single atom of oxygen (O) in a water molecule (H_2O). The gases nitrogen and oxygen, which make up most of the atmosphere, are each made of simple two-atom molecules (N_2 and O_2). In contrast, the double helix of deoxyribonucleic acid (DNA), the basic building block of life, is composed of millions of atoms.

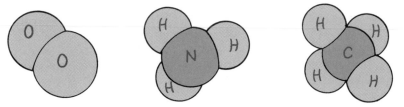

Fig. 17-5 Models of simple molecules. The atoms that compose a molecule are not just mixed together, but are connected in a well-defined way.

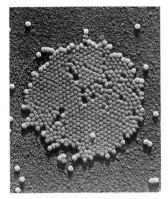

Fig. 17-6 An electron microscope photo of virus molecules.

Not all matter is made of molecules. Metals and rock minerals (including common table salt) are made of atoms that are not joined in molecules. Matter that is a gas or liquid at room temperature tends to be made of molecules.

Matter made of molecules can contain all the same kind of molecule or it can be a mixture of different kinds of molecules. Pure water contains only H_2O molecules, whereas "clean" air contains molecules of several different gases.

Like atoms, molecules are too small to be seen with optical microscopes. More direct evidence of molecules is seen in electron-microscope photographs. The photograph in Figure 17-6 is of virus molecules composed of thousands of atoms. These giant molecules are still too small to be seen with visible light. So we see again that an atom or a molecule that is invisible to light may be visible to a shorter-wavelength electron beam.

We are able to detect molecules through our sense of smell. Noxious gases such as sulphur dioxide, ammonia, and ether are clearly sensed by the organs in our nose. The smell of perfume is the result of molecules that rapidly evaporate from the liquid and jostle around completely haphazardly in the air until some of them accidentally enter our noses. The perfume molecules are certainly not attracted to our noses! They are just a few of the billions of aimlessly jostling molecules that have moved out in all directions from the liquid perfume.

17.6   Compounds

A **compound** is a substance made of atoms of different elements combined in a fixed proportion. The **chemical formula** of the compound tells the proportions of each kind of atom. For example, in the gas carbon dioxide, the formula CO_2 indicates that for every carbon (C) atom there are two oxygen (O) atoms. Water, table salt, and carbon dioxide are all compounds. Air, wood, and salty water are not compounds.

A compound may or may not be made of molecules. Water and carbon dioxide are made of molecules. On the other hand, table salt (NaCl) is made of different kinds of atoms arranged in a regular pattern. Every chlorine atom is surrounded by six sodium atoms (see Figure 17-7). In turn, every sodium atom is surrounded by six chlorine atoms. As a whole, there is one sodium atom for each chlorine atom, but there are no separate sodium-chlorine groups that can be labeled molecules.

Compounds have different properties from the elements from which they are made. At ordinary temperatures, water is a liquid, whereas hydrogen and oxygen are both gases. Salt is an edible solid, whereas chlorine is a poisonous gas.

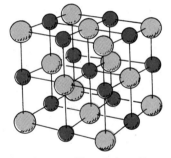

Fig. 17-7   Table salt (NaCl) is a compound that is not made of molecules. The sodium and chlorine atoms are arranged in a repeating pattern. Each atom is surrounded by six atoms of the other kind.

17.7   The Atomic Nucleus

An atom is mostly empty space. Almost all its mass is packed into the central region called the **nucleus**. The New Zealander physicist Ernest Rutherford discovered this in 1909 in his now-famous gold-foil experiment. Rutherford's group directed a beam of electrically-charged particles (alpha particles) from a radio-

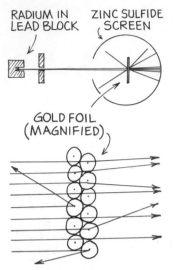

RADIUM IN   ZINC SULFIDE
LEAD BLOCK   SCREEN

GOLD FOIL
(MAGNIFIED)

Fig. 17-8 The occasional
large-angle scattering of
alpha particles from the gold
atoms led Rutherford to the
discovery of the small, very
massive nucleus at their
centers.

active source through a very thin gold foil. They measured the
angles at which the particles were deflected from their straight-
line path as they emerged. This was accomplished by noting
spots of light on a zinc-sulphide screen that nearly surrounded
the gold foil (Figure 17-8). Most particles continued in a more or
less straight-line path through the thin foil. But, surprisingly,
some particles were widely deflected. Some were even scattered
back along their incident paths. It was like firing bullets at a
piece of tissue paper and finding some bullets bouncing backward.

Rutherford reasoned that the particles that were undeflected
traveled through regions of the gold foil that were empty space.
The few particles that were deflected were repelled from the
massive centers of gold atoms that themselves were electrically
charged. Rutherford had discovered the atomic nucleus.

Although the mass of an atom is primarily concentrated in
the nucleus, the nucleus occupies only a few quadrillionths of
the volume of an atom. Atomic nuclei (plural of *nucleus*) are
extremely compact, or equivalently, extremely dense. If bare
atomic nuclei could be packed against each other into a lump
1 cm in diameter (about the size of a large pea), it would weigh
133 000 000 tons!

Huge electrical forces of repulsion prevent such close pack-
ing of atomic nuclei. This is because each nucleus is electrically
charged and repels other nuclei. Only under special circumstances are the nuclei of two or more atoms squashed into contact. When this happens, the violent reaction known as nuclear
fusion takes place. Fusion occurs in the core of the sun and other
stars and in a hydrogen bomb.

The principal building block of the nucleus is the **nucleon**.*
When the nucleon is in its electrically neutral state, it is a **neutron**. When it is in its electrically charged state, it is a **proton**. All
neutrons are identical; they are copies of each other. Similarly,
all protons are identical.

Atoms with the same number of protons all belong to the same
element. There may be a difference in the number of neutrons,
however, in which case we speak of different **isotopes**. The nucleus of the most common hydrogen atom contains only a single
proton. When one proton is accompanied by a single neutron,
we have *deuterium*, an isotope of hydrogen. When there are two
neutrons in the hydrogen nucleus, we have the isotope *tritium*.
Every element has a variety of isotopes. The lighter nuclei have
roughly equal numbers of protons and neutrons. The more massive nuclei have more neutrons than protons.

---

*   Nucleons are composed of still smaller particles called *quarks*, which are
discussed in Chapter 39.

Atoms are classified by their **atomic number**, which is the same as the number of protons in the nucleus. Since the nucleus of a hydrogen atom has a single proton, its atomic number is 1. Helium has two protons, so its atomic number is 2. Lithium has three protons and its atomic number is 3, and so on, in sequence to naturally occurring uranium with atomic number 92.

Electric charge comes in two kinds, positive and negative. You will learn more about electric charge later. For now it is sufficient to know that like kinds of charge repel one another and unlike kinds attract one another. Protons have a positive electric charge and thus repel other protons. Protons are held together within a nucleus, in spite of their mutual repulsion, by the very strong nuclear force, which acts only across tiny distances. (The strong nuclear force is discussed in Chapter 38.)

17.8 | Electrons in the Atom

The number of protons in the nucleus is normally electrically balanced by an equal number of electrons outside the nucleus. These are the electrons that make up the flow of electricity in electric circuits. Electrons have only about $\frac{1}{2000}$ the mass of a nucleon and contribute very little mass to the atom. The negatively charged electrons are attracted to the positive nucleus, but repel other electrons.

When an atom is electrically neutral, it normally does not attract or repel other atoms. But when atoms are close together, the negative electrons on one atom may at times be closer to the positive nucleus of a neighboring atom, which results in a net attraction between the atoms. This is how atoms combine to form molecules.

When the number of electrons do not equal the number of protons in an atom, the atom is electrically charged and is said to be an **ion**. When electrons are knocked off neutral atoms by any means, *positive* ions are produced. This is because the net charge is positive. When neutral atoms gain electrons by any means, *negative* ions are produced. Compounds that are not made of molecules are made of ions. In salt, for example, the sodium atoms are positive ions. The chlorine atoms are negative ions. The compound is held together by the forces between the positive and negative ions.

Just as our solar system is mostly empty space, the atom is mostly empty space. The nucleus and surrounding electrons occupy only a tiny fraction of the atomic volume. If it were not for the electric forces of repulsion between the electrons of neigh-

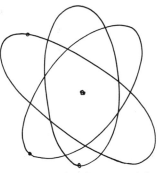

Fig. 17-9 The classic model of the atom consists of a tiny nucleus surrounded by orbiting electrons.

boring atoms, solid matter would be much more dense than it is. We and the solid floor are mostly empty space, because the atoms making up these and all materials are themselves mostly empty space. But we don't fall through the floor. The electric forces of repulsion keep atoms from caving in on each other under pressure. Atoms too close will repel (if they don't combine to form molecules), but when atoms are several atomic diameters apart, the electric forces on each other are negligible.

To explain how atoms of different elements interact to form compounds, scientists have produced the **shell model of the atom**. Electrons are pictured as being in spherical shells around the nucleus. In the innermost shell, there are at most two electrons. In the second shell, there are at most eight. In the third shell, there are at most 18. In the fourth shell, there are at most 32. Consider aluminum as an example. The neutral atom has 13 electrons. The first shell has two, the second shell has eight, and the remaining three are in the third shell.

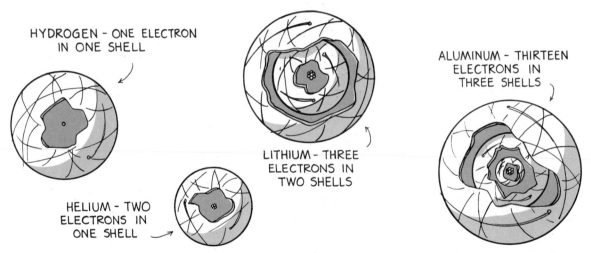

HYDROGEN – ONE ELECTRON IN ONE SHELL

HELIUM – TWO ELECTRONS IN ONE SHELL

LITHIUM – THREE ELECTRONS IN TWO SHELLS

ALUMINUM – THIRTEEN ELECTRONS IN THREE SHELLS

Fig. 17-10   The shell model of the atom pictures the electrons in concentric, spherical shells around the nucleus.

The outermost shell of electrons can never contain more than eight. Thus, potassium, with 19 electrons, has 2, 8, 8, and 1 in four shells. The third shell cannot have 9 electrons when it is the last shell.

The arrangement of electrons about the atomic nucleus dictates whether and how atoms join to become molecules, melting and freezing temperatures, electrical conductivity, as well as the taste, texture, appearance, and color of substances. The electron arrangement quite literally gives life and color to the world.

The **periodic table** is a chart that lists atoms by their atomic number and by their electron arrangements. (See Figure 17-11.)

Periodic Table of the Elements

1A	IIA		IIIB	IVB	VB	VIB	VIIB		VIII		IB	IIB	IIIA	IVA	VA	VIA	VIIA	0
1 H Hydrogen 1.008																		2 He Helium 4.003
3 Li Lithium 6.94	4 Be Beryllium 9.012												5 B Boron 10.81	6 C Carbon 12.011	7 N Nitrogen 14.007	8 O Oxygen 15.999	9 F Fluorine 18.998	10 Ne Neon 20.17
11 Na Sodium 22.990	12 Mg Magnesium 24.305												13 Al Aluminum 26.98	14 Si Silicon 28.09	15 P Phosphorus 30.974	16 S Sulfur 32.06	17 Cl Chlorine 35.453	18 Ar Argon 39.948
19 K Potassium 39.098	20 Ca Calcium 40.08	21 Sc Scandium 44.956	22 Ti Titanium 47.90	23 V Vanadium 50.942	24 Cr Chromium 51.996	25 Mn Manganese 54.938	26 Fe Iron 55.847	27 Co Cobalt 58.933	28 Ni Nickel 58.71	29 Cu Copper 63.546	30 Zn Zinc 65.38	31 Ga Gallium 69.735	32 Ge Germanium 72.59	33 As Arsenic 74.922	34 Se Selenium 78.96	35 Br Bromine 79.904	36 Kr Krypton 83.80
37 Rb Rubidium 85.467	38 Sr Strontium 87.62	39 Y Yttrium 88.906	40 Zr Zirconium 91.22	41 Nb Niobium 92.906	42 Mo Molybdenum 95.94	43 Tc Technetium 98.906	44 Ru Ruthenium 101.07	45 Rh Rhodium 102.91	46 Pd Palladium 106.4	47 Ag Silver 107.868	48 Cd Cadmium 112.41	49 In Indium 114.82	50 Sn Tin 118.69	51 Sb Antimony 121.75	52 Te Tellurium 127.60	53 I Iodine 126.904	54 Xe Xenon 131.30
55 Cs Cesium 132.905	56 Ba Barium 137.33	57–71* Rare Earths	72 Hf Hafnium 178.49	73 Ta Tantalum 180.947	74 W Tungsten 183.85	75 Re Rhenium 186.207	76 Os Osmium 190.2	77 Ir Iridium 192.22	78 Pt Platinum 195.09	79 Au Gold 196.967	80 Hg Mercury 200.59	81 Tl Thallium 204.37	82 Pb Lead 207.2	83 Bi Bismuth 208.98	84 Po Polonium (209)	85 At Astatine (210)	86 Rn Radon (222)
87 Fr Francium (223)	88 Ra Radium 226.03	89–103† Actinides	104 Rf Rutherfordium (261)	105 Ha Hahnium (262)	106 (263)	107 (262)	108 (265)	109 (266)									

*Rare earths (Lanthanide series)

57 La Lanthanum 139.91	58 Ce Cerium 140.12	59 Pr Praseodymium 140.91	60 Nd Neodymium 144.24	61 Pm Promethium (145)	62 Sm Samarium 150.36	63 Eu Europium 151.96	64 Gd Gadolinium 157.25	65 Tb Terbium 158.93	66 Dy Dysprosium 162.50	67 Ho Holmium 164.93	68 Er Erbium 167.26	69 Tm Thulium 168.93	70 Yb Ytterbium 173.04	71 Lu Lutetium 174.967

†Actinide series

89 Ac Actinium 227.028	90 Th Thorium 232.038	91 Pa Protactinium 231.036	92 U Uranium 238.029	93 Np Neptunium 237.048	94 Pu Plutonium (244)	95 Am Americium (243)	96 Cm Curium (247)	97 Bk Berkelium (247)	98 Cf Californium (251)	99 Es Einsteinium (254)	100 Fm Fermium (257)	101 Md Mendelevium (258)	102 No Nobelium (259)	103 Lr Lawrencium (260)

Fig. 17-11 The periodic table of the elements. The atomic number, above the chemical symbol, is equal to the number of protons in the nucleus (and equivalently, the number of electrons that surround the nucleus in a neutral atom). The number below is the atomic mass. Each row in the periodic table corresponds to a different number of electron shells in the atom. Note that the uppermost row consists of only two elements, hydrogen and helium. The electrons of helium complete the innermost shell. Elements are arranged vertically on the basis of similarity in electron arrangement, which dictates similarities in physical and chemical properties of the elements and their compounds.

As you read across from left to right, each element has one more proton than the preceding element. As you go down, each element has one more electron shell than the one above.

Elements in the same column have similar chemical properties. That is, they form compounds with the same elements according to similar formulas. Elements in the same column are said to belong to the same **family** of elements. Elements in the same family have the same number of electrons in the outer shell.

17.9   The States of Matter

Matter exists in four states. You are familiar with the solid, liquid, and gaseous states. In the **plasma** state, matter consists of bare atomic nuclei and free electrons. The plasma state exists only at high temperatures. Although the plasma state is less common to our everyday experience, it is the predominant state of matter in the universe. The sun and other stars as well as much of the intergalactic matter are in the plasma state. Closer to home, the glowing gas in a fluorescent lamp is a plasma.

In all states of matter, the atoms are constantly in motion. In the solid state the atoms and molecules vibrate about fixed positions. If the rate of molecular vibration is increased enough, molecules will shake apart and wander throughout the material, vibrating in non-fixed positions. The shape of the material is no longer fixed but takes the shape of its container. This is the liquid state. If more energy is put into the material and the molecules vibrate at even greater rates, they may break away from one another and assume the gaseous state.

All substances can be transformed from one state to another. We often observe this changing of state in the compound H_2O. When solid, it is ice. If we heat the ice, the increased molecular motion jiggles the molecules out of their fixed positions, and we have water. If we heat the water, we can reach a stage where continued increase in molecular vibration results in a separation between water molecules, and we have steam. Continued heating causes the molecules to separate into atoms. If we heat these to temperatures exceeding 2000°C, the atoms themselves will be shaken apart, making a gas of free electrons and bare atomic nuclei. Then we have a plasma.

The following three chapters treat the solid, liquid, and gaseous states in turn.

17 | Chapter Review

Concept Summary

All matter is made from only about 100 different kinds of atoms.

- Each kind of atom belongs to a different element.
- Most atoms have been recycling through matter even before the solar system came into being.
- Atoms are too small to see with visible light but can be photographed with an electron microscope.

A compound is a substance made of different elements combined in a fixed proportion.

- Some compounds are made of molecules, which are particles made of atoms joined together.
- Other compounds are made of different kinds of atoms arranged in a regular pattern.

The atom is mostly empty space. Its mass is almost entirely in its nucleus.

- The nucleus is made of protons and neutrons.
- The number of protons determines the element to which the atom belongs.
- An electrically neutral atom has electrons outside the nucleus equal in number to the protons inside the nucleus.
- The shell model of the atom pictures electrons in spherical shells around the nucleus.
- The periodic table is a chart of elements arranged according to similar atomic structure and similar properties.

Important Terms

atom (17.1)
atomic number (17.7)
Brownian motion (17.4)
chemical formula (17.6)
compound (17.6)

element (17.1)
family (17.8)
ion (17.8)
isotope (17.7)
molecule (17.5)
neutron (17.7)
nucleon (17.7)
nucleus (17.7)
periodic table (17.8)
plasma (17.9)
proton (17.7)
shell model of the atom (17.8)

Review Questions

1. How many elements are known today? (17.1)

2. Which element has the lightest atoms? (17.1)

3. How does the age of most atoms compare with the age of the solar system? (17.2)

4. What is meant by the statement that you don't "own" the atoms that make up your body? (17.2)

5. How does the approximate number of atoms in the air in your lungs compare to the number of breaths of air in the atmosphere of the whole world? (17.3)

6. How do the sizes of atoms compare to the wavelengths of visible light? (17.3)

7. What causes dust particles to move with Brownian motion? (17.4)

8. Individual atoms cannot be seen with visible light; yet there is a photograph of individual atoms in Figure 17-3. Explain. (17.4)

9. Distinguish between an atom and a molecule. (17.5)

10. a. How many elements compose pure water?
    b. How many individual atoms are there in a water molecule? (17.5)

11. a. Cite an example of a substance that is made of molecules.
    b. Cite a substance that is not made of molecules. (17.5)

12. True or false: We smell things because certain molecules are attracted to our noses. (17.5)

13. a. What is a compound?
    b. Cite the chemical formulas for at least three compounds. (17.6)

14. What did Rutherford discover when he bombarded a thin foil of gold with subatomic particles? (17.7)

15. How does the mass of an atomic nucleus compare to the mass of the whole atom? (17.7)

16. How does the size of an atomic nucleus compare to the size of the whole atom? (17.7)

17. What are the two kinds of nucleons? (17.7)

18. a. What is an isotope?
    b. Give two examples. (17.7)

19. How does the atomic number of an element compare to the number of protons in its nucleus? To the number of electrons that normally surround the nucleus? (17.7)

20. How does the mass of an electron compare to the mass of a nucleon? (17.8)

21. a. What is an ion?
    b. Give two examples. (17.8)

22. At the atomic level, a solid block of iron is mostly empty space. Explain. (17.8)

23. What is the periodic table of elements? (17.8)

24. According to the shell model of the atom, how many electron shells are there in the hydrogen atom? The lithium atom? The aluminum atom? (17.8)

25. What are the four states of matter? (17.9)

Think and Explain

1. Identify which of the following chemical formulas represent pure elements: H_2, H_2O, He, Na, NaCl, Au, U.

2. Which are older, the atoms in the body of an elderly person, or those in the body of a baby?

3. Suppose that your brother comes into the room wearing shaving lotion, which you smell almost immediately. From an atomic point of view, exactly what is happening?

4. In what way does the number of protons in an atomic nucleus dictate the chemical properties of the element?

5. Atoms are mostly empty space, and structures such as a floor are composed of atoms and are therefore also empty space. So why don't you fall through the floor?

6. What element results if you add a proton to the nucleus of carbon? (See periodic table.)

7. What element results if two protons and two neutrons are ejected from a uranium nucleus?

8. You could swallow a capsule of the element germanium without harm. But if a proton were added to each of the germanium nuclei, you would not want to swallow the capsule. Why?

9. Assuming that all the atoms stay in the atmosphere, what are the chances that at least one of the atoms you exhaled in your very first breath will be inhaled in your next breath?

10. What does the addition or subtraction of heat have to do with whether or not a substance is a solid, liquid, gas, or plasma?

18 | Solids

Humans have been classifying solids since the Stone Ages, when they first distinguished between rocks for shelter and rocks for tools. Since then, our knowledge of solids has grown without pause. Before the turn of the century, it was thought that the content of a solid was what determined its characteristics—what made diamonds hard, lead soft, iron magnetic, and copper electrically conducting. It was believed that to change a substance, one merely varied the contents. We have since found that the characteristics of a solid are due to its structure—that is, the arrangement of atoms that make up the material.

This shift in emphasis from the study of the content to the study of the structure of solid materials has changed the role of investigators from being finders and assemblers of materials to actual makers of materials. In today's laboratories people are continually creating new synthetic (humanmade) materials.

18.1 | Crystal Structure

When we look carefully at samples of rock minerals, such as quartz, mica, or galena, we see many smooth, flat surfaces. These flat surfaces are at angles to each other within the mineral. The mineral samples are made of **crystals**, or regular geometric shapes. Each sample is made of many crystals, assembled in various directions. The samples themselves may have very irregular shapes, as if they were tiny cubes or other small units glued together to make a freeform solid sculpture.

Not all crystals are evident to the naked eye. Their existence in many solids was not discovered until the advent of X-ray beams early in the century. The X-ray pattern caused by the crystal structure of common table salt (sodium chloride) is shown in Fig-

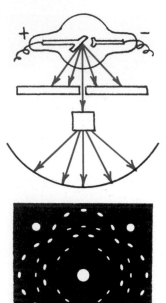

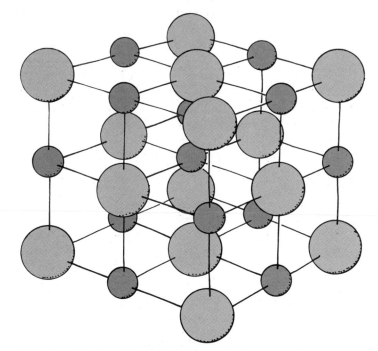

ure 18-1. Rays from the X-ray tube are blocked by a lead screen except for a narrow beam that hits the crystal of sodium chloride. The radiation that penetrates the crystal produces the pattern shown on the photographic film beyond the crystal. The white spot in the center is caused by the main unscattered beam of X rays. The size and arrangement of the other spots indicate the arrangement of sodium and chlorine atoms in the crystal. Every crystalline structure has its own unique X-ray pattern. All crystals of sodium chloride will always produce this same design.

The patterns made by X rays on photographic film show that the atoms in a crystal are in an orderly arrangement. For example, in a sodium chloride crystal, the atoms are arranged like a three-dimensional chess board or a child's jungle gym (Figure 18-2).

Fig. 18-1 X-ray pattern caused by the crystal structure of common table salt (sodium chloride).

Fig. 18-2 A model of a sodium chloride crystal. The large spheres represent chlorine atoms. The small ones represent sodium atoms.

Fig. 18-3 Crystal structure is quite evident on galvanized (zinc-coated) metals.

Metals such as iron, copper, and gold have relatively simple crystal structures. Tin and cobalt are only a little more complex. Metal crystals can be seen if you look carefully at a metal surface that has been cleaned (etched) with acid. You can also see them on the surface of galvanized iron that has been exposed to the weather, or on brass doorknobs that have been etched by the perspiration of hands.

18.2 | Density

One of the properties of solids, as well as liquids and even gases, is the measure of the compactness of the material: **density**. We think of density as the "lightness" or "heaviness" of materials. Density is a measure of how much matter is squeezed into a given space; it is the amount of mass per unit volume:

$$\text{density} = \frac{\text{mass}}{\text{volume}}$$

What happens to the density of a chocolate bar when you break it in two? The answer is, nothing. Each piece may have half the mass, but each piece also has half the volume. Density is not mass and it is not volume. Density is a ratio; it is the amount of mass per unit volume. A pure iron nail has the same density as a pure iron frying pan. The frying pan may have 100 times as many iron atoms and have 100 times as much mass, but they'll take up 100 times as much space. The mass per unit volume for the iron nail and the iron frying pan is the same.

Fig. 18-4 When the loaf of bread is squeezed, its volume decreases and its density increases.

Both the masses of atoms and the spacing between atoms determine the density of materials. Osmium, a hard bluish-white metallic element, is the densest substance on earth, even though the individual osmium atom is less massive than individual atoms of gold, mercury, lead, and uranium. The close spacing of osmium atoms in an osmium crystal gives it the greatest density. More atoms of osmium fit into a cubic centimeter than other more massive and more widely spaced atoms.

The densities of a few materials, in units of grams per cubic centimeter, are given in Table 18-1. Densities vary somewhat with temperature and pressure, so except for water, densities are given at 0°C and atmospheric pressure. Note that water at 4°C has a density of 1.00 g/cm³. The gram is now defined as the mass of a cubic centimeter of water at a temperature of 4°C. Gold, which has a density of 19.3 g/cm³, is 19.3 times more massive than an equal volume of water.

Table 18-1 Densities of a few substances			
Material	Density (in g/cm³)	Material	Density (in g/cm³)
Solids		Liquids	
Osmium	22.6	Mercury	13.6
Platinum	21.4	Glycerin	1.26
Gold	19.3	Sea water	1.03
Uranium	19.0	Water at 4°C	1.00
Lead	11.3	Benzene	0.90
Silver	10.5	Ethyl alcohol	0.81
Copper	8.9		
Brass	8.6		
Iron	7.8		
Steel	7.8		
Tin	7.3		
Diamond	3.5		
Aluminum	2.7		
Graphite	2.25		
Ice	0.92		
Pine wood	0.50		
Balsa wood	0.12		

A quantity known as **weight density** can be expressed by the amount of *weight* a body has compared to its volume:

$$\text{weight density} = \frac{\text{weight}}{\text{volume}}$$

Weight density is commonly used when discussing liquid pressure (see next chapter).*

---

* Weight density is common to British units, where one cubic foot of fresh water (almost 7.5 gallons) weighs 62.4 pounds. So fresh water has a weight density of 62.4 lb/ft³. Salt water is a bit denser, 64 lb/ft³.

The Value of a Simple Computation

One of the reasons gold was used as money is that it is nearly the most dense of all substances and could therefore be easily identified. A merchant who suspected that gold was diluted with a less valuable substance had only to compute its density by measuring its mass and dividing by its volume. The merchant would then compare this value to the density of gold, 19.3 g/cm³.

Consider a gold nugget with a mass of 57.9 g that is measured to have a volume of exactly 3 cm³. Is the nugget pure gold? We compute its density:

$$\text{density} = \frac{\text{mass}}{\text{volume}} = \frac{57.9 \text{ g}}{3 \text{ cm}^3} = 19.3 \text{ g/cm}^3$$

Its density matches that of gold, so the nugget can be presumed to be pure gold. (It is possible to get the same density by mixing gold with a platinum alloy, but this is unlikely since platinum is several times more valuable than gold.)

18.3 Elasticity

When you hang a weight on a spring, the spring stretches. When you add additional weights, the spring stretches still more. When you remove the weights, the spring returns to its original length. We say the spring is **elastic**.

▶ **Answers**

1. The density of *any* amount of water (at 4°C) is 1.00 g/cm³.

2. Any amount of lead always has a greater density than any amount of aluminum. The amount of material is irrelevant.

3. *Any* amount of uranium is more dense than the earth. The density of the earth is actually 5.5 g/cm³, much less than the density of uranium (19.0 g/cm³).

When a batter hits a baseball, he temporarily changes its shape. When an archer shoots an arrow, she first bends the bow, which springs back to its original form when the arrow is released. The spring, the baseball, and the bow are examples of elastic objects. **Elasticity** is that property of a body by which it experiences a change in shape when a deforming force acts on it and by which it returns to its original shape when the deforming force is removed.

Not all materials return to their original shape when a deforming force is applied and then removed. Materials that do not resume their original shape after being distorted are said to be **inelastic**. Clay, putty, and dough are inelastic materials. Lead is also inelastic, since it is easy to distort it permanently.

By hanging a weight on a spring, you are applying a force to the spring. It is found that the stretch or compression is directly proportional to the applied force (Figure 18-6).

Fig. 18-5   The bow is elastic. When the deforming force is removed, it returns to its original shape.

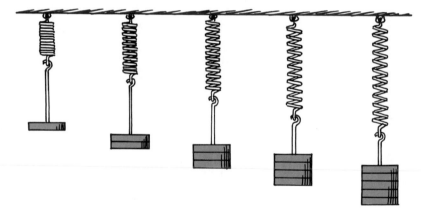

Fig. 18-6   The stretch of the spring is directly proportional to the applied force. If the weight is doubled, the spring stretches twice as much.

This relationship was noted by the British physicist Robert Hooke, a contemporary of Isaac Newton, in the midseventeenth century. It is called **Hooke's law**. The amount of stretch or compression, x, is directly proportional to the applied force F. In shorthand notation,

$$F \sim x$$

If an elastic material is stretched or compressed beyond a certain amount, it will not return to its original state. Instead, it will remain distorted. The distance beyond which permanent distortion occurs is called the **elastic limit**. Hooke's law holds only as long as the force does not stretch or compress the material beyond its elastic limit.

▶ **Questions**

1. When a 20-kg load is hung from the end of a tree branch, the branch is observed to sag a distance of 10 cm. If, instead, a 40-kg load is hung from the same place, by how much will the branch sag? How about if a 60-kg load were hung from the same place? (Assume that none of these loads makes the branch sag beyond its elastic limit.)

2. If a force of 10 N stretches a certain spring 4 cm, how much stretch will occur for an applied force of 15 N?

| 18.4 | **Compression and Stretching** |

Steel is an excellent elastic material. It can be stretched and it can be compressed. Because of its strength and elastic properties, it is used to make not only springs but construction girders as well. Vertical girders of steel used in the construction of tall buildings undergo only slight compression. A typical 25-meter-long vertical girder used in high-rise construction is compressed about a millimeter when it carries a 10-ton load. Most deformation occurs when girders are used horizontally, where the tendency is to sag under heavy loads.

A horizontal beam supported at one or both ends is under stress from the load it supports, including its own weight. It undergoes a stress of both compression and stretching. Consider the beam supported at one end in Figure 18-7. It sags because of its own weight and because of the load it carries at its end.

▶ **Answers**

1. A 40-kg load has twice the weight of a 20-kg load. In accord with Hooke's law, $F \sim x$, two times the applied force will result in two times the stretch, so the branch should sag 20 cm. The weight of the 60-kg load will make the branch sag 3 times as much, or 30 cm. (When the elastic limit is exceeded, then the amount of sag cannot be predicted with the information given.)

2. The spring will stretch 6 cm. By ratio and proportion,

$$\frac{10 \text{ N}}{4 \text{ cm}} = \frac{15 \text{ N}}{x}$$

which is read "10 N is to 4 cm as 15 N is to x." Solving for x gives $x = (15 \text{ N}) \times (4 \text{ cm})/(10 \text{ N}) = 6$ cm. In lab you will learn that the ratio of force to stretch is called the *spring constant k* (in this case $k = 2.5$ N/cm), and Hooke's law is expressed as the equation $F = kx$.

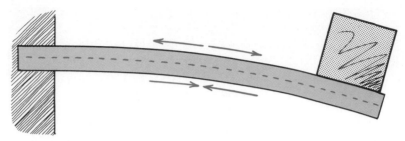

Fig. 18-7   The top part of the beam is stretched and the bottom part is compressed. What happens in the middle portion, between top and bottom?

Can you see that the top part of the beam tends to be stretched? Molecules tend to be pulled apart. The top part is slightly longer because of its deformation. And can you see that the bottom part of the beam is compressed? Molecules there are squeezed. The bottom part is slightly shorter because of the way it is bent. So the top part is stretched, and the bottom part of the beam is compressed. A little thought will show that somewhere in between the top and bottom, there will be a region where the two effects overlap, where there is neither compression nor stretching. This is the *neutral layer*.

Consider the beam shown in Figure 18-8. It is supported at both ends, and carrying a load in the middle. This time the top of the beam is compressed and the bottom is stretched. Again, there is a neutral layer along the middle portion of the length of the beam.

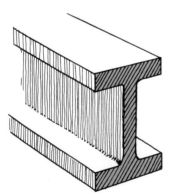

Fig. 18-9   An I-beam is like a solid bar with some of the steel scooped from its middle where it is needed least. The beam is therefore lighter for nearly the same strength.

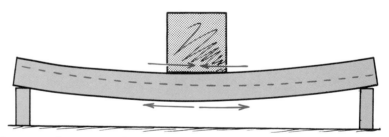

Fig. 18-8   The top part of the beam is compressed and the bottom part is stretched. Where is the neutral layer (the part that is not under stress due to compression or stretching)?

Have you ever wondered why the cross-section of steel girders has the form of the letter *I* (Figure 18-9)? Most of the material in these I-beams is concentrated in the top and bottom bars, whereas the piece joining the bars is of thinner construction. Why is this the shape?

The answer is that the stress is predominantly in the top and bottom bars when the beam is used horizontally in construction. One bar tends to be stretched while the other tends to be compressed. Between the top and bottom bars is a stress-free region that acts principally to connect the top and bottom bars together. This is the neutral layer, where comparatively little material is needed. An I-beam is nearly as strong as if it were a solid bar, and its weight is considerably less.

▶ **Question**

If you had to make a hole horizontally through the tree branch shown, in a location that would weaken it the least, would you bore it through the top, the middle, or the bottom?

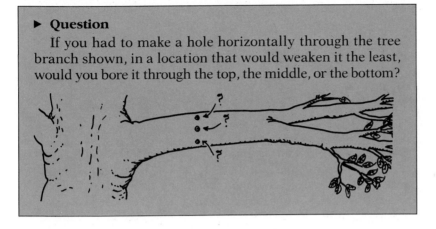

18.5 | Scaling

Did you ever notice how strong an ant is for its size? An ant can carry the weight of several ants on its back, whereas a strong elephant couldn't even carry one elephant on its back. How strong would an ant be if it were scaled up to the size of an elephant? Would this "super ant" be several times stronger than an elephant? Surprisingly, the answer is no. Such an ant would not be able to lift its own weight off the ground. Its legs would be too thin for its greater weight and would likely break.

Ants have thin legs and elephants have thick legs for a reason. The proportions of things in nature are in accord with their size. The study of how size affects the relationship between weight, strength, and surface area is known as **scaling**. As the size of

---

▶ **Answer**

It would be best to drill the hole in the middle, through the neutral layer. Wood fibers in the top part of the branch are being stretched, and if you drilled the hole there, that part of the branch may pull apart. Fibers in the lower part are being compressed, and if you drilled the hole there, that part of the branch might crush under compression. In between, in the neutral layer, the hole will not affect the strength of the branch because fibers there are being neither stretched nor compressed.

a thing increases, it grows heavier much faster than it grows stronger. You can support a toothpick horizontally at its ends and you'll notice no sag. But support a tree of the same kind of wood horizontally at its ends and you'll see a noticeable sag. The tree is much heavier compared to its strength than the toothpick.

Weight depends on volume, and strength comes from the area of the cross-section. To understand this weight-strength relationship, consider a very simple case—a solid cube of matter, 1 centimeter on a side.

A 1-cubic centimeter cube has a cross section of 1 square centimeter. That is, if we sliced through the cube parallel to one of its faces, the sliced area would be 1 square centimeter. Compare this to a cube that has double the linear dimensions, a cube 2 centimeters on each side. Its cross-sectional area will be 2 × 2 (or 4) square centimeters and its volume will be 2 × 2 × 2 (or 8) cubic centimeters. If it has the same density it would be 8 times heavier. Careful investigation of Figure 18-10 shows that for increases of linear dimensions the cross-sectional area (as well as the total area) grows as the square of the increase, whereas volume and weight grow as the cube of the increase.

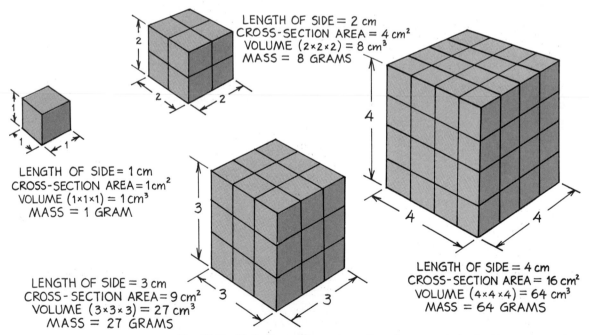

LENGTH OF SIDE = 2 cm
CROSS-SECTION AREA = 4 cm²
VOLUME (2×2×2) = 8 cm³
MASS = 8 GRAMS

LENGTH OF SIDE = 1 cm
CROSS-SECTION AREA = 1 cm²
VOLUME (1×1×1) = 1 cm³
MASS = 1 GRAM

LENGTH OF SIDE = 3 cm
CROSS-SECTION AREA = 9 cm²
VOLUME (3×3×3) = 27 cm³
MASS = 27 GRAMS

LENGTH OF SIDE = 4 cm
CROSS-SECTION AREA = 16 cm²
VOLUME (4×4×4) = 64 cm³
MASS = 64 GRAMS

Fig. 18-10 If the linear dimensions of an object are multiplied by some number, then the area grows by the square of the number, and the volume (and hence the weight) grows by the cube of the number. If the linear dimensions of the cube are increased by 2, the area grows by $2^2 = 4$, and the volume grows by $2^3 = 8$. If the linear dimensions are increased by 3, the area grows by $3^2 = 9$, and the volume grows by $3^3 = 27$.

The volume (and weight) multiplies much more than the corresponding increase of cross-sectional area. Although the figure demonstrates the simple example of a cube, the principle applies to an object of any shape. Consider an athlete who can lift his weight with one arm. Suppose he could somehow be scaled up to twice his size—that is, twice as tall, twice as broad, his bones twice as thick and every linear dimension increased by a factor of 2. Would he be twice as strong? Would he be able to lift himself with twice the ease? The answer to both questions is no. Since his twice-as-thick arms would have 4 times the cross-sectional area, he would be 4 times as strong. At the same time, his volume would be 8 times as great, so he would be 8 times as heavy. Thus, he could lift only half his weight. *In relation to his weight*, he would be weaker than before.

The fact that volume (and weight) grow as the cube of the increase, while strength (and area) grow as the square of the increase is evident in the disproportionally thick legs of large animals compared to small animals. Consider the different legs of an elephant and a deer; or a tarantula and a daddy longlegs.

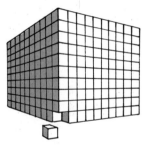

► **Questions**
1. Suppose a cube 1 cm long on each side were scaled up to a cube 10 cm long on each edge. What would be the volume of the scaled-up cube? What would be its cross-sectional surface area? Its total surface area?

2. If an athlete were somehow scaled up proportionally to twice size, would he be stronger or weaker?

So the great strengths attributed to King Kong and other fictional giants cannot be taken seriously. The fact that the consequences of scaling are conveniently omitted is one of the differences between science and science fiction.

► **Answers**
1. The volume of the scaled-up cube would be (length of side)³ = (10 cm)³, or 1000 cm³. Its cross-sectional surface area would be (length of side)² = (10 cm)², or 100 cm². Its total surface area = 6 sides × (length of side)² = 600 cm².

2. The scaled-up athlete would be 4 times as strong, because the cross-sectional area of his twice-as-thick bones and muscles increases by a factor of 4. He could lift a maximum load 4 times as heavy as before. But his own weight is 8 times as much as before, so he would be weaker in relation to his weight. Having 4 times the strength while carrying 8 times the weight gives him a strength-to-weight ratio of only half its former value. This means that if he could just lift his own weight before, he could now lift only half his new weight. In sum, while his actual strength would increase, his strength-to-weight ratio would decrease.

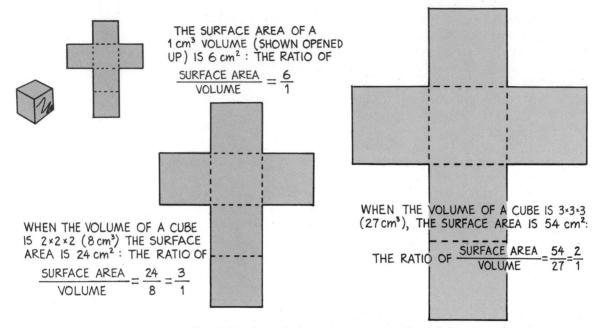

THE SURFACE AREA OF A
1 cm³ VOLUME (SHOWN OPENED
UP) IS 6 cm² : THE RATIO OF

$$\frac{\text{SURFACE AREA}}{\text{VOLUME}} = \frac{6}{1}$$

WHEN THE VOLUME OF A CUBE
IS 2×2×2 (8 cm³) THE SURFACE
AREA IS 24 cm² : THE RATIO OF

$$\frac{\text{SURFACE AREA}}{\text{VOLUME}} = \frac{24}{8} = \frac{3}{1}$$

WHEN THE VOLUME OF A CUBE IS 3×3×3
(27 cm³), THE SURFACE AREA IS 54 cm²:

THE RATIO OF $\dfrac{\text{SURFACE AREA}}{\text{VOLUME}} = \dfrac{54}{27} = \dfrac{2}{1}$

Fig. 18-11   As an object grows proportionally in all directions, there is a greater increase in volume than in surface area. As a result, the ratio of surface area to volume decreases.

Fig. 18-12   The African elephant has less surface area compared to its weight than other animals. It compensates for this with its large ears, which significantly increase its radiating surface area and promote cooling.

Important also is the comparison of total surface area to volume. A study of Figure 18-11 shows that as the linear size of an object increases, the volume grows faster than the total surface area. (Volume grows as the cube of the increase, and both cross-sectional area and total surface area grow as the square of the increase.) So as an object grows, its surface area and volume grow at different rates, with the result that the surface area to volume ratio *decreases*. In other words, both the surface area and the volume of a growing object increase, but the growth of surface area *compared to* the growth of volume decreases. Relatively few people really understand this idea. The following examples may be helpful.

The big ears of elephants are not for better hearing, but for cooling. They are nature's way of making up for the small ratio of surface area to volume for these large animals. The heat that an animal radiates is proportional to its surface area. If an elephant did not have large ears, it would not have enough surface area to cool its huge mass. The large ears of the African elephant greatly increase overall surface area, and enable it to cool off in hot climates.

At the biological level, living cells must contend with the fact that the growth of volume is faster than the growth of surface area. Cells obtain nourishment by diffusion through their surfaces. As cells grow, surface area increases, but not fast enough to keep up with volume. For example, if the surface area increases by four, the corresponding volume increases by eight. Eight times as much mass must be sustained by only four times as much nourishment. This puts a limit on the growth of a living cell. So cells divide, and there is life as we know it. That's nice.

Not so nice is the fate of large animals when they fall. The statement "the bigger they are, the harder they fall" holds true and is a consequence of the small ratio of surface area to weight. Air resistance to movement through the air is proportional to the surface area of the moving object. If you fall out of a tree, even in the presence of air resistance, your speed increases at the rate of very nearly $1g$. You don't have enough surface area compared to your weight—unless you wear a parachute. Small animals, on the other hand, need no parachute. They have plenty of surface area compared to their small weights. An insect can fall from the top of a tree to the ground below without harm. The surface area per weight ratio is in the insect's favor—in a sense, the insect *is* its own parachute.

The fact that small things have more surface area compared to volume, mass, or weight is evident in the kitchen. An experienced cook knows that more skin results when peeling 5 kg of small potatoes than when peeling 5 kg of large potatoes. Smaller objects have more surface area per kilogram. Crushed ice will cool a drink much faster than a single ice cube of the same mass, because crushed ice presents more surface area to the beverage.

The rusting of iron is also a surface phenomenon. Iron rusts when exposed to air, but it rusts much faster and is soon eaten away if it is in the form of small filings.

Chunks of coal burn, while coal dust explodes when ignited. Thin french fries cook faster in oil than fat fries. Flat hamburgers cook faster than meatballs of the same mass. Large raindrops fall faster than small raindrops, and large fish swim faster than small fish. These are all consequences of the fact that differences in volume and differences in area are not in the same proportion to each other.

It is interesting to note that the rate of heartbeat in a mammal is proportional to the size of the mammal. The heart of a tiny shrew beats about twenty times as fast as the heart of an elephant. In general, small mammals live fast and die young; larger animals live at a leisurely pace and live longer. Don't feel bad about a pet hamster that doesn't live as long as a dog. All warmblooded animals have about the same lifespan—not in terms of years, but in the average number of heartbeats (about 800 million). Humans are the exception: we live two to three times longer than other mammals of our size.

18 | Chapter Review

Concept Summary

Many solids are made of crystals.
- The atoms in a crystal are in an orderly arrangement.

One property of solids is density, the amount of mass per unit volume.
- Density is related both to the masses of the atoms and the spacing between atoms.
- Weight density is the amount of weight per unit volume.

A second property of solids is elasticity.
- Elastic materials return to their original shape when a deforming force is applied and removed, as long as they are not deformed beyond their elastic limit.
- According to Hooke's law, the amount of stretch or compression is proportional to the applied force (within the elastic limit.)
- Inelastic materials remain distorted after the force is removed.

Scaling is the study of how size affects the relationship between weight, strength, and surface.

Important Terms

crystal (18.1)
density (18.2)
elastic (18.3)
elastic limit (18.3)
elasticity (18.3)
Hooke's law (18.3)
inelastic (18.3)
scaling (18.5)
weight density (18.2)

Review Questions

1. How does the arrangement of atoms differ in a crystalline and noncrystalline substance? (18.1)

2. What evidence can you cite for the microscopic crystal nature of certain solids? (18.1)

3. What evidence can you cite for the visible crystal nature of certain solids? (18.1)

4. What happens to the density of a uniform piece of wood when you cut it in half? (18.2)

5. Uranium is the heaviest atom. Why is uranium metal not the most dense material? (18.2)

6. Which has the greater density—a heavy bar of pure gold or a pure gold ring? (18.2)

7. Does the *mass* of a loaf of bread change when you squeeze it? Does its *volume* change? Does its *density* change? (18.2)

8. What is the difference between mass density and weight density? (18.2)

9. a. What is the evidence for the claim that steel is elastic?
   b. That putty is inelastic? (18.3)

10. What is Hooke's law? (18.3)

11. What is an elastic limit? (18.3)

12. If the weight of a 2-kg mass stretches a spring by 3 cm, how much will the spring stretch when it supports 6 kg? (Assume the spring has not reached its elastic limit.) (18.3)

13. Is a steel beam slightly shorter when it stands vertically? (18.4)

14. What is the neutral layer in a beam that supports a load? (18.4)

15. Why are the cross sections of metal beams I-shaped, not rectangular? (18.4)

16. What is the weight-strength relationship in scaling? (18.5)

17. a. If the linear dimensions of an object are doubled, by how much does the overall area increase?
    b. By how much does the volume increase? (18.5)

18. True or false: As the volume of an object is increased, its surface area also increases, but the *ratio* of its surface area to volume decreases. Explain. (18.5)

19. Which will cool a drink faster—a 10-gram ice cube or 10 grams of crushed ice? (18.5)

20. a. Which has more skin—an elephant or a mouse?
    b. Which has more skin *per body weight*—an elephant or a mouse? (18.5)

Activity

If you nail 4 sticks together to form a rectangle, they can be deformed into a parallelogram without too much effort. But if you nail 3 sticks together to form a triangle, the shape cannot be changed without actually breaking the sticks or the nails. The triangle is the strongest of all the geometrical figures. Try it and see, and then look at the triangles used in strengthening structures of many kinds.

Think and Explain

1. Which has the greater volume—a kilogram of lead or a kilogram of aluminum?

2. Which has the greater weight—a liter of ice or a liter of water?

3. A certain spring stretches 1 cm for each kilogram it supports. If the elastic limit is not reached, how far will it stretch when it supports a load of 8 kg?

4. Suppose the spring in the preceding question is placed next to an identical spring so that both side-by-side springs support the 8-kg load. By how much will each spring stretch?

5. Compression and stretching stresses occur in a beam that supports loads (even if the load is its own weight). Show by means of a simple sketch how a horizontal load-carrying beam is stretched at the top and compressed at the bottom. Then show a case where the opposite occurs: compression at the top and stretching at the bottom.

6. Metal beams are not "solid" like wooden beams, but are "cut out" in the middle so that their cross section has an I-shape. What are the advantages of this shape?

7. Consider a model steel bridge that is 1/100 the exact scale of the real bridge that is to be built.
   a. If the model bridge weighs 50 N, what will the real bridge weigh?
   b. If the model bridge doesn't appear to sag under its own weight, is this evidence that the real bridge, if built exactly to scale, will not sag either? Explain.

8. If you use a batch of cake batter for cupcakes instead of a cake and bake them for the time suggested for baking a cake, what will be the result?

9. Explain, in terms of scaling, why it is an advantage that natives of the hot African desert tend to be relatively tall and slender, and natives of the Arctic region tend to be short and stout. (*Hint*: A piece of wire will cool faster when stretched out than when rolled up into a ball.)

10. Nourishment is obtained from food through the inner surface area of the intestines. Why is it that a small organism, such as a worm, has a simple and relatively straight intestinal tract, while a large organism, such as a human being, has a complex and many-folded intestinal tract?

19 | Liquids

We live on the only planet in the solar system covered predominantly by a liquid. The earth's oceans are made of H_2O in the liquid state. If the earth were a little closer to the sun, the oceans would turn to vapor. If the earth were a little farther, its surface would be solid ice. It's nice that the earth is where it is.

In the liquid state, molecules can flow. They freely move from position to position by sliding over one another. The shape of a liquid takes the shape of its container.

19.1 | Pressure in a Liquid

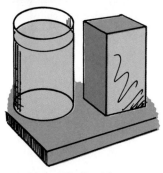

Fig. 19-1 The liquid exerts a pressure against the bottom of its container, just as the block exerts a pressure against the table.

A liquid in a container exerts forces against the walls and bottom of the container. To investigate the interaction between the liquid and the walls, it is useful to discuss the concept of *pressure*. Recall from Chapter 4 that pressure is defined as the force per area on which the force acts.*

$$\text{pressure} = \frac{\text{force}}{\text{area}}$$

The pressure that a block exerts against a table is simply the weight of the block divided by its area of contact. Similarly, for a liquid in a container, the pressure the liquid exerts against the

---

* Pressure may be measured in any unit of force divided by any unit of area. The standard international (SI) unit of pressure, the newton per square meter, is called the pascal (Pa), named after the seventeenth century theologian and scientist Blaise Pascal. A pressure of 1 Pa is very small, approximately the pressure exerted by a dollar bill resting flat on a table. Science types more often use kilopascals (1 kPa = 1000 Pa).

bottom of the container is the weight of the liquid divided by the area of the container bottom.

How much a liquid weighs, and hence how much pressure it exerts, depends on its density. Consider two identical containers, one filled with mercury and the other filled to the same depth with water. For the same depth, the denser liquid exerts the greater pressure. Mercury is 13.6 times as dense as water. So for the same volume of liquid, the weight of mercury is 13.6 times the weight of water. Thus, the pressure of the mercury on the bottom is 13.6 times the pressure of the water.

For liquids of the same density, the pressure will be greater at the bottom of the deeper liquid. Consider the two containers in Figure 19-2. If the liquid in the first container is twice as deep as the liquid in the second container, then like two blocks one atop the other, the pressure of the liquid at the bottom of the first container will be twice that of the second container.

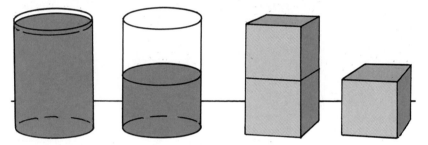

Fig. 19-2   The two blocks exert twice as much pressure as one block against the table. Similarly, the liquid in the first container is twice as deep, so the pressure it exerts on the bottom is twice that exerted by the liquid in the second container.

It turns out that the pressure of a liquid at rest depends only on the density and the depth of the liquid. Liquids are practically incompressible, so except for changes in temperature, the density of a liquid is normally the same at all depths. The pressure of a liquid* is

$$\text{pressure} = \text{weight density} \times \text{depth}.$$

---

*   This relationship is derived from the definitions of pressure and density. Consider an area at the bottom of a container of liquid. Pressure is produced by the weight of the column of liquid directly above this area. From the definition of weight density as weight divided by volume, this weight of liquid can be expressed as weight density times volume. The volume of the column is simply the area multiplied by the depth. Then we get

$$\text{pressure} = \frac{\text{force}}{\text{area}} = \frac{\text{weight}}{\text{area}} = \frac{\text{weight density} \times \text{volume}}{\text{area}}$$

$$= \frac{\text{weight density} \times \text{area} \times \text{depth}}{\text{area}} = \text{weight density} \times \text{depth}$$

At a given depth, a given liquid exerts the same pressure against *any* surface—the bottom *or* sides of its container, or even the surface of an object submerged in the liquid to that depth. The pressure a liquid exerts depends only on the density and depth of the liquid. At twice the depth, the pressure against any surface is twice as great; at three times the depth, pressure is threefold, and so on. Or if the liquid is twice or three times as dense, pressure is correspondingly twice or three times as great for any given depth.

Interestingly enough, the pressure does not depend on the amount of liquid. Neither the volume nor even the total weight of liquid matters. For example, if you sampled water pressure at one meter beneath a large lake surface and one meter beneath a small pool surface, the pressures would be the same.* The dam that must withstand the greater pressure is the dam with the *deepest* water behind it, not the *most* water (Figure 19-3).

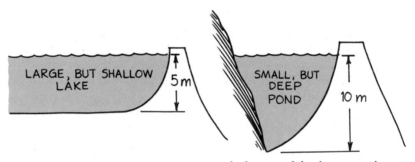

Fig. 19-3   The water pressure is greater at the bottom of the deeper pond, not necessarily the pond with the most water. The dam holding back water twice as deep must withstand twice the average water pressure, regardless of the total volume of water.

The fact that water pressure depends on depth and not on volume is nicely illustrated with "Pascal's vases" (Figure 19-4). Note that the water surface in each of the connected vases is at the same level. This occurs because the pressures at equal depths

---

*   The density of fresh water is 1 gram per cubic centimeter, which is the same as 1000 kilograms per cubic meter. Since the weight (*mg*) of 1000 kilograms is (1000 kg) × (9.8 N/kg) = 9800 N, the weight density of water is 9800 newtons per cubic meter (9800 N/m³). (Seawater is slightly denser than fresh water; its weight density is about 10 000 N/m³.) Water pressure in a pool or lake is simply equal to the weight density of fresh water multiplied by the depth in meters. For example, water pressure is 9800 N/m² at a depth of 1 m , and 98 000 N/m² at a depth of 10 m. In SI units, pressure is measured in pascals (1 Pa = 1 N/m²), so this would be 9800 Pa and 98 000 Pa, respectively; or in kilopascals, 9.8 kPa and 98 kPa, respectively.

beneath the surfaces are the same. At the bottom of each vase, for example, the pressures are equal. If they were not, the liquid would flow until the pressures were equalized. This is the reason that water seeks its own level.

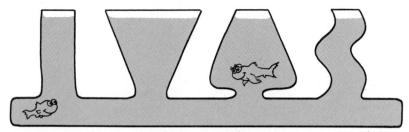

Fig. 19-4   The pressure of the liquid is the same at any given depth below the surface, regardless of the shape of the container.

▶ **Questions**

1. Is there more pressure at the bottom of a bathtub of water 30 cm deep or at the bottom of a pitcher of water 35 cm deep?

2. A brick mason wishes to mark the back of a building at the exact height of bricks already layed at the front of the building. How can he measure the same height using only a transparent garden hose and water?

At any point within a liquid, the forces that produce pressure are exerted equally in all directions. For example, when you are swimming under water, no matter which way you tilt your head, you feel the same amount of water pressure on your ears.

When the liquid is pressing against a surface, there is a net force directed perpendicular to the surface (Figure 19-5 top). If there is a hole in the surface, the liquid initially moves perpendicular to the surface (Fig. 19-5 bottom). Gravity, of course, causes the path of the liquid to curve downward. At greater depths, the net force is greater and the horizontal velocity of the escaping liquid is greater.

Fig. 19-5   (Top) The forces in a liquid that produce pressure against a surface add up to a net force that is perpendicular to the surface. (Bottom) Liquid escaping through a hole initially moves perpendicular to the surface.

▶ **Answers**

1. There is more pressure at the bottom of the pitcher, because the water is deeper in it. The fact that there is more water in the bathtub does not matter.

2. To measure the same height, the brick mason can extend a garden hose from the front to the back of the house, and fill it with water until the water level reaches the height of bricks in the front. Since water seeks its own level, the level of water in the other end of the hose will be the same!

19.2 | Buoyancy

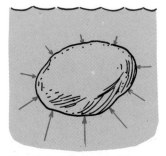

Fig. 19-6 The upward forces against the bottom of a submerged object are greater than the downward forces against the top. There is a net upward force, the buoyant force.

If you have ever lifted a submerged object out of water, you are familiar with **buoyancy**, the apparent loss of weight of objects when submerged in a liquid. It is a lot easier to lift a boulder submerged on the bottom of a river bed than to lift it above the water surface. The reason is that when the boulder is submerged, the water exerts an upward force that is opposite in direction to gravity. This upward force is called the **buoyant force**.

To understand where the buoyant force comes from, look at Figure 19-6. The arrows represent the forces by the liquid that produce pressure against the submerged boulder. The forces are greater at greater depth. The forces acting horizontally against the sides cancel out, so the boulder is not nudged sideways. But the forces acting upward against the bottom are greater than those acting downward against the top. This is simply because the bottom of the boulder is deeper within the water. The difference in upward and downward forces is the buoyant force.

If the weight of a submerged object is greater than the buoyant force, the object will sink. If the weight is equal to the buoyant force, the submerged object will remain at any level, like a fish. If the weight is less than the buoyant force, the object will rise to the surface and float.

To further understand buoyancy, it helps to think some more about what happens when an object is placed in water. If a stone is placed in a container of water, the water level will rise (Figure 19-7). Water is said to be **displaced**, or moved elsewhere, by the stone. A little thought will tell us that the volume—that is, the amount of space taken up or the number of cubic centimeters—of water displaced is equal to the volume of the stone. *A completely submerged object always displaces a volume of liquid equal to its own volume.*

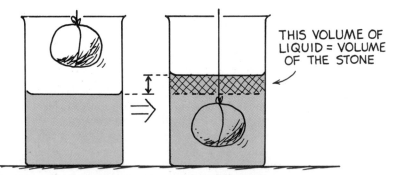

THIS VOLUME OF LIQUID = VOLUME OF THE STONE

Fig. 19-7 When an object is submerged, it displaces a volume of water equal to the volume of the object itself.

This gives us a good way to determine the volume of an irregularly shaped object. Simply submerge it in water in a measuring cup and note the increase in volume of the water. That increase in volume is also the volume of the submerged object.

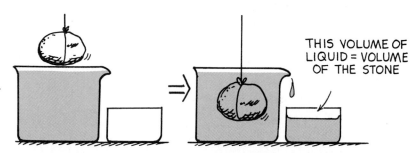

Fig. 19-8 When an object is submerged in a container that is initially brim full, the volume of water that overflows is equal to the volume of the object itself.

19.3 Archimedes' Principle

The relationship between buoyancy and displaced liquid was discovered in ancient times by the Greek philosopher Archimedes (third century B.C.). It is stated as follows:

> An immersed object is buoyed up by a force equal to the weight of the fluid it displaces.

This relationship is called **Archimedes' principle**. It is true of liquids and gases, which are both fluids.

The word *immersed* means *either completely or partially submerged*. For example, if we immerse a sealed one-liter container halfway into water, it will displace a half liter of water and be buoyed up by the weight of a half liter of water. If we immerse it all the way (submerge it), it will be buoyed up by the weight of a full liter of water (9.8 newtons). Unless the completely submerged container is compressed, the buoyant force will equal the weight of one liter of water at any depth. This is because it will displace the same volume of water, and hence same weight of water, at all depths. The weight of this displaced water (not the weight of the submerged object!) is the buoyant force.

A 300-gram block weighs about 3 N in air. Suppose the block displaces 2 N of water when it is submerged (Figure 19-10). The buoyant force on the submerged block will also equal 2 N. The block will seem to weigh less under water than above water. In the water, its apparent weight will be 3 N minus the buoyant force of 2 N, or only 1 N. The apparent weight of a submerged object is its weight in air minus the buoyant force.

Fig. 19-9 A liter of water occupies 1000 cubic centimeters, has a mass of one kilogram, and weighs 9.8 N. Any object with a volume of one liter will experience a buoyant force of 9.8 N when fully submerged in water.

Fig. 19-10 A block weighs less in water than in air. The buoyant force that acts on the submerged block is equal to the weight of the water displaced. So the block appears lighter under water by an amount equal to the weight of water (2 N) that has spilled into the smaller container. The apparent weight of the block under water equals its weight in air minus the buoyant force (3 N − 2 N = 1 N).

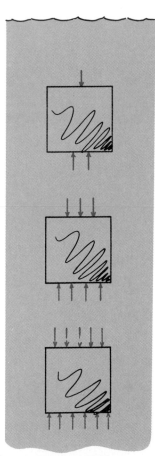

Fig. 19-11 The difference in the upward and downward force acting on the submerged block is the same at any depth.

▶ **Questions**

1. A 1-liter (L) container completely filled with mercury has a mass of 13.6 kg and weighs 133.3 N. If it is submerged in water, what is the buoyant force acting on it?

2. A block is held suspended beneath the water in the three positions, A, B, and C, shown in Figure 19-11. At what position is the buoyant force on it greatest?

3. A stone is thrown into a deep lake. As it sinks deeper and deeper into the water, does the buoyant force on it increase? Decrease? Remain unchanged?

▶ **Answers**

1. The buoyant force is equal to the weight of 1 L of water—about 10 N—because the *volume* of water displaced is 1 L. The mass or weight of mercury is irrelevant; 1 L of *anything* submerged in water will displace 1 L of water and be buoyed upward with a force of 10 N. Or look at it this way: when the container is put in the water, it pushes 1 L of water out of the way by its presence. The water, in action and reaction form, pushes back on the container with a force equal to the weight of 1 L of water. If the container had twice the volume, it would displace 2 L of water and be buoyed up with a force equal to the weight of 2 L of water. (The concepts involved with buoyant force are best dealt with if you *visualize* what is going on—*conceptual physics!*)

2. The buoyant force will be the same at all three positions. Why? Because the amount of water displaced will be the same in each case. The buoyant force will equal the weight of the water displaced, which is the same at positions A, B, and C.

3. Like the previous question, the volume of water displaced is the same at any depth. Water is practically incompressible, so its density is the same at all depths, and equal volumes of water will weigh the same. The buoyant force on the sinking stone will be the same at all depths.

Perhaps your teacher will summarize Archimedes' principle by way of a numerical example to show that the difference between the upward acting and the downward acting forces due to water pressure on a submerged block is numerically identical to the weight of liquid displaced. It makes no difference how deep the block is submerged, for although the pressures are greater with increasing depth, the *difference* in pressures on the bottom and top of the block is the same at any depth (Figure 19-11). Whatever the shape of a submerged object, the buoyant force is equal to the weight of liquid displaced.

19.4 | The Effect of Density on Submerged Objects

You have learned that the buoyant force that acts on a submerged object depends on the volume of the object. Small objects displace small amounts of water and are acted on by small buoyant forces. Large objects displace large amounts of water and are acted on by larger buoyant forces. It is the *volume* of the submerged object—not its *weight*—that determines buoyant force. (A misunderstanding of this idea is at the root of a lot of confusion that you or your friends may have about buoyancy!)

Thus far, the weight of the submerged object has not been taken into account. Now we will consider its role.

Whether an object will sink or float in a liquid has to do with how great the buoyant force is *compared to the object's weight*. Careful thought will show that if the buoyant force is exactly equal to the weight of a completely submerged object, then the weight of the object must be the same as the weight of the water displaced. Since the volumes of the object and of the displaced water are the same, the density of the object must equal the density of water.

This is true for a fish, which has a density equal to the density of water. The fish is "at one" with the water—it doesn't sink and it doesn't float. If the fish were somehow bloated up, it would be less dense than water and it would float to the top. If the fish swallowed a stone and became more dense than water, it would sink to the bottom.

This can be summed up in three simple rules.

1. If an object is denser than the fluid in which it is immersed, it will sink.
2. If an object is less dense than the fluid in which it is immersed, it will float.
3. If an object has a density equal to the density of the fluid in which it is immersed, it will neither sink nor float.

Fig. 19-12 The wood floats because it is less dense than water. The rock sinks because it is more dense than water. The fish does neither because it has the same density as water.

From these rules, what can you say about people who, try as they may, cannot float?* They're simply too dense! To float more easily, you must reduce your density. Since weight density is weight divided by volume, you must either reduce your weight or increase your volume. The purpose of a life jacket is to increase volume while correspondingly adding very little to your weight.

The density of a submarine is controlled by taking water into and out of its ballast tanks. In this way the weight of the submarine can be varied to achieve the desired density. A fish regulates its density by expanding and contracting an air sac that changes its volume. The fish can move upward by increasing its volume (which decreases its density) and downward by contracting its volume (which increases its density). The overall density of a crocodile is increased when it swallows stones. From 4 to 5 kg of stones have been found lodged in the front part of the stomach in large crocodiles. When its density is increased, the crocodile can swim lower in the water, thus exposing less of itself to its prey.

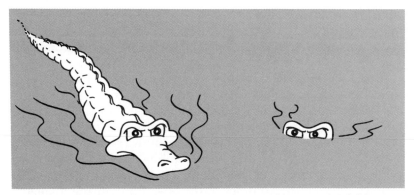

Fig. 19-13   (Left) A crocodile coming toward you in the water. (Right) A crocodile with a bellyful of stones coming toward you in the water.

> ▶ **Question**
>   We know that if a fish makes itself more dense, it will sink; if it makes itself less dense, it will rise. In terms of buoyant force, why is this so?

▶ **Answer**
    When the fish increases its density by decreasing its volume, it displaces less water, so the buoyant force decreases. When the fish decreases its density by expanding, it displaces a greater volume of water, and the buoyant force increases.

---

*   Interestingly enough, the people who can't float are, 9 times out of 10, males. Most males are more muscular and slightly denser than females.

19.5 | Flotation

If you had lived a few hundred years ago and said you were going to make an iron ship, people would have thought you were crazy. They believed that a ship made of iron would sink, and that a ship should be made of a material that itself floats, such as wood. Today it is easy to see how a ship made of iron can float.

Consider a solid 1-ton block of iron. Iron is nearly eight times as dense as water, so when it is submerged, it will displace only $\frac{1}{8}$ ton of water. This is not enough to keep it from sinking. Suppose we reshape the same iron block into a bowl shape, as shown in Figure 19-14. It still weighs 1 ton. If you lower the bowl into a body of water, it displaces a greater volume of water than before. The deeper the bowl is immersed at the surface, the more water is displaced and the greater is the buoyant force exerted on the bowl. When the weight of the displaced water equals the weight of the bowl, it will sink no further. It will float. This is because the buoyant force now equals the weight of the bowl.

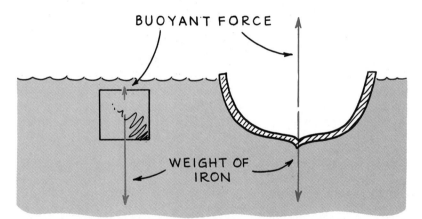

Fig. 19-14 A solid iron block sinks, while the same block shaped to occupy at least 8 times as much volume floats.

Fig. 19-15 The weight of a floating object equals the weight of the water displaced by the submerged part.

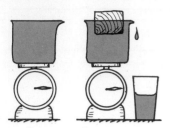

Fig. 19-16 A floating object displaces a weight of liquid equal to its own weight.

This is an example of the **principle of flotation**, which states*:

A floating object displaces a weight of fluid equal to its own weight.

Every ship must be designed to displace a weight of water equal to its own weight. Thus, a 10 000-ton ship must be built wide enough to displace 10 000 tons of water before it sinks too deep below the surface.

Fig. 19-17 The same ship empty and loaded. How does the weight of its load compare to the weight of extra water displaced?

Think about a submarine beneath the surface. If it displaces a weight of water greater than its own weight, it rises. If it displaces less, it falls. If it displaces exactly its weight, it remains at constant depth. Different temperatures of water have slightly different densities, so periodic adjustments must be made. As the next chapter shows, a hot-air balloon obeys the same rules.

▶ **Questions**

Complete the following statements by choosing the correct word for each.

1. The volume of a submerged object is equal to the ? of liquid displaced.

2. The weight of a floating object is equal to the ? of liquid displaced.

▶ **Answers**

1. volume

2. weight

---

* Note that this is a general statement for all fluids rather than just liquids. As the next chapter will show, the same principle applies to gases as well.

19.6 | Pascal's Principle

Push a stick against a wall and you can exert pressure at a distance. Interestingly enough, you can do the same with a fluid. Whenever you change the pressure at one part of a fluid, this change is transmitted to other parts as well. For example, if the pressure of city water is increased at the pumping station by 10 units of pressure, the pressure everywhere in the pipes of the connected system will be increased by 10 units of pressure (providing the water is at rest). This rule is called **Pascal's principle**:

> Changes in pressure at any point in an enclosed fluid
> at rest are transmitted undiminished to all points in
> the fluid and act in all directions.

Pascal's principle was discovered by Blaise Pascal (1623–1662), a French mathematician, physicist, and theologian. The SI unit of pressure, the pascal (1 Pa = 1 N/m²), is named after him.

Pascal's principle is employed in a hydraulic press. If you fill a U-tube with water and place pistons at each end, as shown in Figure 19-18, pressure exerted against the left piston will be transmitted throughout the liquid and against the bottom of the right piston. (The pistons are simply "plugs" that can freely slide snugly inside the tube.) The pressure the left piston exerts against the water will be exactly equal to the pressure the water exerts against the right piston.

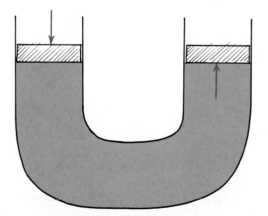

Fig. 19-18 The force exerted on the left piston increases the pressure in the liquid and is transmitted to the right piston.

This is nothing to get excited about. But suppose you make the tube on the right side wider and use a piston of larger area; then the result is impressive. In Figure 19-19 the piston on the left has an area of one square centimeter, and the piston on the right has an area fifty times as great, 50 square centimeters. Suppose there is a one-newton load on the left piston. Then an additional pressure of one newton per square centimeter (1 N/cm²) is transmitted throughout the liquid and up against the larger piston. Here is where the difference between force and pressure comes in. The additional pressure of 1 N/cm² is exerted against *every* square centimeter of the larger piston. Since there are 50 square centimeters, the total extra force exerted on the larger piston is 50 newtons. Thus, the larger piston will support a 50-newton load. This is fifty times the load on the smaller piston!

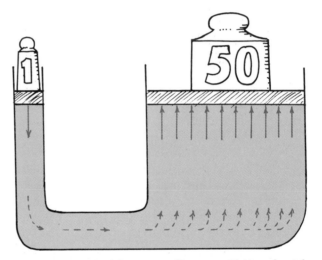

Fig. 19-19 A 1-N load on the left piston will support 50 N on the right piston.

This is quite remarkable, for we can multiply forces with such a device. One newton input, 50 newtons output. By further increasing the area of the larger piston (or reducing the area of the smaller piston), we can multiply forces to any amount. Pascal's principle underlies the operation of the hydraulic press.

The hydraulic press does not violate energy conservation, for the increase in force is compensated for by a decrease in distance moved. When the small piston in the last example is moved downward 10 cm, the large piston will be raised only one-fiftieth of this, or 0.2 cm. Very much like a mechanical lever, the input force multiplied by the distance it moves is equal to the output force multiplied by the distance it moves.

Pascal's principle applies to all fluids, gases as well as liquids. A typical application of Pascal's principle for gases and liquids is the automobile lift seen in many service stations (Figure 19-20). Compressed air exerts pressure on the oil in an underground reservoir. The oil in turn transmits the pressure to a cylinder, which lifts the automobile. The relatively low pressure that exerts the lifting force against the piston is about the same as the air pressure in the tires of the automobile. This is because a low pressure exerted over a relatively large area produces a considerable force.

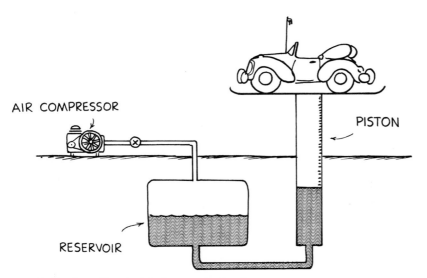

AIR COMPRESSOR

PISTON

RESERVOIR

Fig. 19-20   Pascal's principle in a service station.

▶ **Question**
    As the automobile in Figure 19-20 is being lifted, how does the change in oil level in the reservoir compare to the distance the automobile moves?

▶ **Answer**
    The car moves up a greater distance than the oil level drops, since the area of the piston is smaller than the surface area of the oil in the reservoir.

19   Chapter Review

Concept Summary

Pressure is the force per area on which the force acts.

- Liquids exert pressure at all points within the liquid.
- The forces that produce pressure are exerted equally in all directions.
- The pressure of a liquid at any point is proportional to the density of the liquid times the depth of that point below the liquid surface.

Buoyancy is the apparent loss of weight of an object immersed in a fluid.

- An immersed object displaces the fluid in which it is immersed.
- A completely submerged object always displaces a volume of fluid equal to its own volume.
- According to Archimedes' principle, an immersed object is buoyed up by a force equal to the weight of the fluid it displaces.
- If an object is denser than the fluid in which it is immersed, it will sink.
- If an object is less dense than the fluid in which it is immersed, it will float.
- If an object is as dense as the fluid in which it is immersed, it can remain suspended in the fluid, neither sinking nor floating.
- A floating object displaces a weight of fluid equal to its own weight.

According to Pascal's principle, changes in pressure at any point in an enclosed fluid at rest are transmitted undiminished to all points in the fluid and act in all directions.

- The hydraulic press, which is based on Pascal's principle, multiplies forces.

Important Terms

Archimedes' principle (19.3)
buoyancy (19.2)
buoyant force (19.2)
displaced (19.2)
Pascal's principle (19.6)
principle of flotation (19.5)

Review Questions

1. How does pressure differ from force? (19.1)

2. What is the relationship between liquid pressure and the depth of a liquid? Between liquid pressure and density? (19.1)

3. a. If a diver swims twice as deep in the water, how much more water pressure is exerted on her ears?
   b. If she swims in salt water, will the pressure at the same depth be greater than in fresh water? (19.1)

4. How does water pressure one meter below the surface of a small pond compare to water pressure one meter below the surface of a huge lake? (19.1)

5. If you immerse a tin can with a small hole in it in water so that water spurts through the hole, what will be the direction of water flow where the hole is? (19.1)

6. Why does the buoyant force act upward for an object submerged in water? (19.2)

7. How does the buoyant force that acts on a fish compare to the weight of the fish? (19.2)

8. Why does the buoyant force on submerged objects not act sideways? (19.2)

9. How does the volume of a completely submerged object compare to the volume of water displaced? (19.2)

10. If an object is said to be immersed in water, does this mean it is completely submerged? Does it mean it is partially submerged? Does the word *immersed* apply to either case? (19.2)

11. What is the mass of 1 liter of water? What is its weight in newtons? (19.3)

12. a. Does the buoyant force on a submerged object depend on the weight of the object itself or on the weight of the fluid displaced by the object?
    b. Does it depend on the weight of the object itself or on its volume? Defend your answer. (19.3)

13. If the buoyant force on a submerged object is equal to the weight of the object, how do the densities of the object and water compare? (19.4)

14. If the buoyant force on a submerged object is more than the weight of the object, how do the densities of the object and water compare? (19.4)

15. If the buoyant force on a submerged object is less than the weight of the object, how do the densities of the object and water compare? (19.4)

16. a. How is the density of a submarine controlled?
    b. How is the density of a fish controlled? (19.4)

17. Does the buoyant force on a floating object depend on the weight of the object itself or on the weight of the fluid displaced by the object? Or are these both the same for the special case of floating? (19.5)

18. What is the buoyant force that acts on a 100-ton ship? (To make things simple, give your answer in tons.) (19.5)

19. According to Pascal's principle, what will happen to the pressure in all parts of a confined fluid if you produce an increase in pressure in one part? (19.6)

20. If the pressure in a hydraulic press is increased by an additional 10 N/cm², how much extra load will the output piston support if its cross-sectional area is 50 square centimeters? (19.6)

Activities

1. Note how your veins stand out on the back of your hands when you do hand stands, or when you bend over so your hands are the lowest part of your body. Then note the difference when you hold your hands above your head. This is "pressure depends on depth" in action.

2. Try to float an egg in water. Then dissolve salt in the water until the egg floats. How does the density of an egg compare to that of tap water? To salt water?

3. Make a Cartesian diver (Figure A). Completely fill a large, pliable plastic bottle with water. Partially fill a small pill bottle so that it just barely floats when capped, turned upside down, and placed in the large bottle. (You may have to experiment to get it just right.) Once the pill bottle is barely floating, secure the lid or cap on the large bottle so that it is airtight. When you press the sides of the large bottle, the pill bottle sinks; when you release it, the bottle returns to the top. Experiment by squeezing the bottle different ways to get different results. Can you explain the behavior you see?

Fig. A

4. Punch a couple of holes in the bottom of a water-filled container, and water will spurt out because of water pressure (Figure B left). Now drop the container and note that

as it falls freely the water no longer spurts out (Figure B right)! If your friends don't understand this, can you figure it out and then explain it to them? (What happens to g, and hence weight, and hence weight density, and hence pressure in the reference frame of the falling container?)

Fig. B

Think and Explain

1. Persons confined to bed are less likely to develop bedsores on their bodies if they use a waterbed rather than an ordinary mattress. Why is this so?

2. Why is your blood pressure measured in your upper arm, level with your heart? Is the blood pressure in your legs greater?

3. There is a legend of a Dutch boy who bravely held back the whole Atlantic Ocean by plugging a hole in a dike with his finger. Is this possible and reasonable? Estimate the force if the hole were about one square centimeter in area and one meter below the water level.

4. How much force is needed to push a nearly weightless but rigid one-liter carton beneath the surface of water?

5. Why is it inaccurate to say that heavy objects sink and light objects float?

6. Compared to an empty ship, would a ship loaded with a cargo of Styrofoam sink deeper into water or rise in water? Defend your answer.

7. The density of a rock doesn't change when it is submerged in water, but *your* density does change when you are submerged. Why is this so?

8. A rubber balloon is weighted so that it is barely able to float in water (Figure C). If it is pushed beneath the surface, will it come back to its starting point, stay at the depth to which it is pushed, or sink? Defend your answer. (*Hint*: What change in density, if any, does the balloon undergo?)

Fig. C

9. a. A half-filled bucket of water is on a spring scale. If a fish is placed in it, will the reading on the scale increase?
   b. Would your answer be different if the bucket were initially filled to the brim?

10. a. If the cross-sectional area of the output piston in a hydraulic device is ten times that of the input piston's area, by how much will the device multiply the input force?
    b. How far will the output piston move compared to the distance the input piston is moved? (Is this consistent with $F_1 d_1 = F_2 d_2$, the equation for conservation of energy?)

20 | Gases

Gases are similar to liquids in that they flow; hence both are called *fluids*. The primary difference between gases and liquids is the distance between molecules. In a liquid, the molecules are close together, where they experience forces from the surrounding molecules. These forces strongly affect the motion of the molecules. In a gas, the molecules are far apart and free from forces between molecules. They can move with less restriction.

When molecules in a gas collide with each other or with the walls of a container, they rebound without loss of kinetic energy. A gas expands to fill all space available to it and takes the shape of its container. Only when the quantity of gas is very large, such as the earth's atmosphere or in a star, does gravitation determine the shape of the gas.

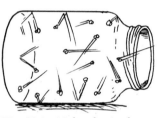

Fig. 20-1 Molecules in the gaseous state are far apart, are in continuous motion, bounce off one another without loss of energy, and fill up and take the shape of their container.

20.1 | The Atmosphere

We don't have to look far to find a sample of gas. We live in an ocean of gas, the atmosphere. Molecules in the air occupy space many kilometers above the earth's surface. The molecules are energized by sunlight and set into continual motion. Without the earth's gravity, they would fly off into outer space. And without the energizing sun, they would be just so much more matter on the ground. Fortunately, there is an energizing sun and there is gravity, so we have an atmosphere.

Unlike the ocean, which has a very definite surface, there is no definite surface for the earth's atmosphere. And unlike the uniform density of a liquid at any depth, the density of the atmosphere varies with altitude. Air is more compressed at sea level than at higher altitudes. Like a huge pile of feathers, those at the bottom are more squashed than those nearer the top. The air gets thinner and thinner (less dense) the higher one goes; it eventually thins out into space.

Even in the vacuous regions of interplanetary space there is a gas density of about one molecule per cubic centimeter. This is primarily hydrogen, the most plentiful element in the universe.

As Figure 20-2 shows, 50% of the atmosphere is below 5.6 kilometers, 75% is below 11 kilometers, 90% is below 17.7 kilometers, and 99% of the atmosphere is below an altitude of about 30 kilometers. Compared to the earth's radius, 30 kilometers is very small. To give you an idea of how small, the "thickness" of the atmosphere compared to the size of the world is like the thickness of condensed breath on a billiard ball. Our atmosphere is quite finite; that's why we should care for it.

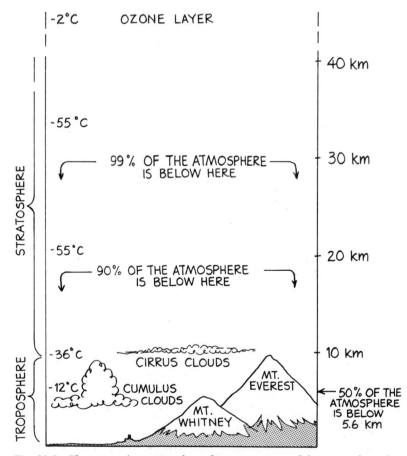

Fig. 20-2 The atmosphere. Note how the temperature of the atmosphere drops as one goes higher.

20.2 | Atmospheric Pressure

We live at the bottom of our ocean of air. The atmosphere, much like water in a lake, exerts a pressure. And just as water pressure is caused by the weight of water, atmospheric pressure is caused by the weight of air. We are so accustomed to the invisible air that we sometimes forget it has weight. Perhaps a fish "forgets" about the weight of water in the same way.

Table 20-1 Densities of Various Gases		
Gas		Density (kg/m³)*
Dry air	0°C	1.29
	10°C	1.25
	20°C	1.21
	30°C	1.16
Helium		0.178
Hydrogen		0.090
Oxygen		1.43

*At sea-level atmospheric pressure and at 0°C (unless otherwise specified)

At sea level, one cubic meter of air at 20°C has a mass of about 1.2 kg. Calculate the number of cubic meters in your room, multiply by 1.2 kg/m³, and you'll have the mass of air in your room. Don't be surprised if it has more mass than your kid sister. Air is heavy if you have enough of it.

Fig. 20-3 It takes more than 1000 kg of additional air to fully pressurize a 747 jumbo jet.

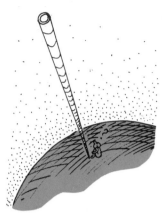

Fig. 20-4 The mass of air that would occupy a bamboo pole that extends to the "top" of the atmosphere is about 1 kg. This air has a weight of 10 N.

Consider a super-long bamboo pole that reaches up through the atmosphere for 30 km. Suppose that the inside cross sectional area of the hollow pole is one square centimeter. If the density of air inside the pole matches the density of air outside, the enclosed mass of air would be about one kilogram. The weight of this much air is about 10 newtons. So air pressure at the bottom of the bamboo pole would be about 10 newtons per square centimeter (10 N/cm²). Of course, the same is true without the bamboo pole.

There are 10 000 square centimeters in one square meter, so a column of air one-square meter in cross section that extends up through the atmosphere has a mass of about 10 000 kilograms. The weight of this air is about 100 000 newtons (10⁵ N). This weight produces a pressure of 100 000 newtons per square meter, or equivalently, 100 000 pascals, or 100 kilopascals. To be more exact, the average atmospheric pressure at sea level is 101.3 kilopascals (101.3 kPa).*

The pressure of the atmosphere is not uniform. Aside from variations with altitude, there are variations in atmospheric pressure at any one locality due to moving air currents and storms. Measurement of changing air pressure is important to meteorologists in predicting weather.

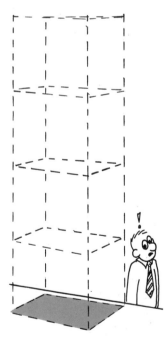

Fig. 20-5 The weight of air that bears down on a one-square-meter surface at sea level is about 100 000 newtons. So atmospheric pressure is about 100 000 newtons per square meter (10⁵ N/m²), or about 100 kPa.

▶ **Questions**

1. About how many kilograms of air occupy a classroom that has a 200-square meter floor area and a 4-meter high ceiling?

2. Why doesn't the pressure of the atmosphere break windows?

▶ **Answers**

1. 950 kg. The volume of air is (200 m²) × (4m) = 800 m³. Each cubic meter of air has a mass of about 1.2 kg, so (800 m³) × (1.2 kg/m³) = 960 kg.

2. The atmospheric pressure doesn't normally break windows because it acts on *both* sides of a window. So no net force is exerted by the atmosphere on the windows.

---

* The average pressure at sea level used to be called one *atmosphere*. This term is still commonly used, but it is no longer acceptable with SI units. In British units, the average atmospheric pressure at sea level is 14.7 pounds/inch².

20.3 | The Simple Barometer

An instrument used for measuring the pressure of the atmosphere is called a **barometer**. A simple mercury barometer is illustrated in Figure 20-6. A glass tube, longer than 76 cm and closed at one end, is filled with mercury and tipped upside down in a dish of mercury. The mercury in the tube runs out of the submerged open bottom until the level falls to about 76 cm. The empty space trapped above, except for some mercury vapor, is a pure vacuum. The vertical height of the mercury column remains constant even when the tube is tilted, unless the top of the tube is less than 76 cm above the level in the dish, in which case the mercury completely fills the tube.

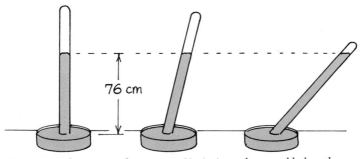

Fig 20-6   A simple mercury barometer. Variations above and below the average column height of 76 cm are caused by variations in atmospheric pressure.

Why does mercury behave this way? The explanation is similar to the reason a simple see-saw will balance when the weights of people at its two ends are equal. The barometer "balances" when the weight of liquid in the tube exerts the same pressure as the atmosphere outside. Whatever the width of the tube, a 76-cm column of mercury weighs the same as the air that would fill a supertall 30-km tube of the same width. If the atmospheric pressure increases, then it pushes the mercury column higher than 76 cm. The mercury is literally pushed up into the tube of a barometer by atmospheric pressure.

Could water be used to make a barometer? The answer is yes, but the glass tube would have to be much longer—13.6 times as long, to be exact. You may recognize this number as the density of mercury compared to that of water. A volume of water 13.6 times that of mercury is needed to provide the same weight as the mercury in the tube (or in the imaginary tube of air outside). So the height of the tube would have to be at least 13.6 times taller than the mercury column. A water barometer would have to be 13.6 × (0.76 m), or 10.3 m high—too tall to be practical.

The operation of a barometer is similar to the process of drinking through a straw. By sucking, you reduce the air pressure in the straw that is placed in a drink. Atmospheric pressure on the drink pushes liquid up into the reduced-pressure region. Strictly speaking, the liquid is not *sucked* up; it is *pushed* up, by the pressure of the atmosphere. If the atmosphere is prevented from pushing on the surface of the drink, as in the party trick bottle with the straw through the air-tight cork stopper, one can suck and suck and get no drink.

Fig. 20-7   You cannot drink soda through the straw unless the atmosphere exerts a pressure on the surrounding liquid.

If you understand these ideas, you can understand why there is a 10.3-meter limit on the height water can be lifted with vacuum pumps. The old fashioned farm-type pump (Figure 20-8) operates by producing a partial vacuum in a pipe that extends down into the water below. The atmospheric pressure exerted on the surface of the water simply pushes the water up into the region of reduced pressure inside the pipe. Can you see that even with a perfect vacuum, the maximum height to which water can be lifted is 10.3 meters?

Fig. 20-8   The atmosphere pushes water from below up into a pipe that is evacuated of air by the pumping action.

20.4 | The Aneroid Barometer

A popular classroom demonstration to illustrate atmospheric pressure is crushing a can with atmospheric pressure. A can with a little water in it is heated until steam forms. Then it is capped securely and removed from the source of heat. There is now less air inside than before it was heated. (Why? Because when the water boils and changes to steam, the steam pushes air out of the can.) When the sealed can cools, the pressure inside is reduced. (This is because steam inside the can liquefies when it cools.) The greater pressure of the atmosphere outside the can then proceeds to crush the can (Figure 20-9). The pressure of the atmosphere is even more dramatically shown when a 50-gallon drum is crushed by the same procedure.

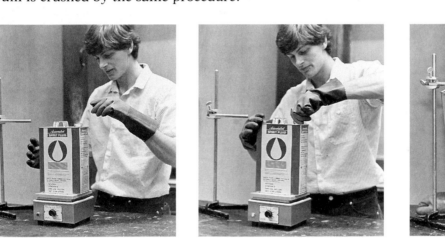

Fig. 20-9   When the air pressure inside is reduced, the greater atmospheric pressure outside crushes the can.

A much more subtle application of atmospheric crushing is used in an **aneroid** ("without liquid") **barometer**. This small portable instrument is more prevalent than the mercury barometer. It uses a small metal box that is partially exhausted of air and has a slightly flexible lid that bends in or out with atmospheric-pressure changes. The pressure difference between the inside and outside is less drastic than that of the crushed can of Figure 20-9. Motion of the lid is indicated on a scale by a mechanical spring-and-lever system. Since atmospheric pressure decreases with increasing altitude, a barometer can be used to determine the elevation. An aneroid barometer calibrated for altitude is called an *altimeter* ("altitude meter"). Some of these instruments are sensitive enough to indicate changes in elevation of less than a meter.

Fig. 20-10   An aneroid barometer.

20.5 | Boyle's Law

The air pressure inside the inflated tires of an automobile is considerably more than the atmospheric pressure outside. The density of air inside is also more than that of the air outside. To understand the relation between pressure and density, think of the molecules of air* inside the tire.

Inside the tire, the molecules behave like tiny table-tennis balls, perpetually moving helter skelter and banging against the inner walls. Their impacts on the inner surface of the tire produce a jittery force that appears to our coarse senses as a steady push. This pushing force averaged over a unit of area provides the pressure of the enclosed air.

Suppose there are twice as many molecules in the same volume (Figure 20-11). Then the air density is doubled. If the molecules move at the same average speed—or, equivalently, if they have the same temperature—then to a close approximation, the number of collisions will be doubled. This means the pressure is doubled. So pressure is proportional to density.

Fig. 20-11 When the density of the air in the tire is increased, the pressure is increased.

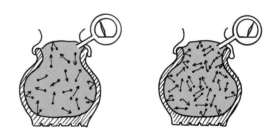

The density of the air can also be doubled by simply compressing the air to half its volume. Consider the cylinder with the movable piston in Figure 20-12. If the piston is pushed downward so that the volume is half the original volume, the density of molecules will be doubled, and the pressure will correspondingly be doubled. Decrease the volume to a third its original value, and the pressure will be increased by three, and so forth.

Fig. 20-12 When the volume of gas is decreased, the density—and therefore pressure—are increased.

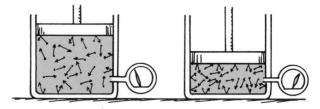

---

*   Air is composed of a mixture of gases—mainly nitrogen, oxygen, and carbon dioxide. When we speak of *molecules of air*, we are referring to any of the different kinds of molecules found in air.

Notice from these examples that the product of pressure and volume is the same. For example, a doubled pressure multiplied by a halved volume gives the same value as a tripled pressure multiplied by a one-third volume. In general, we can say that the product of pressure and volume for a given mass of gas is a constant as long as the temperature does not change. "Pressure × volume" for a quantity of gas at one time is equal to any "different pressure × different volume" at any other time. In shorthand notation:

$$P_1 V_1 = P_2 V_2$$

where P_1 and V_1 represent the original pressure and volume, respectively, and P_2 and V_2 the second, or final, pressure and volume. This relationship is called **Boyle's law**, after Robert Boyle, the seventeenth-century physicist who is credited with its discovery.*

▶ **Questions**

1. If you squeeze a balloon to one third its volume, by how much does the pressure inside increase?

2. A piston in an airtight pump is withdrawn so that the volume of the air chamber is increased 5 times. What is the change in pressure?

3. A scuba diver 10.3 m deep breathes compressed air. If she holds her breath while returning to the surface, by how much does the volume of her lungs tend to increase?

▶ **Answers**

1. The pressure in the balloon is multiplied by 3. No wonder balloons break when you squeeze them!

2. The pressure in the piston chamber is reduced to $\frac{1}{5}$. This is the principle behind a mechanical vacuum pump.

3. Atmospheric pressure can support a column of water 10.3 m high, so the pressure in water due to the weight of the water alone equals atmospheric pressure at a depth of 10.3 m. Taking the pressure of the atmosphere at the water's surface into account, the total pressure at this depth is twice atmospheric pressure. Unfortunately for the scuba diver, her lungs tend to inflate to twice their normal size if she holds her breath while rising to the surface. A first lesson in scuba diving is *not* to hold your breath when ascending. To do so can be fatal.

---

*   A general law that takes temperature changes into account is $P_1 V_1 / T_1 = P_2 V_2 / T_2$, where T_1 and T_2 represent the initial and final *absolute temperatures*, measured in SI units called *kelvins* (Chapter 21).

20.6 Buoyancy of Air

Fig. 20-13 The dirigible and the fish both float for the same reason.

Fig. 20-14 Everything is buoyed up by a force equal to the weight of air it displaces. Why, then, doesn't everything float like this balloon?

In the last chapter you learned about buoyancy in liquids. All the rules for buoyancy were stated in terms of *fluids* rather than liquids. The reason is simple enough: the rules hold for gases as well as liquids. The physical laws that explain a dirigible aloft in the air are the same that explain a fish "aloft" in water. We can state Archimedes' principle for air:

> An object surrounded by air is buoyed up by a force equal to the weight of the air displaced.

Recall that a cubic meter of air at ordinary atmospheric pressure and room temperature has a mass of about 1.2 kg, so its weight is about 12 N. Therefore any one-cubic-meter object in air is buoyed up with a force of 12 N. If the mass of the one-cubic-meter object is greater than 1.2 kg (so that its weight is greater than 12 N), it falls to the ground when released. If this size object has a mass less than 1.2 kg, it rises in the air. Any object that has a mass less than the mass of an equal volume of air will rise in air. Another way to say this is, any object less dense than air will rise in air. Gas-filled balloons that rise in air are less dense than air.

Large dirigible airships are designed to displace a weight of air that is at least as great as their total weight. When in motion, the ship may be raised or lowered by means of horizontal rudders or "elevators."

▶ **Questions**

1. Is there a buoyant force acting on you? If there is, why are you not buoyed up by this force?

2. Two balloons are inflated to the same size, one with air and the other with helium. Which balloon experiences the greater buoyant force? Why does the air-filled balloon sink and the helium-filled balloon float?

▶ **Answers**

1. There *is* a buoyant force acting on you, and you *are* buoyed upward by it. You don't notice it only because your weight is so much greater.

2. Both balloons are buoyed upward with the same buoyant force because they displace the same weight of air. The reason the air-filled balloon sinks in air is because it is heavier than the buoyant force that acts on it, while the helium-filled balloon is lighter than the buoyant force that acts on it. Or put another way, the air-filled balloon is slightly more dense than the surrounding air (principally because it is filled with *compressed* air). Helium, even somewhat compressed, is appreciably less dense than air.

20.7 | Bernoulli's Principle

The discussion of fluid pressure thus far has been confined to stationary fluids. Motion produces an additional influence.

Most people think that atmospheric pressure increases in a gale, tornado, or hurricane. Actually, the opposite is true. High-speed winds may blow the roof off your house but the pressure within the winds is actually less than for still air of the same density. As strange as it may first seem, when the speed of a fluid increases, the internal pressure decreases proportionally. This is true for all fluids—liquids and gases alike.

Daniel Bernoulli, a Swiss scientist of the eighteenth century, studied the relationship of fluid speed and pressure. When a fluid flows through a narrow constriction, its speed increases. This is easily noticed by the increased speed of a brook when it flows through the narrow parts. The fluid must speed up in the constricted region if the flow is to be continuous.

Fig. 20-15 The water speeds up when it flows through the narrow part of the brook. What happens to the water pressure?

Bernoulli wondered how the fluid got the energy for this extra speed. He reasoned that it is acquired at the expense of a lowered internal pressure. His discovery, now called **Bernoulli's principle**, states:

> The pressure in a fluid decreases as the speed of the fluid increases.

Bernoulli's principle is a consequence of the conservation of energy. When a fluid flows, it has kinetic energy because of

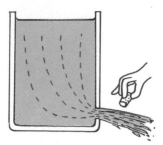

Fig. 20-16 The pressure in the water flowing out the spout is less than that in the water inside the tank at the same level.

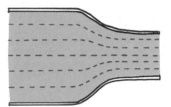

Fig. 20-17 A fluid speeds up when it flows into the narrow region. The constricted streamlines indicate increased speed and decreased internal pressure.

its motion. It also has gravitational potential energy, or stored energy due to the earth's gravitational field. If the fluid picks up speed, or accelerates, it has more kinetic energy than before. Let's suppose that the fluid does not move up or down as it travels through the constricted region. Then its gravitational potential energy does not change. How, then, does the accelerating fluid in the constricted region gain kinetic energy?

The answer is that the surrounding fluid does work on the part that goes through the constricted region. The forces that produce pressure push the accelerating fluid from behind. They do work on the accelerating fluid. The accelerating fluid has to do work on the fluid ahead of it. It turns out that when the fluid is accelerating, *more* work is done *on it* than it does on the fluid ahead. In this way, its energy increases. All through the fluid, some parts are gaining energy while others are losing energy. The net energy of the entire fluid is unchanged.

Bernoulli's principle holds for steady flow. In steady flow, the paths taken by each little region of fluid do not change as time passes. The motion of a fluid in steady flow can be represented with **streamlines**, which are indicated by dashed lines in Figure 20-17 and later figures. Streamlines are the smooth paths of the neighboring regions of fluid. The lines are closer together in the narrower regions, where the flow speed is greater and the pressure within the fluid is less.

If the flow speed is too great, the flow may become turbulent and follow changing, curling paths known as **eddies**. Then Bernoulli's principle will not hold.

20.8 Applications of Bernoulli's Principle

Hold a sheet of paper in front of your mouth, as shown in Figure 20-18. When you blow across the top surface, the paper rises. This is because the moving air pushes against the top of the paper with less pressure than the air that pushes against the lower surface, which is at rest.

Bernoulli's principle accounts for the flight of birds and aircraft. The shape and orientation of the wings insure that air passes slightly faster across the top surface of the wing than against the lower surface. The difference between upward and downward pressures produces a net upward force, appropriately called **lift**. When lift equals weight, horizontal flight is possible.

The lift is greater for higher speeds and larger wing areas. Hence, low-speed gliders have very large wings. The wings of faster-moving fighter aircraft are relatively small.

Fig. 20-18 The paper rises.

We all know that a baseball pitcher can throw a ball in such a way that it will curve off to one side of its trajectory. This is accomplished by imparting a large spin to the ball. Similarly, a tennis player can hit a ball that will curve. A thin layer of air is dragged around the spinning ball by friction, which is enhanced by the baseball's threads or the tennis ball's fuzz. The moving layer produces a crowding of streamlines on one side. Note in Figure 20-20 right that the streamlines are more crowded at B than at A for the direction of spin shown. Air pressure is greater at A, and the ball curves as shown.

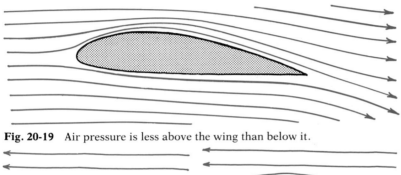

Fig. 20-19 Air pressure is less above the wing than below it.

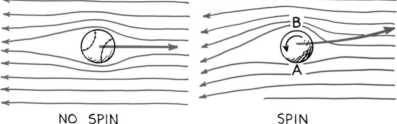

NO SPIN   SPIN

Fig. 20-20 (Left) The streamlines are the same on either side of a nonspinning ball. (Right) A spinning ball produces a crowding of streamlines. It curves as shown.

You can demonstrate Bernoulli's principle quite interestingly in your kitchen sink (Figure 20-21). Tape a table-tennis ball to a string and allow the ball to swing into a stream of running water. You'll see that it will remain in the stream even when tugged slightly to the side, as shown. This is because the pressure of the stationary air next to the ball is greater than the pressure of the moving water. The ball is *pushed* into the region of reduced pressure by the atmosphere.

The same thing happens to a bathroom shower curtain when the shower water is turned on full blast. The pressure in the shower stall is reduced, and the relatively greater pressure outside the curtain pushes it inward. The next time you're taking a shower and the curtain swings in against you, think of Daniel Bernoulli!

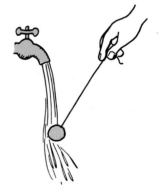

Fig. 20-21 Pressure is greater in the stationary fluid (air) than in the moving fluid (water). The atmosphere pushes the ball into the region of reduced pressure.

20 | Chapter Review

Concept Summary

The earth's atmosphere is an ocean of air extending about 30 km above the earth's surface.
- Air is more compressed at sea level than at higher altitudes.
- Air exerts pressure on everything; the pressure at sea level is about 100 kPa.
- Simple barometers measure atmospheric pressure in terms of how high a column of mercury in a closed tube can be supported by atmospheric pressure.
- Aneroid barometers work without liquids and measure the position of a movable lid against a box with low pressure inside.

Boyle's law states that at constant temperature, the pressure times the volume of an enclosed gas remains constant; when one increases, the other decreases.

Bernoulli's principle states that the pressure of a fluid decreases as the speed of the fluid increases.
- Bernoulli's principle holds only for steady flow, in which the flow paths do not change over time.
- Bernoulli's principle explains the lift forces that enable birds and aircraft to fly.

Important Terms

aneroid barometer (20.4)
barometer (20.3)
Bernoulli's principle (20.7)
Boyle's law (20.5)
eddy (20.7)
lift (20.8)
streamline (20.7)

Review Questions

1. a. What is the energy source for the motion of gases in the atmosphere?
   b. What prevents atmospheric gases from flying off into space? (20.1)

2. How does the density of gases at different elevations in the atmosphere differ from the density of liquids at different depths? (20.1)

3. What causes atmospheric pressure? (20.2)

4. What is the mass of a cubic meter of air at 20°C at sea level? (20.2)

5. a. What is the mass of a column of air that has a cross-sectional area of one square centimeter and that extends from sea level to the top of the atmosphere?
   b. What is the weight of such a column of air? (20.2)

6. What is the pressure at the bottom of the column of air discussed in the last question? (20.2)

7. Is the value for atmospheric pressure at the surface of the earth a constant? Why, or why not? (20.2)

8. How does the pressure at the bottom of the 76-cm column of mercury in a barometer compare to the pressure due to the weight of the atmosphere? (20.3)

9. Why is a water barometer 13.6 times taller than a mercury barometer? (20.3)

10. When you drink liquid through a straw, it is more accurate to say that the liquid is *pushed* up the straw rather than *sucked* up the straw. What exactly does the pushing? Explain. (20.3)

11. Why will a vacuum pump not operate for a well that is deeper than 10.3 m? (20.3)

12. The atmosphere does not ordinarily crush cans. Yet it will crush a can after it has been heated, capped, and cooled. Why? (20.4)

13. Why is it that an aneroid barometer is able to measure altitude as well as atmospheric pressure? (20.4)

14. When air is compressed, what happens to its density? (20.5)

15. When a balloon is squeezed to half its volume, what happens to the air pressure inside it? (20.5)

16. a. How great is the buoyant force on a floating balloon that weighs 1 N?
    b. What happens if the buoyant force decreases? If it increases? (20.6)

17. When the speed of a fluid flowing in a pipe increases, what happens to the internal pressure in the fluid? (20.7)

18. a. What are streamlines?
    b. Is the pressure greater or less in regions where streamlines are crowded? (20.7)

19. Why does a spinning ball curve in its flight? (20.8)

20. Why does a shower curtain swing against you when you take a shower? (20.8)

Activities

1. Try this in the bathtub or when you're washing dishes. Lower a drinking glass, mouth downward, over a small floating object (Figure A). What do you observe? How deep would the glass have to be pushed in order to compress the enclosed air to half its volume? (*Hint*: You won't be able to do this in your bathtub unless it's 10.3 m deep!)

Fig. A

2. Fill a glass with water and hold it partially under water so that its mouth is beneath the surface, as shown in Figure B. Why does the water not run out? How tall would the glass have to be before water ran out? (*Hint*: You won't be able to do this indoors unless you have a ceiling 10.3 m high!)

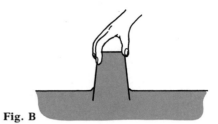

Fig. B

3. Place a card over the open top of a glass filled to the brim with water, and invert it (Figure C left). Why does the card stay intact? Try it sideways (Figure C right).

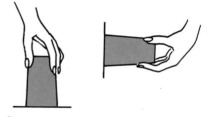

Fig. C

4. Heat a small amount of water to boiling in an aluminum soda-pop can. Using tongs, invert it quickly into a pot of cold water. The result is very dramatic. Be sure to try this one!

5. Lower a narrow glass tube or drinking straw in water, and place your finger over the top of the tube (Figure D). Lift the tube from the water, and then lift your finger from the top of the tube. What happens? (You'll do this often if you enroll in a chemistry lab.)

Fig. D

6. Hold a spoon in a stream of water, as shown in Figure E, and feel the effect of the differences in pressure.

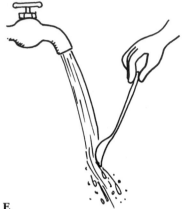

Fig. E

Think and Explain

1. Compared to sea level, would it be slightly more difficult or somewhat easier to drink via a straw at the bottom of a deep mine? At the top of a high mountain? Defend your answer.

2. If there were a liquid twice as dense as mercury, and it were used to make a barometer, how tall would the column be?

3. Which would weigh more—a bottle filled with helium gas, or the same bottle empty?

4. Small bubbles of air are released by a scuba diver deep in the water. As the bubbles rise, do they become larger, smaller, or stay about the same size? Explain.

5. It is easy to breathe when snorkeling with only your face beneath the surface of water, but quite difficult to breathe if you are submerged nearly a meter, and nearly impossible if you are more than a meter deep (even if your snorkel tube reaches to the surface). Can you figure out why?

6. From Table 20-1, which filling would be more effective in making a balloon rise—helium or hydrogen? Why?

7. If an inflated balloon is sufficiently weighted with rocks, it will sink in water. What will happen to the size of the balloon as it sinks? Compared to its volume at the surface, what size will it be when it is 10.3 m below the surface?

8. The buoyant force due to air displacement is considerably greater on an elephant than on a small helium-filled balloon. Why, then, does the elephant remain on the ground, while the balloon rises?

9. The force of the atmosphere against a common window may be about a million newtons. Why does this force not shatter the window? Why will a window sometimes shatter in a strong wind?

10. Why is it that when cars pass each other at high speeds on the road, they tend to be "drawn" to each other?

ON A COLD MORNING, THE PAVEMENT FEELS COLDER THAN THE LAWN TO MY BARE FEET. MORE HEAT ENERGY GOES FROM MY FOOT TO THE PAVEMENT THAN TO THE LAWN. AT FIRST THOUGHT, I TEND TO THINK THE PAVEMENT HAS A LOWER TEMPERATURE THAN THE LAWN---BUT DOES IT? AND ON A HOT AFTERNOON THE PAVEMENT FEELS MUCH HOTTER THAN THE LAWN. DOES IT HAVE A HIGHER TEMPERATURE THAN THE LAWN, OR IS THERE SOME OTHER PHYSICS GOING ON HERE? PERHAPS DIFFERENCES IN HEAT CONDUCTIVITIES? I WONDER ABOUT SUCH THINGS, AND IT'S NICE TO FIND EXPLANATIONS IN MY PHYSICS BOOK. IT'S EVEN NICER TO FIND I CAN UNDERSTAND THEM!

299

21 | Temperature and Heat

All matter—whether solid, liquid, or gas—is composed of continually jiggling molecules or atoms. By virtue of this energetic motion, the molecules or atoms in matter possess kinetic energy. The average kinetic energy of the individual particles is directly related to a property you can sense: how hot something is. Whenever something becomes warmer, the kinetic energy of its particles increases.

It's easy to increase the kinetic energy in matter. Strike a solid penny with a hammer and it becomes warm because the hammer's blow causes the molecules in the metal to jostle faster. Put a flame to a liquid and it too becomes warmer. Rapidly compress air in a tire pump and the air becomes warmer. When a solid, liquid, or gas gets warmer, its molecules or atoms move faster. The atoms or molecules have more kinetic energy.*

21.1 | Temperature

The quantity that tells how warm or cold a body is with respect to some standard is called **temperature**. We express the temperature of matter by a number which corresponds to the degree of hotness on some chosen scale.

Nearly all materials expand when their temperature is raised, and contract when it is lowered. A thermometer is a common

---

* The kinetic energy in matter is sometimes called *thermal energy*. This term is also used to include the potential energy due to the forces between the particles. Because *thermal energy* means different things to different people, this text avoids using the term.

instrument that measures temperature by means of the expansion and contraction of a liquid, usually mercury or colored alcohol.

On the scale commonly used in laboratories, the number 0 is assigned to the temperature at which water freezes, and the number 100 to the temperature at which water boils (at standard atmospheric pressure). The space between is divided into 100 equal parts, called *degrees*. This scale is called the **Celsius scale** in honor of the man who first suggested the scale, the Swedish astronomer Anders Celsius (1701–1744).*

In the everyday scale used in the United States, the number 32 is assigned to the temperature at which water freezes, and the number 212 is assigned to the temperature at which water boils. This scale is called the **Fahrenheit scale**, after the German physicist Gabriel Fahrenheit (1686–1736). The Fahrenheit scale will become obsolete if and when the United States goes metric.

Arithmetic formulas are used for converting from one temperature scale to the other and are popular in classroom exams. Such arithmetic exercises are not really physics, and the probability of your having the occasion to do this task elsewhere is small, so we will not be concerned with it here. Besides, this can be very closely approximated by simply reading the corresponding temperature from the side-by-side scales in Figure 21-1.

Temperature and Kinetic Energy

Temperature is a measure of the motion of the molecules or atoms within a substance; more specifically, it is a measure of the *average* kinetic energy of the molecules or atoms in a substance.

Temperature is *not* a measure of the *total* kinetic energy of the particles within a substance. There is twice as much molecular kinetic energy in two liters of boiling water as in one liter of boiling water. But the temperatures of both amounts of water are the same because the average kinetic energy of molecules in each is the same.

Absolute Zero

The SI temperature scale is called the **Kelvin scale**, after the British physicist Lord Kelvin (1824–1907). On this scale, the number zero is assigned to the lowest possible temperature— **absolute zero**, at which a substance has absolutely no kinetic energy to give up. Temperatures on the Kelvin scale are measured not in degrees but in units called **kelvins**, denoted by K (*not* by °K). On the Celsius scale, absolute zero corresponds to −273°C.

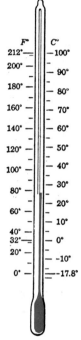

Fig. 21-1 Fahrenheit and Celsius scales on a thermometer.

Fig. 21-2 There is more molecular kinetic energy in the bucketful of warm water than in the small cupful of higher-temperature water.

---

* The Celsius scale used to be called the *centigrade scale*, from *centi* ("hundredth") and *gradus* ("degree").

Units on the Kelvin scale are the same size as degrees on the Celsius scale, so the temperature of melting ice is +273 kelvins. There are no negative numbers on the Kelvin scale.

Scientists favor the Kelvin scale of temperature because it is calibrated not in terms of the freezing and boiling points of some liquid, but in terms of energy itself. Absolute zero corresponds to zero available kinetic energy—temperature zero, no available kinetic energy. Or if a substance has a small amount of available kinetic energy per particle, it has a correspondingly small Kelvin temperature. If the substance has twice the kinetic energy per particle, then it has twice the Kelvin temperature. The available kinetic energy per particle and absolute temperature go hand in hand, as the following question illustrates.

> ▶ **Question**
> A piece of metal has a temperature of 0°C. A second identical piece of metal is twice as hot (has twice the kinetic energy per atom). What is its temperature?

21.2 │ Heat

If you touch a hot stove, energy enters your hand because the stove is warmer than your hand. When you touch a piece of ice, on the other hand, energy passes out of your hand and into the colder ice. The direction of energy transfer is always from a warmer body to a neighboring cooler body. The energy that is transferred from one object to another because of a temperature difference between the objects is called **heat**.

It is common—but incorrect—to think of matter as *containing* heat. Matter contains atomic or molecular kinetic energy, not heat. Heat flows from one thing to another; it is the energy that is being transferred. Once heat has been transferred to an object or substance, it ceases to be heat. It becomes, as we will discuss shortly, *internal energy*.

▶ **Answer**
The piece of metal with twice the kinetic energy per atom has twice the absolute temperature, or two times 273K. This would be 546K, or 273°C. Or look at it this way: The metal at 0°C is 273K above absolute zero. Twice as hot would put it 273K higher, or 273K above 0°C. Since kelvins and Celsius degrees are the same in size, 273K above 0°C is the same as 273 Celsius degrees above 0°C, or simply 273°C.

When it is possible for heat to flow from one object or substance to another with which it is in contact, the objects or substances are said to be in **thermal contact**. For objects or substances in thermal contact, heat tends to flow from the substance at a higher temperature into the substance at a lower temperature. It will not necessarily flow from a substance with more atomic or molecular kinetic energy into a substance with less atomic or molecular kinetic energy. For example, there is more molecular kinetic energy in a large bowl of warm water than there is atomic kinetic energy in a red-hot thumbtack. Yet, if the tack is immersed in the water, heat will not flow from the warm water to the tack. Instead, heat will flow from the hot tack to the relatively cooler water. Heat never flows on its own from a cooler substance into a hotter substance.

Fig. 21-3 Just as water will not flow uphill by itself, regardless of the relative amounts of water in the reservoirs, heat will not flow from a cooler substance into a hotter substance by itself.

21.3 | Thermal Equilibrium

It is interesting to note that a thermometer registers its own temperature. When a thermometer is in thermal contact with a substance, heat will flow between the two until they are both at the same temperature. If we then know the temperature of the thermometer, we also know the temperature of the substance. When two or more substances reach a common temperature, they are said to be in **thermal equilibrium**.

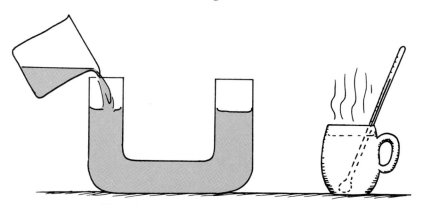

Fig. 21-4 Just as water in the pipes seeks a common level (for which the pressures at the bottom are the same), the thermometer and its immediate surroundings reach a common temperature (at which the average kinetic energy per particle is the same for both).

A thermometer should be small enough that it does not appreciably alter the temperature of the substance being measured. If you are measuring the temperature of room air, then the heat absorbed by your thermometer will not lower the air temperature noticeably; but if you are measuring a drop of water, its temperature after thermal contact may be quite different from its initial temperature.

21.4 | Internal Energy

In addition to the kinetic energy of jostling molecules or atoms in a substance, there is energy in other forms. There is potential energy due to the forces between molecules or atoms. There is also kinetic energy due to movements of atoms within molecules. The grand total of all energies inside a substance is called **internal energy**. A substance does not contain heat—it contains internal energy.

When a substance absorbs or gives off heat, any of these energies may change. Thus, as a substance absorbs heat, this energy may or may not make the molecules or atoms jostle faster. In some cases, as when ice is melting, a substance absorbs heat without an increase in molecular kinetic energy. The substance undergoes a change of state, the subject of Chapter 24.

21.5 | Quantity of Heat

Heat is the internal energy transferred from one body to another by virtue of a temperature difference. The quantity of heat involved in such a transfer is measured by some change—such as the change in temperature of a known amount of water that absorbs the heat.

When a substance absorbs heat, the temperature change depends on the amount of the substance. The amount of heat that can bring a cupful of soup to a boil might raise the temperature of a pot of soup by only a few degrees.

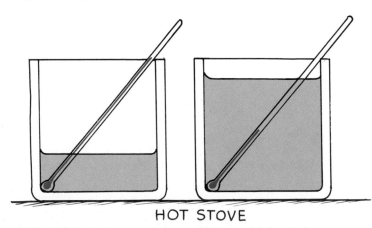

HOT STOVE

Fig. 21-5 Although the same quantity of heat is added to both containers, the temperature of the container with the smaller amount of water increases more.

> ▶ **Question**
> Suppose you apply a flame and add a certain amount of heat to one liter of water, and its temperature rises by 2°C. If you add the same amount of heat to two liters of water, by how much will its temperature rise?

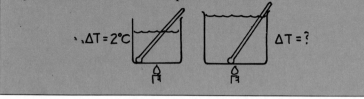

The unit of heat is defined as the heat necessary to produce some standard, agreed-on change. The most commonly used unit for heat is the **calorie**.*

> The calorie is defined as the amount of heat required to raise the temperature of 1 gram of water by 1°C.

The **kilocalorie** is 1000 calories (the heat required to raise the temperature of 1 kilogram of water by 1°C). The heat unit used in rating foods is actually a kilocalorie. To distinguish this unit from the smaller calorie, the food unit is sometimes called a Calorie (written with a capital C).

It is important to remember that the calorie and Calorie are units of energy. These names are historical carry-overs from the early idea that heat was an invisible fluid called *caloric*. This view persisted almost to the nineteenth century. We now know that heat is a form of energy, and we are presently in a transition period to the International System of units (SI), where the quantity of heat is measured in joules, the SI unit for all forms of energy. (The relationship between calories and joules is that one calorie equals 4.187 J.)

The energy value in food is determined by burning the food and measuring the amount of internal energy that is released as heat. Fuels are rated by how much energy a certain amount of the fuel gives off as heat when burned.

Fig. 21-6 To the weight watcher, the peanut contains 10 Calories; to the physicist, it releases 10 000 calories (or 41 870 joules) of energy when burned or digested.

▶ **Answer**
Its temperature will rise by only 1°C, because there are twice as many molecules in two liters of water and each molecule receives only half as much energy on the average. So the average kinetic energy, and thus the temperature, increases by half as much.

---

* Another common unit of heat is the *British thermal unit* (Btu). The Btu is defined as the amount of heat required to change the temperature of 1 pound of water by 1°F.

Computational Example

A woman with an average diet consumes and expends about 2000 Calories per day. The energy used by the body is eventually given off as heat. What is the average heat energy output in joules of the body per second? In other words, what is the body's average thermal power output?

We find this by converting 2000 Calories per day to joules per second. We use the information that 1 Calorie = 4187 joules, 1 day = 24 hours, and 1 hour = 3600 seconds. The conversion is then set up as follows:

$$\frac{2000 \text{ Cal}}{1 \text{ d}} \times \frac{1 \text{ d}}{24 \text{ h}} \times \frac{1 \text{ h}}{3600 \text{ s}} \times \frac{4187 \text{ J}}{1 \text{ Cal}} = 96.9 \text{ J/s}$$
$$= 96.9 \text{ W}$$

Notice that the original quantity (2000 Cal/d) is multiplied by a set of fractions in which the numerator equals the denominator. Since each fraction has the value 1, multiplying by it does not change the value of the original quantity. The rule for choosing which quantity to put in the numerator is that the units should cancel out except for those of the end result.

So, on the average, the body emits heat at the rate of 96.9 J/s, which is the same as 96.9 watts. This is nearly the same as a glowing 100-W lamp! It's easy to see why a crowded room soon becomes warm! (Don't confuse the 96.9 watts given off by the body with the body's internal temperature of 98.6°F. The closeness of the numerical values is a coincidence. A body's temperature and its rate of expending heat are entirely different from each other.)

21.6 Specific Heat

Almost everyone has noticed that some foods remain hotter much longer than others. Boiled onions and squash on a hot dish, for example, are often too hot to eat when mashed potatoes may be eaten comfortably. The filling of hot apple pie can burn your tongue while the crust will not, even when the pie has just been taken out of the oven. The aluminum covering on a frozen dinner can be peeled off with your bare fingers as soon as it is removed from the oven. A piece of toast may be comfortably eaten a few seconds after coming from the hot toaster, whereas we must wait several minutes before eating soup from a stove as hot as the toaster.

Different substances have different capacities for storing internal energy. If we heat a soup pot of water on a stove, we might find that it requires 15 minutes to raise it from room temperature to its boiling temperature. But if we put an equal mass of iron on the same flame, we would find that it would rise through the same temperature range in only about 2 minutes. For silver, the time would be less than a minute. We find that different materials require different quantities of heat to raise the temperature of a given mass of the material by a specified number of degrees.

Different materials absorb energy in different ways. Some of the energy may increase the overall motion of the atoms or molecules, which raises the temperature. Some may increase the kinetic energy of vibrating and/or rotating atoms within the molecules, and some may go into potential energy; such changes do not raise the temperature. Generally, only part of the energy absorbed raises the temperature.

A gram of water requires 1 calorie of energy to raise the temperature 1ºC. It takes only about one-eighth as much energy to raise the temperature of a gram of iron by the same amount. Water absorbs more heat than iron for the same change in temperature. We say water has a higher **specific heat** (sometimes called *specific heat capacity*).

> The specific heat of any substance is defined as the quantity of heat required to raise the temperature of a unit mass of the substance by 1 degree.

Fig. 21-7   You can touch the aluminum pan of the TV dinner soon after it has been taken from the hot oven, but you'll burn your fingers if you touch the food it contains.

▶ **Question**
Which has a higher specific heat—water or sand?

21.7   The High Specific Heat of Water

Water has a much higher capacity for storing energy than all but a few uncommon materials. A relatively small amount of water absorbs a great deal of heat for a correspondingly small temperature rise. Because of this, water is a very useful cooling agent,

---

▶ **Answer**
Water has the higher specific heat. The temperature of water increases less than the temperature of sand in the same sunlight. Sand's low specific heat, as evidenced by how quickly the surface warms in the morning sun and how quickly it cools at night, affects local climates.

and is used in the cooling system of automobiles and other engines. If a liquid of lower specific heat were used in cooling systems, its temperature would rise higher for a comparable absorption of heat. (Of course, if the temperature of the liquid increases to the temperature of the engine, no further cooling takes place.) Water also takes a long time to cool, a fact that explains why hot-water bottles used to be employed on cold winter nights. (Electric blankets have taken their place.)

This tendency on the part of water to resist changes in temperature improves the climate in many places. The next time you are looking at a world globe, notice the high latitude of Europe. If water did not have a high specific heat, the countries of Europe would be as cold as the northeastern regions of Canada, for both Europe and Canada get about the same amount of sunlight per square kilometer. The Atlantic current known as the Gulf Stream carries warm water northeast from the Caribbean. It holds much of its internal energy long enough to reach the North Atlantic off the coast of Europe, where it then cools. The energy released, one calorie per degree for each gram of water that cools, is carried by the westerly winds over the European continent.

Similarly, the climates differ on the east and west coasts of North America. The winds in the latitudes of North America are westerly. On the west coast, air moves from the Pacific Ocean to the land. Because of water's high specific heat, an ocean does not vary much in temperature from summer to winter. The water is warmer than the air in the winter, and cooler than the air in the summer. In winter the water warms the air that moves over and warms the western coastal regions of North America. In summer, the water cools the air and the western coastal regions are cooled. On the east coast, air moves from the land to the Atlantic Ocean. Land, with a lower specific heat, gets hot in the summer but cools rapidly in the winter. As a result of water's high specific heat and the wind directions, the west coast city of San Francisco is warmer in the winter and cooler in the summer than the east coast city of Washington, D.C., which is at about the same latitude.

The central interior of a large continent usually experiences the greatest extremes of temperature. The high summer and low winter temperatures common in Manitoba and the Dakotas, for example, are largely due to the absence of large bodies of water. Europeans, islanders, and people living near ocean air currents should be glad that water has such a high specific heat. San Franciscans are!

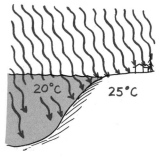

Fig. 21-8 Water has a high specific heat and is transparent, so it takes more energy to heat up than land. Why would its transparency be a factor?

21 | Chapter Review

Concept Summary

Temperature is the quantity that tells how warm or cold something is.
- On the Celsius scale, 0 degrees is the temperature at which water freezes, and 100 degrees is the temperature at which water boils (at standard atmospheric pressure).
- On the Kelvin scale, 0 kelvins is absolute zero, at which a substance has no kinetic energy to give up.
- Temperature is a measure of the average kinetic energy of the molecules or atoms within a substance.

Heat is energy that is transferred between two things because of a temperature difference.
- Heat flows on its own from a hotter to a cooler substance, regardless of the amount of each substance.
- Matter does not contain heat; rather, it contains internal energy.
- Heat is commonly measured in calories, although in SI the joule is preferred.

Specific heat is a measure of how much heat is required to raise the temperature of a unit mass of a substance by one degree.
- Water has a much higher specific heat than almost anything else.
- The high specific heat of water is responsible for the differences in climate between the east and west coasts of large continents.

Important Terms

absolute zero (21.1)
calorie (21.5)
Celsius scale (21.1)
Fahrenheit scale (21.1)
heat (21.2)
internal energy (21.4)
kelvin (21.1)
Kelvin scale (21.1)

kilocalorie (21.5)
specific heat (21.6)
temperature (21.1)
thermal contact (21.2)
thermal equilibrium (21.3)

Review Questions

1. How is temperature commonly measured? (21.1)

2. How many degrees are there between the melting point of ice and the boiling point of water on the Celsius scale? On the Fahrenheit scale? (21.1)

3. How does temperature relate to kinetic energy? (21.1)

4. On the Kelvin scale, what is the temperature of melting ice? Of boiling water? (21.1)

5. Why do scientists favor the kelvin scale? (21.1)

6. Why is it incorrect to say that matter *contains* heat? (21.2)

7. In terms of differences in temperature between objects in thermal contact, in what direction does heat flow? (21.2)

8. What is meant by saying that a thermometer measures its *own* temperature? (21.3)

9. What is meant by thermal equilibrium? (21.3)

10. What is internal energy? (21.4)

11. What is the difference between a calorie and a Calorie? (21.5)

12. What does it mean to say that a certain material has a high or low specific heat? (21.6)

13. Do substances that heat up quickly normally have high or low specific heats? (21.6)

14. How does the specific heat of water compare to that of other common substances? (21.7)

15. Why is the west coast of North America appreciably warmer in winter months and cooler in summer months than the east coast? (21.7)

Think and Explain

1. Why can't you tell whether you have a fever by touching your own forehead?

2. a. Which has the greater amount of internal energy—an iceberg or a hot cup of coffee?
   b. Which has the higher temperature?

3. If you drop a hot rock into a pail of water, the temperature of the rock will decrease and the temperature of the pail of water will increase until the rock and pail of water reach the same temperature. Does the same thing happen if you drop the hot rock into the Pacific Ocean? Explain.

4. If you mix a liter of water at 30°C with a liter of water at 20°C, what will be the final temperature?

5. If you take a bite of hot pizza, the sauce can burn your mouth while the crust, at the same temperature, does not. Explain.

6. In the old days, on a cold winter night it was common to bring a hot object to bed with you. Which would be better—a 10-kilogram iron brick or a 10-kilogram jug of hot water at the same temperature?

7. On a cold winter night, why is it poor economy to allow the bathtub to drain right after taking your bath?

8. Glass bottles have appreciably more mass than aluminum cans. When beverages in glass bottles are cooled, ten times as much heat must be removed as when the same beverages in aluminum cans are cooled. If you were the proprietor of a store and had to pay electric bills to cool drinks, which containers would you prefer—aluminum cans or glass bottles?

9. If you wish to warm 100 kg of water by 15°C for your bath, how much heat is required? (Express your answer in both calories and joules.)

10. Iceland, so named to discourage conquest by expanding empires, is not at all ice-covered like Greenland and Siberia, even though it is nearly on the Arctic Circle. The average winter temperature of Iceland is considerably higher than regions at the same latitude in eastern Greenland and central Siberia. Explain.

22 Thermal Expansion

When the temperature of a substance is increased, its molecules or atoms jiggle faster and tend to move farther apart, on the average. The result is an expansion of the substance. With few exceptions, all forms of matter—solids, liquids, and gases—expand when they are heated and contract when they are cooled.

22.1 Expansion of Solids

If concrete sidewalks and highway paving were laid down in one continuous piece, cracks would appear due to the expansion and contraction brought about by the difference between summer and winter temperatures. To prevent this, the surface is laid in small sections, each one being separated from the next by a small gap which is filled in with a substance such as tar. On a hot summer day, expansion often squeezes this material out of the joints.

Fig. 22-1 This road has buckled because of expansion during the summer heat.

Fig. 22-2 One end of the bridge is fixed, while the end shown rides on rockers to allow for thermal expansion.

The expansion of substances must be allowed for in the construction of structures and devices of all kinds. A dentist uses filling material that has the same rate of expansion as teeth. The aluminum pistons of an automobile engine are just smaller enough in diameter than the steel cylinders to allow for the much greater expansion rate of aluminum. A civil engineer uses reinforcing steel of the same expansion rate as concrete. Long steel bridges commonly have one end fixed while the other rests on rockers (Figure 22-2). The roadway itself is segmented with tongue-and-groove type gaps called expansion joints (Figure 22-3).

Fig. 22-3 This gap is called an expansion joint, and allows the bridge to expand and contract.

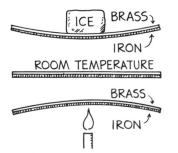

Fig. 22-4 A bimetallic strip. Brass expands (or contracts) more when heated (or cooled) than does iron, so the strip bends as shown.

Different substances expand at different rates. In a **bimetallic strip**, two strips of different metals, say one of brass and the other of iron, are welded or riveted together (Figure 22-4). When the strip is heated, the difference in the amounts of expansion of brass and iron shows up easily. One side of the double strip becomes longer than the other, causing the strip to bend into a curve. On the other hand, when the strip is cooled, it tends to bend in the opposite direction, because the metal that expands the most also shrinks the most. The movement of the strip may be used to turn a pointer, regulate a valve, or close a switch.

A practical application of this is the **thermostat** (Figure 22-5). The back-and-forth bending of the bimetallic coil opens and closes an electric circuit. When the room becomes too cold, the coil bends toward the brass side, and in so doing activates an electrical switch that turns on the heat. When the room becomes too warm, the coil bends toward the iron side, which activates an electrical contact that turns off the heating unit. Refrigerators are equipped with special thermostats to prevent them from becoming either too warm or too cold. Bimetallic strips are used in oven thermometers, electric toasters, automatic chokes on carburetors, and various other devices.

The amount of expansion of a substance depends on the amount of heat it absorbs. If one part of a piece of glass is heated or cooled more rapidly than adjacent parts, the expansion or contraction that results may break the glass. This is especially true for thick glass. Heat-resistant glass (such as Pyrex) is specially formulated to expand very little with increasing temperature.

Fig. 22-5 A thermostat. When the bimetallic coil expands, the mercury rolls away from the electrical contacts and breaks the circuit. When the coil contracts, the mercury rolls against the contacts and completes the electric circuit.

Computational Example: Ratio and Proportion

Steel expands about 1 part in 100 000 for each Celsius degree increase in temperature. This part compared to the whole is a *ratio*, namely

$$\frac{1}{100\ 000}$$

For different lengths of steel, expansion would follow the same proportion. For short lengths of steel, expansion is very small. But consider the expansion of a make-believe snugly-fitting steel pipe that completely circles the earth. How much longer would this 40-million-meter pipe be if its temperature increased by 1°C?

The ratio of its change in length to its full size is the same as the ratio above, so we say

$$\frac{1}{100\ 000} = \frac{?\ \text{meters}}{40\ 000\ 000\ \text{meters}}$$

A little computation will show that the change in length is equal to 400 meters. Here's the interesting part: If such a pipe were elongated by this 400 meters, then there would be a gap between it and the earth's surface. Would the gap be big enough to put this book under? To crawl under? To drive a truck under? How big would this gap be?

We can find the gap by ratio and proportion. The ratio of circumference (C) to diameter (D) for any circle is equal to π (about 3.14). The ratio of the change in circumference (ΔC) to the change in diameter (ΔD) also has the same value. Inserting values, we have

$$\frac{\Delta C}{\Delta D} = \frac{400\ \text{m}}{\Delta D} = 3.14$$

(continued)

Solving for ΔD gives

$$\Delta D = \frac{400 \text{ m}}{3.14} = 127.4 \text{ m}$$

This 127.4 m is the increase in *diameter* of the circular pipe. The increase in *radius* is half this amount, 63.7 m, which is the size of the gap between the earth's surface and the expanded pipe.

So if a world-round steel pipe that fits snugly against the earth were increased in temperature by 1°C, perhaps by people all along its length breathing hard on it, the pipe would expand and stand an amazing 63.7 meters off the ground!

Using ratio and proportion is a powerful way to solve many problems. Another way to solve for the expansion of a material involves a formula ($L = \alpha L_0 \Delta T$). You may encounter this formula in the lab part of your course, but it will not be treated in the text.

▶ **Question**

Why is it advisable to allow telephone lines to sag when stringing them between poles in summer?

22.2 | Expansion of Liquids

Liquids expand appreciably with increases in temperature. When the gasoline tank of a car is filled at a filling station and the car is then parked, the gasoline often overflows the tank after a while. This is because the gasoline is cold when it comes from the underground storage tanks. It then warms up to the temperature of the car as it sits in the car's tank. As the gasoline warms, it expands and overflows the gas tank. Similarly, an automobile radiator filled to the brim with cold water overflows when heated.

▶ **Answer**

Telephone lines are longer in summer, when they are warmer, and shorter in winter, when they are cooler. They therefore sag more on hot summer days than in winter. If they were strung with little sag in summer, in winter they might contract too much and snap.

In most cases the expansion of liquids is greater than the expansion of solids. The gasoline overflowing a car's tank on a hot day is evidence for this. If the tank and contents expanded at the same rate, they would expand together and no overflow would occur. Similarly, if the expansion of the glass of a thermometer were as great as the expansion of the mercury, the mercury would not rise with increasing temperature. The reason the mercury in a thermometer rises with increasing temperature is because the expansion of liquid mercury is greater than the expansion of glass.

Fig. 22-6 When a pot of water filled to the brim is heated on a stove, the water overflows. This shows that the water expands more than its container.

22.3 | Expansion of Water

Almost all liquids will expand when they are heated. Ice-cold water, however, does just the opposite! Water at the temperature of melting ice, 0°C (or 32°F), *contracts* when the temperature is increased. This is most unusual. As the water is heated and its temperature rises, it continues to contract until it reaches a temperature of 4°C. With further increase in temperature, the water then begins to *expand*; the expansion continues all the way to the boiling point, 100°C. The result of this odd behavior is shown graphically in Figure 22-7.

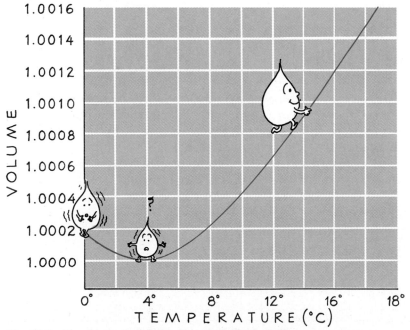

Fig. 22-7 The change in volume of water with increasing temperature.

A given amount of water has its smallest volume—and thus its greatest density—at 4°C. The same amount of water has its largest volume—and smallest density—in its solid form, ice. (Remember, ice floats in water, so it must be less dense than water.) The volume of ice at 0°C is not shown in Figure 22-7. (If it were plotted to the same exaggerated scale, the graph would extend far beyond the top of the page.) After water has turned to ice, further cooling causes it to contract.

The explanation for this behavior of water has to do with the odd crystal structure of ice. The crystals of most solids are arranged in such a way that the solid state occupies a smaller volume than the liquid state. Ice, however, has open-structured crystals (Figure 22-8). These crystals result from the angular shape of the water molecules, plus the fact that the forces binding water molecules together are strongest at certain angles. Water molecules in this open structure occupy a greater volume than they do in the liquid state. Consequently, ice is less dense than water.

Fig. 22-8 Water molecules in their crystal form have an open-structured six-sided arrangement. As a result, water expands upon freezing, and ice is less dense than water.

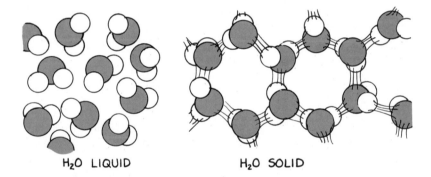

H₂O LIQUID   H₂O SOLID

The reason for the dip in the curve of Figure 22-7 is that two types of volume changes are taking place. The open-structured crystals that make up solid ice are also present, to a much smaller extent, in ice-cold water—a "microscopic slush." At about 10°C all the ice crystals have collapsed. The left-hand graph in Figure 22-9 indicates how the volume of cold water changes due to the collapsing of the microscopic ice crystals.

Fig. 22-9 The collapsing of ice crystals (left) plus increased molecular motion with increasing temperature (center) combine to make water most dense at 4°C (right).

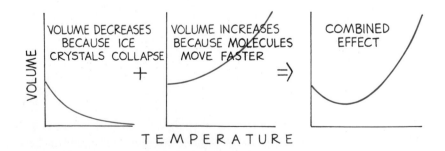

At the same time crystals are collapsing as the temperature increases, increased molecular motion results in expansion. This effect is shown in the center graph in Figure 22-9. Whether ice crystals are in the water or not, increased vibrational motion of the molecules increases the volume of the water.

When we combine the effects of contraction and expansion, the curve looks like the right-hand graph in Figure 22-9 (or Figure 22-7.) This behavior of water is of great importance in nature. Suppose that the greatest density of water were at its freezing point, as is true of most liquids. Then the coldest water would settle to the bottom, and ponds would freeze from the bottom up. Pond organisms would then be destroyed in winter months. Fortunately, this does not happen. The densest water, which settles at the bottom of a pond, is 4 degrees above the freezing temperature. Water at the freezing point, 0°C, is less dense and "floats," so ice forms at the surface, while the pond remains liquid below the ice.

Let's examine this in more detail. Most of the cooling in a pond takes place at its surface, when the surface air is colder than the water. As the surface water is cooled, it becomes more dense, and sinks to the bottom. Water will "float" at the surface for further cooling only if it is equal to or less dense than the water below.

Consider a pond that is initially at, say, 10°C. It cannot possibly be cooled to 0°C without first being cooled to 4°C. And water at 4°C cannot remain at the surface for further cooling unless all the water below has at least an equal density—that is, unless all the water below is at 4°C. If the water below the surface is any temperature other than 4°C, any surface water at 4°C will be denser and sink before it can be further cooled. So before any ice can form, *all* the water in a pond must be cooled to 4°C. Only when this condition is met can the surface water be cooled 3°, 2°, 1°, and 0°C without sinking. Then ice can form.

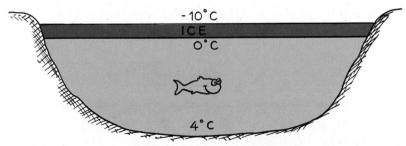

Fig. 22-10   As water is cooled at the surface, it sinks until the entire pond is 4°C. Only then can the surface water cool to 0°C without sinking.

Thus, the water at the surface is first to freeze. Continued cooling of the pond results in the freezing of the water next to the ice, so a pond freezes from the surface downward. In a cold winter the ice will be thicker than in a milder winter.

Very deep bodies of water are not ice-covered even in the coldest of winters. This is because all the water in a lake must be cooled to 4°C before lower temperatures can be reached, and because the winter is not long enough for all the water to be cooled to 4°C. If only some of the water is 4°C, it lies on the bottom. Because of water's high specific heat and poor ability to conduct heat, the bottom of deep lakes in cold regions is a constant 4°C the year round. Fish should be glad that this is so.

> ▶ **Question**
> What was the precise temperature at the bottom of Lake Michigan on New Year's Eve in 1900?

22.4   Expansion and Compression of Gases

Hold an air-filled balloon over a hot stove and you'll easily notice that the size of the balloon increases. This is because the air inside expands with increasing temperature. Gases expand much more than solids and liquids for comparable increases in temperature.

If we warm a volume of air, it will expand. Interestingly enough, if we expand a volume of air, it will not become warmer. Instead, it will become cooler. Warming air and thus making it expand is quite a different process from expanding air and thus making it cool. If you have ever put your hand in the airstream of compressed air, you may have been surprised to find that the air was very cool. A motorist who lets air out of a hot automobile tire is surprised to find the expanding air is cold to the touch.

To increase the temperature of a gas by changing its volume, we must make it not expand but contract. We must compress the gas. If you have ever pumped air into a bicycle tire with a mechanical tire pump, perhaps you noticed that the pump got hot. A tire pump heats up because when you compress the air inside, you make the molecules move faster. The piston in the pump bangs into the molecules of air like a paddle hitting table-tennis balls. The motion of the piston transfers energy to the enclosed air. When you compress air, you heat it.

Interestingly enough, airliners flying at high altitudes, where the air temperature is typically −35°C, are quite comfortable inside—but not because of heaters. The process of compressing outside air to near sea-level cabin pressure would normally heat

Fig. 22-11 Place a dented table-tennis ball in boiling water and you'll remove the dent. Why?

> ▶ **Answer**
> The temperature at the bottom of Lake Michigan, or any deep body of water that experiences freezing temperatures in winter, is 4°C all year round.

the air to a roasting 55°C (131°F). So air conditioners must be used to extract heat from the pressurized air.

> ▶ **Question**
> Would it be correct to say that the reason a gas gets warmer is that its molecules collide more often?

Thus, when you compress air, it gets warmer. When you allow air to expand, it cools. To see why it cools, consider molecules of gas in the midst of a region of expanding air. When molecules collide with others that are approaching at greater speeds, their rebound speeds are increased. When molecules collide with others that are receding, their rebound speeds are lessened. This is easy to see: although a table-tennis ball picks up speed when struck by an approaching paddle, it slows down when it rebounds from a receding paddle. In a region of air that is expanding, molecules will collide, on the average, with more molecules that are receding than are approaching. (This is just the opposite of what happens when air is compressed.) Thus, in expanding air, the average speed of the molecules decreases and the air cools.*

Fig. 22-12 When you compress air, you transfer energy to its molecules as the piston hits them.

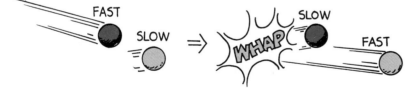

Fig. 22-13 When a molecule collides with a target molecule that is receding, its rebound speed after the collision is less than it was before the collision.

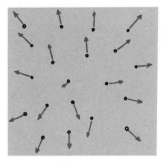

Fig. 22-14 Molecules in a region of expanding air collide primarily with receding molecules rather than with approaching ones. Their rebound speeds therefore tend to decrease and, as a result, the expanding air cools.

▶ **Answer**
No! A common misconception about temperature and molecular motion is that molecules gain energy as they collide against one another—that the more frequently molecules collide, the higher the temperature of the gas should be. This is not true. A pair of molecules bouncing off one another have the same total energy and momentum before and after a collision. Temperature is not a measure of their collision rates but a measure of their kinetic energies. The temperature of a gas would be no different if all the molecules were able to move without colliding with one another. Put this another way: When a gas gets warmer, the molecules collide more often. We can say they collide more often *because* the gas gets warmer. But we can *not* say that the gas gets warmer because the molecules collide more often. The cause produces the effect and not the other way around.

---

* Where does the energy go in this case? It actually goes into the work done on the surrounding air as the expanding air pushes outward.

22.5 | Why Warm Air Rises

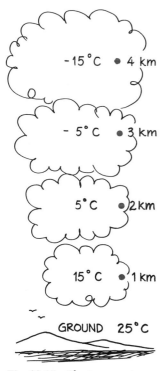

Fig. 22-15 The temperature of a volume of rising dry air decreases by about 10°C for each kilometer of elevation.

We all know that warm air rises. To understand why it rises, think about what you have learned about buoyancy. Something floats when its density is less than that of the surrounding fluid. When air is warmed, it expands and becomes less dense than the surrounding air and is buoyed upward like a balloon. The buoyancy is upward because the air pressure below a region of warmed air is greater than the air pressure above. And the warmed air rises because the buoyant force is greater than its weight.

Because warm air rises, we might at first thought expect the atmosphere to be warmer at higher altitudes. This would mean that mountain tops would be warm and green, and the valleys below cold and snow-covered. But the opposite is the case; the atmosphere is cooler at higher altitudes. Why is this so? The reason is that warm air rises from a region of greater atmospheric pressure at the ground to a region of less pressure above. Because it is moving to a region of less pressure, it *expands*. What happens to the temperature of expanding air? It drops. Measurements show that large volumes of rising dry air cool by about 10°C for each kilometer of elevation (Figure 22-15).

Air flowing over tall mountains or rising in thunderstorms or cyclones may rise several kilometers. If a large volume of dry air at ground level with a comfortable temperature of 25°C were lifted to an altitude of 6 km, the temperature would drop to a frigid −35°C. On the other hand, if air at a typical temperature of −20°C at an altitude of 6 km sinks to the ground, its temperature would be a whopping 40°C. These changes have a pronounced effect on weather, and are produced without external heat input or output. The effect of expansion or compression on gases is quite impressive.

▶ **Answer**
   No! It is true that when you compress air, you heat it; when you expand air, you cool it. But once compressed or expanded, it soon comes to thermal equilibrium with its surroundings. Put your hand on a tank of compressed air and you will find that it has the same temperature as its surroundings. Put your hand in the path of the same air as it escapes and expands from a nozzle and you will find a considerably lower temperature.

22 Chapter Review

Concept Summary

Matter tends to expand when heated and contract when cooled.
- The thermostat is based on a bimetallic coil, made of two metals that expand at different rates.
- Liquids usually expand more than solids.
- Water is highly unusual in that it contracts as it warms from 0°C to 4°C and its solid form (ice) is less dense than its liquid form.
- Gases expand much more than liquids and solids for comparable increases in temperature.

Gases tend to become warmer when compressed and cooler when allowed to expand.
- The temperature changes in a gas during compression or expansion can be explained by changes in the speeds of the molecules during collisions.
- Warm air rises because it is less dense than cool air.
- When air rises over large changes in elevation, its pressure decreases, it expands, and therefore it cools.

Important Terms

bimetallic strip (22.1)
thermostat (22.1)

Review Questions

1. Why are concrete sidewalks made in small sections rather than in one long continuous piece? (22.1)

2. Why does a dentist use fillings that have the same rate of expansion as teeth? (22.1)

3. Why does a bimetallic strip curve when it is heated (or cooled)? (22.1)

4. How is a bimetallic strip or coil used to regulate the temperatures of devices? (22.1)

5. At what temperature is the density of water greatest? (22.3)

6. Ice is less dense than water because of the open spaces in its crystalline structure. But why is water at 0°C less dense than water at 4°C? (22.3)

7. Before water at the surface of a body of water can be cooled to the freezing point, why must all the water below be cooled to at least 4°C? (22.3)

8. Why do lakes and ponds freeze from the top down rather than from the bottom up? (22.3)

9. Why do shallow ponds freeze quickly in winter, and deep ponds often not at all? (22.3)

10. Which expands most for increases in temperature: solids, liquids, or gases? (22.4)

11. What happens to the temperature of a gas that is allowed to expand? (22.4)

12. What happens to the temperature of a gas that is compressed? (22,4)

13. Why does warm air rise? (22.5)

14. Why does warm air generally cool when it rises? (22.5)

15. By about how many degrees Celsius does a large volume of air cool for every kilometer it rises in the atmosphere? (22.5)

Think and Explain

1. Suppose that a metal ball will just barely pass through a metal ring at the same temperature (Figure A).
   a. If the ball is heated and the ring is not, will the heated ball still be able to pass through the ring? Explain.
   b. This time the ring is heated and the ball is not. Will the cool ball be able to pass through the heated ring? Explain.

Fig. A

2. Suppose you cut a small gap in a metal ring (Figure B). If you heat the ring, will the gap become wider or narrower? (Try it and see!)

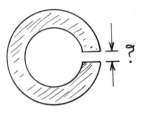

Fig. B

3. Suppose that a metal bar 1 m long expands 0.5 cm when it is heated. How much would it expand if it were 10 m long?

4. Would a bimetallic strip work if the two different metals happened to have the same rates of expansion? Is it important that they expand at different rates? Defend your answer.

5. State whether water at the following temperatures will expand or contract when warmed: 0°C; 4°C; 6°C.

6. If water had a lower specific heat, would lakes be more likely or less likely to freeze in the winter?

7. When you pump a tire with a bicycle pump, the cylinder of the pump becomes hot. Give two reasons why this is so.

8. Everyone knows that warm air rises. So it might seem that the air temperature should be higher at the top of mountains than down below. But the reverse is the case. Explain.

9. a. If a large volume of air initially at 0°C expands while flowing upward alongside a mountain a vertical distance of 1 km, what will its temperature be?
   b. Calculate its temperature when it has risen 5 km.

10. Imagine a giant dry-cleaner's garment bag full of air at a temperature of −10°C floating like a balloon 5 km above the ground with a string hanging from it. Calculate what its approximate temperature would be if you were able to yank it suddenly to the ground.

23 | Transmission of Heat

Heat always tends to pass from warmer objects to cooler objects. If several objects near one another have different temperatures, then those that are warm become cooler, and those that are cool become warmer, until all have a common temperature. This equalization of temperatures is brought about in three ways: by *conduction*, by *convection*, and by *radiation*.

23.1 | Conduction

If you hold an iron rod with one end in a flame, before long the rod becomes too hot to hold. Heat travels through the metal by a process called **conduction**. Conduction of heat takes place within certain materials and from one of these materials to another when they are in direct contact. Materials that are able to conduct heat are known as **conductors** of heat. Metals are the best conductors. Silver and then copper are the very best, and among the common metals, aluminum and then iron are next in order.

Conduction can be explained by the behavior of atoms within the material. In the iron rod example, the flame causes the atoms at the heated end of the rod to vibrate more rapidly. These atoms vibrate against neighboring atoms, which in turn do the same. More importantly, free electrons that can drift through the metal are made to jostle and transfer energy by collisions with atoms and other free electrons within the metal rod.

Materials composed of atoms with "loose" outer electrons are good conductors of heat (as well as electricity). Because metals have the "loosest" outer electrons, they are the best conductors.

Touch a piece of metal and a piece of wood in your immediate vicinity. Which *feels* colder? Which is *really* colder? Your answers

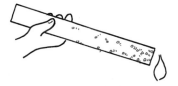

Fig. 23-1 Heat from the flame causes atoms and free electrons in the end of the metal to move faster and jostle against others, which in turn do the same and increase the energy of vibrating atoms down the length of the rod.

323

macroscopic stush particle = ice

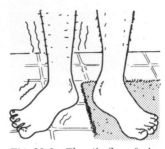

Fig. 23-2 The tile floor feels cold to the bare feet, while the carpet at the same temperature feels warm. This is because tile is a better conductor than carpet.

Fig. 23-3 A "warm" blanket does not provide you with heat; it simply slows the transfer of your body heat to the surroundings.

should be different. If the materials are in the same vicinity, they should have the same temperature: room temperature. Yet, the metal *feels* colder because it is a better conductor; heat easily moves out of your warmer hand and into the cooler metal. Wood, on the other hand, is a poor conductor. Little heat moves out of your hand into the wood, so your hand does not sense that it is touching something cooler. Wood, wool, straw, paper, cork, and polystyrene (Styrofoam) are all poor conductors of heat. Instead, they are called good **insulators** because they delay the transfer of heat. A poor conductor is a good insulator.

Liquids and gases, in general, are good insulators. Air conducts heat very poorly and is a very good insulator. Porous materials that have many small air spaces are good insulators. The good insulating properties of materials such as wool, fur, and feathers are largely due to the air spaces they contain. Be glad that air is a poor conductor, for if it were not, you'd feel quite chilly on a 25°C (77°F) day!

Snowflakes imprison a lot of air in their crystals and are good insulators. Snow slows the escape of heat from the earth's surface, shields Eskimo dwellings from the cold, and provides protection from the cold to animals on cold winter nights. Snow, like any blanket, is not a source of heat; it simply prevents any heat from escaping too rapidly.

Heat is energy that is transferred and is a tangible thing. Cold is not; cold is simply the absence of heat. Strictly speaking, there is no "cold" that passes through a conductor or an insulator. Only heat is transferred. You don't insulate a home to keep the cold out; you insulate to keep the heat in. If the home becomes colder, it is because heat flows out.

It is important to note that no insulator can actually prevent heat from escaping through it. All an insulator can do is slow the rate at which heat penetrates. Even the best insulated warm homes in winter will gradually cool. Insulation delays the transfer of heat.

Fig. 23-4 Snow lasts longest on a well-insulated roof. Thus, the snow patterns reveal the conduction, or lack of conduction, of heat through the roof. How does the insulation of the roof of the center house compare with that of the roofs on either side?

23.2 | Convection

Heat transfer by conduction involves the transfer of energy from atom to atom. Energy moves, but the atoms stay put. Another means of heat transfer is by movement of the heated substance itself. Air in contact with a hot stove ascends and warms the region above. Water heated in a boiler in the basement rises to warm the radiators in the upper floors. This is **convection**, where heating occurs by currents in a fluid.

A simple demonstration illustrates the difference between conduction and convection. With a bit of steel wool, trap a piece of ice at the bottom of a test tube nearly filled with water. Hold the tube by the bottom with your bare hand and place the top in the flame of a Bunsen burner. (See Figure 23-5.) The water at the top will come to a vigorous boil while the ice below remains unmelted. The hot water at the top is less dense and remains at the top. Any heat that reaches the ice must be transferred by conduction, and water is a poor conductor of heat. If you repeat the experiment, only this time hold the test tube at the top by means of tongs and heat the water from below while the ice floats at the surface, the ice will quickly melt. Heat gets to the top by convection, for the hot water rises to the surface, carrying its energy with it to the ice.

Convection occurs in all fluids, whether liquids or gases. Whether we heat water in a pan or heat the air in a room, the process is the same. If the fluid is heated from below, it expands, becomes less dense, and rises. Warm air or warm water rises for the same reason a block of wood rises in water or a helium-filled

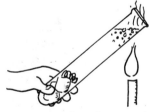

Fig. 23-5 When the test tube is heated at the top, convection is prevented, and heat can reach the ice by conduction only. Since water is a poor conductor, the top water will boil without melting the ice.

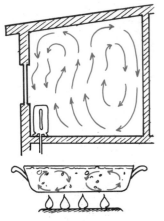

Fig. 23-6 (Top) Convection currents in air. (Bottom) Convection currents in liquid.

balloon rises in air. In effect, convection is an application of Archimedes' principle, for all are buoyed upward by denser surrounding fluid. Cooler fluid then moves to the bottom, and the process continues. In this way, convection currents keep a fluid stirred up as it heats.

▶ **Question**

You can hold your fingers beside the candle flame without harm, but not above the flame. Why?

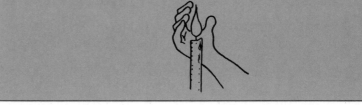

Winds

Winds result from convection currents that stir the atmosphere. Some parts of the earth's surface absorb heat from the sun more readily than others. This causes uneven heating of the air near the surface and creates convection currents.

Convection currents are evident at the seashore. In the daytime, the shore warms more easily than the water. Air over the shore warms faster than air over the water and expands upward. At a given elevation—say 100 m—over the shore, there is now more air pressing down from above than there is over the water at the same elevation (Figure 23-7a). Thus, *at that elevation* the air pressure is greater over the shore than over the water. Air flows from the shore toward the water (Figure 23-7b). This results in more air pressing down on the water from above than is pressing down on the shore. Thus, at ground level the air pressure is greater over the water. So air flows from the water toward the shore (Figure 23-7c). People on the shore experience this as a sea breeze. Cooler air is drawn down from above the water, while the warm air over the shore rises, creating a complete circulation pattern (Figure 23-7d).

At night the process is reversed because the shore cools off more quickly than the water, and then the warmer air is over the sea (Figure 23-7e). Build a fire on the beach and you'll notice that the smoke sweeps inward in the day and seaward at night.

▶ **Answer**

Heat travels upward by air convection. Since air is a poor conductor, very little heat travels sideways.

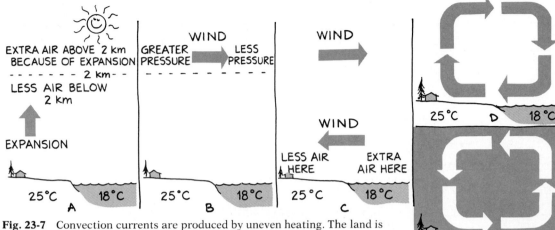

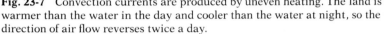

Fig. 23-7 Convection currents are produced by uneven heating. The land is warmer than the water in the day and cooler than the water at night, so the direction of air flow reverses twice a day.

23.3 | Radiation

Heat from the sun is able to pass through the atmosphere and warm the earth's surface. This heat does not pass through the atmosphere by conduction, for air is one of the poorest conductors. Nor does it pass through by convection, for convection begins only after the earth is warmed. We know also that neither convection nor conduction is possible in the empty space between our atmosphere and the sun. Heat must be transmitted by another process. This process is called **radiation**.*

Any energy, including heat, that is transmitted by radiation is called **radiant energy**. Radiant energy is in the form of *electromagnetic waves*. It includes radio waves, microwaves, infrared radiation, visible light, ultraviolet radiation, X rays, and gamma rays. These types of radiant energy are listed in order of wavelength, from longest to shortest.**

---

* This process of heat transmission should not be confused with radioactive radiation, which is given off by the nuclei of radioactive atoms such as uranium and radium.

** Infrared (below-the-red) radiation has longer wavelengths than those of visible light. The longest visible wavelengths are for red light and the shortest are for violet light. Ultraviolet (beyond-the-violet) radiation has shorter wavelengths. Wavelength is treated in more detail in Chapter 25, and electromagnetic waves are covered in more detail in Chapters 27 and 37.

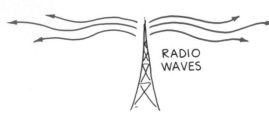

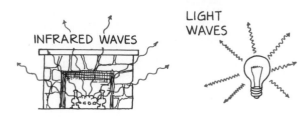

LIGHT WAVES

Fig. 23-8   Types of radiant energy (electromagnetic waves).

Fig. 23-9   Shorter wavelengths are produced when the rope is shaken more vigorously.

Fig. 23-10   Most of the heat from a fireplace goes up the chimney by convection. The heat that warms us comes to us by radiation.

All objects continually give off radiant energy in a mixture of wavelengths. Objects at low temperatures emit long waves, just as long lazy waves are produced when you shake a rope with little energy (Figure 23-9 top). Higher-temperature objects emit waves of shorter wavelengths. Objects of everyday temperatures emit waves mostly in the long-wavelength end of the infrared region, which is between radio and light waves. It is waves in the shorter-wavelength end of the infrared region that our skin experiences as heat. Thus, when we speak of heat radiation, we are speaking of infrared radiation.

When an object is hot enough, some of the radiant energy it emits is in the range of visible light. At a temperature of about 500°C an object begins to emit the longest waves we can see, red light. Higher temperatures produce a yellowish light. At 1200°C all the different waves to which the eye is sensitive are emitted and we see an object as "white hot."

Common sources that give the sensation of heat are the burning embers in a fireplace, a lamp filament, or the sun. All of these emit both infrared radiation and visible light. When this radiant energy falls on other objects, it is partly reflected and partly absorbed. The part that is absorbed increases the internal energy of the objects.

23.4 **Absorption of Radiant Energy**

Absorption and reflection are opposite processes. Therefore, a good absorber of radiant energy reflects very little radiant energy, including the range of radiant energy we call light. So a good absorber appears dark. A perfect absorber reflects no radiant energy and appears perfectly black. The pupil of the eye, for example, allows radiant energy to enter with no reflection and appears perfectly black. (The pink "pupils" that appear in some flash portraits are from light reflected not off the pupil but off the retina at the back of the eyeball.)

Look at the open ends of pipes in a stack. The holes appear black. Look at open doorways or windows of distant houses in the daytime, and they too look black. Openings appear black because the radiant energy that enters is reflected from the inside walls many times and is partly absorbed at each reflection until none remains (Figure 23-12).

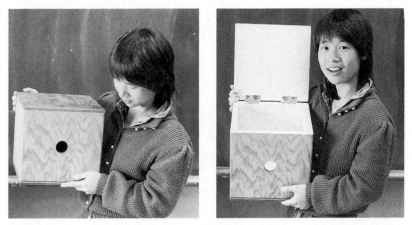

Fig. 23-11 Even though the interior of the box has been painted white, the hole looks black.

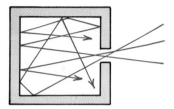

Fig. 23-12 Radiant energy that enters an opening has little chance of leaving before it is completely absorbed.

Good reflectors, on the other hand, are poor absorbers. Light-colored objects reflect more light and heat than dark-colored ones. In summer, light-colored clothing actually keeps people cooler.

23.5 | Emission of Radiant Energy

Good absorbers are also good emitters; poor absorbers are poor emitters. For example, a radio antenna that is constructed to be a good emitter of radio waves will also by its very design be a good receiver of radio waves. A poorly designed transmitting antenna will also be a poor receiver. Interestingly enough, if a good absorber were not also a good emitter, then black objects would all be warmer than lighter-colored objects. But at thermal equilibrium, all objects absorb as much energy as they emit. So we see that good absorbers absorb a lot, and emit a lot as well.

To check this out, find a pair of metal containers of the same size and shape, one having a white or mirrorlike surface and the other a blackened surface (Figure 23-13). Fill the containers with hot water, and place thermometers in the water. You will find that the black container cools faster. The blackened surface is a better emitter. Coffee or tea will stay hot longer in a shiny mirror-like pot than in a blackened one.

Fig. 23-13 When the containers are filled with hot water, the black one cools faster. If filled with cold water, the blackened one warms faster. Why?

If you repeat the same experiment, but this time fill each container with ice water, you will find that the black container warms up faster. A good emitter of radiant energy is also a good absorber. Whether a surface plays the role of emitter or absorber depends on whether its temperature is above or below the surroundings. If the surface is hotter than the surrounding air, for example, it will be a net emitter and will cool. If the surface is colder than the surrounding air, it will be a net absorber and will become warmer.

Buildings that are painted in light colors keep cooler in summer because they reflect much of the incoming radiant energy. They are also poor emitters, so they retain more of their internal energy in the winter. Paint your house a light color.

Fig. 23-14 Anything with a mirrorlike surface reflects most of the radiant energy it encounters. That's why it is a poor absorber of radiant energy.

▶ **Questions**

1. If a good absorber of radiant energy were a poor emitter, how would its temperature compare to its surroundings?

2. Is it more efficient to paint a heating radiator black or silver?

23.6 Newton's Law of Cooling

An object at a different temperature from its surroundings will ultimately come to a common temperature with its surroundings. A hot object will cool as it warms the surroundings, and an object cooler than its surroundings will warm up as the surroundings cool.

The rate of cooling of an object depends on how much hotter the object is than the surroundings. The temperature change per minute of a hot apple pie will be more if the hot pie is put in a cold freezer than if put on the kitchen table. When the pie cools

▶ **Answers**

1. If a good absorber were not also a good emitter, there would be a net absorption of radiant energy and the temperature of good absorbers would be higher than the temperature of the surroundings. A common thermal equilibrium of things around us is possible only because good absorbers are, by their very design, also good emitters.

2. Most of the heat provided by a heating radiator is accomplished by convection, so the color is not really that important. For optimum efficiency, however, the radiators should be painted a dull black so that the contribution by radiation is increased.

in the freezer, the temperature difference is greater. A warm home will leak heat to the cold outside at a greater rate when there is a large difference in the temperature inside and outside. Keeping the inside of your home at a high temperature on a cold day is more costly than keeping it at a lower temperature. If you keep the temperature difference small, the rate of cooling will be correspondingly low.

The rate of cooling of an object—whether by conduction, convection, or radiation—is approximately proportional to the temperature difference ΔT between the object and its surroundings.

$$\text{rate of cooling} \sim \Delta T$$

This is known as **Newton's law of cooling**. (Guess who is credited with discovering this?)

A similar law holds for heating. If an object is cooler than its surroundings, its rate of warming up is also proportional to ΔT. Frozen food will warm up faster in a warm room than in a cold room.

▶ **Question**

Since a hot cup of tea loses heat more rapidly than a lukewarm cup of tea, would it be correct to say that a hot cup of tea will cool to room temperature *before* a lukewarm cup of tea?

23.7 | The Greenhouse Effect

The earth and its atmosphere gain energy when they absorb radiant energy from the sun. This warms the earth. The earth, in turn, emits what is called **terrestrial radiation**, much of which escapes to outer space. The average temperature of the earth is dictated by the difference in radiant energy entering and radiant energy escaping (Figure 23-15). Over the last 500 000 years the average temperature of the earth has fluctuated between 19°C and 27°C and is presently at the high point, 27°C. The earth's temperature increases when either the radiant energy coming in increases, or there is a decrease in the escape of terrestrial radiation.

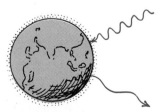

Fig. 23-15 The earth's temperature depends on the energy balance between incoming solar radiation and outgoing terrestrial radiation.

▶ **Answer**

No! Although the rate of cooling is greater for the hotter cup, it has farther to cool to reach thermal equilibrium. The extra time is equal to the time it takes to cool to the initial temperature of the lukewarm cup of tea. Cooling *rate* and cooling *time* are not the same thing.

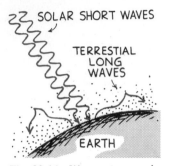

Fig. 23-16 Water vapor and carbon dioxide in the earth's atmosphere absorb terrestrial radiation that would otherwise pass into space.

Radiant energy from the sun, like that from any hot source, is mainly in the range of ultraviolet, visible light, and short-wavelength infrared. These short wavelengths pass rather freely through the atmosphere without being absorbed. They are absorbed at the earth's surface. Radiant energy from the earth, like that from any cooler source, is mainly in the range of long-wavelength infrared. These longer wavelengths are not transmitted as freely through the atmosphere. Much of the terrestrial radiation is absorbed by water vapor and carbon dioxide in the earth's atmosphere. The atmosphere radiates much of this energy back to the earth and keeps the earth's temperature higher than it would be otherwise.

Florists' greenhouses retain energy in a somewhat similar manner, so the overall process is known as the **greenhouse effect** (Figure 23-17). Glass has the property of being transparent to waves of visible light and opaque to ultraviolet and infrared. Glass acts as a sort of one-way valve. It allows visible light to enter but prevents longer waves from leaving. So when radiant energy from the sun in visible-light wavelengths enters through the glass roof, it is absorbed by the interior of the greenhouse—mainly the soil and plants. The soil and plants, in turn, emit infrared radiation. This energy cannot get through the glass and the greenhouse warms up.

Fig. 23-17 Shorter-wavelength radiant energy from the sun enters through the glass roof of the greenhouse. The soil warms up and emits longer-wavelength radiant energy, which is unable to pass through the glass.

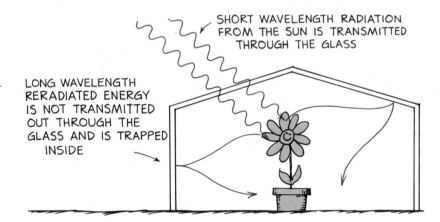

SHORT WAVELENGTH RADIATION FROM THE SUN IS TRANSMITTED THROUGH THE GLASS

LONG WAVELENGTH RERADIATED ENERGY IS NOT TRANSMITTED OUT THROUGH THE GLASS AND IS TRAPPED INSIDE

Interestingly enough, most of the warmth in a greenhouse is due to the role the enclosure plays in preventing convection. Outside the greenhouse, air warmed by contact with the ground rises and is continually being replaced via convection currents. But inside the greenhouse, the soil warms only the air within the enclosure. This limited amount of air warms up easily. So we find that the earth is considerably warmed by the greenhouse effect, while ironically, greenhouses are not.

23	**Chapter Review**

Concept Summary

Heat transfer by conduction takes place within certain materials and from one of these materials to another when they are in direct contact.
- Metals are good conductors.
- Poor conductors, such as wood, cork, polystyrene, and most liquids and gases, are good insulators.

Heat transfer by convection takes place by movement of the heated material itself.
- Convection occurs in all fluids, whether liquids or gases.
- Winds result from convection currents that stir the atmosphere.

Heat transfer by radiation takes place from everything, even when it is surrounded by empty space.
- Energy transmitted by radiation is called radiant energy.
- A good absorber of radiant energy reflects very little radiant energy, including visible light, and thus appears dark.
- Good absorbers of radiant energy are also good emitters.
- According to Newton's law of cooling, the rate of cooling of an object is approximately proportional to the temperature difference between the object and its surroundings.

Important Terms

conduction (23.1)
conductor (23.1)
convection (23.2)
greenhouse effect (23.7)
insulator (23.1)
Newton's law of cooling (23.6)
radiant energy (23.3)
radiation (23.3)
terrestrial radiation (23.7)

Review Questions

1. What is the role of "loose" electrons in heat conductors? (23.1)

2. Why does a piece of room-temperature metal feel cooler to the touch than paper, wood, or cloth? (23.1)

3. What is the difference between a conductor and an insulator? (23.1)

4. Why are materials such as wood, fur, feathers, and even snow good insulators? (23.1)

5. What is meant by saying that cold is not a tangible thing? (23.1)

6. How is heat transferred from one place to another by convection? (23.2)

7. How does Archimedes' principle relate to convection? (23.2)

8. Why does the direction of coastal winds change from day to night? (23.2)

9. A row of dominoes is placed upright, one next to the other. When one is tipped over, it knocks against its neighbor, which does the same in cascade fashion until the whole row collapses. Which of the three means of heat transfer is this most analogous to? (23.1-23.3)

10. What exactly is radiant energy? (23.3)

11. How do the wavelengths of radiant energy vary with the temperature of the radiating source? (23.3)

12. Why does a good absorber of radiant energy appear black? (23.4)

13. Why do eye pupils appear black? (23.4)

14. Is a good absorber of radiation a good emitter or a poor emitter? (23.5)

15. Which will normally cool faster, a black pot of hot tea or a silvered pot of hot tea? (23.5)

16. Which will undergo the greater rate of cooling: a red hot poker in a warm oven or a red hot poker in a cold room (or do both cool at the same rate)? (23.6)

17. Does Newton's law of cooling apply to warming as well as cooling? (23.6)

18. What is terrestrial radiation? (23.7)

19. Solar radiant energy is composed of short waves, yet terrestrial radiation is composed of relatively longer waves. Why? (23.7)

20. a. What does it mean to say that the greenhouse effect is like a one-way valve?
    b. Is the greenhouse effect more pronounced for florists' greenhouses or for the earth's surface? (23.7)

s

the bottom end of a test tube full of water with your bare hand and place top end in a flame, as shown in Figure 23-5. This is even more dramatic if you wedge chunks of ice at the bottom. You'll see the water at the top coming to a boil before the ice melts. What does this show about convection and conduction?

u live where there is snow, do as Benn Franklin did nearly two centuries ago ay samples of light and dark cloth on now. Note the differences in the rate of ng beneath the cloths.

a piece of paper around a thick metal d place it in a flame. Note that the will not catch fire. Can you figure out why? (*Hint*: Paper generally will not ignite until its temperature reaches 233°C.)

Think and Explain

1. At what common temperature will both a block of wood and a piece of metal feel neither hot nor cool when you touch them with your hand?

2. If you stick a metal rod in the snowbank, the end in your hand soon becomes cold. Does cold flow from the snow to your hand?

3. Notice that a desk lamp has small holes near the top of the metal lampshade. How do these holes keep the lamp cool?

4. A cup of tea may be too hot to drink. If the top surface is removed, it will likely then be drinkable. Why?

5. When a space shuttle is in orbit and there appears to be no gravity in the cabin, why can a candle not stay lit?

6. In Montana, the state highway department spreads coal dust on top of snow. When the sun comes out, the snow rapidly melts. Why?

7. Suppose that a person at a restaurant is served coffee before he or she is ready to drink it. In order that the coffee be hottest when the person is ready for it, should cream be added to it right away or just before it is drunk?

8. Will a can of beverage cool just as fast in the regular part of the refrigerator as it will in the freezer compartment? (What physical law do you think about in answering this?)

9. If you wish to save fuel on a cold day, and you're going to leave your warm house for a half hour or so, should you turn your thermostat down a few degrees, down all the way, or leave it at room temperature?

10. If the composition of the upper atmosphere were changed so that it permitted a greater amount of terrestrial radiation to escape, what effect would this have on the earth's climate? Conversely, what would be the effect if the upper atmosphere reduced the escape of terrestrial radiation?

24 Change of State

The matter around us exists in three common **states**—solid, liquid, and gaseous.* Matter can change from one state to another. Ice, for example, is the solid state of H_2O. Add energy, and the rigid molecular structure breaks down to the liquid state, water. Add more energy, and the liquid changes to the gaseous state, vapor, or if hot enough, steam.

The state of matter depends upon its temperature and the pressure that is exerted upon it. Changes of state require a transfer of energy.

24.1 Evaporation

Water in an open container will eventually evaporate, or dry up. The liquid that disappears becomes water vapor in the air. **Evaporation** is a change of state from liquid to gas that takes place at the surface of a liquid.

The temperature of anything is related to the average kinetic energy of the molecules or atoms. In the liquid state, molecules or atoms move in all directions and at different speeds. Many fast-moving molecules have more than the average kinetic energy. If a very fast molecule happens to be moving upward when at or near the surface, it will have enough energy to break free of the liquid. It can leave the surface and fly into the space above the liquid. Now it is a molecule of a gas.

---

* In advanced physics courses, the term *phase* is used instead of *state*.

Fig. 24-1 The cloth covering on the sides of the canteen promotes cooling when it is wet.

It is the faster, more energetic molecules that evaporate, so the average kinetic energy of the molecules remaining in the liquid is lowered. Thus, evaporation is a cooling process. The canteen shown in Figure 24-1 keeps cool because of this fact. The cloth covering on the sides is kept wet so that water can evaporate from it. As the faster-moving water molecules leave the cloth, the temperature of the cloth decreases. Heat moves from the metal canteen to the cloth, and heat then moves from the water inside the canteen to the metal. In this way the water is cooled appreciably below outside air temperature.

When the human body tends to overheat, its sweat glands produce perspiration. The evaporation of perspiration cools us and helps us maintain a stable body temperature. Many animals do not have sweat glands—they must cool themselves by other means (Figures 24-2 and 24-3).

Fig. 24-2 Dogs have no sweat glands (except between the toes). They cool themselves by panting. In this way evaporation occurs in the mouth and within the bronchial tract.

Fig. 24-3 Pigs have no sweat glands and therefore cannot cool by the evaporation of perspiration. That is why they wallow in the mud to cool themselves.

▶ **Question**
Would evaporation be a cooling process if all the molecules in a container of water had the same speed?

▶ **Answer**
Evaporation would not be a cooling process if all the water molecules had the same speed. This is because the average kinetic energy of the molecules would not change as some molecules left the liquid. No change in temperature would take place. In fact, evaporation would not occur at all, as none of the molecules would have enough energy to escape from the liquid.

24.2 | Condensation

The process opposite to evaporation is **condensation**—the changing of a gas into a liquid. The formation of droplets of water on the outside of a chilled soft drink can is an example of condensation.

When gas molecules near the surface of a liquid are attracted to the liquid, they strike the surface with increased kinetic energy, which is absorbed by the liquid. This increases the temperature of the liquid. Thus, condensation is a warming process.

A steam burn, for example, is more damaging than a burn from boiling water of the same temperature. When the steam condenses on the skin, it gives up energy as it changes to a liquid and wets the skin. The fact that steam gives up considerable energy when it changes state is utilized in steam heating systems.

Condensation in the Atmosphere

The air always contains some water vapor. At any given temperature, however, there is a limit to the amount in the air. When this limit is reached, the air is said to be **saturated**. In weather reports, the **relative humidity** indicates how much water vapor is in the air, compared to the limit for that temperature. At a relative humidity of 100%, the air is saturated.

Fig. 24-4 Heat is given up by steam when it condenses inside the radiator.

Fig. 24-5 Molecules of water vapor are more likely to stick together and form a liquid at lower speeds than at higher speeds.

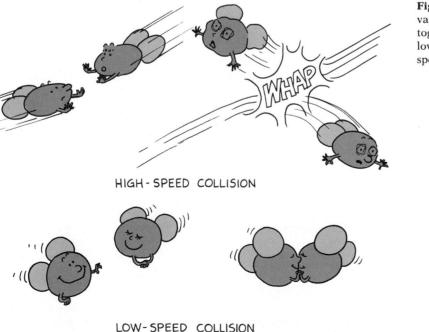

HIGH-SPEED COLLISION

LOW-SPEED COLLISION

More water vapor is required to saturate air when the temperature is high than when it is low. To understand this, think of a fly making a grazing contact with flypaper. At low speed it would surely get stuck, whereas at high speed it stands a greater chance of rebounding into the air. Similarly, when water vapor molecules collide, they are more likely to stick together and become part of a liquid if they are moving at lower speeds (Figure 24-5). At higher speeds, they can bounce apart and remain in the gaseous state. The faster that water vapor molecules move, the less chance there is that they will condense to form droplets. Warmer air therefore can hold a greater number of water molecules in the vapor state than cooler air.

Fog and Clouds

Warm air rises. As it rises, it expands. As it expands, it chills. As it chills, water vapor molecules begin sticking together after colliding rather than bouncing off one another. If there are larger and slower moving particles or ions present, water vapor condenses upon these particles, and we have a cloud.

Fog is basically a cloud that forms near the ground. Flying through a cloud is much like driving through fog. Fog occurs in areas where moist air near the ground cools. For example, moist air that has blown in from over an ocean or lake may pass over cooler land. Some of the water vapor condenses out of the air as it cools, and we have fog.

24.3 Evaporation and Condensation Rates

When you emerge from a shower and step into a dry room, you are likely to feel chilly. This is because evaporation is taking place quickly. If you remain in the shower stall, even with the water off, you do not feel as chilly. This is because when you are in a moist environment, moisture from the air condenses on your skin. This produces a warming effect that counteracts the cooling effect of evaporation. If as much moisture condenses as evaporates, you feel no change in body temperature. That's why you can dry yourself with a towel much more comfortably if you remain in the shower area.

If you leave a dish of water on a table for several days and no apparent evaporation takes place, you might conclude that nothing is happening in the water. You would be mistaken, for much activity is taking place at the molecular level. Evaporation as well as condensation are occuring continuously. In this case, the rates of evaporation and condensation are equal. The number of molecules and amount of energy leaving the liquid's surface

Fig. 24-6 If you feel chilly outside the shower stall, step back inside and be warmed by the condensation of the excess water vapor there.

by evaporation are counteracted by as many molecules and as much energy returning by condensation. The liquid is said to be in **equilibrium**—that is, in a state of balance—since evaporation and condensation have cancelling effects.

Evaporation and condensation normally take place at the same time. If evaporation exceeds condensation, the liquid is cooled. If condensation exceeds evaporation, the liquid is warmed.

24.4 Boiling

Evaporation takes place at the surface of the liquid. A change of state from liquid to gas can also take place beneath the surface of a liquid under the proper conditions. The gas that forms beneath the surface occurs as bubbles. The bubbles are buoyed upward to the surface, where they escape into the surrounding air. This change of state is called **boiling**.

The pressure of the gas within the bubbles in a boiling liquid must be great enough to resist the pressure of the surrounding water and atmospheric pressure. Unless the gas pressure is great enough, the surrounding pressures will collapse any bubbles that may form. At temperatures below the boiling point, the gas pressure would not be great enough, so bubbles do not form until the boiling point is reached.

As the atmospheric pressure is increased, the molecules in the gas are required to move faster to exert a pressure within the bubble that is great enough to counteract the additional atmospheric pressure. So increasing the pressure on the surface of a liquid raises the boiling point of the liquid. Conversely, lowered pressure (as at high altitudes) decreases the boiling point of the liquid. Thus, boiling depends not only on temperature but on pressure as well.

A pressure cooker is based on this fact. A pressure cooker has a tight-fitting lid that does not allow vapor to escape. As the evaporating vapor builds up inside the sealed pressure cooker, pressure on the surface of the liquid is increased, which prevents boiling. This raises the boiling point. The increased temperature of the water cooks the food faster.

It is important to note that it is the high temperature of the water that cooks the food, not the boiling process itself. At high altitudes, water boils at a lower temperature. In Denver, Colorado, the "mile-high city," for example, water boils at 95°C, instead of the 100°C boiling temperature characteristic of sea level. If you try to cook food in boiling water of a lower temperature, you must wait a longer time for proper cooking. A "three-minute"

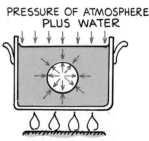

PRESSURE OF ATMOSPHERE PLUS WATER

Fig. 24-7 The motion of molecules in the bubble of steam (much enlarged) creates a gas pressure that counteracts the atmospheric and water pressure against the bubble.

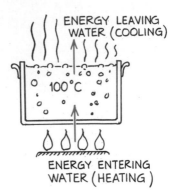

ENERGY LEAVING
WATER (COOLING)

100°C

ENERGY ENTERING
WATER (HEATING)

Fig. 24-8 Heating and boiling are two distinct processes. Heating warms the water, and boiling cools it.

boiled egg in Denver is runny. If the temperature of the boiling water were very low, food would not cook at all.

Boiling, like evaporation, is a cooling process. This is surprising to some people because they associate boiling with heating. Heating water is one thing; boiling is another. When 100°C water at atmospheric pressure is boiling, it is in thermal equilibrium. It is being cooled by boiling as fast as it is being heated by energy from the heat source (Figure 24-8). If cooling did not take place, continued application of heat to a pot of boiling water would result in a continued increase in temperature. A pressure cooker reaches higher temperatures because it prevents boiling, which also prevents cooling.

> ▶ **Question**
>   Since boiling is a cooling process, would it be a good idea to cool your hot and sticky hands by dipping them into boiling water?

24.5   Freezing

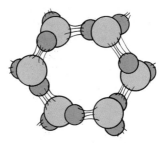

Fig. 24-9 The open structure of pure ice crystals that normally fuse at 0°C. When other kinds of molecules or ions are introduced, crystal formation is interrupted, and the freezing temperature is lowered.

When energy is extracted from water at a temperature of 0°C and at atmospheric pressure, ice is formed. The liquid gives way to the solid state. The change in state from liquid to solid is called **freezing**.

As energy is withdrawn from a liquid, atomic or molecular motion slows until the forces of attraction between the atoms or molecules and atoms cause them to fuse. The atoms or molecules then vibrate about fixed positions and form a solid.

Interestingly enough, if sugar or salt is dissolved in the water, the freezing temperature is lowered. These "foreign" molecules or ions get in the way of water molecules that ordinarily would join together into the six-sided ice-crystal structure. As ice crystals do form, the hindrance is intensified, for the proportion of "foreign" molecules or ions among non-fused water molecules increases. Connections become more and more difficult. In general, dissolving anything in water has this result. Antifreeze is a practical application of this process.

▶ **Answer**
   No, no, no! When we say boiling is a cooling process, we mean that the *water* (not your hands!) is being cooled relative to the higher temperature it would attain otherwise. Because of cooling, it remains at 100°C instead of getting hotter. A dip in 100°C water would be most uncomfortable for your hands!

24.6 | Boiling and Freezing at the Same Time

Suppose that a dish of water at room temperature is placed in a vacuum jar (Figure 24-10). If the pressure in the jar is slowly reduced by a vacuum pump, the water will start to boil. The boiling process takes heat away from the water left in the dish, which cools to a lower temperature. As the pressure is further reduced, more and more of the slower-moving molecules boil away. Continued boiling results in a lowering of temperature until the freezing point of approximately 0°C is reached. Continued cooling by boiling causes ice to form over the surface of the bubbling water. Boiling and freezing are taking place at the same time! This must be witnessed to be appreciated. Frozen bubbles of boiling water are a remarkable sight.

If some drops of coffee are sprayed into a vacuum chamber, they too will boil until they freeze. Even after they are frozen, the water molecules will continue to evaporate into the vacuum until little crystals of coffee solids are left. This is how freeze-dried coffee is made. The low temperature of this process tends to keep the chemical structure of coffee solids from changing. When hot water is added, more of the original flavor of the coffee is retained.

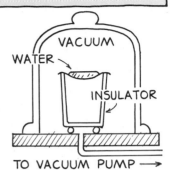

Fig. 24-10   Apparatus to demonstrate that water will freeze and boil at the same time in a vacuum. A gram or two of water is placed in a dish that is insulated from the base by a polystyrene cup.

24.7 | Regelation

The open-structured crystals of ice (Figure 24-9) are readily crushed by the application of pressure. Whereas ice normally melts at 0°C, the application of pressure lowers the melting point. The crystals are simply crushed to the liquid state. At twice standard atmospheric pressure, the melting point is lowered to −0.007°C. Quite a bit more pressure must be applied for an observable effect.

When the pressure is removed, refreezing occurs. This phenomenon of melting under pressure and freezing again when the pressure is reduced is called **regelation**. It is one of the properties of water that make it different from other materials.

You can see regelation in operation if you suspend a fine wire that supports heavy weights over an ice cube, as shown in Figure 24-11. The wire will slowly cut its way through the ice, but its track will refill with ice. You'll see the wire and weights fall to the floor, leaving the ice in a single solid piece!

To make a snowball, you use regelation. When you compress the snow with your hands, you cause a slight melting, which

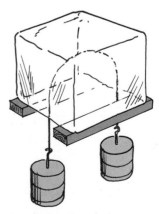

Fig. 24-11   Regelation.

helps to bind the snow into a ball. Making snowballs is difficult in very cold weather because the pressure you can apply may not be enough to melt the snow.

An ice skater skates on a thin film of water between the blade and the ice, which is produced by the blade pressure and friction. As soon as the pressure is released, the water refreezes.

24.8 Energy and Changes of State

If you heat a solid sufficiently, it will melt and become a liquid. If you heat the liquid, it will vaporize and become a gas. Energy must be put into a substance to change its state in the direction from solid to liquid to gas. Conversely, energy must be extracted from a substance to change its state in the direction from gas to liquid to solid (Figure 24-12).

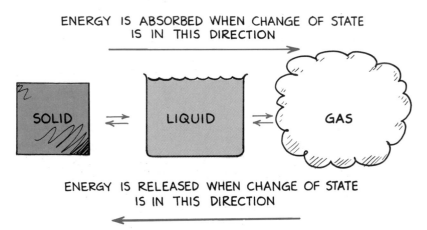

ENERGY IS ABSORBED WHEN CHANGE OF STATE
IS IN THIS DIRECTION

SOLID ⇄ LIQUID ⇄ GAS

ENERGY IS RELEASED WHEN CHANGE OF STATE
IS IN THIS DIRECTION

Fig. 24-12 Energy changes with change of state.

The general behavior of many substances can be illustrated with a description of the changes of state of H_2O. To make the numbers simple, suppose you have a 1-gram piece of ice at a temperature of $-50°C$ in a closed container, and it is put on a stove to heat. A thermometer in the container reveals a slow increase in temperature up to 0°C. At 0°C, the temperature stops rising, yet heat is continually added. This heat melts the ice.

In order for the whole gram of ice to melt, 80 calories of heat energy must be absorbed by the ice. Not until all the ice melts does the temperature again begin to rise. Each additional calorie absorbed by the water increases its temperature by 1°C until it reaches its boiling temperature, 100°C. Again, as heat is added, the temperature remains constant while more and more of the gram of water is boiled away and becomes steam. The water must absorb 540 calories of heat energy to vaporize the whole gram. Finally, when all the water has become steam at 100°C, the temperature begins to rise once more. It will continue to rise as long as you continue to add heat. This process is graphed in Figure 24-13.

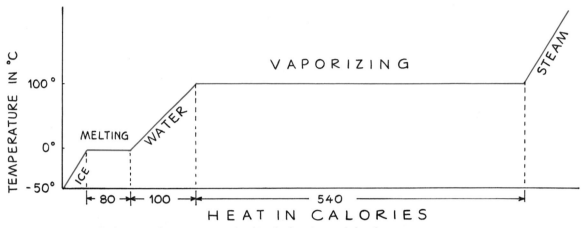

Fig. 24-13   A graph showing the energy involved in the heating and the change of state of 1 g of H₂O.

The change of state sequence is reversible. If the molecules in a gram of steam condense to form boiling water, they liberate 540 calories of heat to the environment. When the water is cooled from 100°C to 0°C, 100 additional calories are liberated to the environment. When ice water fuses to become solid ice, 80 more calories of energy are released by the water. The vaporization of water requires 540 calories per gram, and the fusion (freezing) of water requires 80 calories per gram.[*]

The 540 calories required to vaporize a gram of water is a relatively large amount of energy—much more than would be required to bring a gram of ice at absolute zero to boiling water at 100°C. Although the molecules in steam and boiling water at 100°C have the same average kinetic energy, steam has more potential energy because the molecules are free of each other and are not held together in the liquid. Steam contains a vast amount of energy that can be released during condensation.

---

[*]   These values are known as the *heat of vaporization* and *heat of fusion*, respectively. In SI units, the heat of vaporization of water is 2.26 megajoules per kilogram (mJ/kg), and the heat of fusion of water is 0.335 mJ/kg.

▶ **Questions**
1. How much energy is released when a gram of steam at 100°C condenses to water at 100°C?

2. How much energy is released when a gram of steam at 100°C condenses to ice water at 0°C?

The large value of 540 calories per gram explains why under some conditions hot water will freeze faster than warm water.* This occurs for water hotter than 80°C, and is evident when the surface area that cools by rapid evaporation is large compared to the amount of water involved—as when a car is washed with hot water on a cold winter day, or a skating rink is flooded with hot water to melt and smooth out the rough spots and freeze quickly. The rate of cooling by rapid evaporation is very high because evaporation draws at least 540 calories from each gram of water left behind. In comparison to lukewarm water that cools principally by thermal conduction, the temperature of hot evaporating water takes a nosedive and passes below the temperature of the lukewarm water. Evaporation is truly a cooling process.

Fig. 24-14 When a car is washed on a cold day, hot water will freeze more readily than warm water because of the energy that leaves the hot water during rapid evaporation.

▶ **Answers**
1. One gram of steam at 100°C gives up 540 calories of energy when it condenses to become water at the same temperature.

2. The same steam will give up 640 calories to reach ice water—540 calories to change state to water, and 100 more calories at the rate of 1 calorie per degree to cool to 0°C.

---

*   Hot water will not freeze before cold water, but will freeze before lukewarm water. Water at 100°C, for example, will freeze before water warmer than 60°C, but not before water cooler than 60°C.

> ▶ **Question**
>
> Consider 10 grams of water at 100°C. What will be the temperature of the remaining 9 grams of water if 1 gram rapidly evaporates?

The cooling cycle of a refrigerator neatly employs the energy interchanges that occur with changes of state of a refrigeration fluid—usually Freon. The liquid is pumped into the cooling unit, where it turns into a gas and draws heat from the things stored in the food compartment. The gas is then directed outside the cooling unit to coils located in the back. As the gas condenses in the coils, appropriately called condensation coils, heat is given off to the surrounding air. The liquid returns to the cooling unit, and the cycle starts again. A motor pumps the fluid through the system, where it is made to undergo the cyclic processes of vaporization and condensation. The next time you're near a refrigerator, place your hand near the condensation coils in the back and you will feel the heat that has been extracted from within the cooling unit.

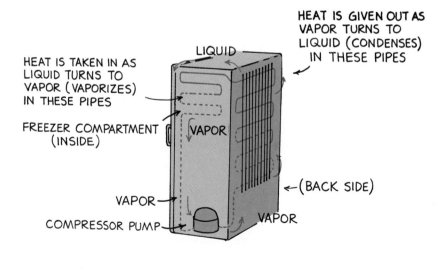

Fig. 24-15 The refrigeration cycle in a common refrigerator.

HEAT IS GIVEN OUT AS VAPOR TURNS TO LIQUID (CONDENSES) IN THESE PIPES

LIQUID

HEAT IS TAKEN IN AS LIQUID TURNS TO VAPOR (VAPORIZES) IN THESE PIPES

FREEZER COMPARTMENT (INSIDE)

VAPOR

←(BACK SIDE)

VAPOR

COMPRESSOR PUMP

VAPOR

▶ **Answer**

40°C, if we assume all the energy for evaporation is supplied by the remaining water. Why? Because 540 calories were taken away by the 1 gram that evaporated. This means that each of the remaining 9 grams gives up 60 calories (since [540 cal] ÷ 9 = 60 cal). Water cools at a rate of 1 degree Celsius per calorie, which means it drops 60 degrees to 40°C (since 100°C − 60°C = 40°C).

In the lab part of your course, if you measure the relationship between heat Q added or subtracted from a substance and its change in temperature ΔT, you'll use the formula $Q = mc\,\Delta T$ as a guide (where m is the mass of the substance and c is its specific heat).

An air conditioner employs the same principles. It simply pumps heat from one part of the unit to another. When the roles of vaporization and condensation are reversed, the air conditioner becomes a heater.

> ▶ **Question**
>
> Suppose you wanted to cool a stuffy kitchen on a hot summer day. Would it be a smart idea simply to turn the refrigerator up full blast and keep its door open?

A way that some people judge the hotness of a clothes iron is to touch it briefly with a finger. This is also a way to burn the finger—unless it is first wet. This is because energy that ordinarily would go into burning the finger goes, instead, into changing the state of the moisture on it. The energy converts the moisture to a vapor, which then provides an insulating layer between the finger and the hot surface.

Similarly, you may have seen news photos or heard stories about people walking barefoot without harm over red-hot coals from firewood. This practice can be highly dangerous if the conditions are not just right. The feet should be moist, perhaps from natural perspiration. The moisture absorbs some of the heat and provides an insulating layer between the soles of the feet and the coals. The wood coals must have a low enough thermal conductivity so that if the feet were in contact very briefly, not much heat could pass through the insulating layer and into the feet. (CAUTION: Never try this on your own; even experienced "firewalkers" have received bad burns when the conditions were not just right.) The same principle is more safely illustrated when a hand is briefly placed in a hot oven. Very little heat is transferred to one's hand because of the low thermal conductivity of air. But touch a metal pan at the same high temperature, and OUCH! Metal is a good conductor. Temperature is one thing; heat transfer is another.

In brief, a solid absorbs energy when it melts; a liquid absorbs energy when it vaporizes. Conversely, a gas releases energy when it liquifies; a liquid releases energy when it solidifies.

▶ **Answer**

Opening the door of the refrigerator would be a senseless idea, for all the device does is pump heat energy from its inside to its outside. When the door is open, both inside and outside are part of the kitchen. Not only would the scheme fail to cool the kitchen, but the operation of the motor would actually heat the kitchen! The kitchen could be cooled only if heat from the condensation coils were discharged to a region outside the kitchen altogether.

24 | Chapter Review

Concept Summary

During evaporation, a liquid changes state at its surface and becomes a gas.
- Evaporation is a cooling process.

During condensation, a gas changes state and becomes a liquid.
- Condensation is a warming process.
- Warmer air holds more water vapor than an equal amount of cooler air.
- Clouds and fog form when air cools and is unable to hold as much water vapor.

When evaporation and condensation occur at the same rate, the liquid is in equilibrium and there is no change in liquid volume.
- A liquid is in equilibrium when the surrounding air is saturated with its vapor.
- In dry air, water evaporates much faster than it condenses; in humid air, it evaporates only slightly faster than it condenses.

During boiling, a liquid changes state at any place within the liquid, and gas bubbles form.
- The boiling temperature of a liquid depends on the pressure on its surface.
- Boiling, like evaporation, is a cooling process.

During freezing, a liquid changes state and becomes a solid.
- The freezing temperature of a liquid is lowered by adding other substances to it.
- During regelation, ice melts under pressure and refreezes when the pressure is removed.

During changes of state, energy is given off or taken in.
- While a substance is changing state, its temperature does not change.
- Much more energy is given off when water vapor condenses than when an equal mass of water freezes.

Important Terms

boiling (24.4)
condensation (24.2)
equilibrium (24.3)
evaporation (24.1)
freezing (24.5)
regelation (24.7)
relative humidity (24.2)
saturated (24.2)
state (24.1)

Review Questions

1. Do all the molecules or atoms in a liquid have about the same speed, or much different speeds? (24.1)

2. Exactly what is evaporation, and why is it a cooling process? What is it that cools? (24.1)

3. Why does a hot dog pant? (24.1)

4. What is condensation, and why is it a warming process? What is it that warms? (24.2)

5. Why is a steam burn more damaging than a burn with boiling water of the same temperature? (24.2)

6. Which will contain more water vapor—warm air or cool air? (24.2)

7. Why does warm moist air form clouds when it rises? (24.2)

8. Why do you feel less chilly if you dry yourself inside the shower stall after taking a shower? (24.3)

9. What evidence would be a clue that the rate of evaporation equaled the rate of condensation? (24.3)

10. What is the difference between evaporation and boiling? (24.4)

11. Why does the temperature at which a liquid boils depend on atmospheric pressure? (24.4)

12. Why is a pressure cooker even more useful when cooking food in the mountains than when cooking at sea level? (24.4)

13. Why does antifreeze or any soluble substance put in water lower its freezing temperature? (24.5)

14. How can water be made both to boil and freeze at the same time? (24.6)

15. What exactly is regelation, and what does it have to do with the open-structured crystals in ice? (24.7)

16. How many calories are needed to
    a. raise the temperature of one gram of water by 1°C?
    b. melt one gram of ice at 0°C?
    c. vaporize one gram of boiling water at 100°C? (24.8)

17. Does a vapor give off or absorb energy when it turns into a liquid? (24.8)

18. In a freezing environment, why does water at 100°C freeze before water at 65°C? (24.8)

19. In a refrigerator, does the food cool when a vapor turns to a liquid, or vice versa? (24.8)

20. Why is it important that a finger be wet before it is touched briefly to a hot clothes iron? (24.8)

Activity

Watch the spout of a teakettle of boiling water. Notice that you cannot see the steam that comes from the spout. The cloud you see farther away from the spout is not steam but condensed water droplets. Steam is invisible. Now hold a candle in the cloud of condensed steam. Can you explain your observations?

Think and Explain

1. You can determine wind direction if you wet your finger and hold it up into the air. Explain.

2. Give two reasons why pouring a hot cup of coffee into a saucer results in faster cooling.

3. At a picnic, why would wrapping a bottle in a wet cloth be a better method of cooling than placing the bottle in a bucket of cold water?

4. Can you ever heat a substance without raising its temperature? Give an example.

5. Why is the constant temperature of boiling water on a hot stove evidence that boiling is a cooling process? (What would happen to its temperature if boiling were not a cooling process?)

6. If you are boiling some potatoes in a pot of water, will they cook faster if the water is boiling vigorously than if the water is boiling gently?

7. Would regelation occur if ice crystals did not have an open structure? Explain.

8. People who live where snowfall is common will attest to the fact that air temperatures are always higher on snowy days than on clear days. Some people get cause and effect mixed up when they say that snowfall cannot occur on very cold days. Explain.

9. If a large tub of water is kept in a small unheated room, even on a very cold day the temperature of the room will not go below 0°C. Why not?

10. On cold winter days the windows of your warm home sometimes get wet on the inside. Why is this so?

IV Sound and Light

ISN'T THIS DISC THE PITS? I MEAN, THERE ARE BILLIONS OF THEM, CAREFULLY INSCRIBED IN AN ARRAY THAT IS SCANNED BY A LASER BEAM MILLIONS OF PITS PER SECOND. DIGITIZED MUSIC! BUT THE BEAUTY OF DISC RECORDINGS IS NOT CONFINED TO SOUND -- JUST LOOK AT THE BRILLIANT SPECTRUM OF COLORS DIFFRACTED BY THE EVENLY SPACED ROWS OF PITS. I FIND IT EVEN MORE BEAUTIFUL WHEN I KNOW *WHY* IT'S SO COLORFUL AND *WHY* IT SOUNDS SO GOOD. THAT'S THE PHYSICS OF IT ALL!

25 Vibrations and Waves

All around us we see things that wiggle and jiggle. Even things too small to see, such as atoms, are constantly wiggling and jiggling. A wiggle in time is called a **vibration**. A vibration cannot exist in one instant, but needs time to move to and fro. Strike a bell and the vibrations will continue for some time before they die down.

A wiggle in space and time is called a **wave**. A wave cannot exist in one place but must extend from one place to another. Light and sound are both forms of energy that move through space as waves. This chapter is about vibrations and waves, and the following chapters continue with the study of sound and light.

25.1 Vibration of a Pendulum

Suspend a stone at the end of a string and you have a simple pendulum. Pendulums swing to and fro with such regularity that they have long been used to control the motion of clocks. Galileo discovered that the time a pendulum takes to swing to and fro through small angles does not depend on the mass of the pendulum or on the distance through which it swings. The time of a to-and-fro swing—called the **period**—depends only on *the length of the pendulum* and *the acceleration of gravity.**

---

* The exact relationship for the period T of a simple pendulum is

$$T = 2\pi \sqrt{L/g}$$

where L is the length of the pendulum, and g is the acceleration of gravity.

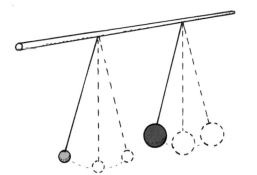

Fig. 25-1 Two pendulums of the same length have the same period as long as they swing through small angles (less than 10° or so). The period does not depend on the mass or the distance moved.

A long pendulum has a longer period than a shorter pendulum; that is, it swings to and fro more slowly than a short pendulum. When walking, we allow our legs to swing with the help of gravity, like a pendulum. In the same way that a long pendulum has a greater period, a person with long legs tends to walk with a slower stride than a person with short legs. This is most noticeable in long-legged animals such as giraffes, horses, and ostriches, which run with a slower gait than do short-legged animals such as dachshunds, hamsters, and mice.

25.2 | Wave Description

Figure 25-2 shows the patterns traced by dark sand that leaks from a pendulum bob moving back and forth in a single plane. When the conveyor belt is not moving (left), the sand traces out a straight line. More interestingly, when the conveyor belt is moving at constant speed (right), the sand traces out a special curve known as a **sine curve**. This curve is a pictorial representation of a wave.

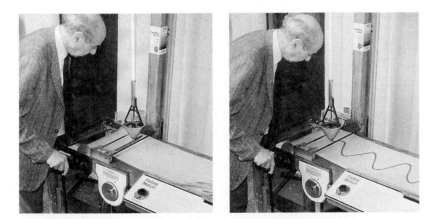

Fig. 25-2 Frank Oppenheimer, founder of the Exploratorium science museum in San Francisco, demonstrates that a pendulum swinging to and fro traces out a straight line over a stationary surface, and a sine curve when the surface moves at constant speed.

The sine curve in Figure 25-3 depicts a wave. The high points are called the **crests**, and the low points are the **troughs**. The straight dashed line represents the "home" position, or midpoint of the vibration. The term **amplitude** refers to the distance from the midpoint to the crest of the wave or, equivalently, from the midpoint to the trough. So the amplitude is equal to the maximum displacement of the pendulum from its position of rest.

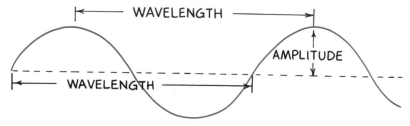

Fig. 25-3 A sine curve.

The **wavelength** of a wave is the distance from the top of one crest to the top of the next one. Or equivalently, the wavelength is the distance between successive identical parts of the wave. The wavelengths of waves at the beach are measured in meters, the wavelengths of ripples in a pond in centimeters, and the wavelengths of light in millionths of a meter (micrometers).

How frequently a vibration occurs is described by its **frequency**. The frequency of a vibrating pendulum specifies the number of to-and-fro vibrations it makes in a given time (usually one second). A complete to-and-fro swing is one vibration. If it occurs in one second, the frequency is one vibration per second. If two vibrations occur in one second, the frequency is two vibrations per second.

The unit of frequency is called the **hertz** (Hz), after Heinrich Hertz, who demonstrated radio waves in 1886. One vibration per second is 1 hertz; two vibrations per second is 2 hertz, and so on. Higher frequencies are measured in kilohertz (kHz), and still higher frequencies in megahertz (MHz). AM radio waves are broadcast in kilohertz, while FM radio waves are broadcast in megahertz. A station at 960 kHz on the AM radio dial, for example, broadcasts radio waves that have a frequency of 960 000 vibrations per second. A station at 101 MHz on the FM dial broadcasts radio waves with a frequency of 101 000 000 hertz. These radio-wave frequencies are the frequencies at which electrons are forced to vibrate in the antenna of a radio station's transmitting tower. The source of all waves is something that vibrates. The frequency of the vibrating source and the frequency of the wave it produces are the same.

If an object's frequency is known, its period can be calculated, and vice versa. Suppose, for example, that a pendulum makes

Fig. 25-4 Electrons in the transmitting antenna of a radio station at 960 kHz on the AM dial vibrate 960 000 times each second and produce 960-kHz radio waves.

two vibrations in one second. Its frequency is 2 Hz. The time needed to complete one vibration—that is, the period of vibration—is $\frac{1}{2}$ second. Or if the vibration period is 3 Hz, then the period is $\frac{1}{3}$ second. The frequency and period are the inverse of each other:

$$\text{frequency} = \frac{1}{\text{period}}$$

or vice versa,

$$\text{period} = \frac{1}{\text{frequency}}$$

> ▶ **Questions**
> 1. What is the frequency in vibrations per second of a 100-hertz wave?
>
> 2. The Sears Building in Chicago sways back and forth at a vibration frequency of about 0.1 Hz. What is its period of vibration?

25.3 | Wave Motion

Most of the information around us gets to us in some form of wave. Sound is energy that travels to our ears in the form of one kind of wave. Light is energy that comes to our eyes in the form of a different kind of wave (an electromagnetic wave). The signals that reach our radio and television sets also travel in the form of electromagnetic waves.

When energy is transferred by a wave from a vibrating source to a distant receiver, there is no transfer of matter between the two points. To see this, think about the very simple wave produced when one end of a horizontally-stretched string is shaken up and down (Figure 25-5). After the end of the string is shaken, a rhythmic disturbance travels along the string. Each part of the string moves up and down while the disturbance moves horizontally along the length of the string. It is the disturbance that moves along the length of the string, not parts of the string itself.

Fig. 25-5 When the string is shaken up and down, a disturbance moves along the length of the string.

▶ **Answers**

1. A 100-hertz wave vibrates 100 times per second.

2. The period is 1/frequency = 1/(0.1 Hz) = 1/(0.1 vibrations/s) = 10 s. Thus, each vibration takes 10 seconds.

Drop a stone in a quiet pond and you'll produce a wave that moves out from the center in expanding circles. It is the disturbance that moves, not the water, for after the disturbance passes, the water is where it was before the wave was produced (Figure 25-6).

When someone speaks to you from across the room, the sound wave is a disturbance in the air that travels across the room. The air molecules themselves do not move along, as they would in a wind. The air, like the rope and the water in the previous examples, is the medium through which wave energy travels. The energy transferred from a vibrating source to a receiver is carried by a *disturbance* in a medium, not by matter moving from one place to another within the medium.

Fig. 25-6 A circular water wave in a still pond.

| 25.4 | **Wave Speed** |

The speed of a wave depends on the medium through which the wave travels. Sound waves, for example, travel at speeds of about 330 m/s to 350 m/s in air (depending on the temperature), and about four times as fast in water. Whatever the medium, the speed of the wave is related to the frequency and wavelength of the wave. You can understand this by considering the simple case of water waves. Imagine that you fix your eyes at a stationary point on the surface of water and observe the waves passing by this point. If you count the number of crests that pass each second (the frequency) and also observe the distance between crests (the wavelength), you can then calculate the horizontal distance a particular crest travels each second.

Fig. 25-7 If the wavelength is 1 m, and one wavelength per second passes the pole, then the speed of the wave is 1 m/s.

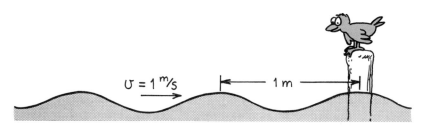

For example, if two crests pass a stationary point each second, and if the wavelength is 3 meters, then 2 × 3 meters of waves pass by in 1 second. The waves therefore travel at 6 meters per second. We can say the same thing this way:*

---

* It is customary to express this equation as $v = f\lambda$, where v is wave speed, f is wave frequency, and λ (Greek letter lambda) is wavelength.

$$\text{wave speed} = \text{frequency} \times \text{wavelength}$$

This relationship holds for all kinds of waves, whether they are water waves, sound waves, or light waves.

Table 25-1 shows some frequencies and corresponding wavelengths of sound in air at the same temperature. Notice that the product of frequency and wavelength is the same for each example—340 m/s in this case. During a concert, you do not hear the high notes in a chord before you hear the low notes. The sounds of all instruments reach you at the same time. Notice that low frequencies have long wavelengths, and high frequencies have relatively short wavelengths. Frequency and wavelength vary inversely to produce the same wave speed for all sounds.

Table 25.1 Sound Waves		
Frequency (Hz)	Wavelength (m)	Wave Speed (m/s)
160	2.13	340
264	1.29	340
396	0.86	340
528	0.64	340

Computational Example

If a train of freight cars, each 10 m long, rolls by you at the rate of 2 cars each second, what is the speed of the train?

This can be seen in two ways, the Chapter 2 way and the Chapter 25 way.

From Chapter 2 recall that

$$v = d/t = [2 \times (10 \text{ m})]/(1 \text{ s}) = 20 \text{ m/s}$$

where d is the length of train that passes you in time t.

Here in Chapter 25 we compare the train to wave motion, where the wavelength corresponds to 10 m, and the frequency is 2 Hz. Then

$$\begin{aligned} \text{wave speed} &= \text{frequency} \times \text{wavelength} \\ &= (2 \text{ Hz}) \times (10 \text{ m}) = 20 \text{ m/s} \end{aligned}$$

One of the nice things about physics is that different ways of looking at things produce the same answer. When this doesn't happen, and there is no error in computation, then the validity of one (or both!) of those ways is suspect.

> ▶ **Questions**
> 1. If a water wave vibrates up and down 2 times each second and the distance between wave crests is 1.5 m, what is the frequency of the wave? What is its wavelength? What is its speed?
>
> 2. What is the wavelength of a 340-Hz sound wave when the speed of sound in air is 340 m/s?

25.5 Transverse Waves

Suppose you create a wave along a rope by shaking the free end up and down (Figure 25-8). In this case the motion of the rope (shown by the up and down arrows) is at right angles to the direction in which the wave is moving. Whenever the motion of the medium (the rope in this case) is at right angles to the direction in which a wave travels, the wave is a **transverse wave**.

Fig. 25-8 A transverse wave.

Waves in the stretched strings of musical instruments and upon the surfaces of liquids are transverse. As Chapter 27 will show, the electromagnetic waves which make up radio waves and light are also transverse.

25.6 Longitudinal Waves

Not all waves are transverse. Sometimes the particles of the medium move to and fro in the same direction in which the wave travels. The particles move *along* the direction of the wave

---

▶ **Answers**
1. The frequency of the wave is 2 Hz; its wavelength is 1.5 m; and its wave speed is

$$\text{frequency} \times \text{wavelength} = (2 \text{ Hz}) \times (1.5 \text{ m}) = 3 \text{ m/s}.$$

2. The wavelength of the 340-Hz sound wave must be 1 m. Then

$$\text{wave speed} = (340 \text{ Hz}) \times (1 \text{ m}) = 340 \text{ m/s}.$$

rather than at right angles to it. This kind of wave is a **longitudinal wave**.

Both transverse and longitudinal waves can be demonstrated with a loose-coiled spring, or "Slinky," as shown in Figure 25-9. A transverse wave is demonstrated by shaking the end of a Slinky up and down. A longitudinal wave is demonstrated by shaking the end of the Slinky in and out. In this case we see that the medium vibrates parallel to the direction of energy transfer. Sound waves are longitudinal waves, and will be discussed in detail in the next chapter.

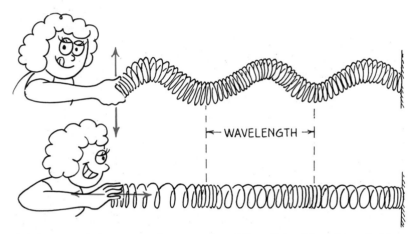

WAVELENGTH

Fig. 25-9 Both waves transfer energy from left to right. When the end of the Slinky is shaken up and down (top), a transverse wave is produced. When it is shaken in and out, a longitudinal wave is produced.

25.7 Interference

A material object such as a rock will not share its space with another rock. But more than one vibration or wave can exist at the same time in the same space. If you drop two rocks in water, the waves produced by each can overlap and form an **interference pattern**. Within the pattern, wave effects may be increased, decreased, or neutralized.

When the crest of one wave overlaps the crest of another, their individual effects add together. The result is a wave of increased amplitude. This is called **constructive interference**, or reinforcement (Figure 25-10 top). When the crest of one wave overlaps the trough of another, their individual effects are reduced. The high part of one wave simply fills in the low part of another. This is called **destructive interference**, or cancellation (Figure 25-10 bottom).

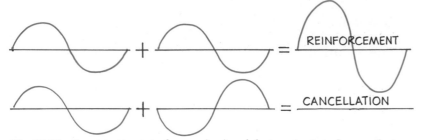

Fig. 25-10   Constructive interference (top) and destructive interference (bottom) in a transverse wave.

Wave interference is easiest to see in water. Figure 25-11 shows the interference pattern made when two vibrating objects touch the surface of water. The gray "spokes" are regions where a crest of one wave overlaps the trough of another to produce regions of zero amplitude. At points along these regions, the waves from the two objects arrive "out of step." We say that they are **out of phase** with one another. The dark and light striped regions are where the crests of one wave overlap the crests of the other, and the troughs overlap as well. In these regions, the two waves arrive "in step." They are **in phase** with each other.

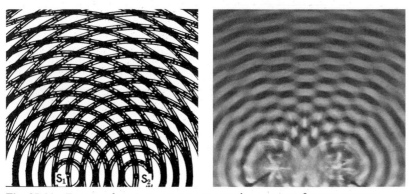

Fig. 25-11   Two overlapping water waves produce an interference pattern.

Interference patterns are neatly illustrated by the overlapping of concentric circles printed on a pair of clear sheets (Figure 25-12). When the sheets overlap with their centers slightly apart, a so-called *moiré pattern* is formed that is very similar to the interference pattern of water waves (or any kind of waves). A slight shift in either of the sheets produces noticeably different patterns. If a pair of such sheets is available, be sure to try this and see the variety of patterns for yourself.

Interference is characteristic of all wave motion, whether the waves are water waves, sound waves, or light waves. The interference of sound is treated in the next chapter, and the interference of light in Chapter 31.

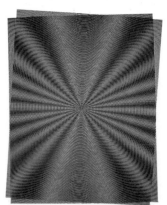

Fig. 25-12   Moiré pattern.

25.8 | Standing Waves

If you tie a rope to a wall and shake the free end up and down, you produce a wave in the rope. The wall is too rigid to shake, so the wave is reflected back along the rope to you. By shaking the rope just right, you can cause the incident (original) and reflected waves to form a **standing wave**, in which parts of the rope, called the **nodes**, remain stationary.

Interestingly enough, you could hold your fingers on either side of the rope at a node, and the rope would not touch them. Other parts of the rope would make contact with your fingers. The positions on a standing wave with the largest amplitudes are known as *antinodes*. Antinodes occur halfway between nodes.

Standing waves are the result of interference. When two waves of equal amplitude and wavelength pass through each other in opposite directions, the waves are always out of phase at the nodes. The nodes are stable regions of destructive interference (Figure 25-13).

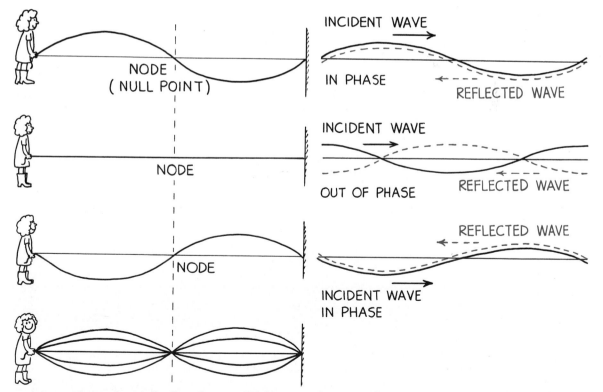

Fig. 25-13 The incident and reflected waves interfere to produce a standing wave. The nodes are regions that remain stationary.

You can produce a variety of standing waves by shaking the rope at different frequencies. The easiest standing wave to produce has one segment (Figure 25-14 top). If you keep doubling the frequency, you'll produce more interesting waves.

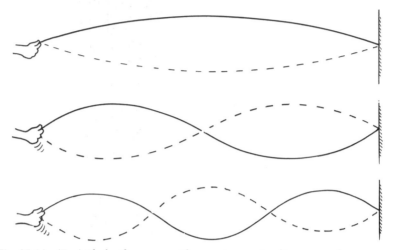

Fig. 25-14   (Top) Shake the rope until you set up a standing wave of one segment (rope length equals ½ wavelength). (Center) Shake with twice the frequency and produce a standing wave with two segments (rope length equals one wavelength). (Bottom) Shake with three times the frequency and produce a standing wave with three segments (rope length equals 1½ wavelengths).

Standing waves are set up in the strings of musical instruments that are plucked, bowed, or struck. They are set up in the air in an organ pipe and the air of a soda-pop bottle when air is blown over the top. Standing waves can be produced in either transverse or longitudinal waves.

▶ **Questions**
1. Is it possible for one wave to cancel another wave so that the combined amplitude is zero?

2. Suppose you set up a standing wave of 3 segments, as shown in Figure 25-14 bottom. If you shake with twice the frequency, how many wave segments will occur in your new standing wave? How many wavelengths will there be?

▶ **Answers**
1. Yes. This is called destructive interference. In a standing wave in a rope, for example, parts of the rope have no amplitude—the nodes.

2. If you impart twice the frequency to the rope, you'll produce a standing wave with twice as many segments. You'll have 6 segments. Since a full wavelength has two segments, you'll have 3 complete wavelengths in your standing wave.

25.9 | The Doppler Effect

Imagine a bug jiggling its legs and bobbing up and down in the middle of a quiet puddle, as shown in Figure 25-15. Suppose the bug is not going anywhere but is merely treading water in a fixed position. The crests of the wave it makes are concentric circles, because the wave speed is the same in all directions. If the bug bobs in the water at a constant frequency, the distance between wave crests (the wavelength) is the same for all successive waves. Waves encounter point A as frequently as they encounter point B. This means that the frequency of wave motion is the same at points A and B, or anywhere in the vicinity of the bug. This wave frequency is the same as the bobbing frequency of the bug.

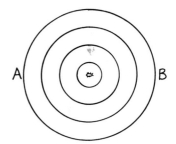

Fig. 25-15   Top view of circular water wave made by a stationary bug jiggling in still water.

Suppose the jiggling bug moves across the water at a speed less than the wave speed. In effect, the bug chases part of the crests it has produced. The wave pattern is distorted and is no longer concentric (Figure 25-16). The center of the outer crest was made when the bug was at the center of that circle. The center of the next smaller crest was made when the bug was at the center of that circle, and so forth. The centers of the circular crests move in the direction of the swimming bug. Although the bug maintains the same bobbing frequency as before, an observer at B would encounter the crests more often. The observer would encounter a *higher* frequency. This is because each successive crest has a shorter distance to travel and therefore arrives at B more frequently than if the bug were not moving toward B.

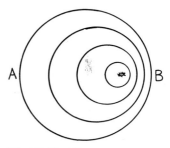

Fig. 25-16   The wave pattern made by a bug swimming in still water.

An observer at A, on the other hand, encounters a *lower* frequency because of the longer time between wave-crest arrivals. This is because each successive crest travels farther to A due to the bug's motion. This change in frequency due to the motion of the source (or receiver) is called the **Doppler effect** (after the Austrian scientist Christian Doppler, 1803–1853). The greater the speed of the source, the greater will be the Doppler effect.

Water waves spread over the flat surface of the water. Sound and light waves, on the other hand, travel in three-dimensional space in all directions like an expanding balloon. Just as circular wave crests are closer together in front of the swimming bug, spherical sound or light wave crests ahead of a moving source are closer together than those behind the source and encounter a receiver more frequently.

The Doppler effect is evident when you hear the changing pitch of a car horn passing you. When it approaches, the pitch is higher than normal (that is, higher on the musical scale). This is because the sound wave crests are encountering you more frequently. And when the car passes and moves away, you hear a drop in pitch because the wave crests are encountering you less frequently.

Fig. 25-17 The pitch of sound increases when the source moves toward you, and decreases when the source moves away.

Police make use of the Doppler effect of radar waves in measuring the speeds of cars on the highway. Radar waves are electromagnetic waves, lower in frequency than light and higher in frequency than radio waves. Police bounce them off moving cars, and a computer built into the radar system calculates the speed of the car relative to the radar unit by comparing the frequency of the radar emitted by the antenna to the frequency of the reflected waves (Figure 25-18).

Fig. 25-18 The police calculate a car's speed by measuring the Doppler effect of radar waves.

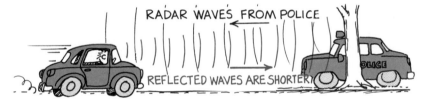

The Doppler effect also occurs for light. When a light source approaches, there is an increase in its measured frequency, and when it recedes, there is a decrease in its frequency. An increase in frequency is called a **blue shift**, because the increase is toward the high-frequency, or blue, end of the color spectrum. A decrease in frequency is called a **red shift**, referring to the low-frequency, or red, end of the color spectrum. The galaxies, for example, show a red shift in the light they emit. A measurement of this shift enables astronomers to calculate their speeds of recession. A rapidly spinning star shows a red shift on the side turning away from us and a relative blue shift on the side turning toward us. This enables a calculation of the star's spin rate.

▶ **Question**
   When a source moves toward you, do you measure an increase or decrease in wave speed?

▶ **Answer**
   Neither! It is the *frequency* of a wave that undergoes a change where there is motion of the source or receiver, not the *wave speed*. Be clear about the distinction between frequency and speed. How frequently a wave vibrates is altogether different from how fast it moves from one place to another.

25.10 | Bow Waves

When the speed of the source is as great as the speed of the waves it produces, something interesting happens. A "wave barrier" is produced. Consider the bug in the previous example when it swims as fast as the wave speed. Can you see that the bug will "keep up" with the wave crests it produces? Instead of the crests getting ahead of the bug, they pile up or superimpose on one another directly in front of the bug (Figure 25-19). The bug encounters a wave barrier. Much effort is required of the bug to swim over this barrier before it can swim faster than wave speed.

The same thing happens when an aircraft travels at the speed of sound. The wave crests overlap to produce a barrier of compressed air on the leading edges of the wings and other parts of the craft. Considerable thrust is required for the aircraft to push through this barrier. Once through, the craft can fly faster than the speed of sound without similar opposition. The craft is then called supersonic—faster than sound. This is like the bug, which once over its wave barrier finds the water ahead relatively smooth and undisturbed.

When the bug swims faster than wave speed, ideally it produces a wave pattern as shown in Figure 25-20. It outruns the wave crests it produces. The crests overlap at the edges, and the pattern made by these overlapping crests is a V shape, called a **bow wave**, which appears to be dragging behind the bug. The familiar bow wave generated by a speedboat knifing through the water is produced by the overlapping of many circular wave crests.

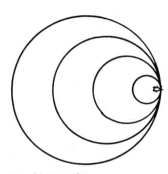

Fig. 25-19 The wave pattern made by a bug swimming at the wave speed.

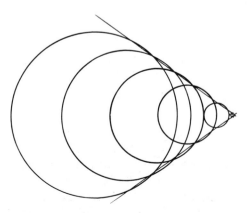

Fig. 25-20 The wave pattern made by a bug swimming faster than the wave speed.

Figure 25-21 shows some wave patterns made by sources moving at various speeds. Note that after the speed of the source exceeds the wave speed, increased speed produces a narrower V shape.

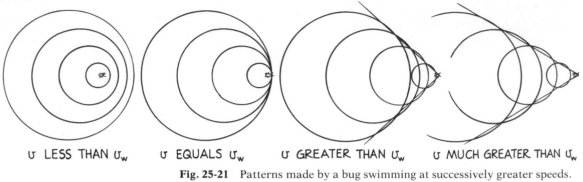

U LESS THAN U$_w$ U EQUALS U$_w$ U GREATER THAN U$_w$ U MUCH GREATER THAN U$_w$

Fig. 25-21 Patterns made by a bug swimming at successively greater speeds. Overlapping at the edges occurs only when the source travels faster than wave speed.

25.11 | Shock Waves

A speedboat knifing through the water generates a two-dimensional bow wave. A supersonic aircraft similarly generates a three-dimensional **shock wave**. Just as a bow wave is produced by overlapping circles that form a V, a shock wave is produced by overlapping spheres that form a cone. And just as the bow wave of a speedboat spreads until it reaches the shore of a lake, the conical shock wave generated by a supersonic craft spreads until it reaches the ground.

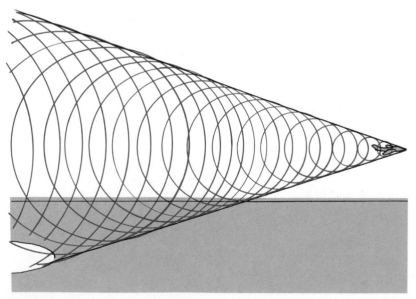

Fig. 25-22 A shock wave from a supersonic aircraft.

The bow wave of a speedboat that passes by can splash and douse you if you are at the water's edge. In a sense, you can say that you are hit by a "water boom." In the same way, when the conical shell of compressed air that sweeps behind a supersonic aircraft reaches listeners on the ground below, the sharp crack they hear is described as a **sonic boom**.

We don't hear a sonic boom from a slower-than-sound, or subsonic, aircraft, because the sound wave crests reach our ears one at a time and are perceived as a continuous tone. Only when the craft moves faster than sound do the crests overlap and encounter the listener in a single burst. The sudden increase in pressure has much the same effect as the sudden expansion of air produced by an explosion. Both processes direct a burst of high-pressure air to the listener. The ear cannot distinguish between the high pressure from an explosion and the high pressure from many overlapping wave crests.

A common misconception is that sonic booms are produced at the moment that an aircraft breaks through the sound barrier—that is, just as the aircraft surpasses the speed of sound. This is equivalent to saying that a boat produces a bow wave only when it overtakes its own waves. This is not so. The fact is that a shock wave and its resulting sonic boom are swept continuously behind an aircraft traveling faster than sound, just as a bow wave is swept continuously behind a speedboat. In Figure 25-23, listener B is in the process of hearing a sonic boom. Listener A has already heard it, and listener C will hear it shortly. The aircraft that generated this shock wave may have broken through the sound barrier hours ago!

It is not necessary that the moving source emit sound for it to produce a shock wave. Once an object is moving faster than the speed of sound, it will *make* sound. A supersonic bullet passing overhead produces a crack, which is a small sonic boom. If the bullet were larger and disturbed more air in its path, the crack would be more boomlike. When a lion tamer cracks a circus whip, the cracking sound is actually a sonic boom produced by the tip of the whip when it travels faster than the speed of sound. Both the bullet and the whip are not in themselves sound sources, but when traveling at supersonic speeds they produce their own sound as waves of air are generated to the sides of the moving objects.

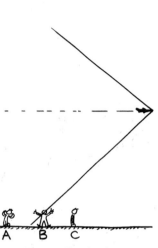

Fig. 25-23 The shock wave has not yet encountered listener C, but is now encountering listener B, and has already reached listener A.

25 Chapter Review

Concept Summary

A vibration is a wiggle in time, and a wave is a wiggle in time and space.

- The period of a wave is the time it takes for one complete to-and-fro vibration.
- The wavelength is the distance between successive identical parts of the wave.
- A wave carries energy from a vibrating source to a receiver without transferring matter from one to the other.
- The frequency, or the number of vibrations in a given time, multiplied by the wavelength equals the speed of the wave.

In a transverse wave, the medium moves at right angles to the direction in which the wave travels.

- Electromagnetic waves, such as light and radio waves, are transverse.

In a longitudinal wave, the medium moves to and fro parallel to the direction in which the wave travels.

- Sound waves are longitudinal.

Interference patterns occur when waves from different sources arrive at the same point at the same time.

- In constructive interference, crest overlaps crest, or trough overlaps trough.
- In destructive interference, a crest overlaps a trough.
- In a standing wave, points of complete destructive interference (at which the medium does not move) remain at the same location.

The Doppler effect is a shift in frequency received due to motion of the vibrating source or receiver.

When a vibrating source moves faster than the speed of waves in the medium, a bow wave or shock wave spreads out behind it.

Important Terms

amplitude (25.2)
blue shift (25.9)
bow wave (25.10)
constructive interference (25.7)
crest (25.2)
destructive interference (25.7)
Doppler effect (25.9)
frequency (25.2)
hertz (25.2)
in phase (25.7)
interference pattern (25.7)
longitudinal wave (25.6)
node (25.8)
out of phase (25.7)
period (25.1)
red shift (25.9)
shock wave (25.11)
sine curve (25.2)
sonic boom (25.11)
standing wave (25.8)
transverse wave (25.5)
trough (25.2)
vibration (25.1)
wave (25.1)
wavelength (25.2)

Review Questions

1. a. What is a wiggle in time called?
   b. What is a wiggle in space and time called? (25.1)

2. What is the period of a pendulum? (25.1)

3. How great is the period of a pendulum that takes one second to make a complete to-and-fro vibration? (25.1)

4. Suppose that a pendulum has a period of 1.5 seconds. Is it a longer or shorter pendulum than the one in Question 3? (25.1)

5. How is a sine curve related to a wave? (25.2)

6. Distinguish among these different parts of a wave: amplitude, crest, trough, and wavelength. (25.2)

7. Distinguish between the *period* and *frequency* of a vibration or a wave. How do they relate to one another? (25.2)

8. Does the medium in which a wave travels move along with the wave itself? Defend your answer. (25.3)

9. How does the speed of a wave relate to its frequency and wavelength? (25.4)

10. As the frequency of sound is increased, does the wavelength increase or decrease? Give an example. (25.4)

11. Distinguish between a *transverse* wave and a *longitudinal* wave. (25.5-25.6)

12. Distinguish between *constructive* interference and *destructive* interference. (25.7)

13. Is interference a property of only some types of waves or all types of waves? (25.7)

14. What causes a standing wave? (25.8)

15. When a wave source moves toward a receiver, does the receiver encounter an increase in wave frequency, wave speed, or both? (25.9)

16. Does the Doppler effect occur for only some types of waves or all types of waves? (25.9)

17. How fast must a wave source move to encounter a "wave barrier"? How fast to produce a bow wave? (25.10)

18. Distinguish between a *bow* wave and a *shock* wave. (25.10-25.11)

19. a. What is a sonic boom?
    b. How fast must an aircraft fly in order to produce a sonic boom? (25.11)

20. If you encounter a sonic boom, is that evidence that an aircraft of some sort exceeded the speed of sound moments ago to become supersonic? Defend your answer. (25.11)

Activities

1. Tie a rubber tube, a spring, or a rope to a fixed support and produce standing waves. See how many nodes you can produce.

2. Repeatedly dip your finger into a wide pan of water to make a circular wave on the surface. What happens to the wavelength if you dip your finger more frequently?

3. Wet your finger and rub it slowly around the rim of a thin-rimmed stemmed glass while you hold its base firmly to a tabletop with your other hand. The friction of your finger will excite standing waves in the glass, much like the wave the friction from a violin bow makes on the strings of a violin.

Think and Explain

1. If you triple the frequency of a vibrating object, what happens to its period?

2. How far, in terms of wavelength, does a wave travel in one period?

3. If a wave vibrates up and down twice each second and travels a distance of 20 m each second, what is its frequency? Its wave speed? (Why is this question best answered by careful reading of the question rather than searching for a formula?)

4. If a wave vibrates to and fro 3 times each second, and its wavelength is 2 m, what is its frequency? Its period? Its speed?

5. Radio waves travel at the speed of light: 300 000 kilometers per second. What is the wavelength of radio waves received at 100 megahertz on your radio dial?

6. Light coming from point *A* at the edge of the sun (see Figure A) is found by astronomers to have a slightly higher frequency than light from point *B* at the opposite side. What do these measurements tell you about the sun's motion?

Fig. A

7. Would it be correct to say that the Doppler effect is the apparent change in the speed of a wave by virtue of motion of the source or receiver? (Why is this question a test of reading comprehension as well as a test of physics knowledge?)

8. Whenever you watch a high-flying aircraft overhead, it seems that its sound comes from behind the craft rather than from where you see it. Why is this?

9. For faster-moving supersonic aircraft, would the conical angle of the shock wave be wider, be more narrow, or remain constant?

10. Why is it that a subsonic aircraft, no matter how loud it may be, cannot produce a sonic boom?

26 Sound

Pretend the whole room is filled with Ping-Pong balls, and in the middle is a big paddle. You shake the paddle back and forth. What happens? The Ping-Pong balls are made to shake back and forth also. Their frequency of shaking back and forth will be the same as the frequency at which the paddle is shaken. Or instead, place a tuning fork in the middle of a room and strike it with a rubber hammer. What happens? The surrounding air molecules are made to vibrate in rhythm with the vibrating prongs of the tuning fork. We hear these vibrations as sound. There is very little difference between the idea of a shaking paddle bumping into Ping Pong balls and a vibrating tuning fork bumping into air molecules. In both cases the vibrations are carried throughout the surrounding medium—the balls or the air.

Fig. 26-1 Vibrate a Ping-Pong paddle in the midst of a lot of Ping-Pong balls, and they will vibrate also.

26.1 The Origin of Sound

All sounds are produced by the vibrations of material objects. In a piano, violin, and guitar, the sound wave is produced by the vibrating strings; in a saxophone, by a vibrating reed; in a flute, by a fluttering column of air blown out the mouthpiece. Your voice results from the vibration of your vocal chords.

Fig. 26-2 The source of all sound waves is something that vibrates.

In each of these cases a vibrating source sends a disturbance through the surrounding medium, usually air, in the form of longitudinal waves. Under ordinary conditions, the frequency of the vibrating source and the frequency of the sound waves produced are the same.

We describe our subjective impression about the frequency of sound by the word **pitch**. A high-pitch sound like that from a piccolo has a high vibration frequency, while a low-pitch sound like that from a fog horn has a low vibration frequency.

The human ear can normally hear pitches corresponding to the range of frequencies between about 20 and 20 000 hertz. As we grow older, the limits of this human hearing range shrink. Sound waves with frequencies below 20 hertz are called **infrasonic**, and those with frequencies above 20 000 hertz are called **ultrasonic**. We cannot hear infrasonic and ultrasonic sound waves.

26.2   The Nature of Sound in Air

Clap your hands and you produce a wave pulse that travels out in all directions. The pulse vibrates the air in the same way that a similar pulse would vibrate a coiled spring or a Slinky. Each particle moves to and fro along the direction of motion of the expanding wave.

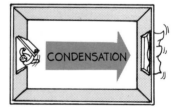

Fig. 26-3   A compression travels along the spring.

For a clearer picture of this process, consider the long room shown in Figure 26-4. At one end is an open window with a curtain over it. At the other end is a door.

When you open the door (top sketch), you can imagine the door pushing the molecules next to it away from their initial positions, and into their neighbors. The neighboring molecules, in turn, push into their neighbors, and so on, like a compression traveling along a spring, until the curtain flaps out the window. A pulse of compressed air has moved from the door to the curtain. This pulse of compressed air is called a **condensation**.

When you close the door (bottom sketch), the door pushes neighboring air molecules out of the room. This produces an area of low pressure next to the door. Neighboring molecules

Fig. 26-4   (Top) When the door is opened, a condensation travels across the room. (Bottom) When the door is closed, a rarefaction travels across the room.

then move into it, leaving a zone of lower pressure behind them. We say this zone of lower-pressure air is *rarefied*. Other molecules farther away from the door, in turn, move into these rarefied regions, and a disturbance again travels across the room. When the lower-pressure air reaches the curtain, it flaps inward. This time the disturbance is a **rarefaction**.

As for all wave motion, it is not the medium itself that travels across the room, but a *pulse* that travels. In both cases the pulse travels from the door to the curtain. We know this because in both cases the curtain moves *after* the door is opened or closed.

If you continually swing the door open and closed in periodic fashion, you can set up a wave of periodic condensations and rarefactions that will make the curtain swing in and out of the window. On a much smaller but more rapid scale, this is what happens when a tuning fork is struck. The vibrations of the tuning fork and the waves it produces are considerably higher in frequency and lower in amplitude than in the case of the swinging door. You don't notice the effect of sound waves on the curtain, but you are well aware of them when they meet your sensitive eardrums.

Consider sound waves in the tube shown in Figure 26-5. For simplicity, only the waves that travel in the tube are depicted. When the prong of the tuning fork next to the tube moves toward the tube, a condensation enters the tube. When the prong swings away, in the opposite direction, a rarefaction follows the condensation. It is like the Ping-Pong paddle moving to and fro in a room packed with Ping-Pong balls. As the source vibrates, a series of compressions and rarefactions is produced.

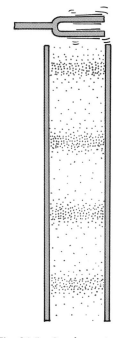

Fig. 26-5 Condensations and rarefactions traveling from the tuning fork through the tube.

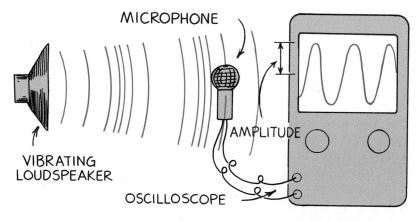

Fig. 26-6 The radio loudspeaker at the left is a paper cone that vibrates in rhythm with an electric signal. The sound that is produced sets up similar vibrations in the microphone (center), which are displayed on the screen of an oscilloscope (right). The shape of the waveform on the oscilloscope reveals information about the sound.

> ▶ **Question**
>   Do condensations and rarefactions in a sound wave travel in the same direction or in opposite directions from one another?

26.3 | Media That Transmit Sound

Most sounds you hear are transmitted through the air. But sound also travels through solids and liquids. Put your ear to the ground as American Indians did, and you can hear the hoofbeats of distant horses before you can hear them in air. More practically, put your ear to a metal fence and have a friend tap it from far away. The sound is more pronounced in the metal than in the air. The metal atoms are more compact and more resilient ("springy") than molecules in the air, so sound is more readily conducted in metals than in air.

Fig. 26-7 Sound can be heard from the ringing bell when air is inside the jar, but when the air is removed by a vacuum pump, no sound can be heard.

Or click a couple of rocks together under water while your ear is submerged. You will hear the clicking sound very clearly. If you've ever been swimming in the presence of motorized boats, you probably noticed that you can hear the motors of the boats much more clearly underwater than above water. Solids and liquids are generally excellent conductors of sound—much better than air.

Sound will not travel in a vacuum (Figure 26-7). The transmission of sound requires a medium. If there is nothing to compress and expand, there can be no sound.

26.4 | The Speed of Sound

Have you ever watched a person at a distance chopping wood or hammering, and noticed that the sound of the blow takes an appreciable time to reach your ears? You see the blow before you hear it. This is most noticeable in the case of lightning. You hear thunder *after* you see a flash of lightning (unless you're at the source). These experiences are evidence for the slower speed of sound compared to light.

▶ **Answer**
They travel in the same direction.

The speed of sound in dry air at 0°C is about 330 meters per second, nearly 1200 kilometers per hour. Water vapor in the air increases this speed slightly. Increased temperature increases the speed of sound also. A little thought will show this makes sense, for the faster-moving molecules in warm air bump into each other more often and therefore can transmit a pulse in less time. For each degree rise in air temperature above 0°C, the speed of sound in air increases by 0.6 m/s. So in air at a normal room temperature of about 20°C, sound travels at about 340 m/s.

The speed of sound in a material depends on the elasticity of the material. An elastic material is one that, when distorted, returns quickly to its initial shape when the distorting force is removed. An elastic substance such as steel (in contrast to putty, which is an inelastic substance) has resilience and can transmit energy with little loss. Steel is "springy" because the atoms in steel are close together and are very elastic. Sound will travel about fifteen times as fast in steel as in air. In water, sound will travel about four times as fast as in air.

> ► **Question**
> About how far away is a thunderstorm when you note a 3-second delay between the flash of lightning and the sound of thunder?

26.5 | Forced Vibration

If you strike an unmounted tuning fork, the sound it makes is faint. Hold the base of the fork on a table top, and the sound is relatively loud. This is because the table is forced to vibrate, and its larger surface sets more air in motion. The table top becomes a sounding board, and can be forced into vibration with forks of various frequencies. This is a case of **forced vibration**.

The mechanism in a music box is mounted on a sounding board to produce the pleasant sounds that are heard. Without the sounding board, the sound the music-box mechanism makes is barely audible. Similarly, stringed musical instruments are made with sounding boards.

Fig. 26-8 The forced vibrations in the sounding board make the sounds audible.

► **Answer**
If you assume that the speed of sound in air is about 340 m/s, the sound of the thunder will travel (340 m/s) × (3 s) = 1020 m. You can assume no measurable time delay for the light, so the storm is slightly more than 1 km away.

26.6 | Natural Frequency

Drop a wrench and a baseball bat on the floor, and you hear distinctly different sounds. This is because both objects vibrate differently when they strike the floor. Tap a wrench, and the vibrations it makes are different from the vibrations of a baseball bat, or of anything else.

Fig. 26-9 The natural frequency of the smaller bell is higher than that of the big bell, and it rings at a higher pitch.

When any object composed of an elastic material is disturbed, it will vibrate at its own special set of frequencies, which together form its special sound. We speak of an object's **natural frequency**, which depends on factors such as the elasticity and shape of the object. Bells and tuning forks, of course, vibrate at their own characteristic frequencies. And interestingly enough, most things—from planets to atoms and almost everything else in between—have a springyness to them and vibrate at one or more natural frequencies. A natural frequency is one at which minimum energy is required to produce forced vibrations. It is also the frequency that requires the least amount of energy for the continuation of vibration.

26.7 | Resonance

When the frequency of forced vibrations on an object matches the object's natural frequency, a dramatic increase in amplitude occurs. This phenomenon is called **resonance**. Literally, resonance means to resound, or sound again. Putty doesn't resonate because it isn't elastic, and a dropped hankerchief is too limp. In order for something to resonate, it needs a force to pull it back to its starting position and enough energy to keep it vibrating.

Fig. 26-10 Pumping a swing in rhythm with its natural frequency produces larger amplitudes.

A common experience illustrating resonance occurs on a swing. When pumping a swing, you pump in rhythm with the natural frequency of the swing. More important than the force with which you pump is the timing. Even small pumps or even small pushes from someone else, if delivered in rhythm with the natural frequency of the swinging motion, produce large amplitudes.

A common classroom demonstration of resonance is illustrated with a pair of tuning forks adjusted to the same frequency and spaced a meter or so apart. When one of the forks is struck, it sets the other fork into vibration. This is a small-scale version of pushing a friend on a swing—it's the timing that's important. When a sound wave impinges on the fork, each condensation gives the prong a tiny push. Since the frequency of these pushes corresponds to the natural frequency of the fork, the pushes will successively increase the amplitude of vibration. This is because the pushes occur at the right time and are repeatedly in the same direction as the instantaneous motion of the fork.

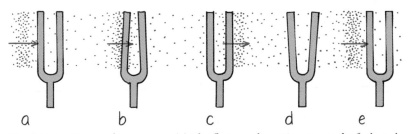

a b c d e

Fig. 26-11 Stages of resonance. (a) The first condensation meets the fork and gives it a tiny and momentary push. The fork bends (b) and then returns to its initial position (c) just at the time a rarefaction arrives. It keeps moving and (d) overshoots in the opposite direction. Just when it returns to its initial position (e), the next condensation arrives to repeat the cycle. Now it bends farther because it is already moving.

If the forks are not adjusted for matched frequencies, the timing of pushes is off and resonance will not occur. When you tune your radio set, you are similarly adjusting the natural frequency of the electronics in the set to match one of the many incoming signals. The set then resonates to one station at a time, instead of playing all stations at once.

Resonance is not restricted to wave motion. It occurs whenever successive impulses are applied to a vibrating object in rhythm with its natural frequency. English cavalry troops marching across a footbridge in 1831 inadvertently caused the bridge to collapse when they marched in rhythm with the bridge's natural frequency. Since then, it is customary to order troops to "break step" when crossing bridges. A bridge disaster in this century was caused by wind-generated resonance (Figure 26-12).

Fig. 26-12 In 1940, four months after being completed, the Tacoma Narrows Bridge in the state of Washington was destroyed by wind-generated resonance. The mild gale produced a fluctuating force in resonance with the natural frequency of the bridge, steadily increasing the amplitude over several hours until the bridge collapsed.

The effects of resonance are all about us. Resonance underscores not only the sound of music, but the color of autumn leaves, the height of ocean tides, the operation of lasers, and a vast multitude of phenomena that add wonder and fascination to the world about us.

26.8 Interference

Sound waves, like any waves, can be made to exhibit interference. Recall that wave interference was discussed in the last chapter. A comparison of interference for transverse waves and longitudinal waves is shown in Figure 26-13. In either case, when the crests of one wave overlap the crests of another wave, there is an increase in amplitude. Or when the crest of one wave overlaps the trough of another wave, there is a decrease in amplitude. In the case of sound, the crest of a wave corresponds to a condensation, and the trough of a wave corresponds to a rarefaction. Interference occurs for both transverse and longitudinal waves.

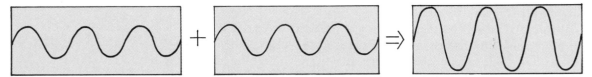

The superposition of two identical transverse waves in phase produces a wave of increased amplitude.

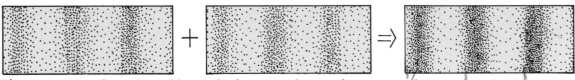

The superposition of two identical longitudinal waves in phase produces a wave of increased amplitude.

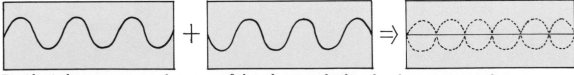

Two identical transverse waves that are out of phase destroy each other when they are superposed.

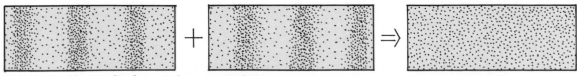

Two identical longitudinal waves that are out of phase destroy each other when they are superposed.

Fig. 26-13   Wave interference for transverse and longitudinal waves.

An interesting case of sound interference is illustrated in Figure 26-14. If you are an equal distance from two sound speakers that each emit an identical tone of constant frequency, the sound is louder because the effects of each speaker add. The condensations and rarefactions of the tones arrive in phase, that is, in step. However, if you move to the side so that the paths from the speakers to you differ by a half-wavelength, then the rarefactions that reach you from one speaker will be filled in by the condensations from the other speaker. This is destructive interference. It is just as if the crest of one water wave exactly filled in the trough of another water wave. If the region is devoid of any reflecting surfaces, little or no sound will be heard!

If the speakers emit a whole range of frequencies, not all wavelengths will destructively interfere for a given difference in path lengths. Interference of this type is usually not a problem, because there is usually enough reflection of sound to fill in cancelled spots. Nevertheless, "dead spots" are sometimes evident in poorly designed theaters or in gymnasiums, where sound waves reflected off walls interfere with unreflected waves to form zones of low amplitude. Moving your head a few centimeters in either direction can make a noticeable difference.

Fig. 26-14 Interference of sound waves. (Top) Waves arrive in phase. (Bottom) Waves arrive out of phase.

26.9 Beats

An interesting and special case of interference occurs when two tones of slightly different frequencies are sounded together. A fluctuation in the loudness of the combined sounds is heard; the sound is loud, then faint, then loud, then faint, and so on. This periodic variation in the loudness of sound is called **beats**.

Beats can be heard when two slightly mismatched tuning forks are sounded together. Because one fork vibrates at a different frequency from the other, the vibrations of the forks will be momentarily in step, then out of step, then in again, and so on. When the combined waves reach your ears in step—say when a condensation from one fork overlaps a condensation from the other—the sound is a maximum. A moment later, when the forks are out of step, a condensation from one fork is met with a rarefaction from the other, resulting in a minimum. The sound that reaches your ears throbs between maximum and minimum loudness and produces a tremolo effect.

If you walk side by side with someone who has a different stride, there will be times when you are both in step, and times when you are both out of step. Suppose, for example, that you take exactly 70 steps in one minute, and your friend takes 72 steps in the same time. Your friend gains two steps per minute

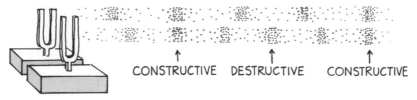

CONSTRUCTIVE   DESTRUCTIVE   CONSTRUCTIVE

Fig. 26-15 The interference of two sound sources of slightly different frequencies produces beats.

on you. A little thought will show that you two will be momentarily in step twice each minute. In general, if two people with different strides walk together, the number of times they are in step each minute is equal to the difference in the frequencies of the steps. This applies also to the pair of tuning forks. If one fork vibrates 264 times per second, and the other fork vibrates 262 times per second, they will be in step twice each second. A beat frequency of 2 hertz will be heard.

Fig. 26-16 The unequal spacings of the combs produce a moiré pattern that is similar to beats.

> ▶ **Question**
>
>   What is the beat frequency when a 262-Hz and a 266-Hz tuning fork are sounded together? A 262-Hz and a 272-Hz fork?

If you overlap two combs of different teeth spacings, you'll see a moiré pattern that is related to beats. The number of beats per length will equal the difference in the number of teeth per length for the two combs (Figure 26-16).

Beats can occur with any kind of wave and are a practical way to compare frequencies. To tune a piano, for example, a piano tuner listens for beats produced between a standard tuning fork and that of a particular string on the piano. When the frequencies are identical, the beats disappear. The members of an orchestra tune up by listening for beats between their instruments and that of a standard tone produced by an oboe or some other instrument.

▶ **Answer**

The 262-Hz and 266-Hz forks will produce 4 beats per second, that is, at 4 Hz (or 266 Hz − 262 Hz). The tone heard will be halfway between, at 264 Hz, as the ear averages the frequencies. The 262-Hz and 272-Hz forks will sound like a tone at 267 Hz beating 10 times per second, or at 10 Hz, which some people will not be able to hear. Beat frequencies greater than 10 Hz are too rapid to be heard normally.

26 | Chapter Review

Concept Summary

Sound waves are produced by the vibrations of material objects.
- A disturbance in the form of a longitudinal wave travels away from the vibrating source.
- High-pitch sounds are produced by sources vibrating at high frequency, while low-pitch sounds are produced by low-frequency sources.

Sound waves consist of traveling pulses of high-pressure zones, or condensations, alternating with traveling pulses of low-pressure zones, or rarefactions.
- Sound can travel through gases, liquids, and solids, but not through a vacuum.
- Sound travels fastest through very elastic materials, such as steel.

Every object vibrates at its own set of natural frequencies.
- When an object such as a sounding board is forced to vibrate by a sound source, the sound becomes louder.
- When an object is forced to vibrate at one of its own natural frequencies, resonance occurs and the sound becomes much louder.

Like any waves, two sound waves can exhibit interference and can make the sound louder or cancel the sound.
- Rapid changes in loudness known as beats occur when two tones very close in frequency are heard at the same time.

Important Terms

beats (26.9)
condensation (26.2)
forced vibration (26.5)
infrasonic (26.1)
natural frequency (26.6)
pitch (26.1)
rarefaction (26.2)
resonance (26.7)
ultrasonic (26.1)

Review Questions

1. What is the source of all sounds? (26.1)

2. How does pitch relate to frequency? (26.1)

3. What is the average frequency range of human hearing? (26.1)

4. Distinguish between *infrasonic* and *ultrasonic* sound. (26.1)

5. a. Distinguish between *condensations* and *rarefactions* of a sound wave.
   b. How is each produced? (26.2)

6. Light can travel through a vacuum, as is evidenced when you see the sun or the moon. Can sound travel through a vacuum also? (26.3)

7. a. Approximately how fast does sound travel in dry air?
   b. Does its speed depend on air temperature? (26.4)

8. How does the speed of sound in air compare to its speed in water and in steel? (26.4)

9. Why does sound travel faster in solids and liquids than in gases? (26.4)

10. Why is sound louder when a vibrating source is held to a sounding board? (26.5)

11. Why do different objects make different sounds when dropped on a floor? (26.6)

12. What does it mean to say that everything has a natural frequency of vibration? (26.6)

13. What is the relationship between forced vibration and resonance? (26.7)

14. Why can a tuning fork or bell be set into resonance, whereas a piece of tissue paper cannot? (26.7)

15. How is resonance produced in a vibrating object? (26.7)

16. How is the process of adjusting the frequency of a tuning fork similar to dialing a station on the radio? (26.7)

17. Is it possible for one sound wave to cancel another? Explain. (26.8)

18. Why does destructive interference occur when the path lengths from two identical sources differ by a half wavelength? (26.8)

19. How does interference of sound relate to beats? (26.9)

20. What is the beat frequency when a 494-Hz tuning fork and a 496-Hz tuning fork are sounded together? (26.9)

Activities

1. Suspend the wire grill of a refrigerator or oven shelf from a string, the ends of which you hold to your ears (Figure A). Let a friend gently stroke the grill with pieces of broom straw and other objects. The effect is best appreciated if you are in a relaxed condition with your eyes closed. Be sure to try this.

Fig. A

2. In the bathtub or when swimming, submerge your head and listen to the sound you make when clicking your fingernails or tapping a pair of stones together. Compare the sound with that you make when both the source and your ears are above water. You'll find it is difficult to hear when the source and your ears are in different media, as when your head is submerged and the sound source is in the air. Most of the sound energy is reflected at the boundary between the two media rather than transmitted through it.

3. If you blow air across the top of a pop bottle, a puff of air (condensation) travels downward, bounces from the bottom, and travels back to the opening. When it arrives (less than a thousandth of a second later), it disturbs the flow of air that you are still producing across the top. This causes a slightly bigger puff of air to start again on its way down the bottle. This happens repeatedly until a very large (and loud) vibration is built up and you hear it as sound. The pitch of the sound depends on the time taken for the back-and-forth trip, which depends on the depth of the bottle. If the bottle is empty, a long wave is reinforced and a relatively low tone is produced. With liquid in the bottle, the bottom is closer to the top and the pitch is higher. With a series of bottles properly filled, you can make your own music.

4. As you pour water into a glass, repeatedly tap the glass with a spoon. As the tapped glass is being filled, does the pitch of the sound increase or decrease? Think about this before you try it and hear for yourself.

Think and Explain

1. If you are in the stands at a track meet and are far from the starter, you'll notice the smoke from the starter's gun before you hear it fire. Why?

2. Why will marchers at the end of a long parade following a band be out of step with marchers nearer the band?

3. You watch a distant farmer driving a stake into the ground with a sledge hammer. He hits the stake at a regular rate of one stroke per second. You hear the sound of the blows exactly synchronized with the blows you see. And then you hear one more blow after you see him stop hammering. How far away is the farmer?   *330 m*

4. When a sound wave propagates past a point in the air, what are the changes that occur in the pressure of air at this point?

5. When a person talks after inhaling helium gas, the voice is high-pitched. This is principally because helium molecules move faster past the vocal chords than do molecules of air. Why do helium molecules move faster? (*Hint*: The helium molecule is significantly lighter than any of the molecules that compose air. The molecules of all gases at the same temperature have the same average kinetic energy. If an elephant and a mouse ran into a barn door with the same kinetic energy, which of the two had the higher speed?)

6. If the handle of a tuning fork is held solidly against a table, the sound becomes louder. Why? How will this affect the length of the time the fork keeps vibrating? Explain, using the law of energy conservation.

7. The sitar, an Indian musical instrument, has a set of strings that vibrate and produce music, even though they are never plucked by the player. These "sympathetic strings" are identical to the plucked strings and are mounted below them. What is your explanation?

8. Suppose three tuning forks of frequency 260 Hz, 262 Hz, and 266 Hz are available. What beat frequencies are possible for pairs of forks sounded together?   *2 4 6*

9. Suppose a piano tuner hears 2 beats per second when listening to the combined sound from her tuning fork and the piano note being tuned. After slightly tightening the string, she hears 1 beat per second. Should she loosen or should she further tighten the string?

10. Do all people in a group hear the same music when they listen to it attentively? (Do all see the same sight when looking at a painting? Do all taste the same flavor when sampling the same cheddar cheese? Do all perceive the same aroma when smelling the same flower? Do all feel the same texture when touching the same fabric? Do all come to the same conclusion when listening to a logical presentation of ideas?)

27 | Light

The only thing we can really see is light. But what *is* light? We know that during the day the primary source of light is the sun, and the secondary source is the brightness of the sky. Other common sources are flames, and since modern times, white-hot filaments in lamps, and glowing gases in glass tubes.

Most objects we see, such as this page, are made visible by the light they reflect from such sources. Some materials, such as water and window glass, allow the passage of light in straight lines. Other materials, such as thin paper, allow the passage of light, but in diffused directions so that you cannot see objects through them. The majority of materials do not allow the passage of light, except when very thin layers of them are used.

Why do things such as water and glass allow light straight through, while things such as wood and steel block light? Before you can answer these questions, you must know something about light itself.

27.1 | Early Concepts of Light

The study of the nature of light has extended over thousands of years. In the fifth century B.C., philosophers such as Socrates and Plato in Greece speculated that light was made up of streamers or filaments emitted by the eye. They believed that seeing takes place when these streamers, acting like antennas, make contact with an object. This view was supported by Euclid, when he asked how else can we explain why we do not see a needle on the floor until our eyes fall upon it. As late as the fifteenth cen-

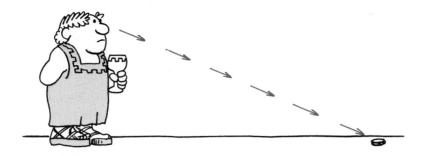

tury, René Descartes, the great French mathematician and philosopher, published a book that explained a similar theory.

Not all the ancients, however, held such views. The Pythagoreans from Greece believed that light traveled from luminous objects to the eye in the form of tiny particles. Another Greek, Empedocles, taught that light traveled in waves. In more recent history, a particle theory of light was championed by Newton and widely accepted by other scientists. The particle theory was supported by the fact that light seemed to move in straight lines instead of spreading out as waves do.

Not everyone in Newton's time believed in the particle theory. One of Newton's contemporaries, the Dutch scientist Christian Huygens, stated that light was a wave. He supported this theory with evidence that under some circumstances light does spread out (this is *diffraction*, which is covered in Chapter 31). Other scientists later found more evidence to support the wave theory. Then in 1905 Einstein published a theory concerning what was called the *photoelectric effect*. According to this theory, light consists of particles—massless bundles of concentrated electromagnetic energy—called **photons**.

Scientists now agree that light has a dual nature, part particle and part wave. This chapter discusses the wave nature of light, and leaves the particle nature of light to Chapter 38 (and your next physics course).

27.2 The Speed of Light

It was not known whether light travels instantaneously or with finite speed until almost the end of the seventeenth century. Galileo had tried to measure the time a light beam takes to travel to a distant mirror and back, but the time interval—if one existed at all—was so short he couldn't begin to measure it. Others tried the experiment at longer distances with lanterns they blinked on and off between distant mountain tops. All they succeeded in doing was measuring their own reaction times.

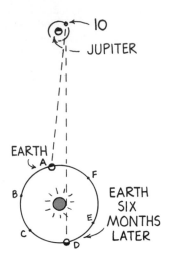

Fig. 27-2 Roemer's method of measuring the speed of light. Light coming from Jupiter's moon Io takes a longer time to reach the earth at position D than at position A. The extra distance that the light travels divided by the extra time it takes gives the speed of light.

Experimental evidence for the first successful measurement of the speed of light was supplied by the Danish astronomer Olaus Roemer about 1675. Roemer made very careful measurements of the periods of Jupiter's moons. The innermost moon, Io, is visible through a small telescope and was measured to revolve around Jupiter in 42.5 hours. Io disappears periodically into the shadow of Jupiter, so this period could be measured with great precision. Roemer was puzzled to find an irregularity in the measurements of the period of Io. He found that when the earth moved away from Jupiter, say from position B to C in Figure 27-2, the measured periods of Io were all somewhat longer than average. When the earth moved toward Jupiter, say from position E to F, the measured periods were shorter than average. Roemer estimated that the cumulative discrepancy between positions A and D amounted to about 22 min. That is, when the earth was at position D, Io would pass into Jupiter's shadow 22 min late with respect to observations at position A.*

The Dutch physicist Christian Huygens correctly interpreted this discrepancy. When the earth was farther away from Jupiter, it was the *light* that was late, not the *moon*. Io passed into Jupiter's shadow at the predicted time, but the light carrying the message did not reach Roemer until it had traveled the extra distance across the diameter of the earth's orbit. There is some doubt as to whether Huygens knew the value of this distance. In any event, this distance is now known to be 300 000 000 km. Using the correct travel time of 1000 s for light to move across the earth's orbit makes the calculation of the speed of light quite simple:

$$\text{speed of light} = \frac{\text{extra distance traveled}}{\text{extra time measured}}$$

$$= \frac{300\ 000\ 000\ \text{km}}{1000\ \text{s}} = 300\ 000\ \text{km/s}$$

The most famous experiment measuring the speed of light was performed by the American physicist Albert Michelson in 1880. Figure 27-3 is a simplified diagram of his experiment. Light from an intense source was directed by a lens to an octagonal mirror initially at rest. The mirror was carefully adjusted so that a beam of light was reflected to a stationary mirror located on a mountain 35 km away, and then reflected back to the octagonal mirror and into the eye of an observer. The distance the light had to travel to the distant mountain was carefully surveyed, so Michelson had only to find the time it took to make a round trip. He accomplished this by spinning the octagonal mirror at a high rate.

---

* Roemer's estimate was not quite correct. The correct value is 16 min, or about 1000 seconds.

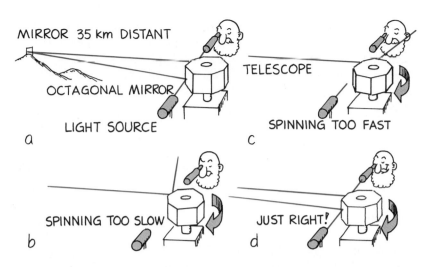

MIRROR 35 km DISTANT

TELESCOPE

OCTAGONAL MIRROR

LIGHT SOURCE

a

SPINNING TOO FAST

c

SPINNING TOO SLOW

b

JUST RIGHT!

d

Fig. 27-3 The mirror arrangement used by Michelson to measure the speed of light. Light is reflected back to the eyepiece when the mirror is at rest (a). Reflected light fails to enter the eyepiece when the mirror is spinning too slowly (b) or too fast (c). When it is rotating at the correct speed (d), light reaches the eyepiece.

When the mirror was spun, the continuous beam of light was chopped up so that only discrete bursts of light reached the mountain mirror to be reflected back to the spinning octagonal mirror. If the rotating mirror made exactly one-eighth rotation in the time the light made the trip to the distant mountain and back, the mirror would be in a position to reflect light into the eyepiece of the observer. If the mirror was rotated too slowly or too quickly, it would not be in a position to reflect light into the eyepiece. When the speed of rotation of the mirror was adjusted so that the light entered the eyepiece, Michelson knew that the time for the light to make the round trip and the time for the octagonal mirror to make one eighth of a rotation was the same. He divided the 70-km round trip distance by this time. Michelson's experimental value for the speed of light was 299 920 km/s, which we round off to 300 000 km/s. Michelson received the 1907 Nobel prize in physics for this experiment. He was the first American scientist to receive the prize.

> ▶ **Question**
> Light entered the eyepiece when Michelson's octagonal mirror made exactly one eighth of a rotation during the time light reflected to the distant mountain and back. Would light enter the eyepiece if the mirror turned one quarter of a rotation in this time?

▶

Answer
Yes, light would enter the eyepiece whenever the octagonal mirror turned in multiples of $\frac{1}{8}$ rotations—$\frac{1}{4}$, $\frac{1}{2}$, 1, etc.—in the time the light made its round trip. What is required is that any of the eight faces is in place when the reflected flash returns from the mountain.

Fig. 27-4 Light would take 30 000 years to reach us from the center of our galaxy. So the center of our galaxy is 30 000 light years distant.

We now know that the speed of light in a vacuum is a universal constant. Light is so fast that if a beam of light could travel around the earth, it would make 7.5 trips in one second. Light takes 8 minutes to travel from the sun to the earth, and 4 years from the next nearest star, Alpha Centauri. The distance light travels in one year is called a **light year**.

So Alpha Centauri is 4 light years away. Our galaxy has a diameter of 100 000 light years, which means that light takes 100 000 years just to travel across the galaxy. Some galaxies are 10 billion light years from earth. If one of those galaxies had exploded 5 billion years ago, this information would not reach earth for another 5 billion years to come. Light is fast and the universe is big!

Computational Example

How far, in kilometers, would a beam of uninterrupted light travel in one year?

The speed of light is a constant, so its instantaneous speed and average speed are the same. From the equation $\bar{v} = d/t$, we can say

$$d = \bar{v}t$$
$$= (300\ 000\ \text{km/s}) \times (1\ \text{yr})$$

By the technique of dimensional analysis we convert 1 year to seconds, and find

$$d = \left(\frac{300\ 000\ \text{km}}{1\ \cancel{s}}\right) \times (1\ \cancel{yr}) \times \left(\frac{365\ \cancel{d}}{1\ \cancel{yr}}\right) \times \left(\frac{24\ \cancel{h}}{1\ \cancel{d}}\right) \times \left(\frac{3600\ \cancel{s}}{1\ \cancel{h}}\right)$$

$$= 9.5 \times 10^{12}\ \text{km}$$

This distance is one light year.

27.3 | Electromagnetic Waves

Light is energy that is emitted by vibrating electric charges in atoms. This energy travels in a wave that is partly electric and partly magnetic. Such a wave is called an **electromagnetic wave**. Light is a small portion of the broad family of electromagnetic waves that includes such familiar forms as radio waves, microwaves, and X rays, which are all radiated by vibrating electrons within the atom. The range of electromagnetic waves, or the **electromagnetic spectrum**, as it is called, is shown in Figure 27-5.

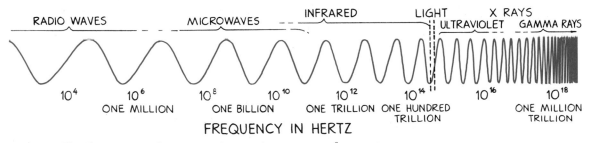

Fig. 27-5 The electromagnetic spectrum is a continuous range of waves extending from radio waves to gamma rays. The descriptive names of the sections are merely a historical classification, for all waves are the same in nature, differing principally in frequency and wavelength; all have the same speed.

The lowest frequency of light we can see with our eyes appears red. The highest visible frequencies are nearly twice the frequency of red and appear violet. Electromagnetic waves of frequencies lower than the red of visible light are called **infrared**. Many heat lamps give off infrared waves. Electromagnetic waves of frequencies higher than those of violet are called **ultraviolet**. These higher-frequency waves are more energetic and are responsible for sunburns.

▶ **Question**

Is it correct to say that a radio wave is a low-frequency light wave? Is a radio wave also a sound wave?

27.4 | Light and Transparent Materials

Light is energy carried in an electromagnetic wave that is made by vibrating electric charges in atoms. When light is incident upon matter, electric charges in the matter are forced into vibration. In effect, vibrations in the emitter are transferred to vibrations in the receiver. This is similar to the way that sound is received by a receiver (see Figure 27-6).

▶ **Answer**

Both a radio wave and a light wave are electromagnetic waves which originate from the vibrations of electrons. Radiowaves have lower frequencies of vibration than visible light waves, so a radio wave may be considered to be a low-frequency light wave. A sound wave, on the other hand, is a mechanical vibration of matter and is not electromagnetic. A sound wave is fundamentally different from an electromagnetic wave. Thus, a radio wave is not a sound wave.

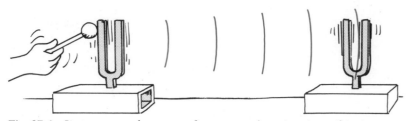

Fig. 27-6 Just as a sound wave can force a sound receiver into vibration, a light wave can force charged particles in materials into vibration.

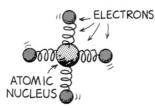

Fig. 27-7 The electrons of atoms in glass can be imagined to be bound to the atomic nucleus as if connected by springs.

Exactly how a receiving material responds when light is incident upon it depends on the frequency of the light and the natural frequency of the electric charges in the material. Visible light vibrates at a very high rate, some 100 trillion times per second (10^{14} hertz). If a charged object is to respond to these ultra-fast vibrations, it must have very little inertia. Electrons have a small enough mass to vibrate this fast.

Glass and water are two materials that allow light to pass through in straight lines. They are said to be **transparent** to light. To understand how light gets through a transparent material such as glass, visualize the electrons in an atom as connected by imaginary springs (Figure 27-7). When a light wave is incident upon them, they are set into vibration.

All materials that are springy or elastic respond more to vibrations at some frequencies than others. Bells ring at a particular frequency, tuning forks vibrate at a particular frequency, and so do the electrons of atoms. The natural vibration frequencies of an electron depend on how strongly it is attached to its atom. Different atoms have different "spring strengths."

Electrons in glass have a natural vibration frequency in the ultraviolet range. When ultraviolet light shines on glass, resonance occurs as the wave builds and maintains a large vibration between the electron and the atomic nucleus, just as pushing someone at the resonant frequency on a swing builds a large vibration. The energy the atom receives can be passed on to neighboring atoms by collisions, or it can be re-emitted as light. If the atom is excited with ultraviolet light (which is at its natural frequency), the atom can hold onto this energy for quite a long time (about 1 million vibrations or 100 millionths of a second). During this time the atom makes many collisions with other atoms and gives up its energy in the form of heat. Glass is not transparent to ultraviolet.

Consider what happens when the electromagnetic wave has a lower frequency than ultraviolet, as does visible light. The electrons of the atom are forced into vibration, but not so strong as before. The atom holds the energy for less time, with less chance of collision with neighboring atoms, and less energy transferred as heat. The energy of the vibrating electrons is re-emitted as light. Glass is transparent to all the frequencies of visible light. The frequency of the re-emitted light that is passed from atom to atom is identical to that of the light that produced the vibration to begin with. The only principal difference is a slight time delay between absorption and re-emission.

This time delay results in a lower average speed of light through a transparent material (see Figure 27-8). Light travels at different average speeds through different materials. In a vacuum the speed of light is a constant 300 000 km/s; we call this speed of light c. Light travels very slightly more slowly than this in the atmosphere, but its speed there is usually rounded off as c. In water light travels at 75 percent of its speed in a vacuum, or $0.75c$. In glass light travels at about $0.67c$, depending on the type of glass. In a diamond light travels at only $0.41c$, less than half its speed in a vacuum. When light emerges from these materials into the air, it travels at its original speed, c.

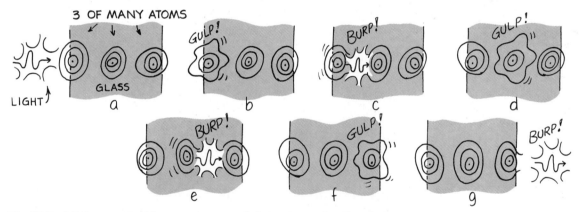

Fig. 27-8   A light wave incident upon a pane of glass sets up vibrations in the atoms that produce a chain of absorptions and re-emissions that pass the light energy through the material and out the other side. Because of the time delay between absorptions and re-emissions, the light travels more slowly in the glass.

Infrared waves, with frequencies lower than visible light, vibrate not only the electrons, but the entire structure of the glass. This vibration of the structure increases the internal energy of the glass and makes it warmer. In sum, glass is transparent to visible light, but not to ultraviolet and infrared light.

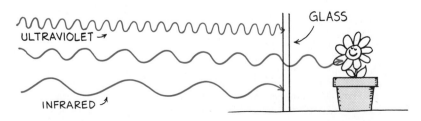

Fig. 27-9   Glass blocks both infrared and ultraviolet, but is transparent to all the frequencies of visible light.

27.5 | Opaque Materials

Most materials absorb light without re-emission and thus allow no light through them; they are said to be **opaque**. Wood, stone, and people are opaque to visible light. In opaque materials, any coordinated vibrations given by light to the atoms and molecules are turned into random kinetic energy—that is, into internal energy. They become slightly warmer.

Metals are also opaque. Interestingly enough, the outer electrons of atoms in metals are not bound to any particular atom. They are free to wander with very little restraint throughout the material (which is why metal conducts electricity and heat so well). When light shines on metal and sets these free electrons into vibration, their energy does not "spring" from atom to atom in the material, but is re-emitted as visible light. This re-emitted light is seen as a reflection. That's why metals are shiny.

Our atmosphere is transparent to visible light and some infrared, but fortunately, quite opaque to high-frequency ultraviolet waves. The small amount of ultraviolet that does get through is responsible for sunburns. If it all got through, we would be fried to a crisp. Clouds are semitransparent to ultraviolet, which is why you can get a sunburn on a cloudy day.

Fig. 27-10 Metals are shiny because light that shines on them forces free electrons into vibration. These electrons then emit their "own" light waves as a reflection.

▶ **Question**

Why is glass transparent to visible light, but opaque to ultraviolet and infrared?

27.6 | Shadows

A thin beam of light is often called a **ray**. Any beam of light—no matter how wide—can be thought of as made of a bundle of rays. When light shines on an object, some of the rays may be

▶ **Answer**

The natural frequency of vibration for electrons in glass matches the frequency of ultraviolet light, so resonance in the glass occurs when ultraviolet waves shine on it. These energetic vibrations of electrons generate heat instead of wave re-emission, so the glass is opaque to ultraviolet. In the range of visible light, the forced vibrations of electrons in the glass are more subtle, and re-emission of light rather than the generation of heat occurs, so that the glass is transparent. Lower-frequency infrared causes the whole structure, rather than electrons, to resonate, and again, heat is generated and the glass is opaque.

stopped while others pass on in a straight-line path. A **shadow** is formed where light rays cannot reach.

Sharp shadows are produced by a small light source nearby or by a larger source farther away. However, most shadows are somewhat blurry. There is usually a dark part on the inside and a lighter part around the edges. A total shadow is called an **umbra**, and a partial shadow a **penumbra**. A penumbra appears where some of the light is blocked, but where other light fills it in. This can happen where light from one source is blocked and light from another source fills in (Figure 27-12). Or a penumbra occurs where light from a broad source is only partially blocked.

Fig. 27-11 A large light source produces a softer shadow than a smaller source.

Fig. 27-12 An object held close to a wall casts a sharp shadow because no light can seep around to form penumbras. As the object is moved farther away, penumbras are formed and cut down on the umbra. When it is very far away, no shadow is evident because all the penumbras mix together into a big blur.

A dramatic example of this occurs when the moon passes between the earth and the sun—during a solar eclipse. Because of the large size of the sun, the rays taper to provide an umbra and a surrounding penumbra (Figure 27-13). The moon's shadow just reaches the earth. If you stand in the umbra part of the shadow, you experience brief darkness during the day. If you stand in the penumbra, you experience a partial eclipse. The sunlight is dimmed and the sun appears as a crescent.*

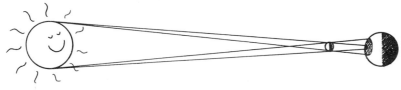

Fig. 27-13 An eclipse of the sun.

---

* People are cautioned not to look at the sun at the time of a solar eclipse because the brightness and ultraviolet radiation of direct sunlight is damaging to the eyes. This good advice is often misunderstood by those who then think that sunlight is more damaging at this special time. But staring at the sun when it is high in the sky is harmful whether or not an eclipse occurs. In fact, staring at the bare sun is more harmful than when part of the moon blocks it! The reason for special caution at the time of an eclipse is simply that more people are interested in looking at the sun during an eclipse.

The earth, like most objects in sunlight, casts a shadow. This shadow extends into space, and sometimes the moon passes into it. When this happens, we have a lunar eclipse. Whereas a solar eclipse can be observed only in a small region of the earth at a given time, a lunar eclipse can be seen by all observers on the nighttime half of the earth (Figure 27-14).

Fig. 27-14   An eclipse of the moon.

> ▶ **Question**
> Why are lunar eclipses more commonly seen than solar eclipses?

Fig. 27-15   A heater at the tip of the submerged J-tube produces convection currents in the water, which are revealed by shadows cast by light that is deflected differently by the water of different temperatures.

Shadows occur when light is bent in passing through a transparent material such as water. In Figure 27-15 shadows are cast by turbulent, rising warm water. Light travels at slightly different speeds in warm and in cold water. The difference bends light, just as layers of warm and cool air in the night sky bend starlight and cause the twinkling of stars. Some of the light gets deflected a bit and leaves darker places on the wall. The shapes of the shadows depend on how the light is bent. Chapter 29 returns to the bending of light.

27.7   Polarization

Light travels in waves. The fact that the waves are transverse—and not longitudinal—is demonstrated by the phenomenon called **polarization**. If you shake the end of a horizontal rope, as in Figure 27-16, a transverse wave travels along the rope. The vibrations are to and fro in one direction, and the wave is said to be *polarized*. If the rope is shaken up and down, a vertically polar-

Fig. 27-16   A vertically polarized wave (left) and a horizontally polarized wave (right).

▶ **Answer**
There are usually two of each every year. However, the shadow of the moon on the earth is very small compared to the shadow of the larger earth on the smaller moon. Only a relatively few people are in the shadow of the moon (solar eclipse), while everybody who views the nighttime sky can see the shadow of the earth on the moon (lunar eclipse).

ized wave is produced; that is, the waves traveling along the rope are confined to a vertical plane. If the rope is shaken from side to side, a horizontally polarized wave is produced.

A single vibrating electron emits an electromagnetic wave that is also polarized. A vertically vibrating electron emits light that is vertically polarized, while a horizontally vibrating electron emits light that is horizontally polarized (Figure 27-17).

A common light source, such as an incandescent or fluorescent lamp, a candle flame, or the sun, emits light that is non-polarized. This is because the vibrating electrons that produce the light vibrate in all different directions. When ordinary light shines on a polarizing filter, such as that from which Polaroid sunglasses are made, the light that is transmitted is polarized. The filter is said to have a *polarization axis* that is in the direction of the vibrations of the polarized light wave.

Light will pass through a pair of polarizing filters when their polarization axes are aligned, but not when they are crossed at right angles. This behavior is very much like the filtering of a vibrating rope that passes through a pair of picket fences (Figure 27-19).

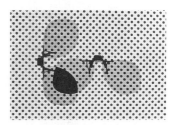

Fig. 27-17   Polarized light lies along the same plane as that of the vibrations of the electron that emits it.

Fig. 27-18   Polaroid sunglasses block out horizontally vibrating light. When the lenses overlap at right angles, no light gets through.

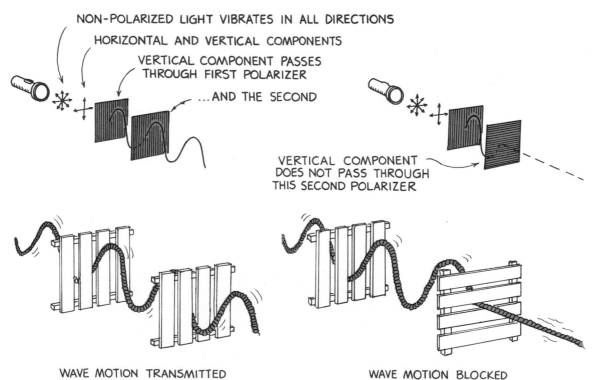

NON-POLARIZED LIGHT VIBRATES IN ALL DIRECTIONS

HORIZONTAL AND VERTICAL COMPONENTS

VERTICAL COMPONENT PASSES THROUGH FIRST POLARIZER

...AND THE SECOND

VERTICAL COMPONENT DOES NOT PASS THROUGH THIS SECOND POLARIZER

WAVE MOTION TRANSMITTED   WAVE MOTION BLOCKED

Fig. 27-19   A rope analogy illustrates the effect of crossed sheets of polarizing material.

Fig. 27-20 Light is transmitted when the axes of the Polaroids are aligned (left), but absorbed when they are at right angles to each other (center). Interestingly enough, when a third Polaroid is sandwiched between the crossed Polaroids (right), light is transmitted. Why? (To answer, you'll have to know more about vectors. See Appendix C, Vector Applications.)

Much of the glare reflected from nonmetallic surfaces, such as from glass, water, or a road surface, is polarized. The reflected light, especially for glancing angles, is primarily made up of light that vibrates in the same plane as the reflecting surface. So the glare from a horizontal surface is polarized horizontally. This is like skipping flat stones across the surface of a pond. When the stones hit the water with their flat sides parallel to the water, they bounce (they are reflected); but when they hit with their flat side at right angles to the surface, they penetrate into the water. Do you see why Polaroid sunglasses are oriented to block horizontal vibrations and transmit vertical vibrations? In this way, it is mainly the glare that is eliminated.

Not all polarizing eyeglasses are made for blocking horizontally polarized light. Three-dimensional slide shows or movies are projected through a pair of projectors fitted with polarizing filters (Figure 27-21). Their polarization axes are at right angles to each other—one is vertical and the other is horizontal. The projectors display a pair of pictures that were taken a short distance apart, just as the eyes are spaced a short distance apart. The pictures are displayed on the same screen and look blurry to the naked eye. To see 3-D, the viewer wears polarizing eyeglasses in which the axes of the two lenses are also at right angles. As a result, each eye sees a separate picture, just as in real life. The brain interprets the two pictures as a single picture with a feeling of depth. (Hand-held stereo viewers produce a similar effect, but the 3-D projector setup allows many people to see the pictures at once.)

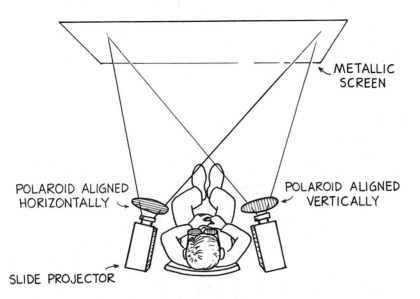

Fig. 27-21   A 3-D slide show using polarizing filters. The left eye sees only polarized light from the left projector; the right eye sees only polarized light from the right projector. Both views merge in the brain to produce an image with depth.

▶ **Question**

Which pair of glasses is best suited for automobile drivers? (The polarization axes are shown by the straight lines.)

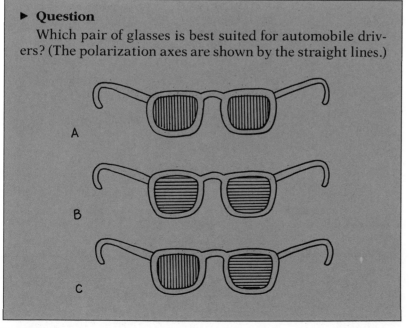

▶ **Answer**

Pair A is best suited because the vertical axis blocks horizontally polarized light that composes much of the glare from horizontal surfaces. (Pair C is suited for viewing 3-D movies.)

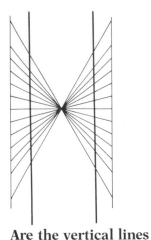

Are the vertical lines
parallel?

Both rectangles are equally bright. Cover the boundary
between them with a pencil and see.

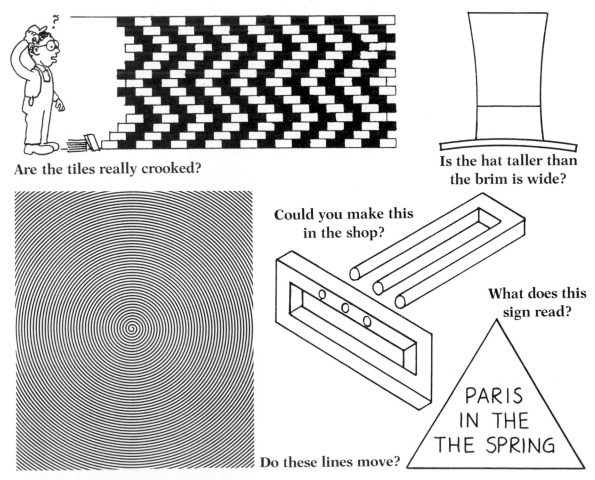

Are the tiles really crooked?

Is the hat taller than
the brim is wide?

Could you make this
in the shop?

What does this
sign read?

PARIS
IN THE
THE SPRING

Do these lines move?

Fig. 27-22 Optical illusions.

27 | Chapter Review

Concept Summary

Light has both a wave and a particle nature, but most everyday phenomena can be explained in terms of the wave nature.

Light has a speed of 300 000 km/s in a vacuum, and lower speeds in matter.

Light is energy that travels in electromagnetic waves within a certain range of frequencies.
- Light is produced by vibrating electric charges in atoms.
- Light passes through materials whose atoms are able to absorb the energy and immediately re-emit it as light.
- Light cannot pass through a material when the energy is changed to random kinetic energy of the atoms.

Because light waves are transverse, they can be polarized so that the vibrations are all in the same direction.
- Polarizing filters transmit components of incident nonpolarized light that are parallel to the polarization axis, and block components vibrating at right angles to the polarization axis. The result is the emergence of polarized light.

Important Terms

electromagnetic spectrum (27.3)
electromagnetic wave (27.3)
infrared (27.3)
light year (27.2)
opaque (27.5)
penumbra (27.6)
photon (27.1)
polarization (27.7)
ray (27.6)
shadow (27.6)
transparent (27.4)
ultraviolet (27.3)
umbra (27.6)

Review Questions

1. a. What is a photon?
   b. Which theory of light is the photon more consistent with—the wave theory or the particle theory? (27.1)

2. How long does it take for light to travel across the diameter of the earth's orbit around the sun? (27.2)

3. How did Michelson know the time that light took to make the round trip to the distant mountain? (27.2)

4. How long does light take to travel from the sun to the earth? From the star Alpha Centauri to the earth? (27.2)

5. How long does light take to travel a distance of one light year? (27.2)

6. What is the source of electromagnetic waves? (27.3)

7. Is the color spectrum simply a small segment of the electromagnetic spectrum? Defend your answer. (27.3)

8. How do the frequencies of infrared, visible, and ultraviolet light compare? (27.3)

9. How does the role of inertia relate to the rate at which electric charges can be forced into vibration? (27.4)

10. Different bells and tuning forks have their own natural vibrations, and emit their own

tones when struck. How is this analogous with atoms, molecules, and light? (27.4)

11. When light encounters a material, it can build up vibrations in the electrons of certain molecules that may be intense enough to last over a long period of time. Will the energy of these vibrations tend to be absorbed and turned into heat, or absorbed and re-emitted as light? (27.4)

12. Will glass be transparent to frequencies of light that match its own natural frequencies? (27.4)

13. Does the time delay between the absorption and re-emission of light affect the average speed of light in a material? Explain. (27.4)

14. Why would you expect the speed of light to be slightly less in the atmosphere than in a vacuum? (27.4)

15. Light incident upon a pane of glass slows down in passing through the glass. Does it emerge at a slower speed or at its initial speed? Explain (27.4).

16. What determines whether or not a material is transparent or opaque? (27.4-27.5)

17. Why are metals shiny in appearance? (27.5)

18. Distinguish between an umbra and a penumbra. (27.6)

19. a. Distinguish between a solar eclipse and a lunar eclipse.
    b. Which type of eclipse is dangerous to your eyes if viewed directly? (27.6)

20. What is the difference between light that is polarized and light that is not? (27.7)

21. Why is light from a common lamp or from a candle flame nonpolarized? (27.7)

22. In what direction is the polarization of the glare that reflects from a horizontal surface? (27.7)

23. How do polarizing filters allow each eye to see separate images in the projection of three-dimensional slides or movies? (27.7)

Think and Explain

1. What evidence can you cite to support the idea that light can travel through a vacuum?

2. If the octagonal mirror in the Michelson apparatus were spun at twice the speed that produced light in the eyepiece, would light still be seen? At 2.1 times the speed? Explain.

3. If the mirror in Michelson's apparatus had had six sides instead of eight, would it have had to spin faster or more slowly to measure the speed of light? Explain.

4. You can get a sunburn on a sunny day and on an overcast day. But you cannot get a sunburn if you are behind glass. Explain.

5. If you fire a bullet through a tree, it will slow down in the tree and emerge at less than its initial speed. But when light shines on a pane of glass, even though it slows down inside, its speed upon emerging is the same as its initial speed. Explain.

6. Short wavelengths of visible light interact more frequently with the atoms in glass than do longer wavelengths. Which do you suppose takes the longer time to get through glass—red light or blue light?

7. Suppose that sunlight is incident upon both a pair of reading glasses and a pair of sunglasses. Which pair would you expect to be warmer, and why?

8. An ideal polarizing filter transmits 50% of the incident nonpolarized light. Explain.

9. What percentage of light would be transmitted by two ideal polarizing filters, one atop the other, with their axes aligned? With their axes crossed at right angles?

28 | Color

Roses are red and violets are blue; colors intrigue artists and physics types too. To the physicist, the colors of things are not in the substances of the things themselves. Color is in the eye of the beholder and is provoked by the frequencies of light emitted or reflected by things. We see red in a rose when light of certain frequencies reaches our eyes. Other frequencies will provoke the sensation of other colors. Whether or not these frequencies of light are actually perceived as colors depends on the eye-brain system. Many organisms, including people with defective color vision, see no red in a rose.

28.1 | The Color Spectrum

Isaac Newton was the first to make a systematic study of color. By passing sunlight through a triangular-shaped glass prism, he was the first to show that sunlight is composed of a mixture of all the colors of the rainbow. The prism cast the sunlight into an elongated patch of colors on the wall (Figure 28–1). Newton called this spread of colors a **spectrum**, and noted that the colors were formed in the order red, orange, yellow, green, blue, and violet.

Sunlight is an example of what is called **white light**. Under white light, white objects appear white and colored objects appear in their individual colors. Newton showed that the colors in the spectrum were a property not of the prism but of white light itself. He demonstrated this when he recombined the colors with a second prism to produce white light again (Figure 28–2). In other words, all the colors, one atop the other, combine to produce white light. Strictly speaking, white is not a color but a combination of all the colors.

Fig. 28–1 Newton passed sunlight through a glass prism to form the color spectrum.

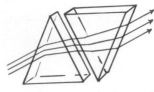

Fig. 28–2 Recombination of colors to produce white light.

Black is similarly not a color itself, but is the absence of light. Objects appear black when they absorb all visible frequencies of light. Carbon soot is an excellent absorber of light and looks very black. The dull finish of black velvet is an excellent absorber also. But even a polished surface may look black under some conditions. For example, highly polished razor blades are not black, but when stacked together and viewed end on, they appear quite black (Figure 28–3). Most of the light that gets between the closely spaced edges of the blades gets trapped and is absorbed after being reflected many times.

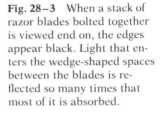

Fig. 28–3 When a stack of razor blades bolted together is viewed end on, the edges appear black. Light that enters the wedge-shaped spaces between the blades is reflected so many times that most of it is absorbed.

Black objects that you can see do not absorb all light that falls on them, for there is always some reflection at the surface. If not, you wouldn't be able to see them.

28.2 Color by Reflection

The colors of most objects around you are due to the way the objects reflect light. Light is reflected from objects in a manner similar to the way sound is "reflected" from a tuning fork when another that is nearby sets it into vibration. A tuning fork can be made to vibrate even when the frequencies are not matched, although at significantly reduced amplitudes. The same is true of atoms and molecules. We can think of atoms and molecules as three-dimensional tuning forks with electrons that behave as tiny oscillators that can vibrate as if attached by invisible springs (Figure 28–4).* Electrons can be forced into vibration by the vibrations of electromagnetic waves (such as light). Like acoustical tuning forks, once vibrating, they send out their own energy waves in all directions.

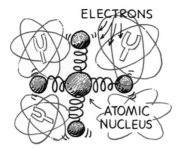

Fig. 28–4   The outer electrons in an atom vibrate as if they were attached to the nucleus by springs. As a result, atoms and molecules behave like tuning forks for light.

Different kinds of atoms and molecules have different natural vibration frequencies. The electrons of one kind of atom can be set into vibration over a range of frequencies different from the range for other kinds of atoms. At the resonant frequencies where the amplitudes of oscillation are large, light is absorbed. But at frequencies below and above the resonant frequencies, light is re-emitted. If the material is transparent, the re-emitted light passes through it. If the material is opaque, the light passes back into the medium from which it came. This is reflection.

Most materials absorb some frequencies and reflect the rest. If a material absorbs most visible frequencies and reflects red, for example, the material appears red. If it reflects all the visible frequencies, like the white part of this page, it will be the same color as the light that shines on it. If a material absorbs all the light that shines on it, it reflects none and is black.

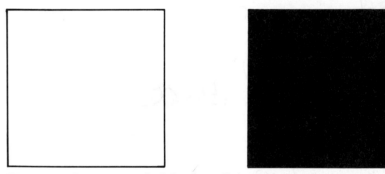

Fig. 28–5   The square on the left *reflects* all the colors illuminating it. In sunlight it is white. When illuminated with blue light, it is blue. The square on the right *absorbs* all the colors illuminating it. In sunlight it is warmer than the white square.

When white light falls on a flower, some of the frequencies are absorbed by the cells in the flower and some are reflected. Cells that contain chlorophyll absorb most of the frequencies and reflect the green part of the light that falls on them, so they appear green. The petals of a red rose, on the other hand, reflect

---

*   The words *oscillate* and *vibrate*, or *oscillator* and *vibrator*, can be used interchangeably; their meanings are the same.

Fig. 28–6 Color depends on the light source.

primarily red light, with a lesser amount of blue. Interestingly enough, it is found that the petals of most yellow flowers, such as daffodils, reflect red and green as well as yellow. Yellow daffodils reflect a broad band of frequencies. The reflected colors of most objects are not pure single-frequency colors, but are composed of a spread of frequencies.

It is important to note that an object can reflect only frequencies that are present in the illuminating light. The appearance of a colored object therefore depends on the kind of light used. A candle flame emits light that is deficient in blue; it emits a yellowish light. Things look yellowish in candlelight. An incandescent lamp emits light that is richer toward the lower frequencies, enhancing the reds. A fluorescent lamp is richer in the higher frequencies, so blues are enhanced when illuminated with fluorescent lamps. In a fabric with a little bit of red, for example, the red will be more apparent when illuminated with an incandescent lamp than with a fluorescent lamp. Colors appear different in daylight than when illuminated with either of these lamps (Figure 28–6). The "true" color of an object is subjective and depends on the light source, although color differences between two objects are most easily detected in bright sunlight.

28.3   Color by Transmission

The color of a transparent object depends on the color of the light it transmits. A red piece of glass appears red because it absorbs all the colors that compose white light, except red, which it transmits. Similarly, a blue piece of glass appears blue because it transmits primarily blue and absorbs the other colors that illuminate it.

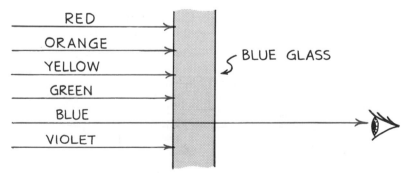

Fig. 28–7 Blue glass transmits only energy of the frequency of blue light; energy of the other frequencies is absorbed and warms the glass.

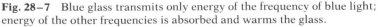

The material in the glass that selectively absorbs colored light is known as a **pigment**. From an atomic point of view, electrons in the pigment atoms are set into vibration by the illuminating light. Some of the frequencies are absorbed by the pigment, and others are re-emitted from atom to atom in the glass. The energy of the absorbed light increases the kinetic energy of the atoms, and the glass is warmed. Ordinary window glass is colorless because it transmits all visible frequencies equally well.

> ▶ **Questions**
>
> 1. When red light shines on a red rose, why do the leaves become warmer than the petals?
>
> 2. When green light shines on a red rose, why do the petals look black?
>
> 3. What color does a daffodil appear when illuminated with red light? With yellow light? With green light? With blue light?

28.4 | Sunlight

White light from the sun is a composite of all the visible frequencies. The brightness of solar frequencies is uneven, as indicated in the graph of brightness versus frequency (Figure 28–8). The graph indicates that the lowest frequencies of sunlight, in the red region, are not as bright as those in the middle-range yellow and green region. Yellow-green is the brightest part of sunlight. (Since humans evolved in the presence of sunlight, it is not surprising that we are most sensitive to yellow-green. That is why it is more and more common for new fire engines to be painted yellow-green, particularly at airports where visibility is vital.

▶ **Answers**

1. The petals appear red because they reflect red light. The leaves absorb rather than reflect red light, so the leaves become warmer.

2. The petals absorb rather than reflect the green light. Since green is the only color illuminating the rose, and green contains no red to be reflected, the rose looks no color at all—black.

3. A daffodil reflects red, yellow, and green light, so when illuminated with any of these colors it reflects that color and appears that color. A daffodil does not reflect blue, so when illuminated with blue light it looks black.

This also explains why at night we see better under the illumination of yellow sodium-vapor lamps than under tungsten lamps of the same brightness.) The blue part of sunlight is not as bright, and the violet is even less bright.

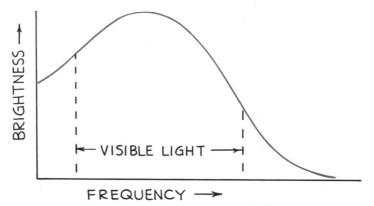

Fig. 28–8   The radiation curve of sunlight is a graph of brightness versus frequency. Sunlight is brightest in the yellow-green region, in the middle of the visible range.

This graphical distribution of brightness versus frequency is called the *radiation curve* of sunlight. Most whites produced from reflected sunlight share this frequency distribution.

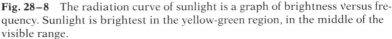

28.5   Mixing Colored Light

All the visible frequencies mixed together produce white. Interestingly enough, white also results from the combination of only red, green, and blue light. Look at Figure 28–9 and Plate 1 of the colored illustrations. (All the colored illustrations are in the four-page color section in this chapter. Many of the effects caused by the physics of color are shown there.) When a combination of only red, green, and blue light of equal brightness is overlapped on a screen, it appears white. Where red and green light alone overlap, the screen appears yellow. Red and blue light alone produce the bluish-red color called *magenta*. Green and blue light alone produce the greenish-blue color called *cyan*.

This is understood if the frequencies of white light are divided into three regions: the lower-frequency red end, the middle-frequency green part, and the higher-frequency blue end (Figure 28–10). The low and middle frequencies combined appear yellow to the human eye. The middle and high frequencies com-

Plate 1

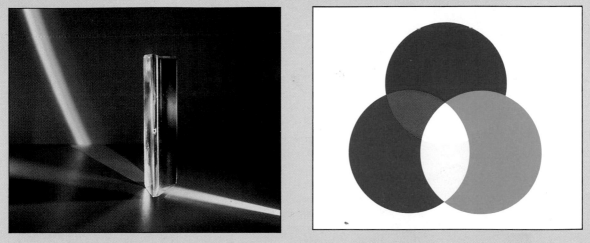

(Left) When sunlight passes through a prism, it separates into a spectrum of all the colors of the rainbow. (Right) When red light, green light, and blue light from three separate projectors overlap on a screen, they add to produce white light. The overlap of red and green light alone produces yellow; the overlap of green and blue light alone produces cyan (greenish blue); the overlap of red and blue light alone produces magenta (bluish red).

Plate 2

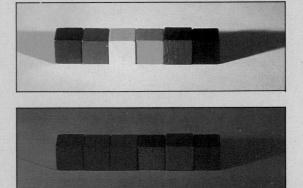

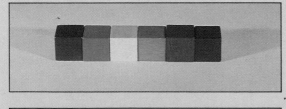

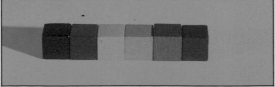

(Top left) Under white light, the colored blocks appear red, orange, yellow, green, blue, and purple, and the shadows are gray. (Top right) When the blocks are lit by red light from the right and green light from the left, the blocks themselves reflect no blue light; the shadow on the left, where red light is absent, appears green; similarly the shadow on the right appears red; the background reflects both red and green light and appears yellow. (Bottom left) The blocks are lit by green light and blue light. (Bottom right) The blocks are lit by red light and blue light. For the last two photos, identify which color of light comes from which direction, and explain the apparent colors of the blocks and the background.

Plate 3

In the printing of all these colored illustrations, only four colors of ink are used: magenta, yellow, cyan, and black. The first three images of the chicken and the colored eggs show how the photo appears when printed separately in magenta, yellow, and cyan ink. The fourth image shows the combination of those three images. The fifth image includes black ink as well.

Plate 4

The blue appearance of distant foliage-covered mountains can be explained in terms of scattering (see Sections 28.8 and 28.9).

Plate 5

(From top to bottom) The continuous spectrum of an incandescent lamp and the line spectra of three elements: hydrogen, sodium, and mercury (see Section 28.11).

Plate 6

A double rainbow. In the inner rainbow, which is brighter, red appears at the outside edge and violet appears at the inside edge. The colors are in reverse order in the outer rainbow, which is due to an extra reflection inside each water drop (see Section 29.11).

Plate 7

The colors that flash from a compact disk (CD) are due to the interference of light waves that reflect from the tiny pits on the surface of the disk, which act like a diffraction grating (see Section 31.4).

Plate 8

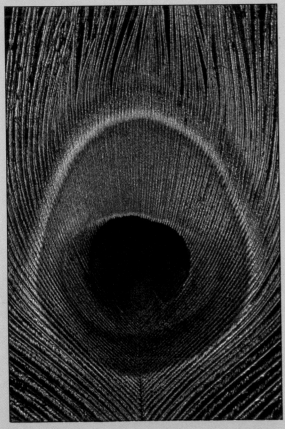

The brilliant colors in a peacock feather are due to iridescence (see Section 31.6).

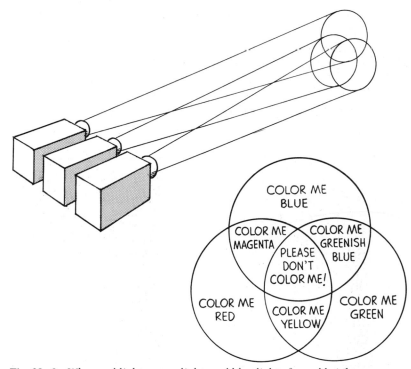

Fig. 28–9   When red light, green light, and blue light of equal brightness are projected on a white screen, the overlapping areas appear different colors. Where all three overlap, white is produced. See Plate 1 for a color version of this illustration.

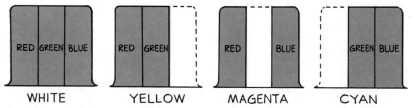

Fig. 28–10   (Left) The lower-frequency, middle-frequency, and high-frequency parts of white light appear *red*, *green*, and *blue*. (Right) To the human eye, red + green = yellow; red + blue = magenta; green + blue = cyan.

bined appear greenish-blue (cyan). The low and high frequencies combined appear bluish-red (magenta).

In fact, almost any color at all can be made by overlapping light of three colors and adjusting the brightness of each color of light. The three colors do not have to be red, green, and blue, although those three produce the highest number of different colors. This amazing phenomenon is due to the way the human eye works.

Color television is based on the ability of the human eye to see combinations of three colors as a variety of different colors. A close examination of the picture on most color television tubes will reveal that the picture is made up of an assemblage of tiny spots, each less than a millimeter across. When the screen is lit, some of the spots are red, some green, and some blue. At a distance the mixtures of these colors provide a complete range of colors, plus white.*

28.6 | Complementary Colors

Two colors of light which when added appear white are called **complementary colors**. For example, red and cyan are complementary colors, since cyan is made from green and blue light (no red), and red, green, and blue light together appear white. Notice in Figure 28–9 that with only two colors of light, the following colors are produced:

$$red + green = yellow$$
$$red + blue = magenta$$
$$blue + green = cyan$$

A little thought will show that:

$$blue + yellow = white$$
$$green + magenta = white$$
$$red + cyan = white$$

Every hue has some complementary color that when added will produce white.

Now, if you begin with white light and *subtract* some color from it, the resulting color will appear as the complement of the one subtracted. Not all the light incident upon an object is reflected. Some is absorbed. The part that is absorbed is in effect subtracted from the incident light. If white light falls on a pigment that absorbs red light, for example, the light reflected appears cyan. A pigment that absorbs blue light will appear yellow; similarly, a pigment that absorbs yellow light will appear blue. Whenever you subtract a color from white light, you end up with the complementary color.

---

*   On a black and white television set, the black you see in the darkest scenes is simply the color of the tube face itself, which is more a light gray than black. Your eyes are sensitive to the contrast with the illuminated parts of the screen and you see the light gray as black. In your mind, you make it black.

28.7 | Mixing Colored Pigments

Every artist knows that if you mix red, green, and blue paint, the result will be not white but a muddy dark brown. Red and green paint certainly do not combine to form yellow as red and green light do. The mixing of paints and dyes is an entirely different process from the mixing of colored light.

Paints and dyes contain finely divided solid particles of pigment that produce their colors by absorbing certain frequencies and reflecting other frequencies of light. Pigments absorb a relatively wide range of frequencies and reflect a wide range as well. In this sense, pigments reflect a mixture of colors.

Blue paint, for example, reflects mostly blue light, but also violet and green; it absorbs red, orange, and yellow light. Yellow paint reflects mostly yellow light, but also red, orange, and green; it absorbs blue and violet light. When blue and yellow paints are mixed, then between them they absorb all the colors except green. The only color they both reflect is green (Figure 28–11), which is why the mixture looks green. This process is called *color mixing by subtraction*, to distinguish it from the effect of mixing colored light, which is called *color mixing by addition*.

So if you cast lights on the stage at a school play, you use the rules of color addition to produce various colors. But if you mix paint, you use the rules of color subtraction.

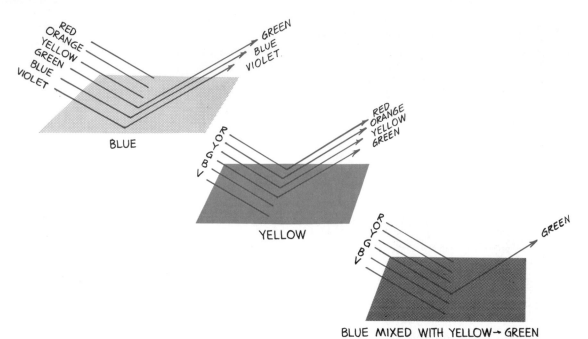

Fig. 28–11   (Left) Blue pigment reflects not only blue light, but also colors to either side of blue—namely, green and violet. It absorbs red, orange, and yellow light. (Center) Yellow pigment reflects not only yellow light, but also red, orange, and green. It absorbs blue and violet light. (Right) When blue and yellow pigments are mixed, the only common color reflected is green. The other colors have been *subtracted* from the incident white light.

You may have learned as a child that you could make any color with crayons or paints of three so-called primary colors: red, yellow, and blue. Actually, the three paint or dye colors that are most useful in color mixing by subtraction are magenta (bluish red), yellow, and cyan (greenish blue). These are the colors used in printing illustrations in full color.*

Color printing is done on a press that prints each page with four differently colored inks (magenta, yellow, cyan, and black) in succession. Each color of ink comes from a different plate, which transfers the ink to the paper. The ink deposits are regulated on different parts of the plate by tiny dots. Examine the colored pictures in this book or in any magazine with a magnifying glass and see how the overlapping dots of three colors plus black give the appearance of many colors. The color section shows the images made by the four plates, separately and combined.

---

*   Note that magenta, yellow, and cyan are used to make other colors by *subtraction*, as when colored paints or dyes are mixed. When colors are mixed by *addition*, as when colored light is mixed, red, green, and blue are the most useful colors to mix.

28.8 Why the Sky Is Blue

If a sound beam of a particular frequency is directed to a tuning fork of similar frequency, the tuning fork will be set into vibration and effectively redirect the beam in multiple directions. The tuning fork **scatters** the sound. A similar process occurs with the scattering of light from atoms and particles that are far apart from one another—as in the atmosphere.

We know that atoms behave like tiny optical tuning forks and re-emit light waves that shine on them. Very tiny particles do the same. The tinier the particle, the higher the frequency of light it will scatter. This is similar to small bells that ring with higher notes than larger bells. The nitrogen and oxygen molecules and the tiny particles that make up the atmosphere are like tiny bells that "ring" with high frequencies when energized by sunlight. Like the sound from bells, the re-emitted light is sent in all directions. It is scattered.

Most of the ultraviolet light from the sun is absorbed by a protective layer of ozone gas in the upper atmosphere. The remaining ultraviolet sunlight that passes through the atmosphere is scattered by atmospheric particles and molecules. Of the visible frequencies, violet is scattered the most, followed by blue, green, yellow, orange, and red, in that order. Red is scattered only a tenth as much as violet. Although violet light is scattered more than blue, our eyes are not very sensitive to violet light. Our eyes are more sensitive to blue, so we see a blue sky.

The blue of the sky varies in different places under different conditions. Where there are a lot of particles of dust and other particles larger than oxygen and nitrogen molecules, the lower frequencies of light are scattered more. This makes the sky less blue, and it takes on a whitish appearance. After a heavy rainstorm, when the particles have been washed away, the sky becomes a deeper blue.

The higher that one goes into the atmosphere, the fewer molecules there are in the air to scatter light. The sky appears darker. When there are no molecules, as on the moon for example, the "sky" is black.

Clusters of water molecules in a variety of sizes make up clouds. The different-size clusters result in a variety of scattered frequencies: low frequencies from larger droplets and high frequencies from tinier clusters of water molecules. The overall result is a white cloud. The electrons in a cluster vibrate together and in step, which results in the scattering of a greater amount of energy than when the same number of electrons vibrate separately. Hence, clouds are bright!

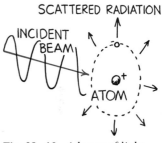

Fig. 28–12 A beam of light falls on an atom and causes the electrons in the atom to vibrate. The vibrating electrons, in turn, re-emit light in various directions. Light is scattered.

Fig. 28-13 When the air is full of particles larger than molecules, lower frequencies as well as blue are scattered. These add to give white, as seen in the haze against distant mountains (above) or in white clouds (below).

28.9 | Why Sunsets Are Red

The lower frequencies of light are scattered the least by nitrogen and oxygen molecules. Therefore red, orange, and yellow light are transmitted through the atmosphere more readily than violet and blue. Red light, which is scattered the least, passes through more atmosphere without interacting with matter than light of any other color. Therefore, when light passes through a thick atmosphere, the lower frequencies are transmitted while the higher frequencies are scattered. At dawn and at sunset, sunlight reaches us through a longer path through the atmosphere than at noon.

At noon sunlight travels through the least amount of atmosphere to reach the earth's surface (Figure 28–14). Only a small amount of high-frequency light is scattered from sunlight. As the day progresses and the sun is lower in the sky, the path through the atmosphere is longer, and more blue is scattered from the sunlight. High-frequency light is scattered, and lower-frequency light predominates in the sunlight that reaches the earth. The sun appears progressively redder, going from yellow to orange and finally to a red-orange at sunset. (The sequence is reversed between dawn and noon.)

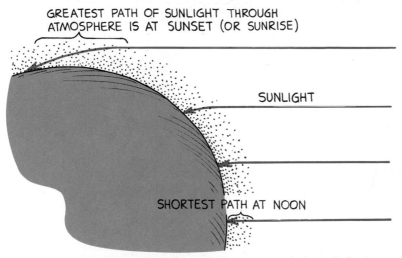

Fig. 28–14 A beam of sunlight must travel a longer path through the atmosphere at sunset than at noon. As a result, more blue is scattered from the beam at sunset than at noon. By the time a beam of white light gets to the ground at sunset, only the lower frequencies survive, producing a red sunset.

The colors of the sun and sky are consistent with our rules for color mixing. When blue is subtracted from white light, the complementary color that is left is yellow. When green is subtracted, magenta is left. The amount of scattering of each frequency depends on atmospheric conditions, which change from day to day and give us a variety of sunsets.

The next time you find yourself admiring a crisp blue sky, or delighting in the shapes of bright clouds, or watching a beautiful sunset, think about all those ultratiny optical tuning forks vibrating; you'll appreciate these everyday wonders of nature even more!

▶ **Questions**
1. If molecules in the sky scattered low-frequency light instead of high-frequency light, how would the colors of the sky and sunsets appear?

2. Distant dark mountains are bluish in color. What is the source of this blueness? (*Hint*: Exactly what is between you and the mountains you see?)

3. Distant snow-covered mountains reflect a lot of light and are bright. But they sometimes look yellowish, depending on how far away they are. Why are they yellow? (*Hint*: What happens to the reflected white light as it travels from the mountain to you?)

28.10 | Why Water Is Greenish Blue

The color of water is not the beautiful deep blue that you often see on the surface of a lake or the ocean. That blue is the reflected color of the sky. The color of water itself, as you can see by looking at a piece of white material under water, is a pale greenish blue.

Water is transparent to nearly all the visible frequencies of light. Water molecules absorb infrared waves. This is because

---

▶ **Answers**
1. If low frequencies were scattered, the noontime sky would appear reddish orange. At sunset more reds would be scattered by the longer path of the sunlight, and the sunlight would be predominantly blue and violet. So sunsets would appear blue!

2. If you look at distant dark mountains, very little light from them reaches you, and the blueness of the atmosphere between you and the mountains predominates. The blueness is of the low-altitude "sky" between you and the mountains. That's why distant mountains look blue!

3. The reason that distant snow-covered mountains often appear a pale yellow is because the blue in the white light from the snowy mountains is scattered on its way to you. What happens to white when blue is scattered from it? The complementary color left is yellow.

   Why do you see the scattered blue when the background is dark, but not when the background is bright? Because the scattered blue is faint. A faint color will show itself against a dark background, but not against a bright background. For example, when you look from the earth's surface at the atmosphere against the darkness of space, the atmosphere is sky blue. But astronauts above who look below through the same atmosphere to the bright surface of the earth do not see the same blueness.

water molecules resonate to the frequencies of infrared. The energy of the infrared waves is transformed into kinetic energy of the water molecules. That is why sunlight warms water.

Water molecules resonate very weakly to the visible-red frequencies. This causes a gradual absorption of red light by water. A 15-m layer of water reduces red light to a quarter of its initial brightness. There is very little red light in the sunlight that penetrates below 30 m of water. When red is taken away from white light, what color remains? Or this question can be asked in another way: What is the complementary color of red? The complementary color of red is cyan—a greenish-blue color. In sea water, the color of everything at these depths looks greenish blue.

It is interesting to note that many crabs and other sea animals that appear black in deep water are found to be red when they are raised to the surface. At great depths, black and red look the same. So both black and red sea animals are hardly seen by predators and prey in deep water. They have survived an evolutionary history while more visible varieties have not.

In summary, the sky is blue because blue from sunlight is re-emitted in all directions by molecules in the atmosphere. Water is greenish blue because red is absorbed by molecules in the water. The colors of things depend on what colors are reflected by molecules, and also by what colors are absorbed by molecules.

28.11 The Atomic Color Code—Atomic Spectra

Every element has its own characteristic color when made to emit light. If the atoms are far enough apart so that their vibrations are not interrupted by neighboring atoms, their true colors are emitted. This occurs when atoms are made to glow in the gaseous state. (In the solid state, as in a lamp filament, where atoms are crowded together, the characteristic colors of the atoms are smudged to produce a continuous spectrum.) Neon gas, for example, glows a brilliant red; mercury vapor glows a bluish violet; and helium glows a pink. The glow of each element is unlike the glow of any other element.

The light from glowing elements can be analyzed with an instrument called a **spectroscope**. This chapter began with a brief account of Newton's investigation of light passing through a prism. The spectrum formed in Newton's first experiment was impure. This was because it was formed by overlapping circular images of the circular hole in his window shutter. He later produced a better spectrum by first passing light through a thin slit and then focusing it with lenses through the prism and onto a white screen (Figure 28–15). If the slit is made narrow, over-lapping is reduced and the colors in the resulting spectrum are much clearer.

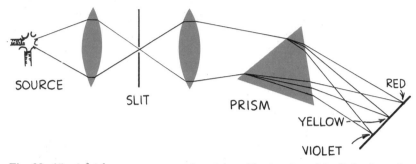

Fig. 28–15 A fairly pure spectrum is produced by passing white light through a thin slit, two lenses, and a prism.

This arrangement of thin slit, lenses, and a prism (or a diffraction grating) is the basis for the spectroscope (Figure 28–16).* A spectroscope displays the spectra of the light from hot gases and other light sources. (*Spectra* is the plural of *spectrum*.) The spectra of light sources are viewed through a magnifying eyepiece.

Fig. 28–16 A spectroscope. Light to be analyzed is placed near the thin slit at the left, where it is focused by lenses onto either a diffraction grating (shown) or a prism on the rotating table in the middle, and then viewed through the eyepiece on the right.

---

* Although a *diffraction grating* works differently from a prism, it too spreads light into a spectrum. It is more commonly used than a prism in spectroscopes. A *spectrometer* is similar to a spectroscope except that it also measures the wave-lengths of a spectrum and records the spectrum (on film for example).

When light from a glowing element is analyzed through a spectroscope, it is found that the colors are the composite of a variety of different frequencies of light. The spectrum of an element appears not as a continuous band of color but as a series of lines (see Figure 28–17 and the color section). Each line corresponds to a distinct frequency of light. Such a spectrum is known as a **line spectrum**. The spectral lines seen in the spectroscope are images of the slit through which the light passes. Note that each colored line appears in the same position as that color in the continuous spectrum.

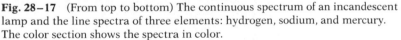

Fig. 28–17   (From top to bottom) The continuous spectrum of an incandescent lamp and the line spectra of three elements: hydrogen, sodium, and mercury. The color section shows the spectra in color.

The light from each different element produces its own characteristic pattern of lines. This is because each element has its own distinct configuration of electrons, and these emit distinct frequencies of light when electrons change from one energy state to another in the atom.* The frequencies of light emitted by atoms in the gaseous state are the "fingerprints" of the elements. Much of the information that physicists have about atomic structure is from the study of atomic spectra. The atomic composition of common materials, the sun, and distant galaxies is revealed in the spectra of these sources. The spectrometer is a very useful and powerful tool.

---

*   Electrons in atoms behave differently when materials glow and are sources of illumination than when materials simply reflect light that shines on them. This distinction won't be treated in detail here, except to say that when materials are made to glow, the electrons in their atoms do more than simply vibrate— they move from one orbit to another in a process called *excitation*. There is more to learn about all this in follow-up physics courses.

28 | Chapter Review

Concept Summary

White light is a combination of light of all visible frequencies.
- Black is the absence of light; objects that appear black absorb all visible frequencies.

The color of an object is due to the color of the light it reflects (if opaque) or transmits (if transparent).
- Light is absorbed when its frequency matches the natural vibration frequencies of the electrons in the atoms of a material illuminated by the light.

Color mixing by addition is the mixing of light of different frequencies.
- The eye sees a combination of red, green, and blue light of equal brightness as white.

Color mixing by subtraction is the mixing of colored paints or dyes, which absorb most frequencies except for the ones that give them their characteristic color.
- When paints or dyes are mixed, the mixture absorbs all the frequencies each paint or dye absorbs.

Scattering of violet and blue frequencies of sunlight in all directions is what gives the sky its blue color.
- When sunlight travels a long path through the atmosphere, as at dawn or sunset, only the lower frequencies of light are transmitted; the higher ones are scattered out.

Atoms of each element have characteristic line spectra that can be used to identify the element.

Important Terms

complementary colors (28.6)
line spectrum (28.11)
pigment (28.3)
scatter (28.8)
spectroscope (28.11)
spectrum (28.1)
white light (28.1)

Review Questions

1. List the order of colors in the color spectrum. (28.1)

2. Are black and white real colors, in the sense that red and green are? Explain. (28.1)

3. A vibrating tuning fork emits sound. What is emitted by the vibrating electrons of atoms? (28.2)

4. What happens to light of a certain frequency that encounters atoms of the same resonant frequency? (28.2)

5. Why does the color of an object look different under a fluorescent lamp than under an incandescent lamp? (28.2)

6. a. What color(s) of light does a transparent red object *transmit*?
   b. What color(s) does it *absorb*? (28.3)

7. What is the function of a pigment? (28.3)

8. Why are more and more fire engines being painted yellow-green instead of red? (28.4)

9. How can yellow be produced on a screen if only red light and green light are available? (28.5)

10. What is the name of the color produced by a mixture of green and blue light? (28.5)

11. What colors of spots are lit on a television tube to give full color? (28.5)

12. What are complementary colors? (28.6)

13. What color is the complement of blue? (28.6)

14. The process of producing a color by mixing pigments is called *color mixing by subtraction*. Why do we say "subtraction" instead of "addition" in this case? (28.7)

15. What colors of ink are used to print full-color pictures in books and magazines? (28.7)

16. What is light scattering? (28.8)

17. a. Do the tiniest particles in the air scatter high or low frequencies of light?
    b. How about larger particles? (28.8)

18. Why is the sky blue? (28.8)

19. Why is the sky sometimes whitish? (28.8)

20. Why are clouds white? (28.8)

21. Why are sunsets red? (28.9)

22. Why is water greenish blue? (28.10)

23. What is a spectroscope, and what is its function? (28.11)

24. Does the red light from glowing neon gas have only one frequency or a mixture of frequencies? (28.11)

25. Why might atomic spectra be considered the "fingerprints" of atoms? (28.11)

Activities

1. Stare at a brightly colored object for a minute or so. The color receptors in your eyeballs become fatigued, so that when you look at a white area afterward, you see an afterimage of the complementary color. This is because the fatigued receptors send a weaker signal to the brain. All the colors produce white, but all the colors minus one produce the complementary to the missing color. Try it and see.

2. Make up a cardboard tube with closed ends of metal foil. Punch a hole in each end with a pencil, one about 3 or so millimeters and the other twice as big. Put your eye to the smaller hole and look through both holes at the colors of things against the black background of the tube. Colors will appear very different from how they appear against ordinary backgrounds.

3. Simulate your own sunset: Add a few drops of milk to a glass of water and look through it to a lit incandescent bulb. The bulb appears to be red or pale orange, while light scattered to the side appears blue. Try it and see.

Think and Explain

1. Why is space beyond the world's atmosphere black?

2. In a dress shop that has only fluorescent lighting, a customer insists on taking a garment into the daylight at the doorway. Is she being reasonable? Explain.

3. What color would a yellow cloth appear if illuminated with sunlight? With yellow light? With blue light?

4. A spotlight is coated so that it won't transmit blue from its white-hot filament. What color is the emerging beam of light?

5. How could you use the spotlights at a play to make the yellow clothes of the performers suddenly change to black?

6. A stage performer stands where beams of red and green light cross.
   a. What is the color of her white shirt under this illumination?
   b. What are the colors of the shadows she casts on the stage floor?

7. Very big particles, such as droplets of water, absorb more radiation than they scatter. How does this fact help to explain why rain clouds appear dark?

8. The only light to reach very far beneath the surface of the ocean is greenish blue. Objects at these depths either reflect greenish blue or reflect no color at all. If a ship that is painted red, green, and white sinks to the bottom of the ocean, how will these colors appear?

9. The element helium is so named because it was discovered in the sun—named "Helios" in Greek—before it was detected on earth. What instrument do you suppose enabled this discovery?

10. A lamp filament is made of tungsten. When made to glow, it emits a continuous spectrum—all the colors of the rainbow. When tungsten *gas* is made to glow, however, the light is a composite of very discrete colors. Why is there a difference in spectra?

29 | Reflection and Refraction

If you shine a beam of light on a mirror, the light doesn't travel through the mirror, but is returned at the surface back into the air. When sound waves strike a canyon wall, they travel back to you as an echo. A transverse wave transmitted along a spring reverses direction when it reaches the wall. In all these situations, waves remain in one medium rather than enter a new medium. These waves are *reflected*.

In other situations, as when light passes from air into water, waves travel from one medium into another. If the waves strike the surface of the medium at an angle, their direction changes in the second medium. These waves are *refracted*.

In most cases, waves are both reflected and refracted when they fall on a transparent medium. When light shines on water, for example, some of the light is reflected and some is refracted. To understand this, let us see why reflection occurs.

29.1 | Reflection

When a wave reaches the boundary between two media, some or all of the wave bounces back into the first medium. This is **reflection**. For example, suppose you fasten a spring to a wall and send a pulse along its length (Figure 29–1). The wall is a very rigid medium compared to the spring. As a result, all the wave energy is transmitted back along the spring rather than into the wall. Waves that travel along the spring are *totally reflected* at the wall.

If the wall is replaced with a less rigid medium, such as the heavy spring shown in Figure 29–2, some energy is transmitted into the new medium. Some of the wave energy is still reflected. These waves are *partially reflected*.

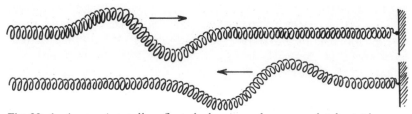

Fig. 29–1 A wave is totally reflected when it reaches a completely rigid boundary.

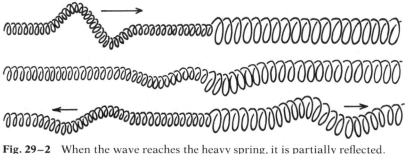

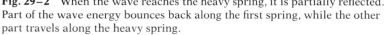

Fig. 29–2 When the wave reaches the heavy spring, it is partially reflected. Part of the wave energy bounces back along the first spring, while the other part travels along the heavy spring.

A metal surface is rigid to light waves that shine upon it. Light energy does not propagate into the metal, and instead is returned in a reflected wave. The wave reflected from a metal surface has almost the full intensity of the incoming wave, apart from small energy losses due to the friction of the vibrating electrons in the surface. This is why metals such as silver and aluminum are so shiny. They reflect almost all the frequencies of visible light. Smooth surfaces of these metals are therefore used as mirrors.

Materials such as glass and water are not as rigid to light waves. Like the different springs of Figure 29–2, wave energy is both reflected and transmitted at the boundary. When light falls perpendicularly on the surface of still water, about 2 percent is reflected and the rest transmitted. When light strikes glass perpendicularly, about 4 percent is reflected. Except for slight losses, the rest is transmitted.

29.2 | The Law of Reflection

In one dimension, reflected waves simply travel back in the direction from which they came. Let a ball drop to the floor, and it bounces straight up, along its initial path. In two dimensions, the situation is a little different.

The direction of incident and reflected waves is best described by straight lines called *rays*. Incident rays and reflected rays make equal angles with a line perpendicular to the surface, called the **normal** (Figure 29–3). The angle made by the incident ray and the normal, called the **angle of incidence**, is equal to the angle made by the reflected ray and the normal, called the **angle of reflection**. That is:

angle of incidence = angle of reflection

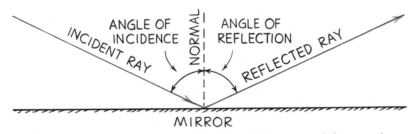

Fig. 29–3   In reflection, the angle between the incident ray and the normal is equal to the angle between the reflected ray and the normal.

This relationship is called the **law of reflection**. The incident ray, the normal, and the reflected ray all lie in the same plane. The law of reflection applies to both partially reflected and totally reflected waves.

29.3 | Mirrors

Consider a candle flame placed in front of a plane (flat) mirror. Rays of light are reflected from its surface in all directions. The number of rays is infinite, and every one obeys the law of reflection. Figure 29–4 shows only two rays that originate at the tip of the candle flame and reflect from the mirror to someone's eye. Note that the rays diverge (spread apart) from the tip of the flame, and continue diverging from the mirror upon reflection. These divergent rays *appear* to originate from a point located behind the mirror. The image of the candle the person sees in the mirror is called a **virtual image**, because light does not actually pass through the image position but behaves virtually as if it did.

Your eye cannot ordinarily tell the difference between an object and its reflected image. This is because the light that enters your eye is entering in exactly the same manner, physically, as it would if there really were an object there. Notice that the image is as far behind the mirror as the object is in front of the mirror. Notice also that the image and object have the same size.

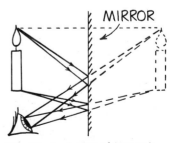

Fig. 29–4   A virtual image is formed behind the plane mirror and is located at the position where the extended reflected rays (broken lines) converge.

When you view yourself in a mirror, for example, the size of your image is the same size your identical twin would appear if located as far behind the mirror as you are in front—as long as the mirror is flat.

Fig. 29–5 For reflection in a plane mirror, object size equals image size and object distance equals image distance.

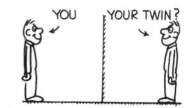

When the mirror is curved, the sizes and distances of object and image are no longer equal. This text will not treat curved mirrors, except to say that the law of reflection still holds. The angle of incidence is equal to the angle of reflection (Figure 29–6). Note that for a curved mirror, unlike a plane mirror, the normals (shown as dashed black lines) at different points on the surface are not parallel to each other.

Fig. 29–6 (a) The virtual image formed by a *convex* mirror (a mirror that curves outward) is smaller and closer to the mirror than the object. (b) When the object is close to a *concave* mirror (a mirror that curves inward like a "cave"), the virtual image is larger and farther away than the object. In any case the law of reflection holds for each ray.

▶ **Questions**

1. If you look at your blue shirt in a mirror, what is the color of its image? What does this tell you about the frequency of light incident upon a mirror compared to its frequency when reflected?

2. If you wish to take a picture of your image while standing 2 m in front of a plane mirror, for what distance should you set your camera to provide sharpest focus?

▶ **Answers**

1. The color of the image will be the same as the color of the object. This is evidence that the frequency of light does not change when it undergoes reflection.

2. You should set your camera for a distance of 4 m. The situation is equivalent to your standing 2 m in front of an open window and viewing your twin standing 2 m in back of the window.

29.4 | Diffuse Reflection

When light is incident on a rough surface, it is reflected in many directions. This is called **diffuse reflection** (Figure 29–7). Although the reflection of each single ray obeys the law of reflection, the many different angles that light rays encounter in striking a rough surface cause reflection in many directions.

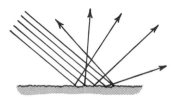

Fig. 29–7 Diffuse reflection from a rough surface.

What constitutes a rough surface for some rays may be a polished surface for others. If the differences in elevations in a surface are small (less than about one-eighth the wavelength of the light that falls on it), the surface is considered polished. A surface therefore may be polished for long wavelengths, but not polished for short wavelengths. The wire-mesh "dish" shown in Figure 29–8 is very rough for light waves, not mirrorlike at all. Yet for long-wavelength radio waves it is polished. It acts as a mirror to radio waves and is an excellent reflector. Whether a surface is a diffuse reflector or a polished reflector depends on the size of the waves it reflects.

Fig. 29–8 The open-mesh parabolic dish acts like a diffuse reflector for light waves but like a polished reflector for long-wavelength radio waves.

Light that reflects from this page is diffuse. The page may be smooth to a long radio wave, but to the short wavelengths of visible light it is rough. This roughness is evident in the microscopic view of an ordinary paper surface (Figure 29–9). Rays of light incident on this page encounter millions of tiny flat surfaces facing in all directions. The light is therefore reflected in all directions. This is very nice, for it allows us to read the page from any direction or position. We see most of the things around us by diffuse reflection.

Fig. 29–9 A microscopic view of the surface of ordinary paper.

Fig. 29–10 (a) If you shine a beam of light on paper, you can see diffusely reflected light at any position. (b) However, your eye must be at the right place to see a reflected beam from a small mirror.

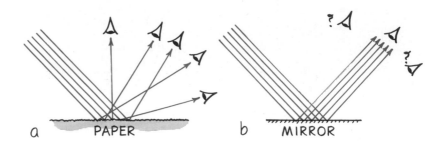

29.5 Reflection of Sound

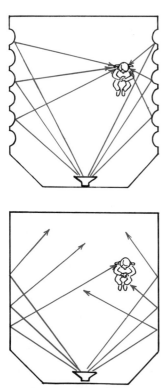

Fig. 29–11 (Top) With grooved walls, sound reflects from many small sections of the wall to a listener. (Bottom) With flat walls, an intense reflected sound comes from only one part of the wall.

An echo is reflected sound. The fraction of sound energy that is reflected from a surface is large if the surface is rigid and smooth, and less if the surface is soft and irregular. Sound energy not reflected is transmitted or absorbed.

Sound reflects from all surfaces—the walls, ceiling, floor, furniture, and people—of a room. Designers of interiors of buildings, whether office buildings, factories, or auditoriums, need an understanding of the reflective properties of surfaces. The study of these properties is the field of *acoustics*.

If the walls of a room, auditorium, or concert hall are too reflective, the sound becomes garbled. This is due to multiple reflections called **reverberations**. On the other hand, if the reflective surfaces are too absorbent, the sound level would be low, and the hall would sound dull and lifeless. Reflection of sound in a room makes it sound lively and full, as you have probably found out while singing in the shower. In the design of an auditorium or concert hall, a balance between reverberation and absorption is desired.

The walls of concert halls are often designed with grooves so that the sound waves are diffused (Figure 29–11 top). In this way a person in the audience receives a small amount of reflected sound from many parts of the wall rather than a larger amount of sound from one part of the wall.

Highly reflective surfaces are often placed behind and above the stage to direct sound out to an audience. The large shiny plastic disks in Figure 29–12 also reflect light. A listener can look up at these reflectors and see the reflected images of the members of the orchestra. (The plastic reflectors are somewhat curved, which increases the field of view.) Both sound and light obey the law of reflection, so if a reflector is oriented so that you can *see* a particular musical instrument, rest assured that you will *hear* it also. Sound from the instrument will follow the line of sight to the reflector and then to you.

Fig. 29–12 The disks above the orchestra in Davies Symphony Hall in San Francisco reflect both light and sound. Adjusting them is quite simple: What you see is what you hear.

29.6 Refraction

Take a pair of wheels off an old toy cart and roll them along the pavement and onto a mowed lawn. They roll more slowly on the lawn because of the interaction of the wheels with the blades of grass. If you roll them at an angle (Figure 29–13), they will be deflected from their straight-line course. The direction of the rolling wheels is shown in the illustration. Note that on meeting the lawn, the left wheel is slowed down first. This is because it meets the grass while the right wheel is still rolling on the pavement. The wheels pivot, and the path is bent toward the normal (the dashed black line perpendicular to the grass-pavement boundary). They then continue across the lawn in a straight line at reduced speed.

Water waves similarly bend when one part of the waves is made to travel more slowly (or faster) than another part. This is **refraction**. Waves travel faster in deep water than in shallow water. Figure 29–14 left shows a view from above of straight wave crests (the bright lines) moving toward the top edge of the photo. They are moving from deep water across a diagonal boundary into shallow water. At the boundary, the wave speed and direction of travel are abruptly altered. Since the wave moves more slowly in shallow water, the crests are closer together. If you look carefully, you'll see that some reflection from the boundary is also taking place.

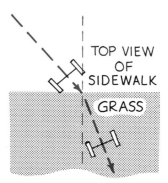

Fig. 29–13 The direction of the rolling wheels changes when one part slows down before the other part.

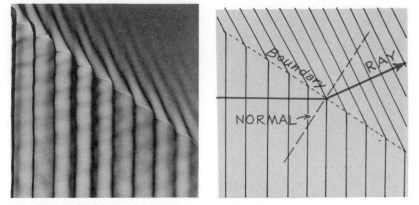

Fig. 29–14   (Left) Photograph of the refraction of a water wave at a boundary where the wave speed changes because the water depth changes. (Right) Diagram of wavefronts and a sample ray. The ray is perpendicular to the wavefront it intersects.

In drawing a diagram of a wave, as in Figure 29–14 right, it is convenient to draw lines that represent the positions of different crests. Such lines are called **wave fronts**.* At each point along a wave front, the wave is moving perpendicular to the wave front. The direction of motion of the wave can thus be represented by rays that are perpendicular to the wave fronts. The rays in Figure 29–14 right show how the water wave changes direction after it crosses the boundary between deep and shallow water. Sometimes we analyze waves in terms of wave fronts, and at other times in terms of rays. Both are useful models for understanding wave behavior.

29.7   Refraction of Sound

Sound waves are refracted when parts of a wave front travel at different speeds. This happens in uneven winds or when sound is traveling through air of uneven temperature. On a warm day, for example, the air near the ground may be appreciably warmer than the air above. Since sound travels faster in warmer air, the speed of sound near the ground is increased. The refraction is not abrupt but gradual (Figure 29–15). Sound waves therefore tend to bend away from warm ground, making it appear that the sound does not carry well.

---

*   Wave fronts also can be considered to represent the positions of different troughs—or any continuous portions of the wave that are all vibrating the same way at the same time.

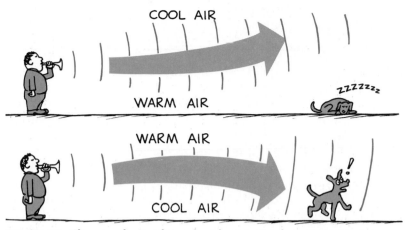

Fig. 29–15 The wave fronts of sound are bent in air of uneven temperature.

On a cold day or at night, when the layer of air near the ground is colder than the air above, the speed of sound near the ground is reduced. The higher speed of the wave fronts above cause a bending of the sound toward the earth. When this happens, sound can be heard over considerably longer distances.

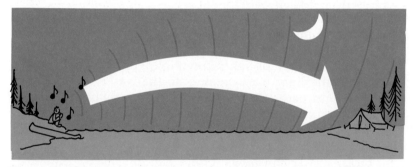

Fig. 29–16 At night, when the air is cooler over the surface of the lake, sound is refracted toward the ground and carries unusually well.

▶ **Question**

Suppose you are downwind from a factory whistle. In which case will the whistle sound louder—if the wind speed near the ground is more than the wind speed several meters above the ground, or if it is less?

▶ **Answer**

You'll hear the whistle better if the wind speed near the ground is less than the wind speed higher up. For this condition, the sound will be refracted toward the ground. If the wind speed were greater near the ground, the refraction would be upward.

29.8 | Refraction of Light

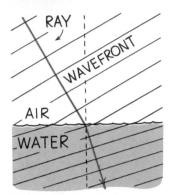

Fig. 29–17 As a light wave passes from air into water, its speed decreases. Note that the refracted ray is closer to the normal than is the incident ray.

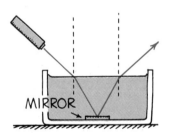

Fig. 29–18 The laser beam bends toward the normal when it enters the water, and away from the normal when it leaves.

A pond or swimming pool both appear shallower than they are. A pencil in a glass of water appears bent, the air above a hot stove shimmers, and the stars twinkle. These effects are caused by the change in the speed of light and hence the change in the direction of light when it passes from one medium to another. In other words, these effects are due to the refraction of light.*

Figure 29–17 shows rays and wave fronts of light that is refracted as it passes from air into water. (The wave fronts would be curved if the source of light were close, just as the wave fronts of water waves near a stone thrown into the water are curved. If we assume that the source of light is the sun, then it is so far away that the wave fronts are practically straight lines.) Note that the left portions of the wave fronts are the first to slow down when they enter the water. The refracted ray of light, which is at right angles to the refracted wave fronts, is closer to the normal than is the incident ray.

Compare the refraction in this case to the bending of the cart wheels in Figure 29–13. When light enters a medium in which its speed decreases, the rays bend toward the normal. But when light enters a medium in which its speed increases—as when light passes from water into air—the rays bend away from the normal (Figure 29–18).

Note that the path of light shown in Figure 29–18 would be the same if the light were shone into the water where it now exits. The light paths are reversible for both reflection and refraction. If you can see somebody by way of a reflective or refractive device, such as a mirror or a prism, then you should know that the person can see you by that device also.

As Figure 29–19 left shows, a thick pane of glass appears to be only two thirds its real thickness when viewed straight on. (For

---

* The ratio n of the speed of light in vacuum to the speed in a given material is called the *index of refraction* of that material.

$$\text{index of refraction } n = \frac{\text{speed of light in vacuum}}{\text{speed of light in material}}$$

The quantitative law of refraction, called *Snell's law*, was first worked out in 1621 by W. Snell, a Dutch astronomer and mathematician. According to Snell's law,

$$n \sin \theta = n' \sin \theta'$$

where n and n' are the indices of refraction of the media on either side of the boundary, and θ and θ' are the respective angles of incidence and refraction. If three of these values are known, the fourth can be calculated from this relationship.

clarity, the diameter of the eye pupil is made larger than true scale.) Similarly, water in a pond or pool appears to be only three quarters its true depth. Look at a fish in water from a bank, and the fish appears to be nearer the surface than it really is (Figure 29–19 right). It will also seem closer. These effects are due to the refraction of light whenever it crosses the boundary between air and another transparent medium.

Fig. 29–19 Because of refraction, the apparent depth of the glass block is less than the real depth (left), and the fish appears to be nearer than it actually is (right).

29.9 Atmospheric Refraction

Although the speed of light in air is only 0.03 percent less than its speed in a vacuum, there are situations in which atmospheric refraction is quite noticeable. One of the most interesting occurs in the **mirage**. On hot days there may be a layer of very hot air in contact with the ground. Since the molecules in hot air are farther apart, the light travels faster through it than through cooler air above. The speeding up of the part of the wave nearest the ground produces a gradual bending of the light rays. The tree in Figure 29–20 would appear upside down (as well as right-side-up) to an observer at the right, just as if it were reflected from a surface of water. But the light is not reflected; it is refracted.

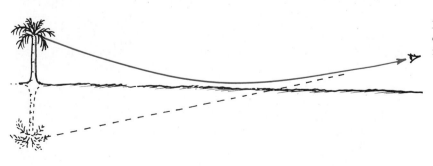

Fig. 29–20 The refraction of light in air produces a mirage.

Wave fronts of light are shown in Figure 29–21. The refraction of light in air in this case is very much like the refraction of sound in Figure 29–15. Undeflected wave fronts would travel at one speed and in the direction shown by the broken lines. Their greater speed near the ground, however, causes the light ray to bend upward as shown.

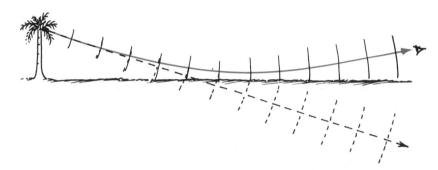

Fig. 29–21   Wave fronts of light travel faster in the hot air near the ground, thereby bending the rays of light upward.

A motorist experiences a similar situation when driving along a hot road that appears to be wet ahead. The sky appears to be reflected from a wet surface but, in fact, light from the sky is being refracted through a layer of hot air. A mirage is not, as some people mistakenly believe, a "trick of the mind." A mirage is formed by real light and can be photographed (see Figure 29–22).

Fig. 29–22   A mirage.

When you see shimmering images in air over a hot pavement or hot stove, you are seeing the effects of atmospheric refraction. The speed of light alters as it travels through varying temperatures of air. The twinkling of stars in the nighttime sky is produced by variations in the speed of light as it passes through unstable layers in the atmosphere.

Whenever you watch the sun set, you see the sun for several minutes after it has really sunk below the horizon. This is because light is refracted by the earth's atmosphere (Figure 29–23). Since the density of the atmosphere changes gradually, the refracted rays bend gradually to produce a curved path. The same thing occurs at sunrise, so our daytimes are about 5 minutes longer because of atmospheric refraction.

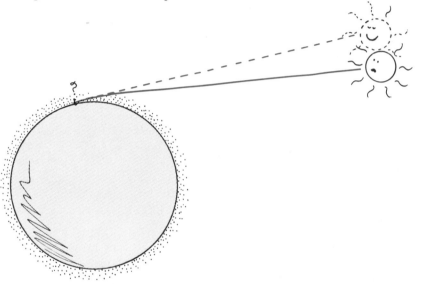

Fig. 29–23 When the sun is already below the horizon, you can still see it.

When the sun (or moon) is near the horizon, the rays from the lower edge are bent more than the rays from the upper edge. This produces a shortening of the vertical diameter and makes the sun (or moon) look elliptical instead of round (Figure 29–24).

Fig. 29–24 Atmospheric refraction produces a "pumpkin" sun.

> ▶ **Question**
>
> If the speed of light were the same in the various temperatures and densities of air, would there still be mirages, slightly longer daytimes, and a pumpkin sun at sunset?

▶ **Answer**
No! There would be no refraction if light traveled at the same speed in air of different temperatures and densities.

29.10 | Dispersion in a Prism

Chapters 27 and 28 discussed how the speed of light is less than *c* in a transparent medium. How much less depends on the medium and the frequency of the light. Light of frequencies closer to the natural frequency of the electron oscillators in a medium travel more slowly in the medium. This is because in the process of absorption and re-emission, there is more interaction with the medium. Since the natural or resonant frequency of most transparent materials is in the ultraviolet part of the spectrum, visible light of higher frequencies travels more slowly than light of lower frequencies. Violet travels about 1 percent more slowly in ordinary glass than red light. The colors between red and violet travel at their own speeds.

Since different frequencies of light travel at different speeds in transparent materials, they will refract differently and bend at different angles. When light is bent twice at nonparallel boundaries, as in a prism, the separation of the different colors of light is quite apparent. This separation of light into colors arranged according to their frequency is called **dispersion** (Figure 29–25).

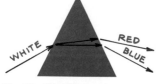
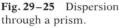

Fig. 29–25 Dispersion through a prism.

29.11 | The Rainbow

A spectacular illustration of dispersion is the rainbow. The conditions for seeing a rainbow are that the sun be shining in one part of the sky and that rain be falling in the opposite part of the sky. When you turn your back to the sun, you see the spectrum of colors in a bow. From an airplane the bow may form a complete circle. The colors are dispersed from the sunlight by thousands of tiny drops that act like prisms.

Fig. 29–26 The rainbow is seen in a part of the sky opposite the sun and is centered around the imaginary line extending from the sun to the observer.

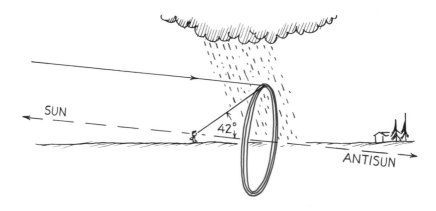

To understand how light is dispersed by raindrops, consider an individual spherical raindrop, as shown in Figure 29–27. Follow the ray of sunlight as it enters the drop near its top surface. Some of the light here is reflected (not shown), and the rest is refracted into the water. At this first refraction, the light is dispersed into its spectral colors. Violet is bent the most and red the least.

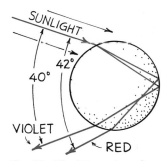

Fig. 29–27 Dispersion of sunlight by a single drop.

The rays reach the opposite part of the drop to be partly refracted out into the air (not shown) and partly reflected back into the water. Part of the rays that arrive at the lower surface of the drop are refracted into the air. This second refraction is similar to that of a prism, where refraction at the second surface increases the dispersion already produced at the first surface.

Each drop disperses a full spectrum of colors. An observer, however, is in a position to see only a single color from each drop (Figure 29–28). If violet light from a single drop enters your eye, red light from the same drop falls below your eye. To see red light you have to look at a drop higher in the sky. You'll see the color red when the angle between a beam of sunlight and the dispersed light is 42°. The color violet is seen when the angle between the sunbeam and dispersed light is 40°.

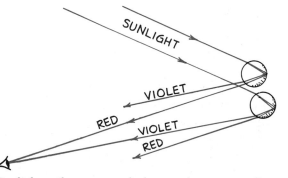

Fig. 29–28 Sunlight strikes two sample drops and emerges as dispersed light. The observer sees red from the upper drop and violet from the lower drop. Millions of drops produce the whole spectrum.

You don't need to look only upward at 42° to see dispersed red light. You could see red by looking sideways at the same angle or anywhere along a circular arc swept out at a 42° angle (Figure 29–29). The dispersed light of other colors is along similar arcs, each at their own slightly different angle. Altogether, the arcs for each color form the familiar rainbow shape.

A secondary rainbow larger than the primary rainbow can often be seen. The colors are reversed for the secondary bow, with violet on the outside and red on the inside. The secondary bow is formed by similar circumstances and is a result of double

Fig. 29–29 Only raindrops along the dashed arc disperse red light to the observer at a 42° angle.

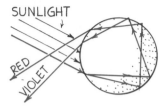

SUNLIGHT

RED

VIOLET

Fig. 29–30 Double reflection in a drop produces a secondary bow.

reflection within the raindrops. Because some light is refracted out the back during the extra reflection, the secondary bow is much dimmer.

> ▶ **Question**
>   If light traveled at the same speed in raindrops as it does in air, would we still have rainbows?

29.12 | Total Internal Reflection

When you're in a physics mood and you're going to take a bath, fill the tub extra deep and bring a waterproof flashlight into the tub with you. Put the bathroom light out. Shine the submerged light straight up and then slowly tip it and note how the intensity of the emerging beam diminishes and how more light is reflected from the water surface to the bottom of the tub.

At a certain angle, called the **critical angle**, you'll notice that the beam no longer emerges into the air above the surface. Instead, it grazes the surface. For water, the critical angle is 48° between the incident ray and the normal to the surface. When the flashlight is tipped beyond the critical angle, you'll notice that the beam cannot enter the air; it is only reflected. The beam is experiencing **total internal reflection**. The only light emerging from the water surface is that which is diffusely reflected from the bottom of the bathtub.

Figure 29–31 shows the refraction and reflection of light for different angles of incidence. The proportions of light refracted and reflected are indicated by the relative lengths of the solid arrows. Note that the light reflected beneath the surface obeys the law of reflection: the angle of incidence is equal to the angle of reflection.

The critical angle for glass is about 43°, depending on the type of glass. This means that within the glass, rays of light that are more than 43° from the normal to a surface will be totally internally reflected at that surface. Rays of light in the glass prisms shown in Figure 29–32, for example, meet the back surface at 45° and are totally internally reflected. They will stay inside the glass until they meet a surface at an angle between 0° (straight on) and 43° to the normal.

> ▶ **Answer**
>   No.

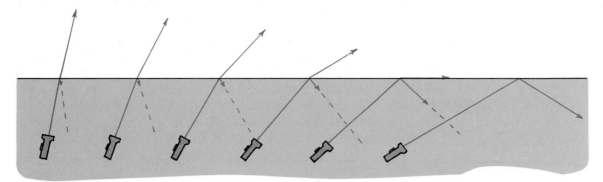

Fig. 29–31 Light emitted in the water at angles below the critical angle is partly refracted and partly reflected at the surface. At the critical angle (second sketch from right), the emerging beam skims the surface. Past the critical angle (far right), there is total internal reflection.

Total internal reflection is as the name implies: total—100%. Silvered or aluminized mirrors reflect only 90 to 95% of incident light, and are marred by dust and dirt; prisms are more efficient. This is the main reason they are used instead of mirrors in many optical instruments.

The critical angle for a diamond is 24.6°, smaller than any other known substance. This small critical angle means that light inside a diamond is most likely to be totally internally reflected. All light rays more than 24.6° from the normal to a surface in a diamond stay inside by total internal reflection. When a diamond is cut as a gemstone, light that enters at one facet is usually totally internally reflected several times, without any loss in intensity, before exiting from another facet in another direction. That's why you see unexpected flashes from a diamond. A small critical angle plus the pronounced refraction because of the unusually low speed of light in diamond produce wide dispersion and a wide array of colors with less overlap between adjacent colors. The colors seen in a diamond are therefore more brilliant.

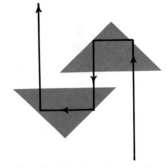

Fig. 29–32 Total internal reflection in glass prisms.

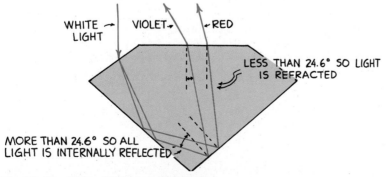

Fig. 29–33 Paths of light in a diamond.

Total internal reflection underlies the usefulness of **optical fibers**, which have been called *light pipes*. As the name implies, these transparent fibers pipe light from one place to another. They do this by a series of total internal reflections, much like the ricocheting of a bullet down a steel pipe. Optical fibers are useful for getting light to inaccessible places. Mechanics and machinists use them to look at the interior of engines, and physicians use them to look inside a patient's body. Light shines down some of the fibers to illuminate the scene and is reflected back along others.

Fig. 29–34   In an optical fiber, light is piped from one end to the other by a succession of total internal reflections.

Optical fibers are important in communications because they are more efficient than copper wire or coaxial cable. In many cities, thin glass fibers have replaced thick, bulky, and expensive copper cables to carry thousands of simultaneous telephone messages between major switching centers. More information can be carried in the short wavelengths of visible light than in the vibrations of electric current. Optical fibers are more and more replacing electric circuits in communications technology.

29 | Chapter Review

Concept Summary

In reflection, a wave reaches the boundary between two media and bounces back into the first medium.

- At a boundary, usually part of a wave is reflected and part passes into the second medium.
- According to the law of reflection, the angle of incidence is equal to the angle of reflection.
- A plane mirror forms a virtual image of an object; the image appears to be as far in back of the mirror as the object is in front of it, and is the same size as the object.
- Light that falls on a rough surface is reflected diffusely.
- The field of acoustics is concerned with how different surfaces reflect sound.

In refraction, a wave reaches the boundary between two media and changes direction as it passes into the second medium.

- Refraction is caused by a difference in the speed of the wave in the two media.
- The speed of light in materials depends on frequency, causing different colors of white light to refract differently and spread out to form a visible spectrum.

In total internal reflection, a wave is incident on a boundary at an angle such that none of the wave can be refracted and there is reflection only.

Important Terms

angle of incidence (29.2)
angle of reflection (29.2)
critical angle (29.12)
diffuse reflection (29.4)

dispersion (29.10)
law of reflection (29.2)
mirage (29.9)
normal (29.2)
optical fiber (29.12)
reflection (29.1)
refraction (29.6)
reverberation (29.5)
total internal reflection (29.12)
virtual image (29.3)
wave front (29.6)

Review Questions

1. What becomes of a wave's energy when the wave is totally reflected at a boundary? When it is partially reflected at a boundary? (29.1)

2. Why do smooth metal surfaces make good mirrors? (29.1)

3. When light strikes the surface of a pane of glass perpendicularly, how much light is reflected and how much is transmitted? (29.1)

4. What is meant by the normal to a surface? (29.2)

5. What is the law of reflection? (29.2)

6. When you view your image in the mirror, how far behind the mirror is your image compared to your distance in front of the mirror? (29.3)

7. Does the law of reflection hold for *curved* mirrors? (29.3)

8. Does the law of reflection hold for diffuse reflection? Explain. (29.4)

9. What is meant by the idea that a surface may be polished for some waves and rough for others? (29.4)

10. Distinguish between an echo and a reverberation. (29.5)

11. Does the law of reflection hold for both sound waves and light waves? (29.5)

12. Distinguish between reflection and refraction. (29.1, 29.6)

13. When a wave crosses a surface at an angle from one medium into another, why does it "pivot" as it moves across the boundary into the new medium? (29.6)

14. What is the orientation of a ray in relation to the wave front of a wave? (29.6)

15. Give an example where refraction is abrupt, and another where refraction is gradual. (29.6–29.7)

16. Does refraction occur for both sound waves and light waves? (29.7–29.8)

17. If light had the same speed in air and in water, would light be refracted in passing from air into water? (29.8)

18. If you can see the face of a friend who is underwater, can your friend also see you? (29.8)

19. Does refraction tend to make objects submerged in water seem shallower or deeper than they really are? (29.8)

20. Is a mirage a result of refraction or reflection? Explain. (29.9)

21. Is daytime a bit longer or a bit shorter because of atmospheric refraction? (29.9)

22. As light passes through glass or water, do the high or low frequencies of light interact more in the process of absorption and re-emission, and therefore lag behind? (29.10)

23. Why does blue light refract at greater angles than red light in transparent materials? (29.10)

24. What conditions are necessary for viewing a rainbow in the sky? (29.11)

25. How is a raindrop similar to a prism? (29.11)

26. What is meant by the *critical angle* in terms of refraction and total internal reflection? (29.12)

27. Why are dispersed colors so brilliant in a diamond? (29.12)

28. Why are optical fibers often called *light pipes*? (29.12)

Activities

1. Stand a pair of mirrors on edge with the faces parallel to each other. Place an object such as a coin between the mirrors, and look at the reflections in each mirror. Neat?

2. What must be the minimum length of a plane mirror in order for you to see a full view of yourself? To find out, stand in front of a mirror and put pieces of tape on the glass: one piece where you see the top of your head, and the other where you see the bottom of your feet. Compare the distance between the pieces of tape to your height. If a full-length mirror is not handy, put pieces of tape on a smaller mirror where you see the top of your head and the bottom of your chin.

3. What effect does your distance from the plane mirror have in the answer to Activity 2? (*Hint*: Move closer and farther from your initial position. Be sure that the top of your head lines up with the top piece of tape. How does the bottom alignment compare? At greater distances, is your image smaller than, larger than, or the same size as the space between the pieces of tape allow? Surprised?)

Think and Explain

1. Why are metals generally shiny?

2. When light strikes glass perpendicularly, about 4% is reflected at each surface. How much light is transmitted through a pane of window glass?

3. Suppose that a mirror and three lettered cards are set up as in Figure A. If a person's eye is at point P, which of the lettered cards will be seen reflected in the mirror?

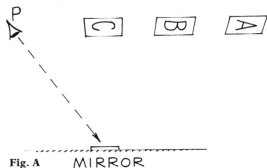

Fig. A MIRROR

4. Does the reflection of a scene in calm water look exactly the same as the scene itself only upside down? (*Hint*: Place a mirror on the floor between you and a table. Do you see the top of the table in the reflected image?)

5. Why is the lettering on the front of some vehicles "backward" (see Figure B)?

AMBULANCE

Fig. B

6. Suppose you walk toward a mirror at one meter per second. How fast do you and your image approach each other? (The answer is *not* one meter per second.)

7. Contrast the types of reflection from a rough road and from the smooth surface of a wet road to explain why it is difficult for a motorist to see the roadway ahead when driving on a rainy night.

8. Cameras with automatic focus bounce a sonar (sound) beam from the object being photographed, and compute distance from the time interval between sending and receiving the signal. Why will these cameras not focus properly for photographs of mirror images?

9. A bat flying in a cave emits a sound and receives its echo in one second. How far away is the cave wall?

10. Why is an echo weaker than the original sound?

11. If you were spearing a fish with a spear, would you aim above, below, or directly at the observed fish to make a direct hit? Would your answer be the same if you used laser light to "spear" the fish? Defend your answer.

12. A rainbow viewed from an airplane may form a complete circle. Will the shadow of the airplane appear at the center of the circle? Explain with the help of Figure 29-26.

30 | Lenses

A light ray bends as it enters glass and bends again as it leaves. The bending, or refraction, is due to the difference in the speed of light in glass and in air. Glass of certain shapes can form images which appear larger, smaller, closer, or farther than the object being viewed. Magnifying glasses have been used for centuries and were well known to the early Greeks and medieval Arabs. Today, eyeglasses allow millions of people to read in comfort, and cameras, projectors, telescopes, and microscopes widen our view of the world.

30.1 | Converging and Diverging Lenses

If a piece of glass has just the right shape, it can bend parallel rays of light so that they all cross—or appear to have crossed—at a single point. A piece of glass that does this is called a **lens**.

The special shape of a lens can be understood by considering a lens to be a large number of portions of triangular prisms (Figure 30–1). When arranged properly, the prisms refract incoming parallel rays so they converge to (or diverge from) a single point. The arrangement shown at the left is thicker in the middle and converges the light. The arrangement at the right, however, is thinner in the middle than at the edges; it diverges the light.

Fig. 30–1 A lens may be thought of as a set of prisms that converge light (left) or diverge light (right).

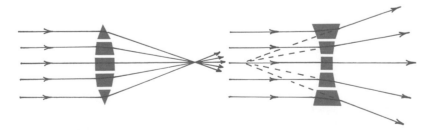

In both arrangements, the greatest net bending of rays occurs at the outermost prisms, for they have the greatest angle between the two refracting surfaces. No net bending occurs in the middle "prism," for its glass faces are parallel to each other, and a ray emerges in its original direction.

Real lenses are made not of prisms, of course, but of solid pieces of glass with surfaces ground usually to a spherical curve. Figure 30–2 shows how smooth lenses refract rays of light and wave fronts. The lens at the left is thicker in the middle and converges parallel rays of light (straight wave fronts). It is called a **converging lens**. The lens at the right is thinner in the middle and diverges parallel rays of light. It is called a **diverging lens**.

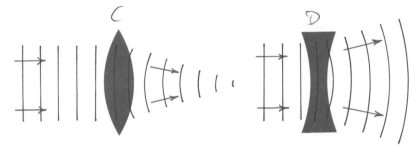

Fig. 30–2   Wave fronts travel more slowly in glass than in air. In the converging lens (left), the wave fronts are retarded more through the center of the lens, and the light converges. In the diverging lens (right), the waves are retarded more at the edges, and the light diverges.

Figure 30–3 illustrates some important terms for a lens. The **principal axis** of a lens is the line joining the centers of curvatures of its surfaces. For a converging lens, the **focal point** is the point at which a beam of parallel light, parallel to the principal axis, converges. Incident parallel beams that are not parallel to the principal axis focus at points above or below the focal point. All such possible points make up a **focal plane**. Since a lens has two surfaces, it has two focal points and two focal planes. When the lens of a camera is set for distant objects, the film is in the focal plane behind the lens in the camera.

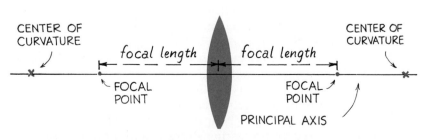

Fig. 30–3   Key features of a converging lens.

For a diverging lens, an incident beam of light parallel to the principal axis is not converged to a point, but is diverged so that the light appears to come from a point in front of the lens. The **focal length** of a lens, whether converging or diverging, is the distance between the center of the lens and its focal point. When the lens is thin, the focal lengths on either side are equal, even when the curvatures on the two sides are not.

30.2   Image Formation by a Lens

With unaided vision, an object far away is seen through a relatively small angle of view, while the same object when closer is seen through a larger angle of view. This wider angle enables the perception of more detail. Magnification occurs when an image is observed through a wider angle with the use of a lens than without the lens. A magnifying glass is simply a converging lens that increases the angle of view.

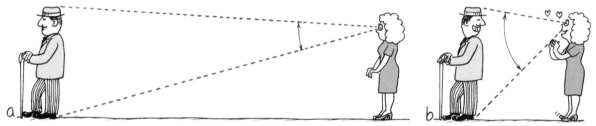

Fig. 30–4   (a) A distant object is viewed through a narrow angle. (b) When the same object is viewed through a wide angle, more detail is seen.

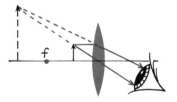

Fig. 30–5   A converging lens can be used as a magnifying glass to produce a virtual image of a nearby object. The image appears larger and farther from the lens than the object.

When you use a magnifying glass, you hold it close to the object you wish to see magnified. This is because a converging lens will magnify only when the object is between the focal point and the lens. The magnified image will be farther from the lens than the object, and it will be right-side-up. If a screen were placed at the image distance, no image would appear on the screen. This is because no light is actually directed to the image position. The rays that reach your eye, however, behave virtually as if they did, so the image is a virtual image.

When the object is far enough away to be beyond the focal point of a converging lens, light from the object does converge and can be focused on a screen (Figure 30–6). An image formed by converging light is called a **real image**. A real image formed by a single converging lens is upside down (inverted). Converging lenses are used for projecting slides and motion pictures on a screen, and for projecting a real image on the film of a camera.

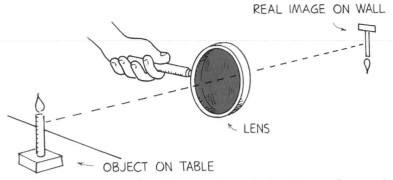

Fig. 30–6   A converging lens forms a real, upside-down image of a more distant object.

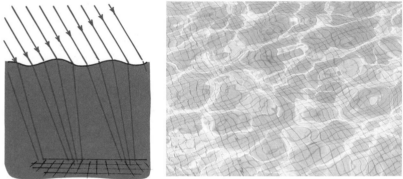

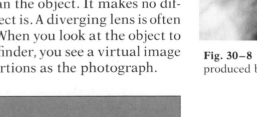

Fig. 30–7   The moving pattern of bright lines on the bottom of a swimming pool results from the uneven surface of water, which behaves as a moving blanket of converging lenses.

When a diverging lens is used alone, the image is always virtual, right-side-up, and smaller than the object. It makes no difference how far or how near the object is. A diverging lens is often used as a viewfinder on a camera. When you look at the object to be photographed through the viewfinder, you see a virtual image that approximates the same proportions as the photograph.

Fig. 30–8   A virtual image produced by a diverging lens.

▶ **Question**
   Why is the greater part of the photograph in Figure 30–8 out of focus?

▶ **Answer**
   Both Jamie and his cat and the virtual image of Jamie and his cat are "objects" for the lens of the camera that took this photograph. Since the objects are at different distances from the camera lens, their respective images are at different distances with respect to the film in the camera. So only one can be brought into focus. The same is true of your eyes. You cannot focus on near and far objects at the same time.

30.3 | Constructing Images Through Ray Diagrams

Ray diagrams, like the one in Figure 30–9, show the principal rays that can be used to determine the size and location of an image. The size and location of the object, its distance from the center of the lens, and the focal length of the lens must be known.* An arrow is used to represent the object (which may be anything from a microbe viewed in a microscope to a galaxy viewed through a telescope). For simplicity, one end of the object is placed right on the principal axis.

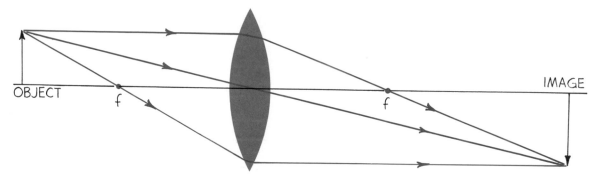

Fig. 30–9   Ray diagram. Three useful rays from the object that converge on the image.

To locate the position of the image, you have to know only the paths of two rays from a point on the object. Any point will work, but it is customary to choose a point at the tip of the arrow.

The path of one refracted ray is known from the definition of the focal point. A ray parallel to the principal axis will be refracted by the lens to the focal point, as shown in Figure 30–9.

Another path is known: through the center of the lens where the faces are parallel to each other. A ray of light will pass through the center with no appreciable change in direction. Therefore a ray from the tip of the arrowhead proceeds in a straight line through the center of the lens.

A third path is known: A ray that passes through the focal point in front of the lens emerges from the lens parallel to the principal axis.

---

*   The mathematical relationship between object distance o, image distance i, and focal length f is given by

$$\frac{1}{o} + \frac{1}{i} = \frac{1}{f}$$

$$\frac{H_i}{H_o} = \frac{D_i}{D_o}$$

This is called the *thin-lens equation*.

All three paths are shown in Figure 30–9, which is a typical ray diagram. The image is located where the three rays intersect. Any two of these three rays is sufficient to locate the relative size and location of the image.

The ray diagram for a converging lens used as a magnifying glass is shown in Figure 30–10. In this case, where the object is within one focal length of the lens, the rays diverge as they leave the lens. They appear to come from a point in front of the lens (same side of the lens as the object). The location of the image is found by extending the rays to the point where they converge. The virtual image is magnified and right-side-up.

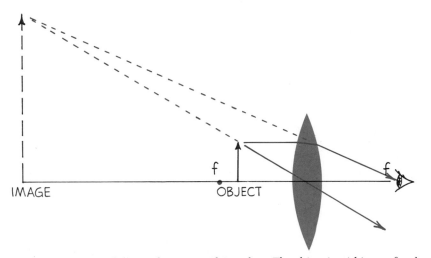

Fig. 30–10 Ray diagram for a magnifying glass. The object is within one focal length of the lens, so the image is virtual, right-side-up, and magnified.

The three rays useful for the construction of a ray diagram are summarized:

1. A ray parallel to the principal axis which passes through the focal point after refraction by the lens.

2. A ray through the center of the lens which does not change direction.

3. A ray through the focal point in front of the lens which emerges parallel to the principal axis after refraction by the lens.

Any two rays are sufficient to locate an image; which particular pair is chosen is merely a matter of convenience.

The ray diagrams in Figure 30–11 show image formation by a converging lens as an object initially at the focal point is moved away from the lens along the principal axis. Since the object is not within one focal length, all the images are real and inverted.

Object position: distance f from lens (at the focal point)
Image position: at infinity

Object position: between f and $2f$ from lens
Image position: beyond $2f$ from lens
Image size: magnified

Object position: distance $2f$ from lens
Image position: distance $2f$ from lens
Image size: same as object

Object position: beyond $2f$ from lens
Image position: between f and $2f$ from lens
Image size: smaller

Object position: at infinity
Image position: distance f from lens (at the focal point)

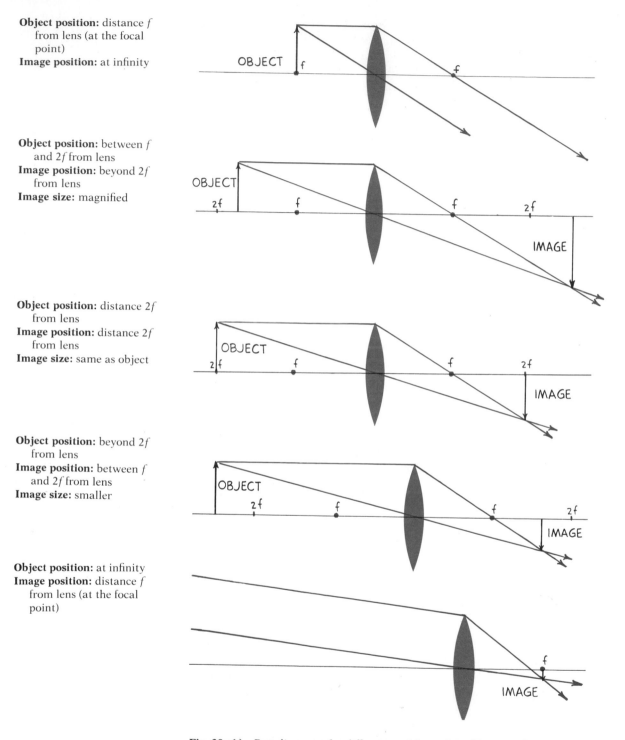

Fig. 30–11 Ray diagrams for different positions of an object in relation to a converging lens of focal length f.

   The method of drawing ray diagrams applies also to diverging lenses (Figure 30–12). A ray parallel to the principal axis from the tip of the arrow will be bent by the lens in the same direction as if it had come from the focal point. A ray through the center goes straight through. A ray that is heading for the focal point on the far side of the lens is bent so that it emerges parallel to the lens.

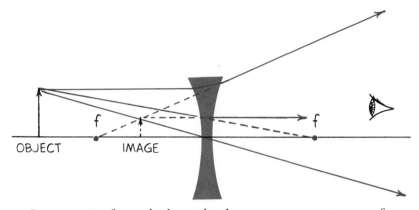

Fig. 30–12 Ray diagram for a diverging lens.

   On emerging from the lens, the three rays appear to come from a point on the same side of the lens as the object. This point defines the position of the virtual image. The image is nearer the lens than the object. It is smaller than the object and right-side-up. Regardless of the object position, the image formed by a diverging lens is always virtual, reduced, and right-side-up.

30.4   Image Formation Summarized

A converging lens is a simple magnifying glass when the object is within one focal length of the lens. The image is then virtual, magnified, and right-side-up.

   When the object is beyond one focal length, a converging lens produces a real, inverted image. The location of the image depends on how close the object is to the focal point. If it is close to the focal point, the image is far away (as with a slide projector or movie projector). If the object is far from the focal point, the image is nearer (as with a camera). In all cases where a real image is formed, the object and the image are on opposite sides of the lens.

   When the object is viewed with a diverging lens, the image is virtual, reduced, and right-side-up. This is true for all locations of the object. In all cases where a virtual image is formed, the object and the image are on the same side of the lens.

> ▶ **Question**
> Where must an object be located so that the image formed by a converging lens will be (a) at infinity? (b) As near the object as possible? (c) Right-side-up? (d) The same size? (e) Inverted and enlarged?

30.5 Some Common Optical Instruments

Many optical instruments use lenses. Among these are the camera, telescope (and binoculars), compound microscope, and projector.

The Camera

A camera consists of a lens and sensitive film mounted in a light-tight box. In many cameras, the lens is mounted in a screw mount which can be moved to or fro to adjust the distance between the lens and film. The lens forms a real, inverted image on the film.

Figure 30–13 shows a camera with a single simple lens. In practice, most cameras make use of compound lenses to minimize distortions called *aberrations*.

The amount of light that gets to the film is regulated by a shutter and a diaphragm. The shutter controls the length of time that the film is exposed to light. The diaphragm controls the opening that light passes through to reach the film. Varying the size of the opening (aperture) varies the amount of light that reaches the film at any instant.

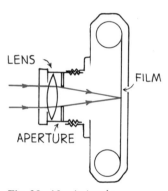

Fig. 30–13 A simple camera.

The Telescope

A **telescope** uses a lens to form a real image of a distant object. The real image is not caught on film but is projected in space to be examined by another lens used as a magnifying glass. The second lens, called the **eyepiece**, is positioned so that the image produced by the first lens is within one focal length. The eyepiece forms an enlarged virtual image of the real image. When you look through a telescope, you are looking at an image of an image.

▶ **Answer**
The object should be (a) one focal length from the lens (at the focal point) (see Figure 30–11); (b) and (c) within one focal length of the lens (see Figure 30–10); (d) at two focal lengths from the lens (see Figure 30–11); (e) between one and two focal lengths from the lens (see Figure 30–11).

Figure 30–14 shows the lens arrangement for an *astronomical telescope*. The image is inverted, which explains why maps of the moon are printed with the moon upside down.

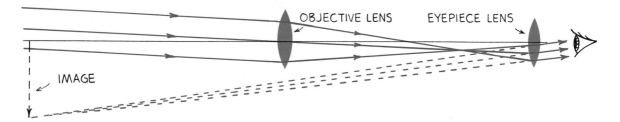

Fig. 30–14   Lens arrangement for an astronomical telescope. (For simplification, the image is shown close here; it is actually located at infinity.)

A third lens or a pair of reflecting prisms is used in the *terrestrial telescope*, which will produce an image that is right-side-up. A pair of these telescopes side by side, each with a pair of prisms to provide four reflecting surfaces to turn images right-side-up, makes up a pair of *binoculars* (Figure 30–15).

Since no lens transmits 100% of the light incident upon it, astronomers prefer the brighter, inverted images of a two-lens telescope to the less bright, right-side-up images that a third lens or prisms would provide. For nonastronomical uses, such as viewing distant landscapes or sporting events, right-side-up images are more important than brightness, so the additional lens or prisms are used.

Fig. 30–15   The arrangement of prisms in binoculars.

The Compound Microscope

A compound microscope uses two converging lenses of short focal length, arranged as shown in Figure 30–16. The first lens, called the **objective lens**, produces a real image of a close object. Since the image is farther from the lens than the object, it is enlarged. A second lens, the eyepiece, forms a virtual image of the first image, further enlarged. The instrument is called a compound microscope because it enlarges an already enlarged image.

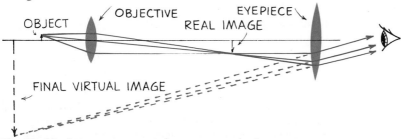

Fig. 30–16   Lens arrangement for a compound microscope.

The Projector

The arrangement of converging lenses for a slide or movie projector is shown in Figure 30–17. A concave mirror reflects light from an intense source back onto a pair of *condenser lenses*. The condenser lenses direct the light through the slide or movie frame to a *projection lens*. The projection lens is mounted in a sliding tube so that it can be positioned to or fro to focus a sharp image on the screen.

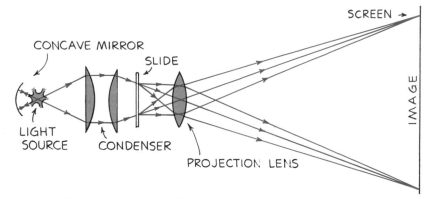

Fig. 30–17   Lens arrangement for a projector.

30.6   The Eye

In many respects the human eye is similar to the camera. The amount of light that enters is regulated by the **iris**, the colored part of the eye which surrounds the opening called the **pupil**.* Light enters through the transparent covering called the **cornea**, passes through the pupil and lens, and is focused on a layer of tissue at the back of the eye—the **retina**—that is more sensitive to light than any artificial detector made. Different parts of the retina receive light from different directions.

The retina is not uniform. There is a small region in the center of our field of view at which we have the most distinct vision. This spot is called the *fovea*. Much greater detail can be seen here than at the side parts of the eye.

There is also a spot in the retina where the nerves carrying all the information run out. This is the *blind spot*. You can demon-

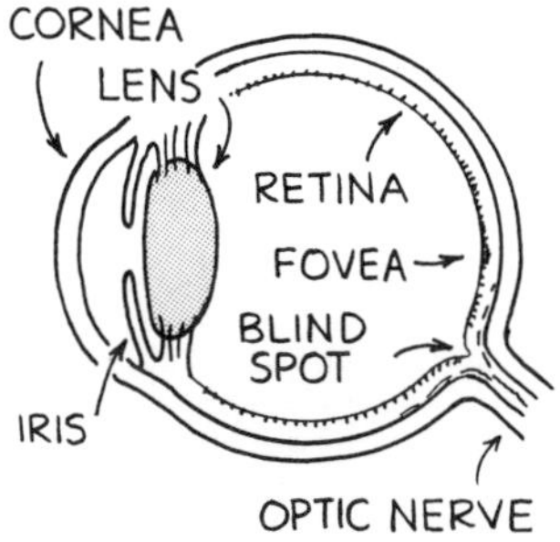

Fig. 30–18   The human eye.

---

*   The hole of the pupil usually looks black because light is going in but not coming out. Sometimes in flash photos, the light from the flashbulb enters the eye at just the right angle to reflect off the retina at the back of the eye. That's why flash photographs sometimes show the pupils to be pinkish.

strate that you have a blind spot in each eye if you hold this book at arm's length, close your left eye, and look at the circle in Figure 30–19 with only your right eye. You can see both the circle and the X at this distance. If you now move the book slowly toward your face, with your right eye still fixed upon the circle, you'll reach a position about 20 to 25 cm from your eye where the X disappears. To establish the blind spot in your left eye, close your right eye and similarly look at the X with your left eye so that the circle disappears. With both eyes opened, you'll find no position where either the X or circle disappears because one eye "fills in" the part of the object to which the other eye is blind. It's nice to have two eyes.

O   **X**

Fig. 30–19   For the blind spot experiment.

In both the camera and the eye, the image is upside down, and this is compensated for in both cases. You simply turn the camera film around to look at it. Your brain has learned to turn around images it receives from your retina!

A principal difference between a camera and the human eye has to do with focusing. In a camera, focusing is accomplished by altering the distance between the lens and the film. In the human eye, most of the focusing is done by the cornea, the transparent membrane at the outside of the eye. Adjustments in focusing of the image on the retina are made by changing the thickness and shape of the lens to regulate its focal length. This is called *accomodation* and is brought about by the action of the *ciliary muscle*, which surrounds the lens.

NORMAL DISTANT VISION   NORMAL CLOSE VISION

Fig. 30–20   The shape of the lens changes to focus light on the retina.

30.7   Some Defects in Vision

If you have what is called normal vision, your eye can accomodate to clearly see objects from infinity (the *far point*) down to 25 cm (the *near point*, which normally recedes for all people with advancing age).

The eyes of a **farsighted** person form images behind the retina (Figure 30–21). The eyeball is too short. Farsighted people have to hold things more than 25 cm away to be able to focus them. The remedy is to increase the converging effect of the eye. This is done by wearing eyeglasses or contact lenses with converging lenses. Converging lenses will converge the rays that enter the eye sufficiently to focus them on the retina instead of behind the retina.

Fig. 30–21 The eyeball of the farsighted eye is too short. A converging lens moves the image closer and onto the retina.

A **nearsighted** person can see nearby objects clearly, but does not see distant objects clearly because they are focused too near the lens, in front of the retina (Figure 30–22). The eyeball is too long. A remedy is to wear corrective lenses that diverge the rays from distant objects so that they focus on the retina instead of in front of it.

Fig. 30–22 The eyeball of the nearsighted eye is too long. A diverging lens moves the image farther away and onto the retina.

Astigmatism of the eye is a defect that results when the cornea is curved more in one direction than the other, somewhat like the side of a barrel. Because of this defect, the eye does not form sharp images. The remedy is cylindrical corrective lenses that have more curvature in one direction than in another.

30.8 Some Defects of Lenses

No lens gives a perfect image. The distortions in an image are called **aberrations**. By combining lenses in certain ways, aberrations can be minimized. For this reason, most optical instruments use compound lenses, each consisting of several simple lenses, instead of single lenses.

Spherical aberration results when light passes through the edges of a lens and focuses at a slightly different place from light passing through the center of the lens (Figure 30–22). This can be remedied by covering the edges of a lens, as with a diaphragm in a camera. Spherical aberration is corrected in good optical instruments by a combination of lenses.

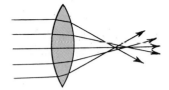

Fig. 30–23   Spherical aberration.

Chromatic aberration is the result of the different speeds of various colors and hence the different refractions they undergo. In a simple lens (as in a prism), red light and blue light do not come to focus in the same place. *Achromatic lenses*, which combine simple lenses of different kinds of glass, correct this defect.

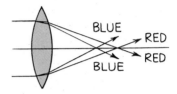

Fig. 30–24   Chromatic aberration.

In the eye, vision is sharpest when the pupil is smallest because light then passes through only the center of the eye's lens, where spherical and chromatic aberrations are minimal. Also, light bends the least through the center of a lens, so minimal focusing is required for a sharp image. An image that is formed by straight lines of light can appear in focus anywhere. You see better in bright light because your pupils are smaller.*

> ► **Question**
> Why is there chromatic aberration in light that passes through a lens, but no chromatic aberration in light that reflects from a mirror?

An option to poor sight in the last five hundred years has been to wear spectacles, and in more recent times an option to wearing spectacles has been to wear contact lenses. It is interesting to note that at the present time there is an option to both spectacles and contact lenses for people with poor eyesight. Experimental and controversial techniques today allow eye surgeons to reshape the cornea of the eye for normal vision. In tomorrow's world, the wearing of eyeglasses and contact lenses may be a thing of the past. We really do live in a rapidly changing world. And that can be nice.

► **Answer**
Different frequencies travel at different speeds in a transparent medium, and therefore refract at different angles, which produces chromatic aberration. The angles at which light *reflects*, on the other hand, have nothing to do with the frequency of light. One color reflects the same as any other. Mirrors are therefore preferable to lenses in telescopes because there is no chromatic aberration with reflection.

---

*   If you wear glasses and ever misplace them, or find it difficult to read small print, as in a telephone book, hold a pinhole (in a piece of paper or whatever) in front of your eye close to the page. You'll see the print clearly, and because you're close, it will seem magnified. Try it and see!

30 | Chapter Review

Concept Summary

A lens refracts parallel rays of light so that they cross—or appear to cross—at a focal point.

- A converging lens is thicker in the middle; a diverging lens is thinner in the middle.
- A converging lens forms virtual, magnified images when the object is within one focal length of the lens.
- A converging lens forms real images when the object is beyond one focal length from the lens.
- A diverging lens always forms virtual, reduced images.
- Optical instruments that use lenses include the camera, telescope, compound microscope, and projector.
- The human eye refracts light and focuses it on the retina (with the help of corrective lenses if necessary).

Important Terms

aberration (30.8)
astigmatism (30.7)
converging lens (30.1)
cornea (30.6)
eyepiece (30.5)
diverging lens (30.1)
farsighted (30.7)
focal length (30.1)
focal plane (30.1)
focal point (30.1)
iris (30.6)
lens (30.1)
nearsighted (30.7)
objective lens (30.5)
principal axis (30.1)
pupil (30.6)
ray diagram (30.3)
real image (30.2)
retina (30.6)
telescope (30.5)

Review Questions

1. Distinguish between a *converging* lens and a *diverging* lens. (30.1)

2. Distinguish between the focal *point* and focal *plane* of a lens. (30.1)

3. Distinguish between a *virtual* image and a *real* image. (30.2)

4. There are three convenient rays commonly used in ray diagrams to estimate the position of an image. Describe these three rays in terms of their orientation with respect to the principal axis and focal points. (30.3)

5. How many of the rays in Question 4 are necessary for estimating the position of an image? (30.3)

6. Do ray diagrams apply only to converging lenses, or to diverging lenses as well? (30.3)

7. Explain what is meant by saying that in a telescope one looks at the image of an image? (30.5)

8. In what two ways does an astronomical telescope differ from a terrestrial telescope? (30.5)

9. How does a compound microscope differ from a telescope? (30.5)

10. Which instrument—a telescope, a compound microscope, or a camera—is most similar to the eye? (30.5–30.6)

11. Why do you not normally see a blind spot when you look at your surroundings? (30.6)

12. Distinguish between *farsighted* and *nearsighted* vision. (30.7)

13. What is astigmatism, and how can it be corrected? (30.7)

14. Distinguish between *spherical* aberration and *chromatic* aberration, and cite a remedy for each. (30.8)

Activities

1. Make a pinhole camera, as illustrated in Figure A. Cut out one end of a small cardboard box, and cover the end with tissue or onionskin paper. Make a clean-cut pinhole at the other end. (If the cardboard is thick, place a piece of metal foil over an opening in the cardboard, and make the hole in the foil.) Aim the camera at a bright object in a darkened room, and you will see an upside-down image on the translucent tissue paper. If in a dark, windowless room you replace the tissue paper with unexposed photographic film, cover the back so it is light-tight, and cover the pinhole with a removable flap, you are now ready to take a picture. Exposure times differ depending mostly on the kind of film and the amount of light. Try different exposure times, starting with about 3 seconds. Also try boxes of various lengths. You'll find everything in focus in your photographs, but the pictures will not have clear-cut sharp outlines. The principal difference between your pinhole camera and a commercial one is the glass lens, which is larger than the pinhole and therefore admits more light in less time. It is because a lens camera is so fast that the pictures it takes are called "snapshots."

Fig. A

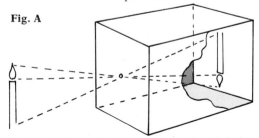

2. Look at the reflections of overhead lights from the two surfaces of eyeglasses, and you will see two fascinatingly different images. Why are they different?

3. Determine the magnification power of a lens by focusing on the lines of a ruled piece of paper (Figure B). Count the spaces between the lines that fit into one magnified space, and you have the magnification power of the lens. For example, if three spaces fit into one magnified space, then the magnification power of the lens is 3. You can do the same with binoculars and a distant brick wall. Hold the binoculars so that only one eye looks at the bricks through the eyepiece while the other eye looks directly at the bricks. The number of bricks, as seen with the unaided eye, that will fit into one magnified brick gives the magnification of the instrument.

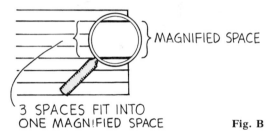

MAGNIFIED SPACE

3 SPACES FIT INTO ONE MAGNIFIED SPACE **Fig. B**

Think and Explain

1. a. What condition must exist for a converging lens to produce a virtual image?
   b. What condition must exist for a diverging lens to produce a real image?

2. How could you prove that an image was indeed a real image?

3. Why do you suppose that a magnifying glass has often been called a "burning glass"?

4. In terms of focal length, how far is the camera lens from the film when very distant objects are being photographed?

5. Can you photograph yourself in a mirror and focus the camera on both your image and the mirror frame? Explain.

6. If you take a photograph of your image in a plane mirror, how many meters away should you set your focus if you are 2 m in front of the mirror?

7. Copy the three drawings in Figure C. Then use ray diagrams to find the image of each arrow.

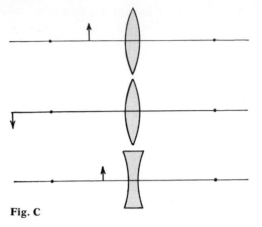

Fig. C

8. Why do you have to put slides into a slide projector upside down?

9. Maps of the moon are actually upside down. Why is this so?

10. What is responsible for the rainbow-colored fringe commonly seen at the edges of a spot of white light from the beam of a slide projector?

11. Why do older people who do not wear glasses read a book farther from their eyes than younger people?

12. Would telescopes and microscopes magnify if light had the same speed in glass as in air? Explain.

31 Diffraction and Interference

Reflection and refraction of light can be understood in terms of either a particle model or a wave model of light. Whether light travels in straight lines as little particles called photons, or as waves that spread out from a source, either model can explain reflection and refraction. This chapter investigates properties of light—diffraction and interference—that can be understood only by a wave model. These properties are closely related.

31.1 Huygens' Principle

In the late 1600s a Dutch mathematician-scientist, Christian Huygens, proposed a very interesting idea about waves. Huygens stated that light waves spreading out from a point source may be regarded as the overlapping of tiny secondary wavelets, and that every point on any wave front may be regarded as a new point source of secondary waves (Figure 31–1). In other words, wave fronts are made up of tinier wave fronts. This idea is called **Huygens' principle**.

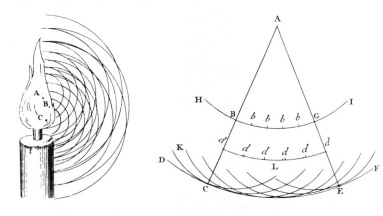

Fig. 31–1 These drawings are from from Huygens' book *Treatise on Light*. Light from A (left) expands in wavefronts, every point of which (right) behaves as if it were a new source of waves. Secondary wavelets starting at *b,b,b,b* form a new wave front (*d,d,d,d*); secondary wavelets starting at *d,d,d,d* form still another new wave front (DCEF).

457

Look at the spherical wave front in Figure 31–2. Each point along the wave front AA' is the source of a new wavelet that spreads out in a sphere from that point. Only a few of the infinite number of wavelets are shown in the figure. The new wave front BB' can be regarded as a smooth surface enclosing the infinite number of overlapping wavelets that started from AA' a short time earlier.

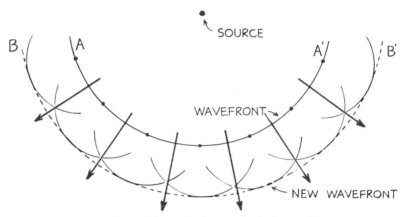

Fig. 31–2   Huygens' principle applied to a spherical wave front.

As a wave front spreads, it appears less curved. Very far from the original source, the wave fronts seem to form a plane. A good example is the plane waves that arrive from the sun. A Huygens' wavelet construction for plane waves is shown in Figure 31–3. (In a two-dimensional drawing, the planes are shown as straight lines.)

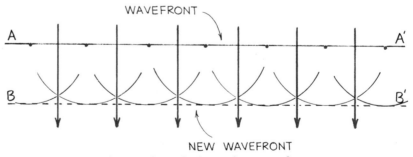

Fig. 31–3   Huygens' principle applied to a plane wave front.

The laws of reflection and refraction are illustrated via Huygens' principle in Figure 31–4.

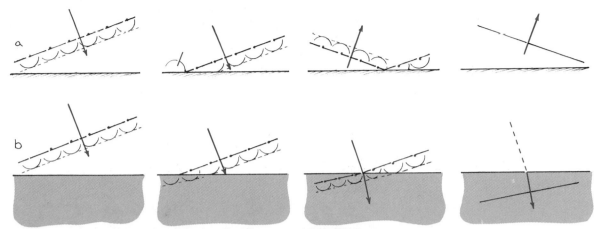

Fig. 31–4 Huygens' principle applied to (a) reflection and (b) refraction.

You can observe Huygens' principle in water waves that are made to pass through a narrow opening. A wave with straight wave fronts can be generated in water by successively dipping a stick lengthwise into the water (Figure 31–5). A ruler works well. When the straight wave fronts pass through the opening in a barrier, interesting wave patterns result.

Fig. 31–5 Making plane waves in a tank of water and watching the pattern they produce when they pass though an opening in a barrier.

When the opening is wide, you'll see straight wave fronts pass through without change—except at the corners, where the wave fronts are bent into the "shadow region" in accord with Huygens' principle. If you narrow the width of the opening, less of the wave gets through, and the spreading into the shadow region is more pronounced. When the opening is small compared to the wavelength of the waves, Huygens' idea that every part of a wave front can be regarded as a source of new wavelets becomes quite apparent. As the waves move into the narrow opening, the water sloshing up and down in the opening is easily seen to act as a point source of circular waves that fan out on the other side of the barrier. The photos in Figure 31–6 are top views of water waves generated by a vibrating stick. Note how the waves fan out as the hole through which they pass becomes smaller.

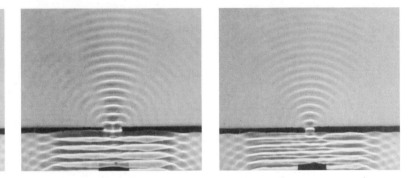

Fig. 31–6 Straight waves passing through openings of various sizes. The smaller the opening, the greater the bending of the waves at the edges.

31.2   Diffraction

Any bending of a wave by means other than reflection or refraction is called **diffraction**. The photos of Figure 31–6 show the diffraction of straight water waves through various openings. When the opening is wide compared to the wavelength, the spreading effect is small. As the opening becomes narrower, the spreading of waves is more pronounced. The same occurs for all kinds of waves, including light waves.

Fig. 31–7 Light casts a sharp shadow when the opening is large compared to the wavelength of the light. It casts a fuzzy shadow because of diffraction when the opening is extremely narrow.

When light passes though an opening that is large compared to the wavelength of light, it casts a rather sharp shadow (Figure 31–7). When light passes through a thin razor slit in a piece of opaque material, it casts a fuzzy shadow, for the light fans out like the water through the narrow opening in Figure 31–6. The light is diffracted by the thin slit.

LIGHT SOURCE

WIDE WINDOW

SHADOW

LIGHT SOURCE

NARROW SLIT

SHADOW

Diffraction is not confined to the spreading of light through narrow slits or to openings in general. Diffraction occurs to some degree for all shadows. On close examination, even the sharpest shadow is blurred at the edge (Figure 31–8). The fuzzy edges of most shadows are diffraction patterns too fine ordinarily to be seen.

Diffraction occurs for radio waves. Longer wavelengths diffract more readily around buildings—and thus can reach more places—than shorter wavelengths. AM radio waves have longer wavelengths than FM waves, so AM radio reception comes in loud and clear in many localities where FM reception is poor. TV waves are really short-wavelength radio waves, so diffraction is not as pronounced, and antennas must be put on rooftops in many localities. Diffraction aids radio reception.

Diffraction is not an asset in viewing things through a microscope. The shadows of small objects become less and less well defined as the size of the object approaches the wavelength of the light illuminating it. If the object is smaller than the wavelength of light, no structure can be seen. Any image is lost due to diffraction. No amount of magnification or perfection of microscope design can defeat this fundamental diffraction limit.

To minimize this problem, microscopists illuminate tiny objects with shorter wavelengths. It turns out that a beam of electrons has a wavelength associated with it. This wavelength is very much shorter than the wavelengths of visible light. Microscopes that use beams of electrons to illuminate tiny things are called *electron microscopes*. The diffraction limit of an electron microscope is much less than that of an optical microscope.

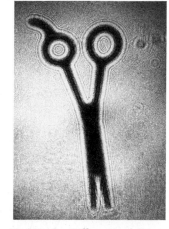

Fig. 31–8 Diffraction fringes are evident in the shadows of laser light, which is of a single frequency. These fringes would be filled in by multitudes of other fringes if the source were white light.

Fig. 31–9 Minimal diffraction by a very-short-wavelength electron beam in an electron microscope produces extraordinary detail.

> ▶ **Question**
> Why is blue light used to view tiny objects in an optical microscope?

31.3 | Interference

The idea of wave interference was introduced in Chapter 25, and applied to sound in Chapter 26. The idea is important enough to summarize here before applying it to light waves.

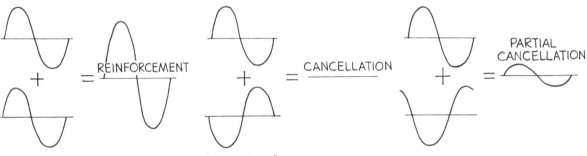

Fig. 31–10   Interference.

Fig. 31–11   Interference of water waves produced by two stones dropped into water.

If you drop a couple of stones into water at the same time, the two sets of waves that result cross each other and produce what is called an *interference pattern*. Within the pattern, wave effects may be increased, decreased, or neutralized. When the crest of one wave overlaps the crest of another, their individual effects add together; this is *constructive interference*. When the crest of one wave overlaps the trough of another, their individual effects are reduced; this is *destructive interference*.

Water waves can be produced in shallow tanks of water known as *ripple tanks* under more carefully controlled conditions. Interesting patterns are produced when two sources of waves are placed side by side. Small spheres are made to vibrate at a controlled frequency in the water while the wave patterns are photographed from above (see Figure 31–12). The gray "spokes" are regions of destructive interference. The dark and light striped

> ▶ **Answer**
> Less diffraction results from the short wavelengths of blue light compared to other longer wavelengths.

regions are regions of constructive interference. The greater the frequency of the vibrating spheres, the closer together the stripes (and the shorter the wavelength). Note how the number of regions of destructive interference depends on the wavelength and on the distance between the wave sources.

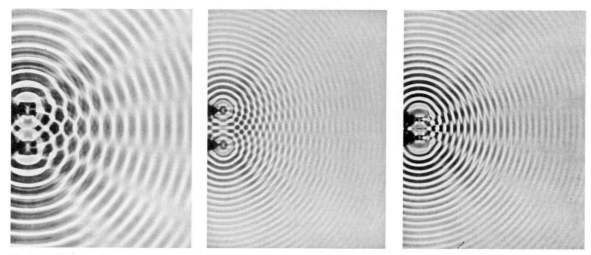

Fig. 31–12   Interference patterns of overlapping water waves from two vibrating sources.

31.4   Young's Interference Experiment

In 1801 the British physicist and physician Thomas Young performed an experiment that was to make him famous.* Young discovered that when **monochromatic** light—light of a single color—was directed through two closely spaced pinholes, fringes of brightness and darkness were produced on a screen behind. He realized that the bright fringes of light resulted from light waves from both holes arriving crest to crest (constructive interference—more light). Similarly, the dark areas resulted from light waves arriving trough to crest (destructive interference— no light). Young had convincingly demonstrated the wave nature of light that Huygens had proposed earlier.

---

*   Young read fluently at the age of two; by four, he had read the Bible twice; by fourteen, he knew eight languages. During his adult life he contributed to an understanding of fluids, work and energy, and the elastic properties of materials. He was also the first person to make progress in deciphering Egyptian hieroglyphics. No doubt about it: Thomas Young was smart—very smart.

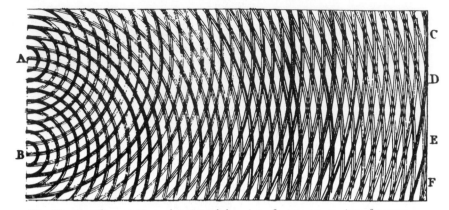

Fig. 31–13 Thomas Young's original drawing of a two-source interference pattern. The crests lie atop each other in the right half. Letters *C*, *D*, *E*, and *F* mark regions of destructive interference.

Young's experiment is now done with two closely spaced slits instead of pinholes, so the fringes are straight lines. A sodium vapor lamp provides a good source of monochromatic light, and a laser is even better. The arrangement is shown in Figure 31–14. Note the similarity of this to the arrangement of sound speakers back in Figure 26–14 in Chapter 26. The effects are similar.

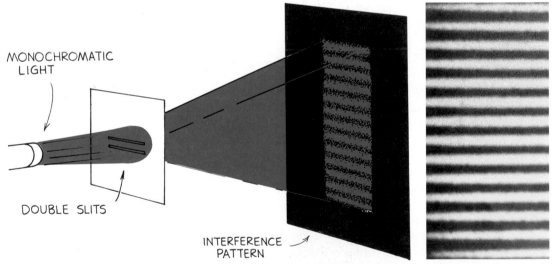

Fig. 31–14 When monochromatic light is passed through two closely spaced slits, a striped interference pattern is produced.

Figure 31–15 shows how the series of bright and dark lines results from the different path lengths from the slits to the screen. A bright fringe occurs when waves from both slits arrive in phase. Dark regions occur when waves arrive out of phase.

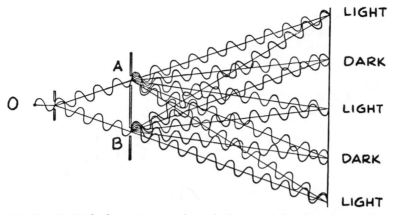

Fig. 31–15   Light from *O* passes through slits *A* and *B* and produces an interference pattern on the screen at the right.

> ► **Questions**
>
> 1. Why is it important that monochromatic (single-frequency) light be used in Young's interference experiment?
>
> 2. If the double slits were illuminated with monochromatic blue light, would the fringes be closer together or farther apart than those produced when monochromatic red light is used?

Interference patterns are not limited to double slits. A multitude of closely-spaced parallel slits makes up a **diffraction grating**. Many spectrometers use diffraction gratings rather than prisms to disperse light into colors. Whereas a prism separates the colors of light by refraction, a diffraction grating separates colors by interference. More common diffraction gratings are seen in reflective materials used in items such as costume jewelry and automobile bumper stickers. These materials are ruled with

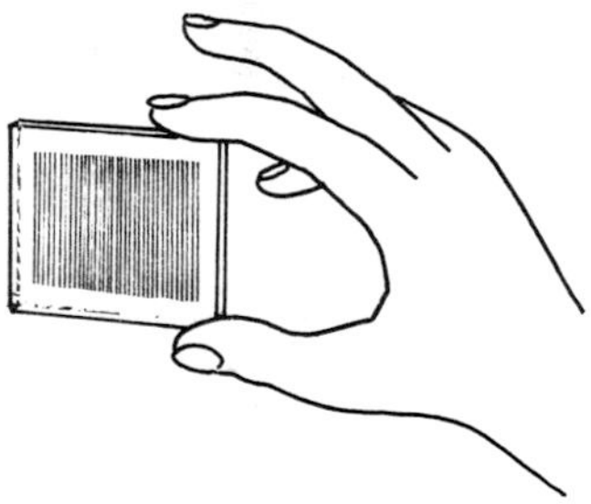

Fig. 31–16   A diffraction grating disperses light into colors by interference. It may be used in place of a prism in a spectrometer.

► **Answers**

1. If light of a variety of wavelengths were diffracted by the slits, dark fringes for one wavelength would be filled in with bright fringes for another, resulting in no distinct fringe pattern. This is similar to the person listening to the pair of speakers back in Chapter 26, Figure 26–14. If the path difference equals one-half wavelength for one frequency, it cannot also equal one-half wavelength for any other frequency. Different frequencies will "fill in" the fringes.

2. The wavelength of blue light is shorter than (nearly half) that of red light. Investigate the differences in the number of fringes for the water waves in Figure 31–12. The fringes of shorter wavelengths are closer together than those of longer wavelengths. So blue fringes would be closer together than red fringes.

tiny grooves that diffract light into a brilliant spectrum of colors. The pits on the reflective surface of digital audio laser discs not only provide high-fidelity music but diffract light spectacularly into its component colors.

31.5 | Single-Color Interference from Thin Films

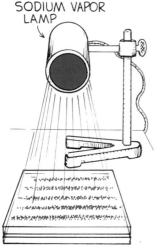

SODIUM VAPOR LAMP

Fig. 31–17 Interference fringes can be produced when monochromatic light is reflected from two plates of glass.

Interference fringes can be produced by the reflection of light from two surfaces very close together. If you shine monochromatic light onto two plates of glass, one atop the other as shown in Figure 31–17, you'll see dark and bright bands.

The cause of these bands is the interference between the waves reflected from the glass on the top and bottom surfaces of the air space between the plates. This is shown in Figure 31–18. The light reflected from point *P* comes to the eye by two different paths. The light that hits the lower glass surface has slightly farther to go to reach your eye. If this extra distance results in light from the upper and lower reflections getting to your eye one-half wavelength out of phase, then destructive interference will occur and a dark region will be seen. Nearby, the path differences will not result in destructive interference, and a light region will be seen.

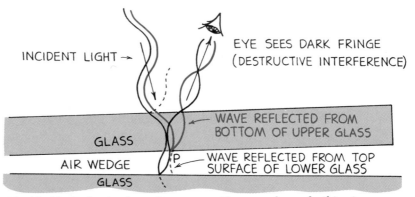

INCIDENT LIGHT →

EYE SEES DARK FRINGE (DESTRUCTIVE INTERFERENCE)

GLASS

← WAVE REFLECTED FROM BOTTOM OF UPPER GLASS

AIR WEDGE   P — WAVE REFLECTED FROM TOP SURFACE OF LOWER GLASS

GLASS

Fig. 31–18 Reflection from the upper and lower surfaces of a thin air space results in constructive and destructive interference.

A practical use of interference fringes is the testing of precision lenses. When a lens to be tested is placed on a perfectly flat piece of glass and illuminated from above with monochromatic light, light and dark fringes are seen (Figure 31–19). Irregular fringes indicate an irregular surface. When a lens is polished until smooth and concentric, the interference fringes will be concentric and regularly spaced.

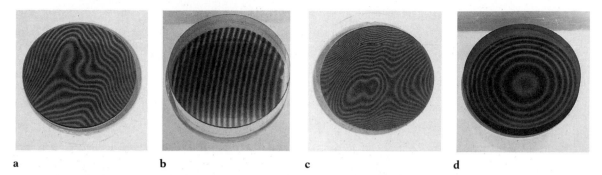

a   b   c   d

Fig. 31–19   The flatness or curvature of a surface can be tested by placing the surface on a very flat piece of glass and observing the interference pattern. (a) A rough surface; (b) a flat surface; (c) a poorly polished lens; (d) a precision lens.

31.6 Iridescence from Thin Films

Everyone who has seen soap bubbles or gasoline spilled on a wet street has noticed the beautiful spectrum of colors reflected from them. Some types of bird feathers have colors that seem to change hue as the bird moves. All these colors are produced by the interference of light waves of mixed frequencies in thin films, a phenomenon known as **iridescence**. Iridescence is illustrated in the color section found between pages 404 and 405.

A thin film, such as a soap bubble, has two closely spaced surfaces. Light that reflects from one surface may cancel light that reflects from the other surface. For example, the film may be just the right thickness in one place to cause the destructive interference of, say, blue light. If the film is illuminated with white light, then the light that reflects to your eye will have no blue in it. What happens when blue is taken away from white light? The answer is, the complementary color will appear. And for the cancellation of blue, that is yellow. So the soap bubble will appear yellow wherever blue is cancelled.

In a thicker part of the film, where green is cancelled, the bubble will appear magenta. The different colors correspond to the cancellations of their complementary colors by different thicknesses of the film.

Figure 31–20 illustrates interference for a thin layer of gasoline on a layer of water. Light reflects from both the upper gasoline-air surface and the lower gasoline-water surface. Suppose that the incident beam is monochromatic blue, as in the illustration. If the gasoline layer is just the right thickness to cause cancellation of light of that wavelength, then the gasoline surface appears dark to the eye. If the incident beam is white sunlight, on the other hand, then the gasoline surface appears yellow to the eye. This is because blue is subtracted from the white, leaving the complementary color, yellow.

The different colors you see in gasoline on a wet street, then, correspond to different thicknesses of the thin film. The colors provide a vivid "contour map" of microscopic differences in surface "elevations."

Fig. 31–20 The thin film of gasoline is just the right thickness so that monochromatic blue light reflected from the top surface of the gasoline is cancelled by light of the same wavelength reflected from the water.

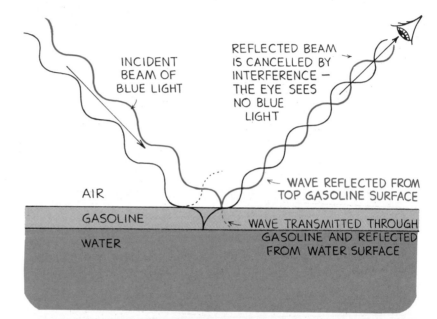

► **Questions**

1. What color will appear to be reflected from a soap bubble in sunlight when its thickness is such that red light is canceled?

2. Many camera lenses have on their surfaces a thin clear coating that makes the lens appear bluish. What does this tell you about the thickness of the coating?

► **Answers**

1. You will see the color cyan, which is the complementary color of red.

2. The thickness of the coating is just right for producing the destructive interference of yellow light, since you see the complementary color, blue. For cancellation of yellow, part of a wave must reflect directly from the coating and part must continue through the coating material and reflect off the glass lens surface beneath, rejoining the directly-reflected light one-half (or an odd-number multiple of one-half) wavelength out of phase. (The purpose of the coating is to cancel yellow light that doubly reflects from inside the camera and arrives at the film out of focus.)

The beautiful colors reflected from some types of seashells are produced by interference of light from their thin transparent coatings. So are the sparkling colors from fractures within opals. Interference colors can even be seen in the thin film of detergent left when dishes are not properly rinsed.

Interference provides the principal method for measuring the wavelengths of light. Wavelengths of other regions of the electromagnetic spectrum are also measured with interference techniques. Extremely small distances (millionths of a centimeter) are measured with instruments called *interferometers*, which make use of the principle of interference. These instruments are sensitive enough to detect the displacement at the end of a long, several-centimeters-thick solid steel bar when you gently twist the other end with your hand. They are among the most accurate measuring instruments known.

The next two sections describe the laser and what is perhaps the most exciting illustration of interference—the *hologram*.

31.7 | Laser Light

Light emitted by a common lamp is **incoherent**. That is, the light has many phases of vibration (as well as many frequencies). The light is as incoherent as the footsteps on an auditorium floor when a mob of people are chaotically rushing about. Incoherent light is chaotic. Interference within a beam of incoherent light is rampant, and a beam spreads out after a short distance, becoming wider and wider and less intense with increased distance.

Fig. 31–21   Incoherent white light contains waves of many frequencies and wavelengths that are out of phase with each other.

Even if a beam is filtered so that it is monochromatic (has a single frequency), it is still incoherent, for the waves are out of phase and interfere with one another. The slightest differences in their directions results in a spreading with increased distance.

Fig. 31–22   Light of a single frequency and wavelength is still out of phase.

A beam of light that has the same frequency, phase, and direction is said to be **coherent**. There is no interference of waves within the beam. Only a beam of coherent light will not spread and diffuse.

Fig. 31–23 Coherent light. All the waves are identical and in phase.

Coherent light is produced by a **laser** (whose name comes from <u>l</u>ight <u>a</u>mplification by <u>s</u>timulated <u>e</u>mission of <u>r</u>adiation).* Within a laser, a light wave emitted from one atom stimulates the emission of light from a neighboring atom so that the crests of each wave coincide. These waves stimulate the emission of others in cascade fashion, and a beam of coherent light is produced. This is very different from the random emission of light from atoms in common sources.

Fig. 31–24 A helium-neon laser. A high voltage applied to a mixture of helium and neon gas energizes helium atoms to a prolonged energy state. Before the helium can emit light, it gives up its energy by collision with neon, which is boosted to an otherwise hard-to-come-by matched energy state. Light emitted by neon stimulates other energized neon atoms to emit matched-frequency light. The process cascades, and a coherent beam of light is produced.

---

*   A word constructed from the initials of a phrase is called an *acronym*.

The laser is not a source of energy. It is simply a converter of energy, taking advantage of the process of stimulated emission to concentrate a certain fraction of the energy input (commonly 1 percent) into a thin beam of coherent light. Like all devices, a laser can put out no more energy than is put in.

Lasers come in many types and find broad applications in fields like the construction industry, communications, medicine, and energy research. Grocery store cash registers read product codes with laser light, and videodiscs use laser light as a type of optical "record needle." A most impressive product of laser light is the hologram.

Fig. 31–25 A product's code is read by laser light that reflects from the bar pattern and is converted to an electrical signal that is fed into a computer. The signal is high when light is reflected from the white spaces and low when reflected from a dark bar.

31.8 The Hologram

Holo- comes from the Greek word for "whole," and *gram* comes from the Greek for "message" or "information." A **hologram** is a three-dimensional version of a photograph that contains the whole message or entire picture in every portion of its surface. To the naked eye it appears to be an imageless piece of transparent film, but on its surface is a pattern of microscopic fringes. Light diffracted from these fringes produces an image that is extremely realistic.

A hologram is produced by the interference between two laser light beams on photographic film. The two beams are part of one beam. One part illuminates the object and is reflected from the object to the film. The second part, called the *reference beam*, is reflected from a mirror to the film (Figure 31–26). Interference between the reference beam and light reflected from the different points on the object produces a pattern of microscopic fringes on the film. Light from nearer parts of the object travel shorter paths than light from farther parts of the object. The different distances traveled will produce slightly different interference patterns with the reference beam. In this way information about the depth of an object is recorded.

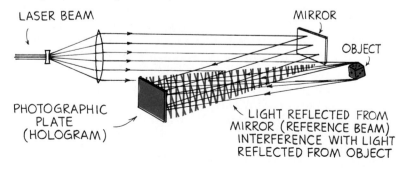

LASER BEAM

MIRROR

OBJECT

PHOTOGRAPHIC PLATE (HOLOGRAM)

LIGHT REFLECTED FROM MIRROR (REFERENCE BEAM) INTERFERENCE WITH LIGHT REFLECTED FROM OBJECT

Fig. 31–26 A simplified arrangement for making a hologram. The laser light that exposes the photographic film is made up of two parts: one part is reflected from the object and one part is reflected from the mirror. The waves of these two parts interfere to produce microscopic fringes on the film. When developed, it is then a hologram.

When laser light (or, in some cases, white light) falls on a hologram, it is diffracted through the fringed pattern to produce wave fronts identical in form to the original wave fronts reflected by the object. The diffracted wave fronts produce the same effect as the original reflected wave fronts. You look through the hologram and see a realistic three-dimensional image as though you were viewing the original object through a window. Parallax is evident when you move your head to the side and see down the sides of the object, or when you lower your head and look underneath the object. Holographic pictures are extremely realistic.

Fig. 31–27 When a hologram is illuminated with coherent light, the diverging diffracted light produces a three-dimensional *virtual* image that can be seen when looking *through* the hologram, like looking through a window. You refocus your eyes to see near and far parts of the image, just as you do when viewing a real object. Converging diffracted light produces a *real* image in front of the hologram, which can be projected on a screen. Since the image has depth, you cannot see near and far parts of the image in sharp focus for any single position on a flat screen.

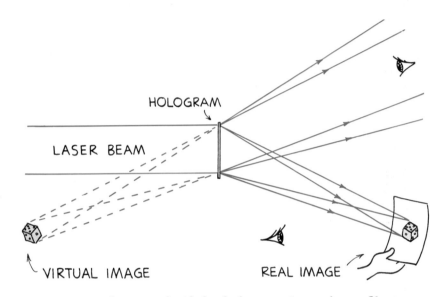

HOLOGRAM

LASER BEAM

VIRTUAL IMAGE

REAL IMAGE

Interestingly enough, if the hologram is made on film, you can cut it in half and still see the entire image. And you can cut one of the pieces in half again and again and see the entire image, just as you can put your eye to any part of a window to see outdoors. Every part of the hologram has received and recorded light from the entire object.

Even more interesting is holographic magnification. If holograms are made using short-wavelength light and viewed with light of a longer wavelength, the resulting image is magnified in the same proportion as the wavelengths. Holograms made with X rays would be magnified thousands of times when viewed with visible light and appropriate viewing arrangements. X-ray holograms have not been made as this book is being written. Technological growth is fast these days. Are X-ray holograms a reality as you are reading this?

Light is interesting—especially when it is diffracted through the interference fringes of that super-sophisticated diffraction grating, the hologram!

31 | Chapter Review

Concept Summary

Diffraction of light is the bending of light by means other than reflection or refraction.

- Huygens' principle, which states that every point on a wave front acts like a point source of secondary wavelets, can be used to understand diffraction.
- Diffraction of light is visible when light passes through an opening comparable in size to the wavelength of the light.

Interference of light is the combining of single-frequency light from two parts of the same beam in a way such that crest overlaps crest or crest overlaps trough.

- The colors seen in soap bubbles or thin films of gasoline on water are due to destructive interference of different frequencies at different thicknesses.
- Holograms are three-dimensional pictures created through the interference of two parts of a laser beam.

Important Terms

coherent (31.7)
diffraction (31.2)
diffraction grating (31.4)
hologram (31.8)
Huygens' principle (31.1)
incoherent (31.7)
iridescence (31.6)
laser (31.7)
monochromatic (31.4)

Review Questions

1. What is Huygens' principle? (31.1)

2. a. Waves spread out when they pass through an opening. Does this spreading become more pronounced or less pronounced for narrower openings?
   b. What name is given to this spreading of waves? (31.1–31.2)

3. Does diffraction aid or hinder radio reception? (31.2)

4. Does diffraction aid or hinder the viewing of images in a microscope? (31.2)

5. Is it possible for a wave to be cancelled by another wave? Defend your answer. (31.3)

6. Does wave interference occur for waves in general, or does it occur only for light waves? Give examples to support your answer. (31.3–31.4)

7. What was Thomas Young's discovery? (31.4)

8. What is the cause of the fringes of light in Young's experiment? (31.4)

9. What is a diffraction grating? (31.4)

10. What is required for part of the light that reflects from a surface to be cancelled by another part reflected from a second surface? (31.5)

11. What is the cause of the bright and dark fringes visible in lenses that rest on flat plates of glass (as shown in Figure 31–19)? (31.5)

12. What is iridescence, and what phenomenon is it related to? (31.6)

13. If the thickness of a soap bubble is sufficient to cancel yellow by interference, what color will the bubble appear when illuminated by white light? (31.6)

14. Why is gasoline that is spilled on a wet surface so colorful? (31.6)

15. What is an interferometer, and on what physics principle is it based? (31.6)

16. How does light from a laser differ from light from an ordinary lamp? (31.7)

17. Is a laser capable of putting out more energy than is put in? (Would you have to know more about lasers to answer this question? Why not?) (31.7)

18. What is a hologram, and on what physics principle is it based? (31.8)

19. How does the image of a hologram differ from that of a common photograph? (31.8)

20. What would be the advantage of making holograms with X rays? (31.8)

Activities

1. Use a razor blade to cut a thin slit in a card and look at a light source through it. You can vary the size of the opening by bending the card slightly. Can you see diffraction fringes? Repeat with two closely spaced slits.

2. Make some slides for a slide projector by sticking crumpled cellophane onto pieces of slide-sized polarizing material. Also try strips of cellophane tape, overlapping at different angles. (Experiment with different brands of tape.) Project the slides onto a large screen or white wall and rotate a second, slightly larger piece of polarizing material in front of the projector lens in rhythm with your favorite music. You'll have your own light and sound show.

3. Do this one at your kitchen sink. Dip a dark-colored coffee cup (dark makes the best background for viewing interference colors) in dishwashing detergent and then hold it sideways and look at the reflected light from the soap film that covers its mouth. Swirling colors appear as the soap runs down to form a wedge that grows thicker at the bottom with time. The top becomes thinner, so thin that it appears black. This happens when the film is thinner than ¼ the wavelength of the shortest waves of visible light. The film soon becomes so thin that it pops.

Think and Explain

1. In our everyday environment, diffraction is much more evident for sound waves than for light waves. Why is this so?

2. Why do radio waves diffract around buildings while light waves do not?

3. Why are TV broadcasts in the VHF range more easily received in areas of marginal reception than broadcasts in the UHF range? (*Hint*: UHF has higher frequencies than VHF.)

4. Suppose a pair of loudspeakers a meter or so apart emit pure tones of the same frequency and loudness. When a listener walks past in a path parallel to the line that joins the loudspeakers, the sound is heard to alternate from loud to soft. What is going on?

5. In the preceding question, suggest a path along which the listener could walk so as not to hear alternate loud and soft sounds.

6. When monochromatic light illuminates a pair of thin slits, an interference pattern is produced on a wall behind. How will the distance between the fringes of the pattern for red light differ from that for blue light?

7. When Thomas Young performed his interference experiment, his monochromatic light passed through a single narrow opening before it reached the double openings. Explain how this procedure made the light coherent so that the interference fringes could be seen.

8. Seashells, butterfly wings, and the feathers

of some birds often change color as you look at them from different positions. Explain this phenomenon in terms of light interference.

9. If you notice the interference patterns of a thin film of oil or gasoline on water, you'll note that the colors form complete rings.

How are these rings similar to the lines of equal elevation on a map that shows the contours of terrain?

10. Why are interference colors seen only for thin films and not for thick films, such as plates of glass, for example?

Ⅴ Electricity and Magnetism

THIS SIMPLE ELECTRIC CIRCUIT ILLUSTRATES SOME REALLY NEAT PHYSICS. THE BATTERY PROVIDES *VOLTAGE*, A SORT OF ELECTRIC PRESSURE, THAT PUSHES ELECTRONS THROUGH THE WIRE AND LAMP. ELECTRONS FLOW EASILY THROUGH THE RELATIVELY THICK WIRE, BUT WITH DIFFICULTY THROUGH THE LAMP FILAMENT. THE FILAMENT HAS A HIGH *RESISTANCE* TO ELECTRON FLOW. *CURRENT* SQUEEZED THROUGH IT SHAKES THE ATOMS SO VIGOROUSLY THAT THEY GLOW. THAT'S WHY THE FILAMENT EMITS *LIGHT* WHILE THE CONNECTING WIRE DOESN'T. EVEN THE LIGHT IS *ELECTRICAL* IN NATURE --- *MAGNETIC* TOO, AS UNIT 5 WILL SHOW.

WHICH PEOPLE ARE MOST AFRAID OF ELECTRICITY, THOSE WHO HAVE SOME UNDERSTANDING OF IT, OR THOSE WHO DON'T?

32 Electrostatics

Electricity in one form or another underlies just about everything around you. It's in the lightning from the sky; it's in the spark beneath your feet when you scuff across a rug; and it's what holds atoms together to form molecules. The control of electricity is evident in technological devices of many kinds, from lamps to computers. In this technological age it is important to have an understanding of how the basics of electricity can be manipulated to give people a prosperity that was unknown before recent times.

This chapter is about **electrostatics**, or electricity at rest. Electrostatics involves electric charges, the forces between them, and their behavior in materials. The next chapter is about the aura that surrounds electric charges—the *electric field*. Chapters 34 and 35 cover moving electric charges, or *electric currents*; the voltages that produce them; and the ways that currents can be controlled. Finally, Chapters 36 and 37 cover the relationship of electric currents to magnetism, and how electricity and magnetism can be controlled to operate motors and other electrical devices.

An understanding of electricity requires a step-by-step approach, for one concept is the building block for the next, and so on. So please put in extra care in the study of this material. It is a good idea at this time to lean more heavily on the laboratory part of your course, for *doing* physics is better than only studying physics. If you're hasty, the physics of electricity and magnetism can be difficult, confusing, and lead to frustration. But with careful effort, it can be comprehensible and rewarding.

32.1 Electrical Forces and Charges

You are familiar with the force of gravity. It attracts you to the earth, and you call it your weight. Now consider a force acting on you that is billions upon billions of times stronger. Such a force should compress you to a size about the thickness of a piece of paper. But suppose that in addition to this enormous force there were a repelling force that is also billions upon billions of times stronger than gravity. The two forces acting on you would balance each other and have no noticeable effect at all. It so happens that there is a pair of such forces acting on you all the time—**electrical forces**.

Electrical forces arise from particles in atoms. In the simple model of the atom proposed about 70 years ago by Ernest Rutherford and Niels Bohr, a nucleus containing protons and neutrons is surrounded by electrons (Figure 32–2). The protons in the nucleus attract the electrons and hold them in orbit, just as the sun holds the planets in orbit. Electrons are attracted to protons, but electrons repel other electrons. This attracting and repelling behavior is attributed to a property called **charge**.* By convention (general agreement), electrons are *negatively* charged and protons *positively* charged. Neutrons have no charge, and are neither attracted nor repelled by charged particles.

Some important facts about atoms are:

1. Every atom has a positively charged nucleus surrounded by negatively charged electrons.

2. All electrons are identical; that is, each has the same mass and the same quantity of negative charge.

3. The nucleus is composed of protons and neutrons. (The

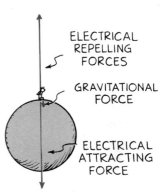

Fig. 32–1 The enormous attractive and repulsive electrical forces between the charges in the earth and the charges in your body balance out, leaving the relatively weaker force of gravity, which attracts only. Hence your weight is due only to gravity.

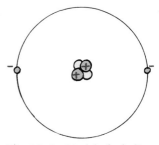

Fig. 32–2 Model of a helium atom. The helium nucleus is composed of two protons and two neutrons. The positively charged protons attract two negative electrons.

---

* Why don't protons pull the oppositely charged electrons into the nucleus? According to Bohr's model of the atom, the electrons are not pulled into the nucleus for the same reason that the earth is not pulled into the sun by gravitational force. That is, the electrons are in motion and overshoot the nucleus but are held in orbit by the pull of the protons. This explanation is very oversimplified but serves as a starter for understanding the electrical nature of the atom. You will learn a very different explanation if you continue your study of physics. You will learn to visualize the electron as a wave rather than an orbiting particle. This is *quantum physics*, which is discussed very briefly in Chapter 38.

Why is it that the protons in the nucleus do not mutually repel and fly apart? What holds the nucleus together? The answer is that in addition to electrical forces in the nucleus, there are even greater forces that are nonelectrical in nature. These are *nuclear forces* and are discussed in Chapter 39.

common form of hydrogen, which has no neutrons, is the only exception.) All protons are identical; similarly, all neutrons are identical. A proton has nearly 2000 times the mass of an electron but its positive charge is equal in magnitude to the negative charge of the electron. A neutron has slightly greater mass than a proton and has no charge.

4. Atoms usually have as many electrons as protons, so the atom has zero *net* charge.

Nobody knows *why* electrons repel electrons and are attracted to protons. At this point we simply say that this is nature as we find it, and since we don't have a deeper understanding of it, we say it is fundamental, or basic. The fundamental rule at the base of all electrical phenomena is:

Like charges repel; opposite charges attract.

The old saying that opposites attract, usually referring to people, was first popularized by public lecturers who traveled about by horse and wagon, entertaining people by demonstrating the scientific marvels of electricity. An important part of these demonstrations was the charging and discharging of pith balls. Pith is a light, spongy plant tissue that resembles Styrofoam, and balls of it were coated with aluminum paint so their surfaces would conduct electricity. When suspended from a silk thread, such a ball would be attracted to a rubber rod just rubbed with cat's fur, but when the two made contact, the force of attraction would change to a force of repulsion. Thereafter, the ball would be repelled by the rubber rod but attracted to a glass rod that had just been rubbed with silk. A pair of pith balls charged in different ways exhibited both attraction and repulsion forces (Figure 32–3). The lecturer pointed out that nature provides two kinds of charge, just as it provides two sexes.

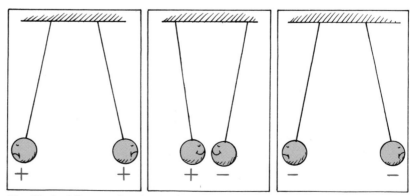

Fig. 32–3 Likes repel and opposites attract.

> ▶ **Questions**
> 1. Beneath the complexities of electrical phenomena, there lies a fundamental rule from which nearly all other effects stem. What is this fundamental rule?
>
> 2. How does the charge of an electron differ from the charge of a proton?

32.2 | Conservation of Charge

Electrons and protons have electric charge. In a neutral atom, there are as many electrons as protons, so there is no net charge. The total positive charge balances the total negative charge exactly. If an electron is removed from an atom, the atom is no longer neutral. The atom has one more positive charge (proton) than negative charge (electron) and is said to be positively charged.

A charged atom is called an *ion*. A *positive ion* has a net positive charge; it has lost one or more electrons. A *negative ion* has a net negative charge; it has gained one or more extra electrons.

Matter is made of atoms, and atoms are made of electrons and protons (and neutrons as well). An object that has equal numbers of electrons and protons has no net electric charge. But if there is an imbalance in the numbers, the object is then electrically charged. An imbalance comes about by adding or removing electrons from an object.

Although the innermost electrons in an atom are bound very tightly to the oppositely charged atomic nucleus, the outermost electrons of many atoms are bound very loosely and can be easily dislodged. How much force is required to tear an electron away from an atom varies for different substances. The electrons are held more firmly in rubber than in fur, for example. Hence, when a rubber rod is rubbed by a piece of fur, electrons transfer from the fur to the rubber rod. The rubber then has an excess of electrons and is negatively charged. The fur, in turn, has a deficiency of electrons and is positively charged. If you rub a glass or plastic rod with silk, you'll find that the rod becomes positively charged. The silk has a greater affinity for electrons than the glass

---

▶ **Answers**

1. Like charges repel; opposite charges attract.

2. The charge of an electron is equal in magnitude, but opposite in sign.

Fig. 32–4 Electrons are transferred from the fur to the rod. The rod is then negatively charged. Is the fur charged? Positively or negatively?

or plastic rod. Electrons are rubbed off the rod and onto the silk. In summary:

> An object that has unequal numbers of electrons and protons is electrically charged. If it has more electrons than protons, the object is negatively charged. If it has fewer electrons than protons, then it is positively charged.

Notice that electrons are neither created nor destroyed but simply transferred from one material to another. Charge is conserved. In every event, whether large-scale or at the atomic and nuclear level, the principle of **conservation of charge** applies. No case of the creation or destruction of net electric charge has ever been found. The conservation of charge is a cornerstone in physics, and ranks with the conservation of energy and momentum.

Any object that is electrically charged has an excess or deficiency of some number of electrons. This means that the charge of the object is therefore a whole-number multiple of the charge of an electron. It cannot have a charge equal to the charge of $1\frac{1}{2}$ or $1000\frac{1}{2}$ electrons, for example.* All charged objects to date have a charge that is a whole-number multiple of the charge of a single electron.

▶ **Question**

If you scuff electrons from your feet while walking across a rug, are you negatively or positively charged?

32.3 | Coulomb's Law

Recall from Newton's law of gravitation that the gravitational force between two objects of mass m_1 and mass m_2 is proportional to the product of the masses and inversely proportional to the square of the distance d between them:

▶ **Answer**

You have fewer electrons after you scuff your feet, so you are positively charged (and the rug is negatively charged).

---

* Within the atomic nucleus, however, elemental charges equal to $\frac{1}{3}$ the electron charge are associated with elemental particles called *quarks*. The charge of three quarks equals the charge of one proton or electron. Since quarks have never been found separated, the whole-number-multiple rule of electron charge holds for nuclear processes as well.

$$F = G\,\frac{m_1 m_2}{d^2}$$

where G is the universal gravitational constant.

The electrical force between any two objects obeys a similar inverse-square relationship with distance. This relationship was discovered by the French physicist Charles Coulomb (1736–1806) in the eighteenth century. **Coulomb's law** states that for charged particles or objects that are small compared to the distance between them, the force between the charges varies directly as the product of the charges and inversely as the square of the distance between them. The role that charge plays in electrical phenomena is much like the role that mass plays in gravitational phenomena. Coulomb's law can be expressed as:

$$F = k\,\frac{q_1 q_2}{d^2}$$

where d is the distance between the charged particles; q_1 represents the quantity of charge of one particle and q_2 the quantity of charge of the other particle; and k is the proportionality constant.

The SI unit of charge is the **coulomb**, abbreviated C. Common sense might say that it is the charge of a single electron, but it isn't. For historical reasons, it turns out that a charge of 1 C is the charge of 6.25 billion billion (6.25×10^{18}) electrons. This might seem like a great number of electrons, but it represents only the amount of charge that passes through a common 100-W light bulb in about one second.

The proportionality constant k in Coulomb's law is similar to G in Newton's law of gravitation. Instead of being a very small number like G, the electrical proportionality constant k is a very large number. Rounded off, it equals

$$k = 9\ 000\ 000\ 000\ \text{N·m}^2/\text{C}^2$$

or, in scientific notation, $k = 9.0 \times 10^9\ \text{N·m}^2/\text{C}^2$. The units N·m²/C² convert the right-hand side of the equation to the unit of force, the newton (N), when the charges are in coulombs (C) and the distance is in meters (m). Note that if a pair of charges of 1 C each were 1 m apart, the force of repulsion between the two charges would be 9 billion newtons.* That would be about ten times the weight of a battleship! Obviously, such amounts of *net* charge do not exist in our everyday environment.

---

* Contrast this to the gravitational force of attraction between two masses of 1 kg each a distance 1 m apart: 6.67×10^{-11} N. This is an extremely small force. For the force to be 1 N, two masses 1 m apart would have to be about 8 million kilograms each! Gravitational forces between ordinary objects are much too small to be detected except in delicate experiments. Electrical forces (noncancelled) between ordinary objects are large enough to be commonly experienced.

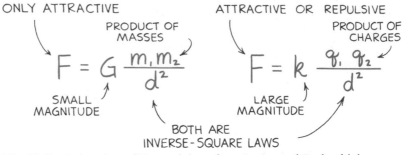

Fig. 32–5   Comparison of Newton's law of gravitation and Coulomb's law.

So Newton's law of gravitation for masses is similar to Coulomb's law for electric charges.* Whereas the gravitational force of attraction between a pair of one-kilogram masses is extremely small, the electrical force between a pair of one-coulomb charges is extremely large. The greatest difference between gravitation and electrical forces is that while gravity only attracts, electrical forces may either attract or repel.

Because most objects have equal numbers of electrons and protons, electrical forces usually balance out. Any electrical

▶ **Questions**
1. What is the chief significance of the fact that G in Newton's law of gravitation is a small number and k in Coulomb's law is a large number?

2. a. If an electron at a certain distance from a charged particle is attracted with a certain force, how will the force compare at twice this distance?
   b. Is the charged particle in this case positive or negative?

▶ **Answers**
1. The small value of G indicates that gravity is a weak force; the large value of k indicates that the electrical force is enormous in comparison.

2. a. In accord with the inverse-square law, at twice the distance the force will be one fourth as much.
   b. Since there is a force of attraction, the charges must be opposite in sign, so the charged particle is positive.

---

*   The similarities between these two forces have made some physicists think they may be different aspects of the same thing. Albert Einstein was one of these people; he spent the later part of his life searching with little success for a "unified field theory." In recent years, the electrical force has been unified with the *weak force*, which plays a role in radioactive decay. Physicists are still looking for a way to unify electrical and gravitational forces.

forces between the earth and the moon, for example, are balanced. In this way the much weaker gravitational force, which attracts only, is the predominant force between astronomical bodies.

Although electrical forces balance out for astronomical and everyday objects, at the atomic level this is not always true. The negative electrons of one atom may at times be closer to the positive protons of a neighboring atom than to the electrons of the neighbor. Then the attractive force between these charges is greater than the repulsive force. When the net attraction is sufficiently strong, atoms combine to form molecules. The chemical bonding forces that hold atoms together to form molecules are electrical forces acting in small regions where the balances of attractive and repelling forces is not perfect. It makes good sense for anyone planning to study chemistry to know something about electricity.

Computational Example

The hydrogen atom has the simplest structure of all atoms. Its nucleus is a proton (mass 1.7×10^{-27} kg), outside of which there is a single electron (mass 9.1×10^{-31} kg) at an average separation distance of 5.3×10^{-11} m. Compare the electrical and gravitational forces between the proton and the electron in a hydrogen atom.

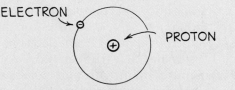

ELECTRON

PROTON

To solve for the electrical force, simply substitute the appropriate values in Coulomb's law.

$$\text{distance } d = 5.3 \times 10^{-11} \text{ m}$$
$$\text{proton charge } q_p = +1.6 \times 10^{-19} \text{ C}$$
$$\text{electron charge } q_e = -1.6 \times 10^{-19} \text{ C}$$

The electric force F_e is:

$$F_e = k \frac{q_e q_p}{d^2}$$

$$= (9.0 \times 10^9 \text{ N·m}^2/\text{C}^2) \frac{(1.6 \times 10^{-19} \text{ C})^2}{(5.3 \times 10^{-11} \text{ m})^2}$$

$$= 8.2 \times 10^{-8} \text{ N}$$

(continued)

The gravitational force F_g between them is:

$$F_g = G \frac{m_e m_p}{d^2}$$

$$= (6.7 \times 10^{-11} \text{ N·m}^2/\text{kg}^2) \frac{(9.1 \times 10^{-31} \text{ kg}) (1.7 \times 10^{-27} \text{ kg})}{(5.3 \times 10^{-11} \text{ m})^2}$$

$$= 3.7 \times 10^{-47} \text{ N}$$

A comparison of the two forces is best shown by their ratio:

$$\frac{F_e}{F_g} = \frac{8.2 \times 10^{-8} \text{ N}}{3.7 \times 10^{-47} \text{ N}} = 2.2 \times 10^{39}$$

The electrical force is more than 10^{39} times greater than the gravitational force. In other words, the electric forces which subatomic particles exert on one another are so much stronger than their mutual gravitational forces that gravitation can be completely neglected.

32.4   Conductors and Insulators

Electrons are more easily moved in some materials than in others. Outer electrons of the atoms in a metal are not anchored to the nuclei of particular atoms, but are free to roam in the material. Such materials are good **conductors**. Metals are good conductors for the motion of electric charges for the same reason they are good conductors of heat: Their electrons are "loose."

Electrons in other materials—rubber and glass, for example— are tightly bound and remain with particular atoms. They are not free to wander about to other atoms in the material. These materials are poor conductors of electricity, for the same reason they are generally poor conductors of heat. Such materials are good **insulators**.

All substances can be arranged in order of their ability to conduct electric charges. Those at the top of the list are the conductors, and those at the bottom are the insulators. The ends of the list are very far apart. The conductivity of a metal, for example, can be more than a million trillion times greater than the conductivity of an insulator such as glass. In a power line, charge flows much more easily through hundreds of kilometers of metal wire than through the few centimeters of insulating material that separates the wire from the supporting tower. In a common appliance cord, charges will flow through several meters of wire to the appliance, and then through its electrical network, and then back through the return wire rather than flow directly across from one wire to the other through the tiny thickness of rubber insulation.

Whether a substance is classified as a conductor or an insulator depends on how tightly the atoms of the substance hold their electrons. Some materials, such as germanium and silicon, are good insulators in their pure crystalline form but increase tremendously in conductivity when even one atom in ten million is replaced with an impurity that adds or removes an electron from the crystal structure. These materials can be made to behave sometimes as insulators and sometimes as conductors. Such materials are **semiconductors**. Thin layers of semiconducting materials sandwiched together make up *transistors*, which are used in a variety of electrical applications. How transistors and other semiconductor devices work will not be covered in this book.

At temperatures near absolute zero, certain metals acquire near infinite conductivity. That is, charge flows through them completely freely. These metals are **superconductors**. Once electric current is established in a superconductor, the electrons will flow indefinitely. Why certain materials become superconducting at low temperatures has to do with the wave nature of matter (quantum mechanics) and is still being studied.

Fig. 32–6 It is easier for electric charge to flow through hundreds of kilometers of metal wire than through a few centimeters of insulating material.

32.5 | Charging by Friction and Contact

We are all familiar with the electrical effects produced by friction. We can stroke a cat's fur and hear the crackle of sparks that are produced, or comb our hair in front of a mirror in a dark room and see as well as hear the sparks of electricity. We can scuff our shoes across a rug and feel the tingle as we reach for the doorknob, or do the same when sliding across plastic seat covers while parked in an automobile (Figure 32–7). In all these cases electrons are being transferred by friction when one material rubs against another.

Fig. 32–7 Charging by friction and then by contact while parked at a drive-in movie.

Electrons can be transferred from one material to another by simply touching. When a charged rod is placed in contact with a neutral object, some charge will transfer to the neutral object. This method of charging is simply called *charging by contact*. If the object is a good conductor, the charge will spread to all parts of its surface because the like charges repel each other. If it is a poor conductor, it may be necessary to touch the rod at several places on the object in order to get a more or less uniform distribution of charge.

32.6 | Charging by Induction

If we bring a charged object *near* a conducting surface, even without physical contact, electrons will move in the conducting surface. Consider the two insulated metal spheres, A and B, in Figure 32–8. In sketch *a*, the uncharged spheres touch each other, so in effect they form a single noncharged conductor. In sketch *b*, a negatively charged rod is near sphere A. Electrons in the metal are repelled by the rod, and excess negative charge has moved onto sphere B, leaving sphere A with excess positive charge. The charge on the two spheres has been redistributed. A charge is said to have been **induced** on the spheres. In sketch *c*, spheres A and B are separated while the rod is still present. In sketch *d*, the rod has been removed. The spheres are charged equally and oppositely. They have been charged by **induction**. Since the charged rod never touched them, it retains its initial charge.

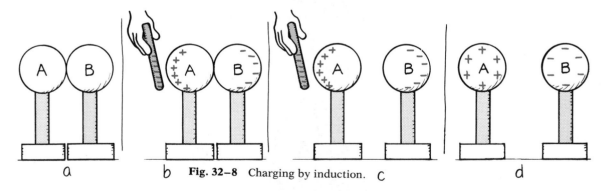

Fig. 32–8 Charging by induction.

A single sphere can be charged similarly if we touch it when the charges are separated by induction. Consider a metal sphere that hangs from a nonconducting string, as shown in Figure 32–9. In sketch *a*, the net charge on the metal sphere is zero. In sketch *b*, a charge redistribution is induced by the presence of the charged rod. The net charge on the sphere is still zero. In sketch *c*, touching the negative side of the sphere removes electrons by contact. In sketch *d*, the sphere is left positively charged. In sketch *e*, the sphere is attracted to the negative rod; it swings over to it and touches it. Now electrons move onto the sphere from the rod. The sphere has been negatively charged by contact. In sketch *f*, the negative sphere is repelled by the negative rod.

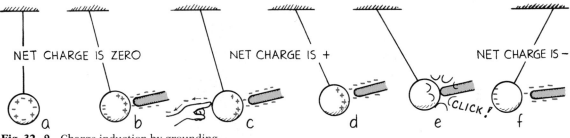

NET CHARGE IS ZERO   NET CHARGE IS +   NET CHARGE IS −

a   b   c   d   e   f

Fig. 32−9   Charge induction by grounding.

When we touch the charged part of the metal surface with a finger (sketch *c*), charges that repel each other have a conducting path to a practically infinite reservoir for electric charge—the ground. When we allow charges to move off (or onto) a conductor by touching it, it is common to say that we are **grounding** it. Chapter 34 returns to this idea of grounding in the discussion of electric currents.

▶ **Questions**

1. Would the charges induced on spheres A and B of Figure 32−8 necessarily be exactly equal and opposite?

2. Why does the negative rod in Figure 32−8 have the same charge before and after the spheres are charged, but not when charging takes place as in Figure 32−9?

▶ **Answers**

1. The charges must be equal and opposite on both spheres, because each single positive charge on sphere A is the result of a single electron being taken from A and moved to B. This is like taking bricks from the surface of a brick road and putting them all on the sidewalk. The number of bricks on the sidewalk will be exactly matched by the number of holes in the road. Similarly, the number of extra electrons on sphere B will exactly match the number of "holes" (positive charges) left in sphere A. Remember that a positive charge is the absence of an electron.

2. In the charging process of Figure 32−8, no contact was made between the negative rod and either of the spheres. In the charging process of Figure 32−9, however, the rod touched the sphere when it was positively charged. A transfer of charge by contact reduced the negative charge on the rod.

Fig. 32-10 The bottom of the negatively-charged cloud induces a positive charge at the surface of the ground below.

Charging by induction occurs during thunderstorms. The negatively charged bottoms of clouds induce a positive charge on the surface of the earth below. Benjamin Franklin was the first to demonstrate this in his famous kite-flying experiment, in which he proved that lightning is an electrical phenomenon.* Most lightning is an electrical discharge between oppositely charged parts of clouds. The kind we are most familiar with is the electrical discharge between the clouds and the oppositely charged ground below.

Franklin also found that charge leaks off sharp points, and fashioned the first lightning rod. If the rod is placed above a structure connected to the ground, charge leaks off instead of building up due to induction. This continual leaking of charge prevents a charge buildup that would otherwise lead to a sudden discharge between the cloud and the ground. The primary purpose of the lightning rod, then, is to prevent a lightning discharge from occuring. If for any reason sufficient charge does not leak off the rod, and lightning strikes anyway, it will be attracted to the rod and shortcircuited to the ground, thereby sparing the structure.

32.7 | Charge Polarization

Charging by induction is not restricted to conductors. When a charged rod is brought near an insulator, there are no free electrons to migrate throughout the insulating material. Instead, there is a rearrangement of the positions of charges within the atoms and molecules themselves (Figure 32-11 left). One side of the atom or molecule is induced to be slightly more positive (or negative) than the opposite side. The atom or molecule is said to be **electrically polarized**. If the charged rod is negative, say, then the positive side of the atom or molecule is toward the rod, and the negative side of the atom or molecule away from it. The atoms or molecules near the surface all become aligned this way (Figure 32-11 right).

---

* Benjamin Franklin was most fortunate that he was not electrocuted as were others who attempted to duplicate his experiment. In addition to being a great statesman, Franklin was a first-rate scientist. He introduced the terms *positive* and *negative* as they relate to electricity and established the "one-fluid theory" of electric currents. He also contributed to our understanding of grounding and insulation. As Franklin approached the height of his scientific career, a more compelling task was presented to him—helping to form the system of government of the newly independent United States. A less important undertaking would not have kept him from spending more of his energies on his favorite activity—the scientific investigation of nature.

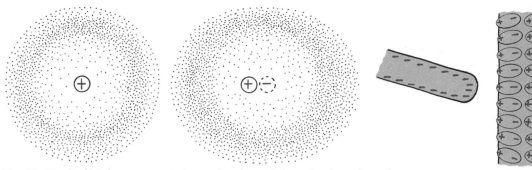

Fig. 32–11   (Left) When an external negative charge is brought closer from the left, the charges within a neutral atom or molecule rearrange so that the left-hand side is slightly more positive and the right-hand side is slightly more negative. (Right) All the atoms or molecules near the surface become electrically polarized.

This explains why electrically neutral bits of paper are attracted to a charged object. Molecules are polarized in the paper, with the oppositely-charged sides of molecules closest to the charged object. Closeness wins, and the bits of paper experience a net attraction. Sometimes they will cling to the charged object and suddenly fly off. This indicates that charging by contact has occurred; the paper bits have acquired the same sign of charge as the charged object and are then repelled.

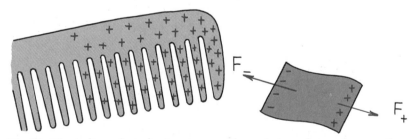

Fig. 32–12   A charged comb attracts an uncharged piece of paper because the force of attraction for the closer charge is greater than the force of repulsion for the farther charge. Closeness wins, and there is a net attraction.

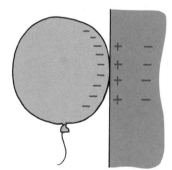

Fig. 32–13   The negatively charged balloon polarizes molecules in the wooden wall and creates a positively charged surface, so the balloon sticks to the wall.

Rub an inflated balloon on your hair and it becomes charged. Place the balloon against the wall and it sticks. This is because the charge on the balloon induces an opposite surface charge on the wall. Closeness wins, for the charge on the balloon is slightly closer to the opposite induced charge than to the charge of same sign (Figure 32–13).

Many molecules—H_2O for example—are electrically polarized in their normal states. The distribution of electric charge is not perfectly even. There is a little more negative charge on one side of the molecule than on the other (Fig. 32–14). Such molecules are said to be *electric dipoles*.

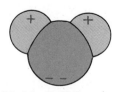

Fig. 32–14   An H_2O molecule is an electric dipole.

32   Chapter Review

Concept Summary

All electrons have the same amount of negative charge; all protons have a positive charge equal in magnitude to the negative charge on the electron.

- Electrical forces arise because of the way that like charges repel and unlike charges attract.
- Electric charge is conserved.
- According to Coulomb's law, the electrical force between two charged objects is proportional to the product of the charges and inversely proportional to the square of the distance between them.

Electrons move easily in good conductors and poorly in good insulators.

- Objects become charged when electrons move onto them or off them.
- Charging by friction occurs when electrons are transferred by rubbing.
- Charging by contact occurs when electrons are transferred by direct contact.
- Charging by induction occurs in the presence of a charge without physical contact.
- Charge polarization occurs in insulators that are in the presence of a charged object.

Important Terms

charge (32.1)
conductor (32.4)
conservation of charge (32.2)
coulomb (32.3)
Coulomb's law (32.3)
electrical force (32.1)
electrically polarized (32.7)
electrostatics (32.1)
grounding (32.6)
induced (32.6)
induction (32.6)
insulator (32.4)
semiconductor (32.4)
superconductor (32.4)

Review Questions

1. Which force—gravitational or electrical—repels as well as attracts? (32.1)

2. Gravitational forces depend on the property called *mass*. What comparable property underlies electrical forces? (32.1)

3. How do protons and electrons differ in their electric charge? (32.1)

4. Is an electron in a hydrogen atom the same as an electron in a uranium atom? (32.1)

5. Which has more mass—a proton or an electron? (32.1)

6. In a normal atom, how many electrons are there compared to protons? (32.1)

7. a. How do like charges behave?
   b. How do unlike charges behave? (32.1)

8. How does a negative ion differ from a positive ion? (32.2)

9. a. If electrons are rubbed from cat's fur onto a rubber rod, does the rod become positively or negatively charged?
   b. How about the cat's fur? (32.2)

10. What does it mean to say that charge is conserved? (32.2)

11. a. How is Coulomb's law similar to Newton's law of gravitation?
    b. How are the two laws different? (32.3)

12. The SI unit of mass is the kilogram. What is the SI unit of charge? (32.3)

13. The proportionality constant k in Coulomb's law is huge compared to the proportionality constant G in Newton's law of gravitation. What does this mean in terms of the relative strengths of these two forces? (32.3)

14. Why does the weaker force of gravity dominate over electrical forces for astronomical objects? (32.3)

15. Why do electrical forces dominate between atoms that are close together? (32.3)

16. What is the difference between a good conductor and a good insulator? (32.4)

17. a. Why are metals good conductors?
    b. Why are materials such as rubber and glass good insulators? (32.4)

18. What is a semiconductor? (32.4)

19. What is a superconductor? (32.4)

20. a. What are the three main methods of charging objects?
    b. Which method involves no touching? (32.5–32.6)

21. What is lightning? (32.6)

22. What is the function of a lightning rod? (32.6)

23. What does it mean to say an object is electrically polarized? (32.7)

24. When a charged object polarizes another, why is there an attraction between the objects? (32.7)

25. What is an electric dipole? (32.7)

Activities

1. Charge a comb by rubbing it through your hair, especially if the weather is dry, and bring it near tiny bits of paper. Explain your observations.

2. Place a charged comb near a thin stream of running water from a faucet. Is the stream of water charged?

Think and Explain

1. Electrical forces between charges are enormous in comparison to gravitational forces. Yet, we normally don't sense electrical forces between us and our environment, while we do sense our gravitational interaction with the earth. Why is this so?

2. By how much is the electrical force between a pair of ions reduced when their separation distance is doubled? Tripled?

3. If you scuff electrons from a rug onto your shoes, are you positively or negatively charged? How about the rug?

4. An electroscope is a simple device. It consists of a metal ball that is attached by a conductor to two fine gold leaves that are protected from air disturbances in a jar, as shown in Figure A. When the ball is touched by a charged object, the leaves that normally hang straight down spring apart. Why? (Electroscopes are useful not only as charge detectors, but also for measuring the amount of charge: the more charge transferred to the ball, the more the leaves diverge.)

Fig. A

5. Would it be necessary for a charged object to actually touch the leaves of an electroscope (see Question 4) for the leaves to diverge? Defend your answer.

6. Give an example and an explanation of charging an object positively with only the help of a negatively charged object.

7. If a glass rod that is rubbed with a plastic dry cleaner's bag acquires a certain charge, why does the plastic bag have exactly the same opposite charge?

8. Why is a good conductor of electricity also a good conductor of heat?

9. Explain how an object that is electrically neutral can be attracted to an object that is charged.

10. The leaves of a charged electroscope (see Question 4) collapse after a while. At higher altitudes they collapse more readily. Speculate an explanation for this behavior. (*Hint*: The existence of cosmic rays was first indicated by this observation.)

33 Electric Fields and Potential

The space around a strong magnet is different from how it would be if the magnet were not there. Put a paper clip in the space and you'll see the paper clip move. The space around a black hole is different from how it would be if the black hole were not there. Put yourself in the space, and that will be the last thing you do. Similarly, the space around a concentration of electric charge is different from how it would be if the charge were not there. If you walk by the charged dome of an electrostatic machine—a Van de Graaff generator, for example—you can sense the charge. Hair on your body stands out—just a tiny bit if you're more than a meter away, and more if you're closer. The space that surrounds each of these things—the magnet, the black hole, and the electric charge—is altered. The space is said to contain a *force field*.

Fig. 33–1 You can sense the force field that surrounds a charged Van de Graaff generator.

33.1 Electric Fields

The force field that surrounds a mass is a gravitational field. If you throw a ball into the air, it follows a curved path. Earlier chapters showed that it curves because there is an interaction between the ball and the earth—between their centers of gravity, to be exact. Their centers of gravity are quite far apart, so this is "action at a distance."

The idea that things not in contact could exert forces on one another bothered Isaac Newton and many others. The concept of a force field eliminates the distance factor. The ball is in contact with the field all the time. We can say the ball curves because it interacts with the earth's gravitational field. It is common to think of distant rockets and space probes as interacting with gravitational fields rather than with the masses of the earth and other astronomical bodies that are responsible for the fields.

Just as the space around the earth and every other mass is filled with a gravitational field, the space around every electric charge is filled with an **electric field**—a kind of aura that extends through space. In Figure 33–2, a gravitational force holds a satellite in orbit about a planet, and an electrical force holds an electron in orbit about a proton. In both cases there is no contact between the objects, and the forces are "acting at a distance." Putting this in terms of the field concept, we can say that the orbiting satellite and electron interact with the force fields of the planet and the proton and are everywhere in contact with these fields. In other words, the force that one electric charge exerts on another can be described as the interaction between one charge and the electric field set up by the other.

Fig. 33–2 The satellite and the electron both experience forces; they are both in force fields.

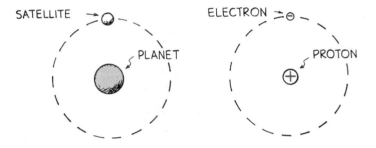

An electric field has both magnitude and direction. Its magnitude (strength) can be measured by its effect on charges located in the field. Imagine a small positive "test charge" that is placed in an electric field. Where the force is greatest on the test charge, the field is strongest. Where the force on the test charge is weak, the field is small.*

The direction of the electric field at any point, by convention, is the direction of the electrical force on a small *positive* test charge placed at that point. Thus, if the charge that sets up the field is positive, the field points away from that charge. If the charge that sets up the field is negative, the field points toward that charge. (Be sure to distinguish between the hypothetical small test charge and the charge that sets up the field.)

---

*   The strength of an electric field is a measure of how great a force will be exerted on a small test charge placed in the field. (The test charge must be small enough that it doesn't push the original charge around and thus alter the field we are trying to measure). If a test charge q experiences a force F at some point in space, then the electric field E at that point is

$$E = \frac{F}{q}$$

Electric field strength can be measured in units of newtons per coulomb (N/C). The next chapter will show that it has equivalent units of volts per meter (V/m).

33.2 | Electric Field Lines

Since an electric field has both magnitude and direction, it is a *vector quantity* and can be represented by vectors. The negatively charged particle in Figure 33–3 top is surrounded by vectors that point toward the particle. (If the particle were positively charged, the vectors would point away from the particle. The vectors always point in the direction of the force that would act on a positive test charge.) The magnitude of the field is indicated by the length of the vectors. The electric field is greater where the vectors are long than where the vectors are short. To represent a complete electric field by vectors, you would have to show a vector at every point in the space around the charge. Such a diagram would be totally unreadable!

A more useful way to describe an electric field is with electric *field lines*, also called *lines of force* (Figure 33-3 bottom). Where the lines are farther apart, the field is weaker. For an isolated charge, the lines extend to infinity, while for two or more opposite charges, the lines emanate from a positive charge and terminate on a negative charge. Some electric field configurations are shown in Figure 33–4.

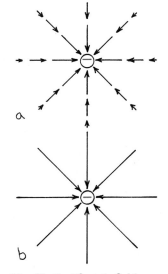

Fig. 33–3 Electric field representations about a negative electric charge. (a) A vector representation; (b) a lines-of-force representation.

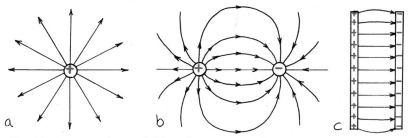

Fig. 33–4 Some electric field configurations. (a) Field lines around a single positive charge. (b) Field lines for a pair of equal but opposite charges. Note that the lines emanate from the positive charge and terminate on the negative charge. (c) Evenly spaced field lines between two oppositely charged parallel plates.

Photographs of field patterns are shown in Figure 33–5. The photographs show bits of thread that are suspended in an oil bath surrounding charged conductors. The ends of the bits of thread are charged by induction and tend to line up end-to-end with the field lines, like iron filings in a magnetic field. Notice that between the plates the threads are aligned parallel to each other. The field has a constant strength between the plates. Also, notice the threads inside the cylinder are unaligned. There is no electric field in the space inside a conductor. The conductor shields the space from the field outside.

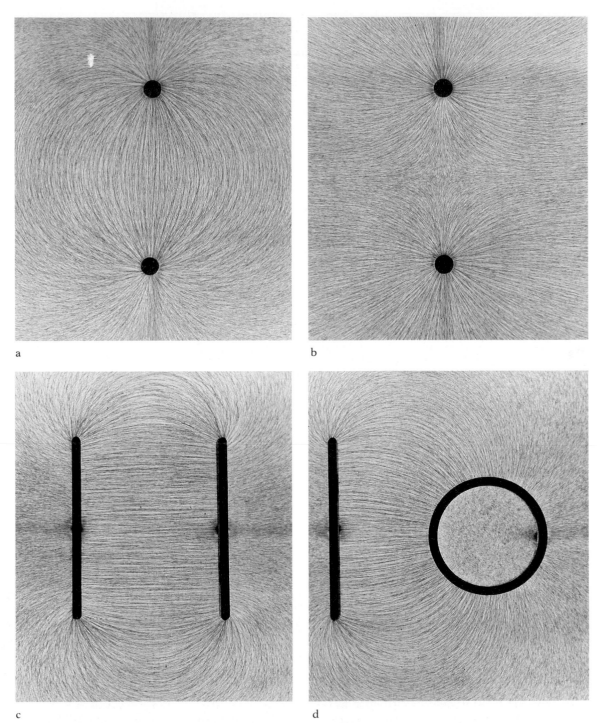

a   b   c   d

Fig. 33–5 Bits of fine thread suspended in an oil bath surrounding charged conductors line up end-to-end along the direction of the field. (a) Equal and opposite charges. (b) Equal like charges. (c) Oppositely charged plates. (d) Oppositely charged cylinder and plate.

If our only concern for electrical forces were limited to isolated point charges, the electric field concept would be of limited use. The force between point charges is adequately described by Coulomb's law. But charges most often are spread out over a wide variety of surfaces. Charges also move. This motion is communicated to neighboring charges by changes in the electric field that emanate at the speed of light.

As later chapters will show, the electric field is a storehouse of energy. The energy can be transported over long distances by an electric field, which may be directed through and guided by metal wires or directed through empty space.

> ▶ **Question**
>
> Suppose that a beam of electrons is produced at one end of a glass tube and lights up a phosphor screen on the inner surface of the other end. When the beam is straight, it produces a spot of light in the middle of the screen. If the beam passes through the electric field of a pair of oppositely charged plates, it is deflected, say to the left. If the charge on the plates is reversed, in what direction would deflection occur?

33.3 | Electric Shielding

The dramatic photo in Figure 33–6 shows a car being struck by lightning. Yet the occupant inside the car is completely safe. This is because the electrons that shower down upon the car are mutually repelled to the outer metal surface before moving to the ground. The configuration of electrons on the car's surface at any moment is such that the electric field contributions inside the car cancel to zero. This is true of any charged conductor. The electric field inside a conductor is normally zero.

This internal field cancellation is more easily understood by considering a charged metal sphere (Figure 33–7). Because of mutual repulsion, the electrons have spread as far apart from one another as possible. They distribute themselves uniformly over

Fig. 33–6 Electrons from the lightning bolt mutually repel to the outer metal surface. Although the electric field they set up may be great *outside* the car, the overall electric field *inside* the car cancels to zero.

▶ **Answer**
    When the charge on the plates is reversed, the electric field will be in the opposite direction, so the electron beam would be deflected to the right. If the field is made to oscillate, the beam will be swept back and forth. With a second set of plates and further refinements it could sweep a picture onto the screen and be a television set.

Fig. 33–7 The forces on a test charge located inside a charged hollow sphere cancel to zero.

the surface of the sphere. A positive test charge located exactly in the middle of the sphere would feel no force. The electrons on the left side of the sphere, for example, would tend to pull the test charge to the left, but the electrons on the right side of the sphere would tend to pull the test charge to the right equally hard. The net force on the test charge would be zero. Thus, the electric field is also zero. Interestingly enough, complete cancellation will occur *anywhere* inside the conducting sphere. Why this is true involves some geometry and will not be treated in this text.

If the conductor is not spherical, then the charge distribution will not be uniform. If it is a cube, for example, then most of the charge is located at the corners. The neat thing is this: The exact charge distribution over the surfaces and corners of a conducting cube is such that the electric field everywhere inside the cube is zero. Look at it this way: If there were an electric field inside a conductor, then free electrons inside the conductor would be set in motion. How far would they move? Until equilibrium is established, which is to say, when the positions of all the electrons produce a zero field inside the conductor.

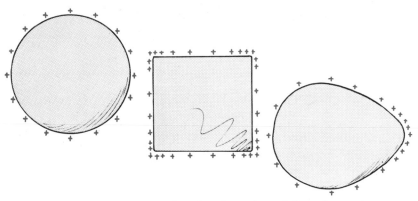

Fig. 33–8 The charges are distributed on the surface of all conductors in a way such that the electric field inside the conductor is zero.

Fig. 33–9 The metal cover shields the internal electrical components from external electric fields. Similarly, the metal cover also shields the coaxial cable.

There is no way to shield gravity, because gravity only attracts. There are no repelling parts of gravity to offset attracting parts. Shielding electric fields, however, is quite simple. Surround yourself or whatever you wish to shield with a conducting surface. Put this surface in an electric field of whatever field strength. The free charges in the conducting surface will arrange themselves on the surface of the conductor in a way such that all field contributions inside cancel one another. That's why certain electronic components are encased in metal boxes, and why certain cables have a metal covering—to shield them from all outside electrical activity.

> ▶ **Question**
> It is said that a gravitational field, unlike an electric field, cannot be shielded. But the gravitational field at the center of the earth cancels to zero. Isn't this evidence that a gravitational field *can* be shielded?

I Love You

33.4 | **Electric Potential Energy**

Recall from Chapter 8 the relation between work and potential energy. Work is done when a force moves something in the direction of the force. An object has potential energy by virtue of its location, say in a force field. For example, if you lift an object, you apply a force equal to its weight. When you raise it through some distance, you are doing work on the object. You are also increasing its gravitational potential energy. The greater the distance it is raised, the greater is the increase in its gravitational potential energy. Doing work increases its gravitational potential energy (Figure 33–10).

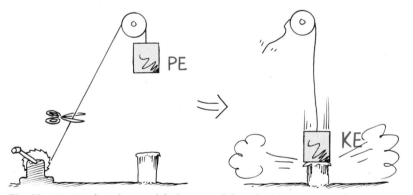

Fig. 33–10 Work is done to lift the ram of the pile driver against the gravitational field of the earth. In an elevated position, the ram has gravitational potential energy. When released, this energy is transferred to the pile below.

▶ **Answer**
No. Gravity can be cancelled inside a planet or between planets, but it cannot be *shielded* by a planet or by any arrangement of masses. During a lunar eclipse, for example, when the earth is directly between the sun and the moon, there is no shielding of the sun's field to affect the moon's orbit. Even a very slight shielding would accumulate over a period of years and show itself in the timing of subsequent eclipses. Shielding requires a combination of repelling and attracting forces, and gravity only attracts.

In a similar way, a charged object can have potential energy by virtue of its location in an electric field. Just as work is required to lift an object against the gravitational field of the earth, work is required to push a charged particle against the electric field of a charged body. (It may be more difficult to visualize, but the physics of both the gravitational case and the electrical case is the same.) The electric potential energy of a charged particle is increased when work is done to push it against the electric field of something else that is charged.

Figure 33–11 top shows a small positive charge located at some distance from a positively charged sphere. If we push the small charge closer to the sphere (Figure 33–11 bottom), we will expend energy to overcome electrical repulsion. Just as work is done in compressing a spring, work is done in pushing the charge against the electric field of the sphere. This work is equivalent to the energy gained by the charge. The energy the charge now possesses by virtue of its location is called **electric potential energy**. If the charge is released, it will accelerate in a direction away from the sphere, and its electric potential energy will transform to kinetic energy.

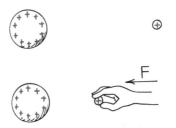

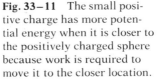

Fig. 33–11 The small positive charge has more potential energy when it is closer to the positively charged sphere because work is required to move it to the closer location.

33.5   Electric Potential

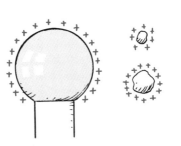

Fig. 33–12 An object of greater charge has more electric potential energy in the field of the charged dome than an object of less charge, but the *electric potential* of any amount of charge at the same location is the same.

If in the preceding discussion we push two charges instead, we do twice as much work. The two charges in the same location will have twice the electric potential energy as one; three charges will have three times the potential energy; a group of ten charges will have ten times the potential energy; and so on.

Rather than deal with the total potential energy of a group of charges, it is convenient when working with electricity to consider the *electric potential energy per charge*. The electric potential energy per charge is the total electric potential energy divided by the amount of charge. At any location the potential energy *per charge*—whatever the amount of charge—will be the same. For example, an object with ten units of charge at a specific location has ten times as much energy as an object with a single unit of charge. But it also has ten times as much charge, so the potential energy per charge is the same. The concept of electric potential energy per charge has a special name, **electric potential**.

$$\text{electric potential} \quad = \frac{\text{electric potential energy}}{\text{charge}}$$

> ▶ **Question**
> If there were twice as much charge on one of the charged
> objects near the charged sphere in Figure 33–12, would the
> electric potential energy of the object in the field of the
> charged sphere be the same or would it be twice as great?
> Would the electric potential of the object be the same or
> would it be twice as great?

The SI unit of measurement for electric potential is the **volt**,
named after the Italian physicist Allesandro Volta (1745–1827).
The symbol for volt is V. Since potential energy is measured in
joules and charge is measured in coulombs,

$$1 \text{ volt} = 1 \; \frac{\text{joule}}{\text{coulomb}}$$

Thus, a potental of 1 volt equals 1 joule of energy per coulomb of
charge; 110 volts equals 110 joules of energy for every coulomb
of charge. If a conductor has a potential of 110 volts, then each
coulomb of charge on the conductor has an electric potential energy of 110 joules. In other words, 110 joules of work must
have been expended in storing each coulomb of charge on the
conductor.*

Since electric potential is measured in volts, it is commonly
called **voltage**. In this book the names will be used interchangeably. The significance of voltage is that a definite value for it can
be assigned to a location whether or not a charge exists at that
location. We can speak about the voltages at different locations
in an electric field whether or not any charges occupy those
locations.

▶ **Answer**
Twice as much charge would cause the object to have twice as much electric
potential energy, because it would have taken twice as much work to put the object at that location. But the electric potential would be the same, because the
electric potential is total electric potential energy divided by total charge. In
this case, twice the energy divided by twice the charge gives the same value as
the original energy divided by the original charge. Electric potential is not the
same thing as electric potential energy. Be sure you understand this before you
study further.

---

*   It is common practice to assign a zero electric potential to places infinitely
far away from any charges. Then if an object is said to be charged to, say, 10 000
volts, it would take 10 000 joules of work per coulomb to assemble the charges by
pushing against their mutual repulsions from zero starting points. As the next
chapter discusses, for electric currents the value zero is assigned to the ground.

Fig. 33–13 Although the voltage of the charged balloon is high, the electric potential energy is low because of the small amount of charge.

Rub a balloon on your hair and the balloon becomes negatively charged, perhaps to several thousand volts! If the charge on the balloon were one coulomb, it would take several thousand joules of energy to give the balloon that voltage. However, one coulomb is a relatively large amount of charge; the charge on a balloon rubbed on hair is typically much less than a millionth of a coulomb. Therefore, the amount of energy associated with the charged balloon is very, very small—about a thousandth of a joule. A high voltage requires great energy only if a great amount of charge is involved. This example highlights the difference between electric potential energy and electric potential.

33.6 | The Van de Graaff Generator

A common laboratory device for building up high voltages is the *Van de Graaff generator*. This is the lightning machine that "evil scientists" used in old science fiction movies. A simple model of the Van de Graaff generator is shown in Figure 33–14. A large hollow metal sphere is supported by a cylindrical insulating stand. A motor-driven rubber belt inside the support stand moves past a comb-like set of metal needles that are maintained at a

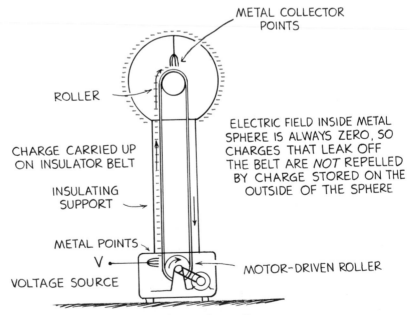

Fig. 33–14 A simple model of a Van de Graaff Generator.

Fig. 33–15   The physics enthusiasts and the dome of the Van de Graaff generator are charged to a high voltage. Why does their hair stand out?

high electric potential. A continuous supply of electrons is deposited on the belt through electric discharge by the points of the needles and is carried up into the hollow metal sphere. The electrons leak onto metal points (which act like tiny lightning rods) attached to the inner surface of the sphere. Because of mutual repulsion, the electrons move to the outer surface of the conducting sphere. (Remember, static charge on any conductor is on the outside surface.) This leaves the inside surface uncharged and able to receive more electrons as they are brought up the belt. The process is continuous, and the charge builds up to a very high electric potential—on the order of millions of volts.

A sphere with a radius of 1 m can be raised to a potential of 3 million volts before electrical discharge occurs through the air (because breakdown occurs in air when the electric field strength is about 3×10^6 V/m). The voltage can be further increased by increasing the radius of the sphere or by placing the entire system in a container filled with high-pressure gas. Van de Graaff generators can produce voltages as high as 20 million volts. These devices accelerate charged particles used as projectiles for penetrating the nuclei of atoms. Touching one can be a hair-raising experience (Figure 33–15).

Before electricity became commonplace, it was regarded by most people as a terrifying phenomenon. To help dispel this fear, Nikola Tesla (who was the chief proponent of alternating current when Thomas Edison was promoting direct current) would sit indifferently in the midst of a high-voltage sparking demonstration (Figure 33–16).

Fig. 33–16   Time exposure of Nikola Tesla quietly reading while sparks leap between conductors around him.

33 | Chapter Review

Concept Summary

An electric field fills the space around every electric charge.
- The field is strongest where it would exert the greatest electrical force on a test charge.
- The direction of the field at any point is the direction of the electrical force on a positive test charge.
- An electric field can be represented by electric field lines.
- The electric field inside a conductor is zero; any static net charge is all on the outside surface.

A charged object has electric potential energy by virtue of its location in an electric field.
- The electric potential, or voltage, at any point in an electric field is the electric potential energy per charge for a charged object at that point.

Important Terms

electric field (33.1)
electric potential energy (33.4)
electric potential (33.5)
volt (33.5)
voltage (33.5)

Review Questions

1. What is meant by the expression *action at a distance*? (33.1)

2. How does the concept of a field eliminate the idea of action at a distance? (33.1)

3. How are a gravitational and an electric field similar? (33.1)

4. Why is an electric field considered a vector quantity? (33.2)

5. a. What are electric field lines?
   b. How do their directions compare with the direction of the force that acts on a positive test charge in the same region? (33.2)

6. How is the strength of an electric field indicated with field lines? (33.2)

7. How do the electric field lines appear when the field has the same strength at all points in a region? (33.2)

8. Why will an occupant inside a car struck by lightning be safe? (33.3)

9. What is the size of the electric field inside any charged conductor? (33.3)

10. a. Can gravity be shielded?
    b. Can electric fields be shielded? (33.3)

11. What is the relationship between the amount of work you do on an object and its potential energy? (33.4)

12. How can the electric potential energy of a charged particle in an electric field be increased? (33.4)

13. What will happen to the electric potential energy of a charged particle in an electric field when the particle is released and free to move? (33.4)

14. Clearly distinguish between *electric potential energy* and *electric potential*. (33.5)

15. If you do more work to move more charge a

certain distance against an electric field, and increase the electric potential energy as a result, why do you not also increase the electric potential (33.5)?

16. The SI unit for electric potential energy is the joule. What is the SI unit for electric potential? (33.5)

17. Charge must be present at a location in order for there to be electric potential energy. Must charge also be present at a location for there to be electric potential? (33.5)

18. How can electric potential be high when electric potential energy is relatively low? (33.5)

19. How does the amount of charge on the inside surface of the sphere of a charged Van de Graaff generator compare to the amount on the outside? (33.6)

20. About how large a voltage can be built up on a Van de Graaff generator before electrical discharge occurs through the air? (33.6)

Think and Explain

1. How is an electric field different from a gravitational field?

2. The vectors for the gravitational field of the earth point *toward* the earth; the vectors for the electric field of a proton point *away* from the proton. Explain.

3. If a "free" electron and "free" proton were placed in an electric field, how would their accelerations and directions of travel compare?

4. Suppose that the strength of the electric field about an isolated point charge has a certain value at a distance of 1 m. How will the electric field strength compare at a distance of 2 m from the point charge? What law guides your answer?

5. When a conductor is charged, the charge moves to the outer surface of the conductor. Why is this so?

6. Suppose that a metal file cabinet is charged. How will the charge concentration at the corners of the cabinet compare to the charge concentration on the flat parts of the cabinet? Defend your answer.

7. a. If you do 12 J of work to push 0.001 C of charge into an electric field, by how much will its voltage increase?
   b. When the charge is released, what will be its kinetic energy as it flies past its starting position? What principle guides your answer?

8. a. If you do 24 J of work to push 0.002 C of charge into an electric field, by how much will its voltage increase?
   b. When the charge is released, what will be its kinetic energy as it flies past its starting position?

9. Is it correct to say that an object with twice the electric potential as another has twice the electric potential energy? Defend your answer. (*Hint*: Is it correct to say that an object with twice the temperature as another has twice the internal energy?)

10. Why does your hair stand out when you are charged by a device such as a Van de Graaff generator?

34 Electric Current

The last chapter discussed the concept of electric potential, or voltage. This chapter will show that voltage is an "electrical pressure" that can produce a flow of charge, or *current*, within a conductor. The flow is restrained by the *resistance* it encounters. When the flow takes place along one direction, it is called *direct current* (dc); when it flows to and fro, it is called *alternating current* (ac). The rate at which energy is transferred by electric current is *power*. You'll note here that there are many terms to be sorted out. This is easier (and more meaningful) to do when you have some understanding of the ideas these terms represent. In turn, the ideas are better understood if you know how they relate to one another. Let's begin with the flow of electric charge.

34.1 Flow of Charge

Recall in your study of heat and temperature that heat flows through a conductor when a difference in temperature exists across its ends. Heat flows from the end of higher temperature to the end of lower temperature. When both ends reach the same temperature, the flow of heat ceases.

In a similar way, when the ends of an electrical conductor are at different electric potentials, charge flows from the higher potential to the lower potential. Charge flows when there is a **potential difference**, or difference in potential (voltage), across the ends of a conductor. The flow of charge will continue until both ends reach a common potential. When there is no potential difference, no flow of charge will occur.

As an example, if one end of a wire were connected to the ground and the other end placed in contact with the sphere of a

Van de Graaff generator charged to a high potential, a surge of charge would flow through the wire. The flow would be brief, however, for the sphere would quickly reach a common potential with the ground.

To attain a sustained flow of charge in a conductor, some arrangement must be provided to maintain a difference in potential while charge flows from one end to the other. The situation is analagous to the flow of water from a higher reservoir to a lower one (Figure 34–1 left). Water will flow in a pipe that connects the reservoirs only as long as a difference in water level exists. (This is implied in the saying, "Water seeks its own level.") The flow of water in the pipe, like the flow of charge in the wire that connects the Van de Graaff generator to the ground, will cease when the pressures at each end are equal. In order that the flow be sustained, there must be a suitable pump of some sort to maintain a difference in water levels (Figure 34–1 right). Then there will be a continual difference in water pressures and a continual flow of water. The same is true of electric current.

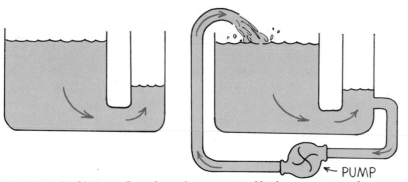

Fig. 34-1   (Left) Water flows from the reservoir of higher pressure to the reservoir of lower pressure. The flow will cease when the difference in pressure ceases. (Right) Water continues to flow because a difference in pressure is maintained with the pump.

34.2 | Electric Current

Electric current is simply the flow of electric charge. In solid conductors, it is the electrons that carry the charge through the circuit. This is because the electrons are free to move throughout the atomic network. These electrons are called *conduction electrons*. Protons, on the other hand, are bound inside atomic nuclei that are more or less locked in fixed positions. In fluids, however, positive and negative ions as well as electrons may compose the flow of electric charge.

Electric current is measured in **amperes**, symbol A.* An ampere is the flow of one coulomb of charge per second. (Recall that one coulomb, the standard unit of charge, is the electric charge of 6.25 billion billion electrons.) In a wire that carries a current of 5 amperes, for example, 5 coulombs of charge pass any cross section in the wire each second. So that's a lot of electrons! In a wire that carries 10 amperes, twice as many electrons pass any cross section each second.

Note that a current-carrying wire does not have a *net* electric charge. Negative electrons swarm through the atomic network that is composed of positively charged atomic nuclei. Under ordinary conditions, the number of electrons in the wire is equal to the number of positive protons in the atomic nuclei. So the net charge of the wire is normally zero at every moment.

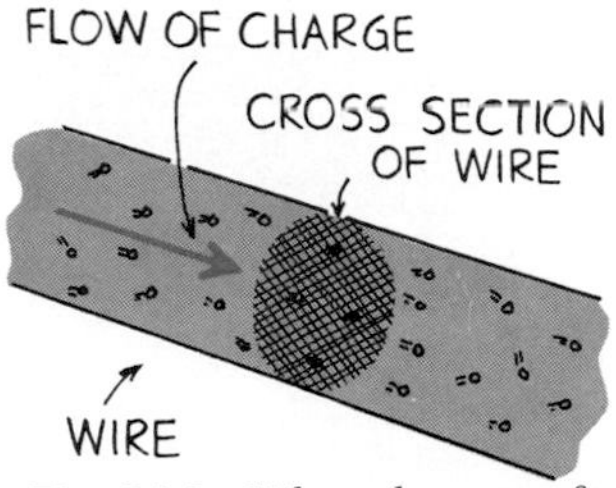

Fig. 34-2 When the rate of flow of charge past any cross section is one coulomb (6.25 billion billion electrons) per second, the current is one ampere.

34.3   Voltage Sources

Charges do not flow unless there is a potential difference. A sustained current requires a suitable "electrical pump" to provide a sustained potential difference. Something that provides a potential difference is known as a **voltage source**. If you charge a rubber rod by rubbing it with fur, you can develop a large voltage between the rod and the fur. This voltage source is not a good electrical pump because when the rod and the fur are connected by a conductor, the potentials equalize in a single brief surge of moving charges. It is not practical. Dry cells, wet cells, and generators, however, are capable of maintaining a steady flow. (A battery is just two or more cells wired together.)

Dry cells, wet cells, and generators supply energy that allows charges to move. In dry cells and wet cells, energy released in a chemical reaction that takes place inside the cell is converted to electric energy.** Generators convert mechanical energy to electric energy, as discussed in Chapter 37. The electric potential energy produced by whatever means is available at the terminals of the cell or generator. The potential energy per coulomb of charge available to electrons that move from one terminal to the other equals the potential difference (voltage) that provides the "electrical pressure" to move electrons through a circuit joined to these terminals.

---

*   The SI symbol for ampere is A. However, an older symbol still in common usage is amp. People often speak of a current of, say, "5 amps."
** A description of the chemical reactions inside dry cells and wet cells can be found in almost any chemistry textbook.

Fig. 34-3 Each coulomb of charge that is made to flow in a circuit that connects the ends of this 1.5-volt flashlight cell is energized with 1.5 joules.

Power utilities use electric generators to provide the 120 volts that is delivered to home outlets. The potential difference between the two holes in the outlet is 120 volts. When the prongs of a plug are inserted into the outlet, an electrical "pressure" of 120 volts is placed across the circuit connected to the prongs. This means that 120 joules of energy is supplied to each coulomb of charge that is made to flow in the circuit.

There is often some confusion between charge flowing *through* a circuit and voltage being impressed *across* a circuit. To distinguish between these ideas, consider a long pipe filled with water. Water will flow *through* the pipe if there is a difference in pressure *across* or between its ends. Water flows from the high-pressure end to the low-pressure end. Only the water flows, not the pressure. Similarly, you say that charges flow *through* a circuit because of an applied voltage *across* the circuit.* You don't say that voltage flows through a circuit. Voltage doesn't go anywhere, for it is the charges that move. Voltage causes current.

34.4 Electrical Resistance

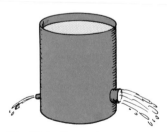

Fig. 34-4 For a given pressure, more water passes through a large pipe than a small one. Similarly, more electric current passes through a large-diameter wire than a small one.

The amount of current that flows in a circuit depends on the voltage provided by the voltage source. It also depends on the resistance that the conductor offers to the flow of charge, or the **electrical resistance**. This is similar to the rate of water flow in a pipe, which depends not only on the pressure behind the water but on the resistance offered by the pipe itself. The resistance of a wire depends on the *conductivity* of the material (that is, how well it conducts) and also on the thickness and length of the wire.

Electrical resistance is less in thick wires. The longer the wire, of course, the greater the resistance. In addition, electrical resistance depends on temperature. The greater the jostling about of atoms within the conductor, the greater resistance the conductor offers to the flow of charge. For most conductors, increased temperature means increased resistance.** The resistance of some metals approaches zero at very low temperatures. These are the superconductors discussed briefly in Chapter 32.

---

* It is conceptually simpler to say that current flows through a circuit, but don't say this around somebody who is "picky" about grammar, for the expression "current flows" is redundant. More properly, charge flows, which *is* current.
** Carbon is an interesting exception. At high temperatures, electrons are shaken from the carbon atom, which increases electric current. Carbon's resistance, in effect, lowers with increasing temperature. This behavior, along with its high melting temperature, accounts for the use of carbon in arc lamps.

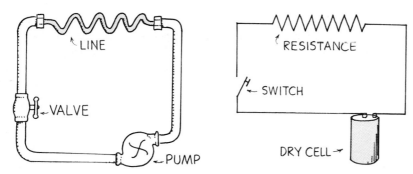

Fig. 34-5 Analogy between a simple hydraulic circuit and an electric circuit.

Electrical resistance is measured in units called ohms,* after Georg Simon Ohm, a German physicist who tested different wires in circuits to see what effect the resistance of the wire had on the current.

34.5 Ohm's Law

Ohm discovered that the amount of current in a circuit is directly proportional to the voltage impressed across the circuit, and is inversely proportional to the resistance of the circuit. In short,

$$\text{current} = \frac{\text{voltage}}{\text{resistance}}$$

This relationship between voltage, current, and resistance is called **Ohm's law**.**

The relationship between the units of measurement for these three quantities is:

$$1 \text{ ampere} = 1\,\frac{\text{volt}}{\text{ohm}}$$

So for a given circuit of constant resistance, current and voltage are proportional. This means that you'll get twice the current for twice the voltage. The greater the voltage, the greater the current. But if the resistance is doubled for a circuit, the current will be half what it would be otherwise. The greater the resistance, the less the current. Ohm's law makes good sense.

---

* The Greek letter omega (Ω) is usually used as a symbol for ohm.
** Many texts use V for voltage, I for current, and R for resistance, and express Ohm's law as $V = IR$. It then follows that $I = V/R$, or $R = V/I$, so if any two variables are known, the third can be found.

Fig. 34-6 Resistors. The stripes are color coded to indicate the resistance in ohms.

Using specific values, a potential difference of 1 volt impressed across a circuit that has a resistance of 1 ohm will produce a current of 1 ampere. If a voltage of 12 volts is impressed across the same circuit, the current will be 12 amperes.

The resistance of a typical lamp cord is much less than 1 ohm, while a typical light bulb has a resistance of about 100 ohms. An iron or electric toaster has a resistance of 15 to 20 ohms. The low resistance permits a large current, which produces considerable heat. Inside electrical devices such as radio and television receivers, the current is regulated by circuit elements called *resistors*, whose resistance may range from a few ohms to millions of ohms.

> ▶ **Questions**
> 1. What is the resistance of an electric frying pan that draws 12 amperes of current when connected to a 120-volt circuit?
>
> 2. How much current is drawn by a lamp that has a resistance of 100 ohms when a voltage of 50 volts is impressed across it?

34.6 | Ohm's Law and Electric Shock

What causes electric shock in the human body—current or voltage? The damaging effects of shock are the result of current passing through your body. From Ohm's law, we can see that this current depends on the voltage applied, and also on the electrical resistance of the human body.

The resistance of your body depends on its condition and ranges from about 100 ohms if you're soaked with salt water to about 500 000 ohms if your skin is very dry. If you touched the

---

▶ **Answers**
1. The resistance is 10 ohms.

   resistance = (voltage)/(current) = (120 volts)/(12 amperes) = 10 ohms

   An electrical device is said to *draw* current when voltage is impressed across it, just as water is said to be drawn from a well or a faucet. In this sense, to draw is not to attract, but to *obtain*.

2. The current is 0.5 ampere.

   current = (voltage)/(resistance) = (50 volts)/(100 ohms) = 0.5 ampere

two electrodes of a battery with dry fingers, the resistance your body would normally offer to the flow of charge would be about 100 000 ohms. You usually would not feel 12 volts, and 24 volts would just barely tingle. If your skin were moist, on the other hand, 24 volts could be quite uncomfortable. Table 34–1 describes the effects of different amounts of current on the human body.

Table 34–1 Effect of Various Electric Currents on the Body	
Current in amperes	Effect
0.001	Can be felt
0.005	Painful
0.010	Involuntary muscle contractions (spasms)
0.015	Loss of muscle control
0.070	Through the heart; serious disruption; probably fatal if current lasts for more than 1 second.

▶ **Questions**

1. If the resistance of your body were 100 000 ohms, how much current would be produced in your body if you touched the terminals of a 12-volt battery?

2. If your skin were very moist so that your resistance was only 1000 ohms, and you touched the terminals of a 24-volt battery, how much current would you draw?

Many people are killed each year by current from common 120-volt electric circuits. If you touch a faulty 120-volt light fixture with your hand while your feet are on the ground, there is a 120-volt "electric pressure" between your hand and the ground. Resistance to current flow is usually greatest between your feet and the ground, so the current is usually not enough to do serious harm. But if your feet and the ground are wet, there is a low-resistance electrical bond between you and the ground.

▶ **Answers**

1. The current in your body would be:

current = (voltage)/(resistance) = (12 V)/(100 000 Ω) = 0.00012 A

2. You would draw (24 V)/(1000 Ω), or 0.024 A, a dangerous amount of current!

Fig. 34-7 Handling a wet hair dryer can be like sticking your fingers into a live socket.

Fig. 34-8 The bird can stand harmlessly on one wire of high potential, but it had better not reach over and grab a neighboring wire! Why not?

Fig. 34-9 The third prong connects the body of the appliance directly to ground. Any charge that builds up on an appliance is therefore conducted to the ground.

Your overall resistance is so lowered that the 120-volt potential difference across your body may produce a harmful current in your body.

Drops of water that collect around the on-off switch of devices such as a hair dryer can conduct current to the user. Although distilled water is a good insulator, the ions in ordinary water greatly reduce the electrical resistance. Dissolved materials, especially small amounts of salt, reduce the resistance even more. There is usually a layer of salt left from perspiration on your skin, which when wet lowers your skin resistance to a few hundred ohms or less. Handling electrical devices while taking a bath is extremely dangerous.

You have seen birds perched on high-voltage wires. Every part of their bodies is at the same high potential as the wire, and they feel no ill effects. For the bird to receive a shock, there must be a *difference* in electric potential between one part of its body and another part. Current will then pass along the path of least electrical resistance connecting these two points.

Suppose you fell from a bridge and managed to grab onto a high-voltage power line, halting your fall. So long as you touch nothing else of different potential, you will receive no shock at all. Even if the wire is several thousand volts above ground potential and even if you hang by it with two hands, no charge will flow from one hand to the other. This is because there is no difference in electric potential between your hands. If, however, you reach over with one hand and grab onto a wire of different potential, **ZAP!!**

Mild shocks occur when the surfaces of electrical appliances are at a different electric potential from the surfaces of other nearby devices. If you touch surfaces of different potentials, you become the pathway to equilibrium. Sometimes the effect is more than mild. To prevent this problem, the outsides of electrical appliances are connected to a ground wire, which is connected to the round third prong of a three-wire electrical plug (Figure 34–9). All ground wires in all plugs are connected together through the wiring system of the house. The two flat prongs are for the current-carrying double wire, part of which is live and the other neutral. If the live wire accidentally comes in contact with the metal surface of an appliance, the current will be directed to ground rather than shocking you if you handle it.

Electric shock overheats tissues in the body or disrupts normal nerve functions. It can upset the nerve center that controls breathing. In rescuing victims, the first thing to do is clear them from the electric supply with a wooden stick or some other nonconductor so that you don't get electrocuted yourself. Then apply artificial respiration.

> ▶ **Question**
> What causes electric shock—current or voltage?

34.7 | Direct Current and Alternating Current

Electric current may be *dc* or *ac*. By *dc*, we mean **direct current**, which refers to a flow of charge that is *always in one direction*. A battery produces direct current in a circuit because the terminals of the battery always have the same sign of charge. Electrons always move through the circuit in the same direction, from the repelling negative terminal and toward the attracting positive terminal. Even if the current moves in unsteady pulses, so long as it moves in one direction only, it is dc.

Alternating current (ac) acts as the name implies. Electrons in the circuit are moved first in one direction and then in the opposite direction, alternating to and fro about relatively fixed positions. This is accomplished by alternating the polarity of voltage at the generator or other voltage source. Nearly all commercial ac circuits in North America involve voltages and currents that alternate back and forth at a frequency of 60 cycles per second. This is 60-hertz current. In some places, 25-hertz, 30-hertz, or 50-hertz current is used.

Fig. 34-10 Direct current (dc) does not change direction over time. Alternating current (ac) cycles back and forth.

The popularity of ac arises from the fact that electric energy in the form of ac can be transmitted great distances with easy voltage step-ups that result in lower heat losses in the wires. Why this is so will be discussed in Chapter 37.

The primary use of electric current, whether dc or ac, is to transfer energy quietly, flexibly, and conveniently from one place to another.

▶ **Answer**
Electric shock *occurs* when current is produced in the body, which is *caused* by an impressed voltage.

34.8 | The Speed of Electrons in a Circuit

When you flip on the light switch on your wall and the circuit is completed, the light bulb appears to glow immediately. When you make a telephone call, the electrical signal carrying your voice travels through the connecting wires at seemingly infinite speed. This signal is transmitted through the conductors at nearly the speed of light. It is *not* the electrons that move at this speed but the signal.

At room temperature, the electrons inside a metal wire have an average speed of a few million kilometers per hour due to their thermal motion. This does not produce a current because the motion is random. There is no flow in any one direction. But when a battery or generator is connected, an electric field is established inside the wire. It is the electric field that travels through a circuit at nearly the speed of light. The electrons continue their random motions while simultaneously being nudged through the wire by the electric field.

The conducting wire acts as a guide or "pipe" to the electric field lines that are established at the voltage source (Figure 34–11). If the voltage source is dc, like the battery shown in Figure 34–11, the electric field lines are maintained in one direction in the conductor.

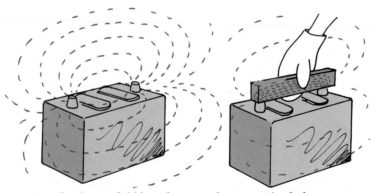

Fig. 34-11   The electric field lines between the terminals of a battery are directed through a conductor, which joins the terminals. A metal bar is shown here, but the conductor is usually an electric circuit.

Conduction electrons are accelerated by the field in a direction parallel to the field lines. Before they gain appreciable speed, they "bump into" the anchored metallic ions in their paths and lose some of their kinetic energy to them. This is why current-carrying wires become hot. These collisions interrupt the motion of the electrons so that their actual *drift speed*, or *net speed* through the wire due to the field, is extremely low. In a typical dc circuit, in the electrical system of an automobile, for example, electrons have a net average drift speed of about 0.01 cm/s. At this rate it would take more than three hours for an electron to travel through 10 meters of wire.

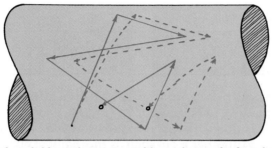

Fig. 34-12   The solid lines depict a possible random path of an electron bouncing off atomic nuclei in a conductor. Instantaneous speeds are about 1/200 the speed of light. The dashed lines show an exaggerated view of how this path may be altered when an electric field is applied. The electron drifts toward the right with an average speed much less than a snail's pace.

In an ac circuit, the conduction electrons don't go anywhere. They oscillate rhythmically to and fro about relatively fixed positions. When you talk to your friend on the telephone, it is the *pattern* of oscillating motion that is carried across town at nearly the speed of light. The electrons already in the wires vibrate to the rhythm of the traveling pattern.

34.9   The Source of Electrons in a Circuit

Some people think that the electrical outlets in the walls of their homes are a source of electrons. They think that electrons flow from the power utility through the power lines and into the wall outlets of their homes. This is not true. The outlets in homes are ac. Electrons do not travel through a wire in an ac circuit, but instead vibrate to and fro about relatively fixed positions.

When you plug a lamp into an outlet, *energy* flows from the outlet into the lamp, not electrons. Energy is carried by the electric field and causes vibratory motion of the electrons that already exist in the lamp filament. If 120 volts are impressed on a lamp, then 120 joules of energy are given to each coulomb of charge that is made to vibrate. Most of this electrical energy is transformed into heat while some of it takes the form of light. Power utilities do not sell electrons. They sell *energy.* *You* supply the electrons.

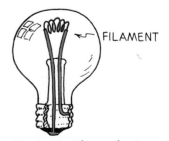

Fig. 34-13   The conduction electrons that surge to and fro in the filament of the lamp do not come from the voltage source. They are in the filament to begin with. The voltage source simply provides them with surges of energy.

Thus, when you are jolted by an electric shock, the electrons making up the current in your body originate in your body. Electrons do not come out of the wire and through your body and into the ground. Energy does. The energy simply causes free electrons in your body to vibrate in unison. Small vibrations tingle. Large vibrations can be fatal.

34.10 | Electric Power

When a charge moves in a circuit, it does work. Usually this results in heating the circuit or in turning a motor. The rate at which work is done, that is, the rate at which electric energy is converted into another form such as mechanical energy, heat, or light is called **electric power**. Electric power is equal to the product of current and voltage.*

$$\text{electric power} = \text{current} \times \text{voltage}$$

If the voltage is expressed in volts and the current in amperes, then the power is expressed in watts. So in units form,

$$1 \text{ watt} = (1 \text{ ampere}) \times (1 \text{ volt})$$

If a lamp rated at 120 watts operates on a 120-volt line, you can see that it will draw a current of 1 ampere, since 120 watts = (1 ampere) × (120 volts). A 60-watt lamp draws 0.5 ampere on a 120-volt line. This relationship becomes a practical matter when you wish to know the cost of electrical energy, which varies from 1 cent to 10 cents per kilowatt-hour depending on locality.

A *kilowatt* is 1000 watts, and a *kilowatt-hour* represents the amount of energy consumed in 1 hour at the rate of 1 kilowatt.**

---

*   Note that this follows from the definitions of current and voltage:

$$\text{current} \times \text{voltage} = \frac{\cancel{\text{charge}}}{\text{time}} \times \frac{\text{energy}}{\cancel{\text{charge}}} = \frac{\text{energy}}{\text{time}} = \text{power}$$

** Since power = (energy)/(time), simple rearrangement gives energy = power × time; hence, energy can be expressed in units of kilowatt-hours.

Physicists measure energy in *joules*, but utility companies customarily sell energy in units of *kilowatt-hours* (kW·h), where 1 kW·h = 3.6 × 10⁶ J. This duplication of units added to an already long list of units unfortunately makes the study of physics more difficult. It will be enough for you to become familiar with and be able to distinguish between the units *coulombs, volts, ohms, amperes, watts, kilowatts,* and *kilowatt-hours* here. Mastering them requires laboratory work and the help of more advanced textbooks. An understanding of electricity takes considerable time and effort, so be patient with yourself if you find this material difficult.

Therefore, in a locality where electric energy costs 5 cents per kilowatt-hour, a 100-watt electric light bulb can be run for 10 hours at a cost of 5 cents, or a half cent for each hour. A toaster or iron, which draws more current and therefore more power, costs several times as much to operate for the same time.

Fig. 34-14 The power and voltage on the light bulb read "60 W 120 V." How much current in amperes will flow though the bulb?

▶ **Questions**

1. How much power is used by a calculator that operates on 8 volts and 0.1 ampere? If it is used for one hour, how much energy does it use?

2. Will a 1200-watt hairdryer operate on a 120-volt line if the current is limited to 15 amperes by a safety fuse? Can two hairdryers operate on this line?

▶ **Answers**

1. Power = current × voltage = (0.1 A) × (8 V) = 0.8 W. If it is used for one hour, then energy = power × time = (0.8 W) × (1 h) = 0.8 watt-hour, or 0.0008 kilowatt-hour.

2. One 1200-watt hairdryer can be operated because the circuit can provide (15 A) × (120 V) = 1800 watts. But there is inadequate power to operate two hairdryers of combined power 2400 watts. This can be seen also from the amount of current involved. Since 1 watt = (1 ampere) × (1 volt), note that (1200 watts)/(120 volts) = 10 amperes; so the hair dryer will operate when connected to the circuit. But two hairdryers on the same plug will require 20 amperes and blow the fuse.

34 | Chapter Review

Concept Summary

Electric current is the flow of electric charge that occurs when there is a potential difference across the ends of an electrical conductor.
- The flow continues until both ends reach a common potential.
- Dry cells, wet cells, and electric generators are voltage sources that maintain a potential difference in a circuit.

The amount of current that flows in a circuit depends on the voltage and the electrical resistance that the conductor offers to the flow of charge.
- An increased temperature or a longer wire increases resistance.
- Increasing the thickness of the wire decreases resistance.

Ohm's law states that the amount of current is directly proportional to the voltage and inversely proportional to the resistance.
- Resistors are used in many electrical devices to control the considerable heat formed by a large current.
- Electric shock is caused by the electric current that passes through the body when there is a voltage difference between two parts of the body.

Direct current (dc) is electric current in which the charge flows in one direction only; electrons in an alternating current (ac) alternate their direction of flow.
- Batteries produce direct current.
- Alternating current allows electrical transmission across great distances.

Electric fields travel through circuits at nearly the speed of light, but the electrons themselves do not.
- In a dc circuit, electrons have a low drift speed within the electric field.

- In ac circuits, energy, not electrons, flows from the outlet; the electrons vibrate rhythmically in fixed positions.

Electric power, the rate at which electric energy is converted into another form of energy, is equal to the product of current and voltage.

Important Terms

alternating current (34.7)
ampere (34.2)
direct current (34.7)
electrical resistance (34.4)
electric current (34.2)
electric power (34.10)
ohm (34.4)
Ohm's law (34.5)
potential difference (34.1)
voltage source (34.3)

Review Questions

1. What condition is necessary for the flow of heat? What analogous condition is necessary for the flow of charge? (34.1)

2. What is meant by the term potential? Potential difference? (34.1)

3. What condition is necessary for the sustained flow of water in a pipe? What analogous condition is necessary for the sustained flow of charge in a wire? (34.1)

4. What is electric current? (34.2)

5. What is an ampere? (34.2)

6. What two devices commonly separate charges to provide an "electrical pressure"? (34.3)

7. How many joules per coulomb are given to charges that flow in a 120-volt circuit? (34.3)

8. Does charge flow through a circuit or into a circuit? (34.3)

9. Does voltage flow through a circuit, or is voltage established across a circuit? (34.3)

10. What is electrical resistance? (34.4)

11. Is electrical resistance greater in a short fat wire or a long thin wire? Explain. (34.4)

12. What is Ohm's law? (34.5)

13. If the voltage impressed across a circuit remains constant while the resistance doubles, what change occurs in the current? (34.5)

14. If the resistance of a circuit remains constant while the voltage across the circuit decreases to half its former value, what change occurs in the current? (34.5)

15. How does wetness affect the resistance of your body? (34.6)

16. Is water usually an insulator or a conductor? Does it vary? (34.6)

17. Why is it that a bird can perch without harm on a high voltage wire? (34.6)

18. What is the function of the third prong in a household electric plug? (34.6)

19. Distinguish between dc and ac. Which is produced by a battery and which is usually produced by a generator? (34.7)

20. Exactly what is it that travels at the speed of light in an electric circuit? (34.8)

21. What is a typical "drift" speed of electrons that make up a current in a typical dc circuit? In a typical ac circuit? (34.8)

22. From where do the electrons originate that flow in a typical electric circuit? (34.9)

23. What is power? (34.10)

24. Which of these is a unit of power and which is a unit of electrical energy: a watt, a kilowatt, or a kilowatt-hour? (34.10)

25. How many amperes flow through a 60 watt bulb when 120 volts are impressed across it? (34.10)

Activity

Batteries are made up of electric cells, which are composed of two unlike pieces of metal separated by a conducting solution. A simple 1.5-volt cell, equivalent to a flashlight cell, can be made by placing a strip of copper and a strip of zinc in a moist vegetable or piece of fruit as shown in Figure A. A lemon or banana works fine. Hold the ends of the strips close together, but not touching, and place the ends on your tongue. The slight tingle you feel and the metallic taste you experience result from a slight current of electricity pushed by the cell through the metal strips when your moist tongue closes the circuit. Try this and compare the results for different metals and different fruits and vegetables.

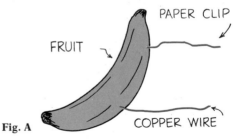

Fig. A

Think and Explain

1. Do an *ampere* and a *volt* measure the same thing, or different things? What are those things, and which is a flow and which is the cause of the flow?

2. Ten coulombs of charge pass a point in 5 seconds. What is the current at that point?

3. A battery does 18 joules of work on 3 cou-

lombs of charge. What voltage does it supply?

4. How much voltage is required to make 1 ampere flow through a resistance of 8 ohms?

5. What is the effect on current in a circuit if both the voltage and the resistance are doubled? If both are halved? Explain.

6. How much current moves through your fingers (resistance = 1200 ohms) when you place them against the terminals of a 6-volt battery?

7. Why do wires heat up when they carry electric current?

8. Why are thick wires rather than thin wires used to carry large currents?

9. Would you expect to find dc or ac in the dome lamp in an automobile? In a lamp in your home? Explain.

10. How many amperes flow in a 60-watt bulb that is rated for 120 volts when it is connected to a 120-volt circuit? How many amperes would flow if it were connected to 240 volts?

11. Use the relationship power = current × voltage to find out how much current is drawn by a 1200-watt hair dryer when it operates on 120 volts. Then use Ohm's law to find the resistance of the hair dryer.

12. The useful life of an automobile battery without recharging is given in terms of ampere-hours. A typical 12-volt battery has a rating of 60 ampere-hours, which means that a current of 60 amperes can be drawn for 1 hour, 30 amperes can be drawn for 2 hours, 15 amperes can be drawn for 4 hours, and so forth. Suppose you forget to turn off the headlights in a parked automobile. If each of the two headlights draws 3 amperes, how long will it be before the battery is "dead"?

35 Electric Circuits

Mechanical things seem to be easier to figure out for most people than electrical things. Maybe this is because most people have had experience playing with blocks and mechanical toys when they were children. If you are among the many who have had far less direct experience with the inner workings of electrical devices, as compared to mechanical gadgets, you are encouraged to put extra effort into the laboratory part of this course. This is because an understanding of electric circuits is helped by hands-on experience. This experience can be fun.

35.1 A Battery and a Bulb

Take apart a flashlight, the ordinary kind shown in Figure 35-1. If you don't have any spare pieces of wire around, cut some strips from some aluminum foil that you probably have in one of your kitchen drawers. Try to light up the bulb with a single battery* and a couple of pieces of wire or foil.

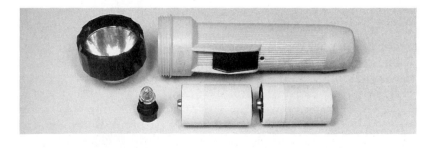

Fig. 35-1 A flashlight taken apart.

---

*   Strictly speaking, a battery consists of two or more cells. What most people call a flashlight battery is more properly called a flashlight dry cell. To conform with popular usage, this chapter uses the term *battery* to mean either a single cell or series of cells.

Some of the ways you *can* light the bulb and some of the ways you *can't* light it are shown in Figure 35-2. The important thing to note is that there must be a complete path, or **circuit**, from the positive terminal at the top of the battery to the negative terminal, which is the bottom of the battery. Electrons flow from the negative part of the battery through the wire or foil to the bottom (or side) of the bulb, through the filament inside the bulb, and out the side (or bottom) and through the other piece of wire or foil to the positive part of the battery. The current then passes through the interior of the battery to complete the circuit.

Fig. 35-2   (a) Unsuccessful ways to light a bulb. (b) Successful ways to light a bulb.

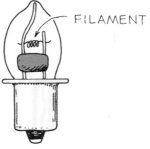

FILAMENT

Fig. 35-3   Electrons do not pile up inside a bulb, but instead flow through its filament.

It is a bit misleading to say that electrons flow "out of" the battery, or "into" the bulb; a better description is to say they flow *through* these devices. The flow of charge in a circuit is analogous to the flow of water in a closed system of pipes. The battery is analogous to a pump, the wires to the pipes, and the bulb to any device that operates when the water is flowing. The water flows through both the pump itself and the circuit it connects. It doesn't "squash up" and concentrate in certain regions, but flows continuously. Electric current behaves the same way.

35.2   Electric Circuits

Any path along which electrons can flow is a circuit. For a continuous flow of electrons, there must be a complete circuit with no gaps. A gap is usually provided by an electric switch that can be opened or closed to either cut off or allow electron flow.

   The water analogy is quite useful for gaining a conceptual understanding of electric circuits, but it does have some limitations. An important one is that a break in a water pipe results in water spilling from the circuit, whereas a break in an electric circuit results in a complete stop in the flow of electricity. Another difference has to do with turning current off and on. When you *close* an electrical switch that connects the circuit, you allow current to flow in much the same way as you allow water to flow by *opening* a faucet. Opening a switch stops the flow of electricity. An electric circuit must be closed for electricity to flow. Opening a water faucet, on the other hand, starts the flow of water. Except for these and some other differences, thinking of electric current in terms of water current is a useful way to study electric circuits.

Most circuits have more than one device that receives electrical energy. These devices are commonly connected in a circuit in two ways, *series* or *parallel*. When connected **in series**, they form a single pathway for electron flow between the terminals of the battery, generator, or wall socket (which is simply an extension of these terminals). When connected **in parallel**, they form branches, each of which is a separate path for the flow of electrons. Both series and parallel connections have their own distinctive characteristics. This chapter briefly treats circuits with these two types of connections.

35.3 Series Circuits

Figure 35-4 shows three lamps connected in series with a battery. This is an example of a simple **series circuit**. When the switch is closed, a current exists almost immediately in all three lamps. The current does not "pile up" in any lamp but flows *through* each lamp. Electrons that make up this current leave the negative terminal of the battery, pass through each of the resistive filaments in the lamps in turn, and then return to the positive terminal of the battery (the same amount of current passes through the battery). This is the only path of the electrons through the circuit. A break anywhere in the path results in an open circuit, and the flow of electrons ceases. Burning out of one of the lamp filaments or simply opening the switch could cause such a break.

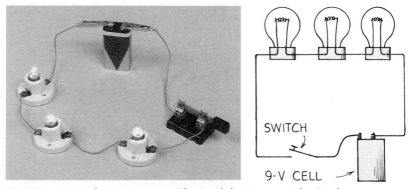

Fig. 35-4 A simple series circuit. The 9-volt battery provides 3 volts across each lamp.

The circuit shown in Figure 35-4 illustrates the following important characteristics of series connections:

1. Electric current has but a single pathway through the circuit. This means that the current passing through the resistance of each electrical device is the same.
2. This current is resisted by the resistance of the first device, the resistance of the second, and the third also, so that the total resistance to current in the circuit is the sum of the individual resistances along the circuit path.
3. The current in the circuit is numerically equal to the voltage supplied by the source divided by the total resistance of the circuit. This follows Ohm's law.
4. The *voltage drop*, or potential difference, across each device is proportional to its resistance. This follows from the fact that more energy is used to move a unit of charge through a large resistance than through a small resistance.
5. The total voltage impressed across a series circuit divides among the individual electrical devices in the circuit so that the sum of the voltage drops across each individual device is equal to the total voltage supplied by the source. This follows from the fact that the amount of energy used to move each unit of charge through the entire circuit equals the sum of the energies used to move that unit of charge through each electrical device in turn.

▶ **Questions**

1. What happens to current in other lamps if one lamp in a series circuit burns out?

2. What happens to the light intensity of each lamp in a series circuit when more lamps are added to the circuit?

It is easy to see the main disadvantage of a series circuit: if one device fails, current in the whole circuit ceases. Some cheap Christmas tree lights are connected in series. When one lamp burns out, it's "fun and games" (or frustration) trying to find which bulb to replace.

▶ **Answers**

1. If one of the lamp filaments burns out, the path connecting the terminals of the voltage source will break and current will cease. All lamps will go out.

2. The addition of more lamps in a series circuit results in a greater circuit resistance. This decreases the current in the circuit and therefore in each lamp, which causes dimming of the lamps. Energy is divided among more lamps so the voltage drop across each lamp will be less.

Most circuits are wired so that it is possible to operate electrical devices independently of each other. In your home, for example, a lamp can be turned on or off without affecting the operation of other lamps or electrical devices. This is because these devices are connected not in series but in parallel to one another.

35.4 | Parallel Circuits

Figure 35-5 shows three lamps connected to the same two points A and B. This is an example of a simple **parallel circuit**. Electrical devices connected in parallel are connected to the same two points of an electric circuit. Electrons leaving the negative terminal of the battery need travel through only *one* lamp filament before returning to the positive terminal of the battery. In this case, current branches into three separate pathways from A to B. A break in any one path does not interrupt the flow of charge in the other paths. Each device operates independently of the other devices.

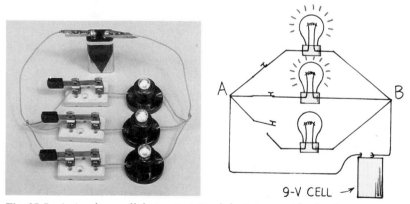

Fig. 35-5   A simple parallel circuit. A 9-volt battery provides 9 volts across each lamp.

The circuit shown in Figure 35-5 illustrates the following major characteristics of parallel connections:

1. Each device connects the same two points A and B of the circuit. The voltage is therefore the same across each device.
2. The total current in the circuit divides among the parallel branches. Current passes more readily into devices of low resistance, so the amount of current in each branch is inversely proportional to the resistance of the branch. This follows Ohm's law.

3. The total current in the circuit equals the sum of the currents in its parallel branches.

4. As the number of parallel branches is increased, the overall resistance of the circuit is *decreased*. Overall resistance is lowered with each added path between any two points of the circuit. This means the overall resistance of the circuit is less than the resistance of any one of the branches.

► **Questions**

1. What happens to the current in other lamps if one of the lamps in a parallel circuit burns out?

2. What happens to the light intensity of each lamp in a parallel circuit when more lamps are added in parallel to the circuit?

35.5 | Schematic Diagrams

Electric circuits are frequently described by simple diagrams, called **schematic diagrams**, that are similar to those of the last two figures. Some of the symbols used to represent certain circuit elements are shown in Figure 35-6. Resistance is shown by a zigzag line, and ideal resistanceless wires are shown with solid straight lines. A single cell battery is represented with a set of short and long parallel lines. The convention is to represent the

---

► **Answers**

1. If one lamp burns out, the other lamps will be unaffected. The current in each branch, according to Ohm's law, is equal to (voltage)/(resistance), and since neither voltage nor resistance is affected in the branches, the current in those branches is unaffected. The total current in the overall circuit (the current through the battery) however, is decreased by an amount equal to the current drawn by the lamp in question before it burned out. But the current in any other single branch is unchanged.

2. The light intensity for each lamp is unchanged as other lamps are introduced (or removed). Only the total resistance and total current in the total circuit changes, which is to say, the current in the battery changes. (There is resistance in a battery also, which we assume is negligible here.) As lamps are introduced, more paths are available between the battery terminals, which effectively decreases total circuit resistance. This decreased resistance is accompanied by an increased current, the same increase that feeds energy to the lamps as they are introduced. Although changes of resistance and current occur for the circuit as a whole, no changes occur in any individual branch in the circuit.

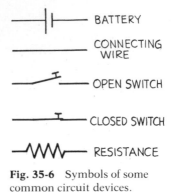

Fig. 35-6 Symbols of some common circuit devices.

positive terminal of the battery with a long line and the negative terminal with a short line. A two-cell battery is represented with a pair of such lines, a three-cell with three, and so on. Figure 35-7 shows schematic diagrams for the circuits of Figures 35-4 and 35-5.

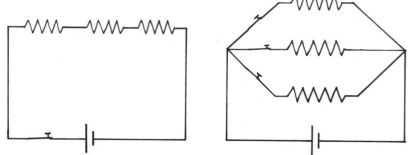

Fig. 35-7 Schematic diagrams. (Left) The circuit of Figure 35-4, with three lamps in series. (Right) The circuit of Figure 35-5, with three lamps in parallel.

35.6   Combining Resistors in a Compound Circuit

Sometimes it is useful to know the *equivalent resistance* of a circuit that has several resistors in its network. The equivalent resistance is the value of the single resistor that would comprise the same load to the battery or power source. The equivalent resistance can be found by the rules for adding resistors in series and parallel. For example, the equivalent resistance for a pair of 1-ohm resistors in series is simply 2 ohms.

The equivalent resistance for a pair of 1-ohm resistors in parallel is 0.5 ohm. (The equivalent resistance is *less* because the current has "twice the path width" when it takes the parallel path. In a similar way, the more doors that are open in an auditorium of people trying to exit, the *less* will be the resistance to migration.) The equivalent resistance for a pair of equal resistors in parallel is half the value of either resistor.

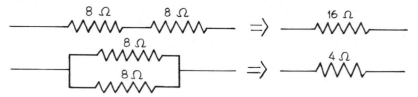

Fig. 35-8 (a) The equivalent resistance of two 8-ohm resistors in series is 16 ohms. (b) The equivalent resistance of two 8-ohm resistors in parallel is 4 ohms.

Figure 35-9 shows a combination of three 8-ohm resistors. The two in parallel are equivalent to a single 4-ohm resistor, which is in series to the 8-ohm resistor and adds to produce an equivalent resistance of 12 ohms. If a 12-volt battery were connected to these resistors, can you see from Ohm's law that the current through the battery would be 1 ampere? (In practice it would be less, for there is resistance inside the battery as well, called the *internal resistance*.)

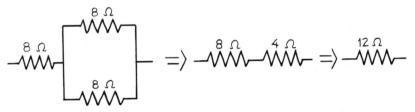

Fig. 35-9   The equivalent resistance of the circuit is found by combining resistors in successive steps.

Two more complex combinations are broken down in successive equivalent combinations in Figures 35-10 and 35-11. It's like a game: combine resistors in series by adding; combine a pair of equal resistors in parallel by halving.* The value of the single resistor left is the equivalent resistance of the combination.

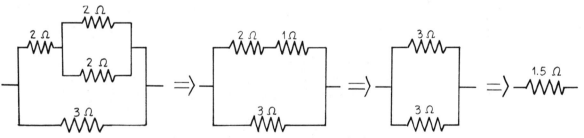

Fig. 35-10   The equivalent resistance of the top branch is 3 ohms, which is in parallel with the 3-ohm resistance of the lower branch. The overall equivalent resistance is 1.5 ohms.

---

*   For a pair of non-equal resistors in parallel, the equivalent resistance is found by taking the product of the pair and dividing by the sum of the pair. That is:

$$R_{equivalent} = \frac{R_1 R_2}{R_1 + R_2}$$

This rule of "product divided by sum" holds only for two resistors in parallel. For three or more parallel resistors, you can do a pair at a time (as is done in Figures 35-10 and 35-11), or use the more general formula:

$$1/R_{equivalent} = 1/R_1 + 1/R_2 + 1/R_3 \text{ and so on}$$

Details can be found in other physics textbooks.

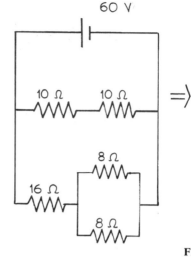 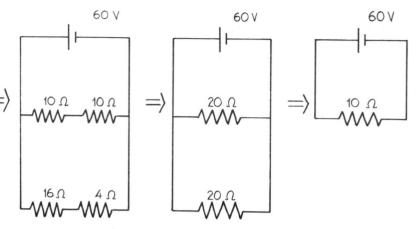

Fig 35-11 Schematic diagrams for an arrangement of various electric devices. The equivalent resistance of the circuit is 10 ohms.

▶ **Questions**
The following questions are based on the schematic diagrams in Figure 35-11.
1. What is the current in amperes through the battery? (Neglect the internal resistance of the battery.)

2. What is the current in amperes through the pair of 10-ohm resistors?

3. What is the current in amperes through each of the 8-ohm resistors?

4. How much power is provided by the battery?

▶ **Answers**
1. The current in the battery (or total current in the circuit) is 6 A. You can get this from Ohm's law: current = (voltage)/(resistance) = (60 V)/(10 ohms) = 6 A. You know that the equivalent resistance of the circuit is 10 ohms from step (d) in the figure.

2. Half the total circuit current, 3 A, will flow through the pair of 10-ohm resistors. You know this because you can see in step (c) of the figure that both branches have equal resistances. This means that the total circuit current will divide equally between the upper and lower branches.

3. The current through the pair of 8-ohm resistors is 3 A, and the current through each is therefore 1.5 A. This is because the 3-A current divides equally through these equal resistances.

4. The battery supplies 360 watts. This is from the relationship

power = current × voltage = (6 A) × (60 V) = 360 watts

This power will be dissipated among all the resistors in the circuit.

35.7 | Parallel Circuits and Overloading

Electricity is usually fed into a home by way of two lead wires called *lines*. These lines are very low in resistance and are connected to wall outlets in each room. About 110 to 120 volts are impressed on these lines by generators at the power utility. This voltage is applied to appliances and other devices that are connected in parallel by plugs to these lines.

As more devices are connected to the lines, more pathways are provided for current. What effect do the additional pathways produce? The answer is, a lowering of the combined resistance of the circuit. Therefore, a greater amount of current occurs in the lines. Lines that carry more than a safe amount of current are said to be *overloaded*. The resulting heat may be sufficient to melt the insulation and start a fire.

You can see how overloading occurs by considering the circuit in Figure 35-12. The supply line is connected to an electric toaster that draws 8 amperes, to an electric heater that draws 10 amperes, and to an electric lamp that draws 2 amperes. When only the toaster is operating and drawing 8 amperes, the total line current is 8 amperes. When the heater is also operating, the total line current increases to 18 amperes (8 amperes to the toaster and 10 amperes to the heater). If you turn on the lamp, the line current increases to 20 amperes. Connecting any more devices increases the current still more.

To prevent overloading in circuits, *fuses* are connected in series along the supply line. In this way the entire line current must pass through the fuse. The fuse shown in Figure 35-13 is constructed with a wire ribbon that will heat up and melt at a given current. If the fuse is rated at 20 amperes, it will pass 20 amperes, but no more. A current above 20 amperes will melt the fuse, which "blows out" and breaks the circuit. Before a blown fuse is replaced, the cause of overloading should be determined and remedied. Often, insulation that separates the wires in a circuit wears away and allows the wires to touch. This effectively shortens the path of the circuit, and is called a *short circuit*. A short circuit draws a dangerously large current because it by-passes the normal circuit resistance.

Circuits may also be protected by *circuit breakers*, which use magnets or bimetallic strips to open the switch. Utility companies use circuit breakers to protect their lines all the way back to the generators. Circuit breakers are used instead of fuses in modern buildings because they do not have to be replaced each time the circuit is opened. Instead, the switch can simply be moved back to the "on" position after the problem has been corrected.

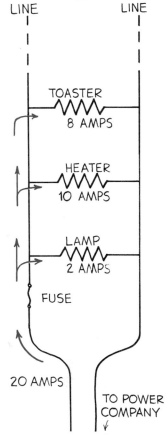

Fig. 35-12 Circuit diagram for appliances connected to a household supply line.

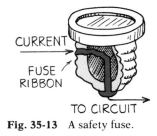

Fig. 35-13 A safety fuse.

35 | Chapter Review

Concept Summary

Any path along which electrons can flow is a circuit.
- A complete circuit is needed to maintain a continuous electron flow.

In a series circuit, electrical devices form a single pathway for electron flow.
- A break anywhere in the path stops the electron flow in the entire circuit.
- The total resistance is equal to the sum of individual resistances along the current path.
- The current is equal to the voltage divided by the total resistance.
- The voltage drop across each device is proportional to its resistance.
- The sum of voltage drops across the resistance of each individual device is equal to the total voltage.

In a parallel circuit, electrical devices form branches, each of which provides a separate path for the flow of electrons.
- Each device connects the same two points of the circuit; the voltage is the same across each device.
- The amount of current in each branch is inversely porportional to the resistance of the branch.
- The total current is equal to the sum of the currents in each branch.

Electric circuits are often described by schematic diagrams, in which each element of the circuit is represented by a symbol.

In a circuit with several resistors, the equivalent resistance is the value of the single resistor that would comprise the same load to the battery or power source.
- For resistors in series, the equivalent resistance is the sum of their values.

- For resistors in parallel, the equivalent resistance is less than the value of any individual resistor.

Lines carrying an unsafe amount of current are overloaded.
- To prevent overloading, fuses are connected in series. Any current above the rating of the fuse will "blow out" the fuse and break the circuit.
- A short circuit is often caused by faulty wire insulation.

Important Terms

circuit (35.1)
in parallel (35.2)
in series (35.2)
parallel circuit (35.4)
schematic diagram (35.5)
series circuit (35.3)

Review Questions

1. Do electrons flow from a battery into a circuit or through both the battery and the circuit it connects? (35.1)

2. Why must there be no gaps in an electric circuit? (35.1)

3. Distinguish between a series circuit and a parallel circuit. (35.2)

4. If three lamps are connected in series to a 6-volt battery, how many volts are impressed across each lamp? (35.3)

5. If one of three lamps blows out when connected in series, what happens to the current in the other two? (35.3)

6. If three lamps are connected in parallel to a 6-volt battery, how many volts are impressed across each lamp? (35.4)

7. If one of three lamps blows out when connected in parallel, what happens to the current in the other two? (35.4)

8. a. In which case will there be more current in each of three lamps—if they are connected to the same battery in series or in parallel?
   b. In which case will there be more voltage across each lamp? (35.4)

9. What happens to the total circuit resistance when more devices are added to a series circuit? To a parallel circuit? (35.6)

10. What is the equivalent resistance of a pair of 8-ohm resistors in series? In parallel? (35.6)

11. Why does the total circuit resistance decrease when more devices are added to a parallel circuit? (35.6)

12. What does it mean to say that lines in a home are overloaded? (35.7)

13. What is the function of fuses in a circuit? (35.7)

14. Why will too many electrical devices operating at one time often blow a fuse? (35.7)

15. What is meant by a short circuit? (35.7)

Think and Explain

1. Sometimes you hear someone say that a particular appliance "uses up" electricity. What is it that the appliance actually "uses up," and what becomes of it?

2. Why are household appliances almost never connected in series?

3. As more and more lamps are connected in series to a flashlight battery, what happens to the brightness of the lamps?

4. As more and more lamps are connected in parallel to a battery, and if the current does not produce heating inside the battery, what happens to the brightness of the lamps?

5. When charge flows in a battery, heating occurs, and battery resistance increases. This lowers the voltage that is supplied to the external circuit. If too many lamps are connected in parallel across a battery, will their brightness diminish? Explain.

6. Consider the combination series and parallel circuit shown in Figure A.
   a. Identify the parallel part of the circuit. What is the equivalent resistance of this part? In other words, what single resistance could take their place and not change the total current from the battery?
   b. What is the equivalent resistance of all the resistors? In other words, what single resistance could take their place without changing the current from the battery?

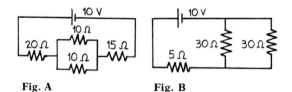

Fig. A   Fig. B

7. What is the current in the battery of the circuit shown in Figure B? (What must you find before you can calculate the current?)

8. The rear window defrosters on automobiles are made up of several strips of heater wire connected in parallel. Consider the case of four wires, each of 6 ohms resistance, connected to 12 volts.
   a. What is the equivalent resistance of the four wires? (Consider the wires to be two groups of two.)
   b. What is the total current drawn?

9. How does the line current compare to the total currents of all devices connected in parallel?

10. Why should you not use a copper penny in place of a safety fuse that blows out?

36 Magnetism

Magnets are fascinating. Bring a pair close together and they snap together and stick. Turn one of the magnets around and they will repel each other. A magnet will stick to a refrigerator door, but it won't stick to an aluminum pan. Magnets come in all shapes and sizes. They are popular as toys, are utilized as compasses, and are essential elements in electric motors and generators. Magnetism is very common to all that you see, for it is an essential ingredient of light itself.

The term *magnetism* stems from certain metallic rocks called *lodestones* found by the early Greeks more than 2000 years ago in the region of Magnesia. In the twelfth century, the Chinese were using them for navigating ships. In the eighteenth century, the French physicist Charles Coulomb demonstrated that they obeyed the inverse-square law.

Magnetism was thought to be independent of electricity until 1820 when a Danish physics professor named Hans Christian Oersted made a remarkable discovery. Oersted discovered in a classroom demonstration that an electric current deflects a compass needle. He was the first to announce that magnetism was related to electricity.* This discovery ushered in a whole new technology, including electric power, radio, and television.

---

* We can only speculate about how often such relationships become evident when they "aren't supposed to" and are dismissed as "something wrong with the apparatus." Oersted, however, had the insight characteristic of a good physicist to see that nature was revealing another of its secrets.

Fig. 36-1 Which interaction has the greater strength—the gravitational attraction between the scrap iron and the earth, or the magnetic attraction between the magnet and the scrap iron?

36.1 | Magnetic Poles

Magnets exert forces on one another. They are similar to electric charges, for they can both attract and repel without touching one another, depending on which ends of the magnets are held near one another. Like electric charges also, the strength of their interaction depends on the distance of separation of the two magnets. Whereas electric charges produce electrical forces, regions called **magnetic poles** produce magnetic forces.

If you suspend a bar magnet from its center by a piece of string, it will act as a compass. The end that points northward is called the *north-seeking pole*, and the end that points southward is called the *south-seeking pole*. More simply, these are called the *north* and *south poles*. All magnets have both a north and a south pole. For a simple bar magnet these are located at the two ends. The common horseshoe magnet is a bar magnet that has been bent, so its poles are also at its two ends.

If the north pole of one magnet is brought near the north pole of another magnet, they repel. The same is true of a south pole

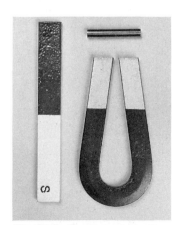

Fig. 36-2 Common magnets.

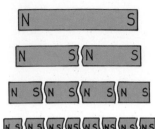

Fig. 36-3 Break a magnet in half and you have two magnets. Break these in half and you have four magnets, each with a north and south pole. Keep breaking the pieces further and further and you find the same results. Magnetic poles exist in pairs.

near a south pole. If opposite poles are brought together, however, attraction occurs.*

Like poles repel; opposite poles attract.

Magnetic poles behave similarly to electric charges in some ways, but there is a very important difference. Whereas electric charges can be isolated, magnetic poles cannot. Electrons and protons are entities by themselves. A cluster of electrons need not be accompanied by a cluster of protons, and vice versa. But a north magnetic pole never exists without the presence of a south pole, and vice versa. The north and south poles of a magnet are like the head and tail of the same coin.

If you break a bar magnet in half, each half still behaves as a complete magnet. Break the pieces in half again, and you have four complete magnets. You can continue breaking the pieces in half and never isolate a single pole. Even when your piece is one atom thick, there are two poles. This suggests that atoms themselves are magnets.

▶ **Question**

Does every magnet necessarily have a north and south pole?

36.2 | Magnetic Fields

Place a sheet of paper over a bar magnet and sprinkle iron filings on the paper. The filings will tend to trace out an orderly pattern of lines that surround the magnet. The space around a magnet, in which a magnetic force is exerted, is filled with a **magnetic field**. The shape of the field is revealed by *magnetic field lines*. Magnetic field lines spread out from one pole, curve around the magnet, and return to the other pole.

▶ **Answer**

Yes, just as every coin has two sides, a "head" and a "tail." Some "trick" magnets, however, have more than one pair of poles, but nevertheless, poles occur in pairs.

---

*   The force of interaction between magnetic poles is given by $F \sim pp'/d^2$, where p and p' represent magnetic pole strengths, and d represents the separation distance between them. Note the similarity of this relationship with Coulomb's law.

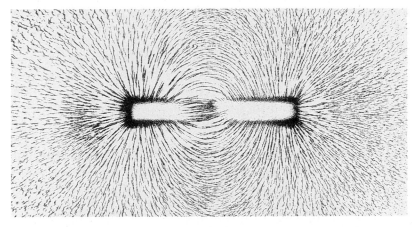

Fig. 36-4   Iron filings trace
out a pattern of magnetic
field lines in the space surrounding the magnet.

The direction of the field outside the magnet is from the north
to the south pole. Where the lines are closer together, the field
strength is greater. We see the magnetic field strength is greater
at the poles. If we place another magnet or a small compass anywhere in the field, its poles will tend to line up with the magnetic field.

Fig. 36-5   Like the iron filings, the compasses line up
with the magnetic field lines.

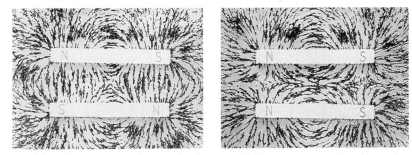

Fig. 36-6   The magnetic field patterns for a pair of magnets when (left) opposite
poles are parallel to each other, and (right) like poles are parallel to each other.

36.3 | The Nature of a Magnetic Field

Magnetism is very much related to electricity. Just as an electric charge is surrounded by an electric field, the same charge is
also surrounded by a magnetic field if it is moving. This is due
to "distortions" in the electric field caused by motion, and was
explained by Albert Einstein in 1905 in his theory of special relativity. This text will not go into the details, except to acknowledge that a magnetic field is a relativistic byproduct of the elec-

tric field. Charges in motion have associated with them both an electric and a magnetic field. A magnetic field is produced by the motion of electric charge.*

Where is the motion of electric charges in a common bar magnet? Although the magnet as a whole may be stationary, it is composed of atoms whose electrons are in constant motion. Electrons behave as if they move in orbits about the atomic nuclei. This moving charge constitutes a tiny current and produces a magnetic field. More important, electrons spin about their own axes like tops. A spinning electron constitutes a charge in motion and thus creates another magnetic field. In most materials, the field due to spinning is predominant over the field due to orbital motion.

Every spinning electron is a tiny magnet. A pair of electrons spinning in the same direction makes up a stronger magnet. A pair of electrons spinning in opposite directions, however, has the opposite effect. The magnetic fields of each cancel the other. This is why most substances are not magnets. In most atoms, the various fields cancel each other because the electrons spin in opposite directions. In materials such as iron, nickel, and cobalt, however, the fields do not cancel each other entirely. Each iron atom has four electrons whose spin magnetism is uncanceled. Each iron atom, then, is a tiny magnet. The same is true to a lesser degree for the atoms of nickel and cobalt.**

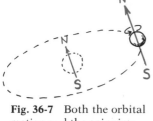

Fig. 36-7 Both the orbital motion and the spinning motion of every electron in an atom produce magnetic fields. These fields combine constructively or destructively to produce the magnetic field of the atom. The resulting field is greatest for iron atoms.

36.4 | Magnetic Domains

The magnetic field of individual iron atoms is so strong that interactions among adjacent iron atoms cause large clusters of them to line up with each other. These clusters of aligned atoms are called **magnetic domains**. Each domain is perfectly magnetized, and is made up of billions of aligned atoms. The domains are microscopic (Figure 36-8), and there are many of them in a crystal of iron.

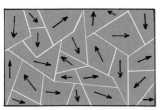

Fig. 36-8 A microscopic view of magnetic domains in a crystal of iron. Each domain consists of billions of aligned iron atoms.

---

* Interestingly enough, since motion is relative, the magnetic field is relative. For example, when a charge moves by you, there is a definite magnetic field associated with the moving charge. But if you move along with the charge so that there is no motion relative to you, you will find no magnetic field associated with the charge. Magnetism is relativistic.
** Most common magnets are made from alloys containing iron, nickel, cobalt, and aluminum in various proportions. In these the electron spin contributes virtually all the magnetic properties. In the rare-earth metals, such as gadolinium, the orbital motion is more significant.

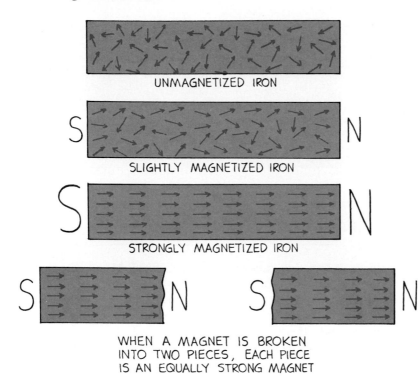

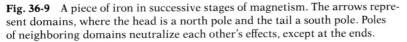

Fig. 36-9 A piece of iron in successive stages of magnetism. The arrows represent domains, where the head is a north pole and the tail a south pole. Poles of neighboring domains neutralize each other's effects, except at the ends.

The difference between a piece of ordinary iron and an iron magnet is the alignment of domains. In a common iron nail, the domains are randomly oriented. When a strong magnet is brought nearby, two effects take place. One is a growth in size of domains that are oriented in the direction of the magnetic field. This growth is at the expense of domains that are not aligned. The other effect is a rotation of domains as they are induced into alignment. The domains become aligned much as electric charges in a piece of paper are aligned in the presence of a charged rod. When you remove the nail from the magnet, ordinary thermal motion causes most or all of the domains in the nail to return to a random arrangement.

Permanent magnets are made by simply placing pieces of iron or certain iron alloys in strong magnetic fields. Alloys of iron differ; soft iron is easier to magnetize than steel. It helps to tap the iron to nudge any stubborn domains into alignment. Another way of making a permanent magnet is to stroke a piece of iron with a magnet. The stroking motion aligns the domains in the iron. If a permanent magnet is dropped or heated, some of the domains are jostled out of alignment and the magnet becomes weaker.

Fig. 36-10 The iron nails become induced magnets.

> ▶ **Questions**
> 1. How can a magnet attract a piece of iron that is not magnetized?
>
> 2. The iron filings sprinkled on the paper that covers the magnet in Figure 36-4 were not initially magnetized. Why, then, do they line up with the magnetic field of the magnet?

36.5 Electric Currents and Magnetic Fields

A moving charge produces a magnetic field. Many charges in motion—an electric current—also produce a magnetic field. The magnetic field that surrounds a current-carrying conductor can be demonstrated by arranging an assortment of magnetic compasses around a wire (Figure 36-11) and passing a current through it. The compasses line up with the magnetic field produced by the current and show it to be a pattern of concentric circles about the wire. When the current reverses direction, the compasses turn completely around, showing that the direction of the magnetic field changes also.

Fig. 36-11 (Left) When there is no current in the wire, the compasses align with the earth's magnetic field. (Right) When there is a current in the wire, the compasses align with the stronger magnetic field about the wire. The magnetic field about the wire forms concentric circles.

▶ **Answers**
1. Domains in the unmagnetized piece of iron are induced into alignment by the magnetic field of the nearby permanent magnet. See the similarity of this with Figure 32-12 back in Chapter 32. Like the pieces of paper, pieces of iron will jump to a strong magnet when it is brought nearby. But unlike the paper, they are not repelled. Can you think of the reason why?

2. Domains align in the individual filings, causing them to act like tiny compasses. The poles of each "compass" are pulled in opposite directions, producing a torque that twists each filing into alignment with the external magnetic field.

If the wire is bent into a loop, the magnetic field lines become bunched up inside the loop (Figure 36-12). If the wire is bent into another loop, overlapping the first, the concentration of magnetic field lines inside the double loop is twice as much as in the single loop. It follows that the magnetic field intensity in this region is increased as the number of loops is increased. The magnetic field intensity is appreciable for a current-carrying coil of wire with many loops.

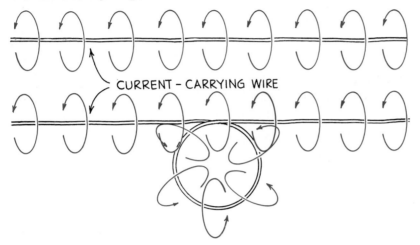

CURRENT – CARRYING WIRE

Fig. 36-12   Magnetic field lines about a current-carrying wire crowd up when the wire is bent into a loop.

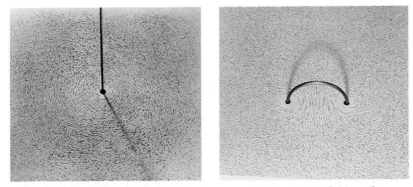

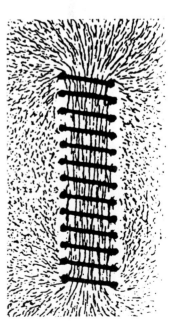

Fig. 36-13   Iron filings sprinkled on paper reveal the magnetic field configurations about (left) a current-carrying wire, (center) a current-carrying loop, and (right) a coil of loops.

If a piece of iron is placed in a current-carrying coil of wire, the magnetic domains in the iron are induced into alignment. This further increases the magnetic field intensity, and we have an **electromagnet**.

Powerful electromagnets, some cooled to temperatures low enough for them to become superconducting, are used for research purposes. These magnets are used to control charged particle beams in high-energy accelerators. More impressively, electromagnets can levitate and propel high-speed trains.

Fig. 36-14 Conventional trains vibrate as they ride on rails at high speeds. This Japanese magnetically levitated train is capable of vibration-free high speeds, even in excess of 200 km/h.

36.6 | Magnetic Forces on Moving Charged Particles

A charged particle at rest will not interact with a static magnetic field. But if the charged particle *moves* in a magnetic field, the magnetic character of its motion becomes evident. It experiences a deflecting force.* The force is greatest when the particle moves in a direction perpendicular to the magnetic field lines. At other angles, the force is less; it becomes zero when the particle moves parallel to the field lines. In any case, the direction of the force is always perpendicular to both the magnetic field lines and the velocity of the charged particle (Figure 36-15). So a moving charge is deflected when it crosses magnetic field lines but not when it travels parallel to the field.

---

* When particles of electric charge q and speed v move perpendicular to a magnetic field of strength B, the force F on each particle is simply the product of the three variables: $F = qvB$. For non-perpendicular angles, v in this relationship must be the component of velocity perpendicular to the field.

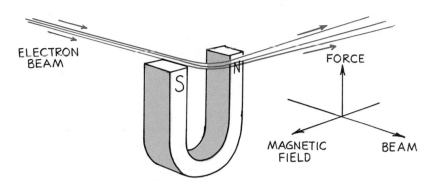

Fig. 36-15 A beam of electrons is deflected by a magnetic field.

This sideways deflecting force is very different from the forces that occur in other interactions, such as the force of gravitation between masses, the electrostatic force between charges, and the force between magnetic poles. The force that acts on a moving charged particle does not act in a direction between the sources of interaction, but instead acts perpendicular to both the magnetic field and the electron beam.

It's nice that charged particles are deflected by magnetic fields, for this fact is employed to spread electrons onto the inner surface of a TV tube and provide a picture. More importantly, charged particles from outer space are deflected by the earth's magnetic field. The harmful cosmic rays bombarding the earth's surface would be much more intense otherwise.

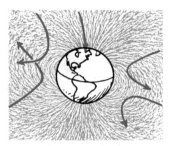

Fig. 36-16 The magnetic field of the earth deflects many charged particles that make up cosmic radiation.

36.7 Magnetic Forces on Current-Carrying Wires

Simple logic tells you that if a charged particle moving through a magnetic field experiences a deflecting force, then a current of charged particles moving through a magnetic field also experiences a deflecting force. If the particles are trapped inside a wire when they respond to the deflecting force, the wire will also move (Figure 36-17).

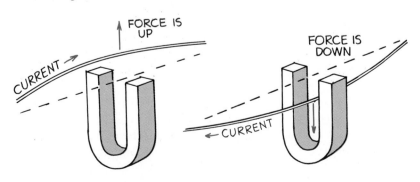

Fig. 36-17 A current-carrying wire experiences a force in a magnetic field. (Can you see that this is a simple extension of Figure 36-15?)

If the direction of current is reversed, the deflecting force acts in the opposite direction. The force is maximum when the current is perpendicular to the magnetic field lines. The direction of force is along neither the magnetic field lines nor the direction of current. The force is perpendicular to both field lines and current. It is a sideways force.

So just as a current-carrying wire will deflect a magnetic compass, as discovered by Oersted in his high school classroom in 1820, a magnet will deflect a current-carrying wire. Both cases show different effects of the same phenomenon. This discovery created much excitement, for almost immediately people began harnessing this force for useful purposes—with great sensitivity in electric meters, and with great force in electric motors.

36.8   Meters to Motors

The simplest meter to detect electric current is shown in Figure 36-18. It consists of a magnetic needle on a pivot at the center of a number of loops of insulated wire. When an electric current passes through the coil, each loop produces its own effect on the needle so that a very small current can be detected. A current-indicating instrument is called a *galvanometer*.

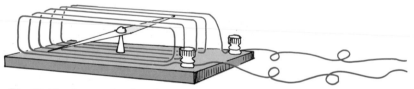

Fig. 36-18   A very simple galvanometer.

A more common design is shown in Figure 36-19. It employs more loops of wire and is therefore more sensitive. The coil is mounted for movement and the magnet is held stationary. The coil turns against a spring, so the greater the current in its loops, the greater its deflection.

A galvanometer may be calibrated to measure current (amperes), in which case it is called an *ammeter*. Or it may be calibrated to measure electric potential (volts), in which case it is called a *voltmeter*.

If the design of the galvanometer is slightly modified, you have an electric motor. The principal difference is that the current is made to change direction every time the coil makes a half revolution. After it has been forced to rotate one half revolution,

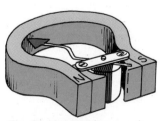

Fig. 36-19   A common galvanometer design.

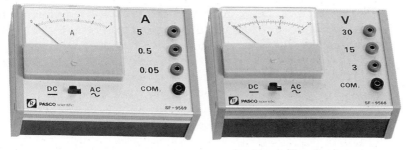

Fig. 36-20   Both the ammeter and the voltmeter are basically galvanometers. (The electrical resistance of the instrument is made to be very low for the ammeter, and very high for the voltmeter.)

it overshoots just in time for the current to reverse, whereupon it is forced to continue another half revolution, and so on in cyclic fashion to produce continuous rotation.

A simple dc motor is shown in bare outline in Figure 36-21. A permanent magnet is used to produce a magnetic field in a region where a rectangular loop of wire is mounted so that it can turn about an axis as shown. When a current passes through the loop, it flows in opposite directions in the upper and lower sides of the loop. (It has to do this because if charge flows into one end of the loop, it must flow out the other end.) If the upper portion of the loop is forced to the left, then the lower portion is forced to the right, as if it were a galvanometer. But unlike a galvanometer, the current is reversed during each half revolution by means of stationary contacts on the shaft. The parts of the wire that brush against these contacts are called *brushes*. In this way, the current in the loop alternates so that the forces in the upper and lower regions do not change directions as the loop rotates. The rotation is continuous as long as current is supplied.

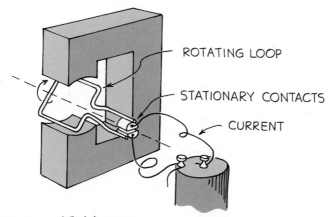

ROTATING LOOP

STATIONARY CONTACTS

CURRENT

Fig. 36-21   A simplified dc motor.

Larger motors, dc or ac, are usually made by replacing the permanent magnet by an electromagnet that is energized by the power source. Of course, more than a single loop is used. Many loops of wire are wound about an iron cylinder, called an *armature*, which then rotates when energized with electric current.

Needless to say, the advent of electric motors saw the replacement of enormous human and animal toil the world over. Electric motors have greatly changed the way that people live.

> ▶ **Question**
> How is a galvanometer similar to a simple electric motor? How is each fundamentally different?

36.9 The Earth's Magnetic Field

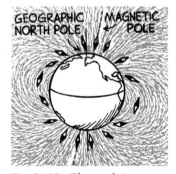

Fig. 36-22 The earth is a magnet.

A compass points northward because the earth itself is a huge magnet. The compass aligns with the magnetic field of the earth. The magnetic poles of the earth, however, do not coincide with the geographical poles, nor are they very close to the geographical poles. The magnetic pole in the Northern Hemisphere, for example, is located nearly 1800 kilometers from the geographical North Pole, somewhere in the Hudson Bay region of northern Canada. The other magnetic pole is located south of Australia (Figure 36-22). This means that compasses do not generally point to true north. The discrepancy between the orientation of a compass and true north is known as the *magnetic declination*.

It is not known exactly why the earth itself is a magnet. The configuration of the earth's magnetic field is like that of a strong bar magnet placed near the center of the earth. But the earth is not a magnetized chunk of iron like a bar magnet. It is simply too hot for individual atoms to remain aligned.

A better candidate for the earth's magnetic field are the currents in the molten part of the earth's core. About 2000 km below the outer rocky mantle (which itself is almost 3000 km thick) lies

---

▶ **Answer**
Both a galvanometer and a motor are similar in that they both employ coils positioned in a magnetic field. When a current passes through the coils, forces on the wires rotate the coils. The fundamental difference is that the maximum rotation of the coil in a galvanometer is one half turn, whereas in a motor the coil (armature) rotates through many complete turns. In the armature of a motor, the current is made to alternate with each half turn of the armature.

the molten part, which surrounds the solid center. Most earth scientists think that moving charges looping around within the earth create the magnetic field. Because of the earth's great size, the speed of moving charges would have to be less than one millimeter per second to account for the field.

Another candidate for the earth's magnetic field is convection currents that result from heat rising from the central core (Figure 36-23). Perhaps such convection currents combined with the rotational effects of the earth produce the earth's magnetic field. A firmer explanation awaits more study.

Whatever the cause, the magnetic field of the earth is not stable, but has wandered throughout geological time. Evidence of this comes from analysis of the magnetic properties of rock strata. Iron atoms in a molten state tend to align themselves with the earth's magnetic field. When the iron solidifies, the direction of the earth's field is recorded by the orientation of the domains in the rock. The slight magnetism that results can be measured with sensitive instruments. As samples of rock from different strata formed throughout geologic time are tested, the magnetic field of the earth for different periods can be charted. The evidence from the rock shows that there have been times when the magnetic field of the earth has diminished to zero and then reversed itself.

More than twenty reversals have taken place in the past 5 million years. The most recent occurred 700 000 years ago. Prior reversals happened 870 000 and 950 000 years ago. Studies of deep sea sediments indicate that the field was virtually switched off for 10 000 to 20 000 years just over 1 million years ago. This was the time that modern humans emerged.

We cannot predict when the next reversal will occur because the reversal sequence is not regular. But there is a clue in recent measurements that show a decrease of over 5% of the earth's magnetic field strength in the last 100 years. If this change is maintained, we may well have another magnetic field reversal within 2000 years.

Fig. 36-23 Convection currents in the molten parts of the earth's interior may produce the earth's magnetic field.

36   Chapter Review

Concept Summary

Magnetic forces are produced by north and south magnetic poles.
- Like poles repel; opposite poles attract.
- North and south poles always occur in pairs.

A magnetic field is produced by the motion of electric charge.
- In magnetic substances such as iron, the magnetic fields created by spinning electrons do not cancel each other out; large clusters of magnetic atoms align to form magnetic domains.
- In nonmagnetic substances, electron pairs within the atoms spin in opposite directions; there is no net magnetic field.

An electric current produces a magnetic field.
- Bending a current-carrying wire into coils intensifies the magnetic field.
- Placing a piece of iron into a current-carrying coil creates an electromagnet.

A charged particle may be deflected by a magnetic field.
- Deflection is greatest for particles moving perpendicular to the magnetic field, and zero for particles moving parallel to the field.

An electric current is also deflected by a magnetic field.
- The force is maximum when the current is perpendicular to the field.
- Galvanometers, ammeters, voltmeters, and electric motors are based on this effect.

The earth itself is a magnet; its magnetic poles are almost 2000 km from its geographical poles.

Important Terms

electromagnet (36.5)
magnetic domain (36.4)
magnetic field (36.2)
magnetic pole (36.1)

Review Questions

1. What do electric charges have in common with magnetic poles? (36.1)

2. What is a major difference between electric charges and magnetic poles? (36.1)

3. What is a magnetic field, and what is its source? (36.2)

4. Every spinning electron is a tiny magnet. Since all atoms have spinning electrons, why are not all atoms tiny magnets? (36.3)

5. What is so special about iron that makes each iron atom a tiny magnet? (36.3)

6. What is a magnetic domain? (36.4)

7. Why do some pieces of iron behave as magnets, and other pieces of iron do not? (36.4)

8. How can a piece of iron be induced into becoming a magnet? For example, if you place a paper clip near a magnet, it will itself become a magnet. Why? (36.4)

9. Why will dropping or heating a magnet weaken it? (36.4)

10. What is the shape of the magnetic field that surrounds a current-carrying wire? (36.5)

11. If a current-carrying wire is bent into a loop, why is the magnetic field stronger inside the loop than outside? (36.5)

12. What must a charged particle be doing in order to experience a magnetic force? (36.6)

13. With respect to an electric and magnetic field, how does the direction of a magnetic force on a charged particle differ from the direction of the electrical force? (36.6)

14. What role does the earth's magnetic field play in cosmic ray bombardment? (36.6)

15. How does the direction in which a current-carrying wire is forced when in a magnetic field compare to the direction that moving charges are forced? (36.7)

16. How do the concepts of force, field, and current relate to a galvanometer? (36.8)

17. Why is it important that the current in the armature of a motor periodically change direction? (36.8)

18. What is meant by magnetic declination? (36.9)

19. According to most geophysicists, what may be the probable causes of the earth's magnetic field? (36.9)

20. What are magnetic pole reversals, and what is the evidence that the earth's magnetic field has undergone pole reversals throughout its history? (36.9)

Think and Explain

1. What kind of force field surrounds a stationary electric charge? A moving electric charge?

2. Why can iron be made to behave as a magnet and wood not?

3. Since iron filings are not themselves magnets, by what mechanism do they align themselves with the field of a magnet, as shown in Figure 36-6?

4. A strong magnet and a weak magnet attract each other. Which magnet exerts the stronger force—the strong one or the weak one? (Could you have answered this way back in Chapter 5?)

5. Why will the magnetic field strength be further increased inside a current-carrying coil if a piece of iron is placed in the coil?

6. Magnetic fields can be used to trap plasmas in "magnetic bottles," but the plasma must be moving. Why?

7. A cyclotron is a device for accelerating charged particles in ever-increasing circular orbits to high speeds. The charged particles are subjected to both an electric field and a magnetic field. One of these fields increases the speed of the particles, and the other field holds them in a circular path. Which field performs which function?

8. A magnetic field can deflect a beam of electrons, but it cannot do work on them to speed them up. Why? (*Hint:* Consider the direction of the force relative to the direction in which the electrons move.)

9. In what direction relative to a magnetic field does a charged particle travel in order to experience maximum magnetic force? Minimum magnetic force?

10. Pigeons have multiple-domain magnetite magnets within their skulls that are connected with a large number of nerves to the pigeon's brain. How does this aid the pigeon in navigation? (Magnetic material also exists in the abdomens of bees; none, however, has been found in humans.)

11. What changes in cosmic ray intensity at the earth's surface would you expect during periods in which the earth's magnetic poles passed through a zero phase while undergoing reversals? (A widely held theory supported by fossil evidence, is that the periods of no protective magnetic field may have been as effective in changing life forms as X rays have been in the famous heredity studies of fruit flies.)

37 Electromagnetic Induction

The discovery that magnetism could be produced with electrical wires was a turning point in physics and the technology that followed. The question arose as to whether electricity could be produced from magnetism. At that time in the early nineteenth century, the only current-producing devices were voltaic cells, which produced small currents by dissolving expensive metals in acids. These were the forerunners of our present-day batteries. A major alternative to these crude devices was discovered independently in 1831 by two physicists: Michael Faraday in England and Joseph Henry in the United States. Their discovery was to change the world by making electricity so commonplace that it would power industries by day and light up cities by night.

37.1 Electromagnetic Induction

Faraday and Henry both discovered that electric current could be produced in a wire by simply moving a magnet in or out of a wire coil (Figure 37-1). No battery or other voltage source was needed—only the motion of a magnet in a coil or in a single wire loop. They discovered that voltage was induced by the relative motion between a wire and a magnetic field.

The production of voltage depends only on the relative motion between the conductor and the magnetic field. Voltage is induced whether the magnetic field of a magnet moves past a stationary conductor, or the conductor moves through a stationary magnetic field (Figure 37-2). The results are the same whether either or both move.

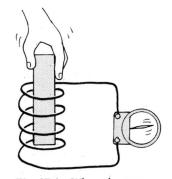

Fig. 37-1 When the magnet is plunged into the coil, charges in the coil are set in motion; voltage is induced in the coil.

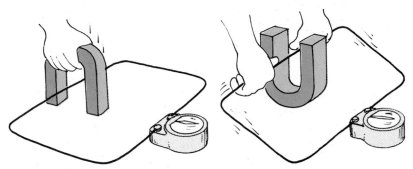

Fig. 37-2 Voltage is induced in the wire loop whether the magnetic field moves past the wire or the wire moves through the magnetic field.

The amount of voltage induced depends on how quickly the magnetic field lines are traversed by the wire. Very slow motion produces hardly any voltage at all. Quick motion induces a greater voltage.

The greater the number of loops of wire that move in a magnetic field, the greater the induced voltage and the greater the current in the wire (Figure 37-3). Pushing a magnet into twice as many loops will induce twice as much voltage; pushing it into ten times as many loops will induce ten times as much voltage; and so on.

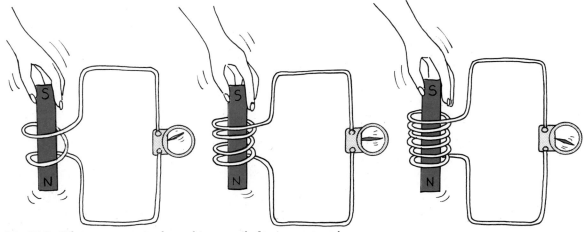

Fig. 37-3 When a magnet is plunged into a coil of twice as many loops as another, twice as much voltage is induced. If the magnet is plunged into a coil with three times as many loops, then three times as much voltage is induced.

Fig. 37-4 It is more difficult to push the magnet into a coil with more loops because the magnetic field of each current loop resists the motion of the magnet.

It may seem that we get something (energy) for nothing by simply increasing the number of loops in a coil of wire, but we don't. It is more difficult to push the magnet into a coil with more loops. Think of it this way: Each additional current loop is an additional electromagnet to resist the motion of your magnet. There is a repulsion between the magnet and the electromagnet that is induced. You do more work to induce more voltage (Figure 37-4).

The amount of voltage induced depends on how quickly the magnetic field changes. Very slow movement of the magnet into the coil produces hardly any voltage at all. Quick motion induces a greater voltage.

It doesn't matter which moves—the magnet or the coil. It is the relative motion between the coil and the magnetic field that induces voltage. It so happens that any change in the magnetic field around a conductor induces a voltage. This phenomenon of inducing voltage by changing the magnetic field around a conductor is called **electromagnetic induction**.

37.2 Faraday's Law

Electromagnetic induction can be summarized in a statement that is called **Faraday's law**:

> The induced voltage in a coil is proportional to the product of the number of loops and the rate at which the magnetic field changes within those loops.

Voltage is one thing, current is another. The amount of current produced by electromagnetic induction depends not only on the induced voltage, but on the resistance of the coil and the circuit that it connects.* For example, you can plunge a magnet in and out of a closed rubber loop and in and out of a closed loop of copper. The voltage induced in each is the same, providing each intercepts the same number of magnetic field lines. But the current in each is quite different—a lot in the copper but almost none in the rubber. The electrons in the rubber sense the same voltage as those in the copper, but their bonding to the fixed atoms prevents the movement of charge that occurs so freely in the copper.

---

* Current also depends on the "reactance" of the coil. Reactance is similar to resistance and is important in ac circuits; it depends on the number of loops in the coil and on the frequency of the ac source, among other things. This complication will not be treated in this book.

> ▶ **Question**
>
> If you push a magnet into a coil, as shown in Figure 37-4, you'll feel a resistance to your push. Why is this resistance greater in a coil with more loops?

37.3   Generators and Alternating Current

If a magnet is plunged in and out of a coil of wire, the induced voltage alternates in direction. As the magnetic field strength inside the coil is increased (magnet entering), the induced voltage in the coil is directed one way. When the magnetic field strength diminishes (magnet leaving), the voltage is induced in the opposite direction. The greater the frequency of field change, the greater the induced voltage. The frequency of the induced alternating voltage is equal to the frequency of the changing magnetic field within the loop.

Rather than moving the magnet, it is more practical to move the coil. This is best accomplished by rotating the coil in a stationary magnetic field (Figure 37-5). This arrangement is called a **generator**. It is essentially a motor running backward. Whereas a motor converts electric energy into mechanical energy, a generator converts mechanical energy into electric energy.

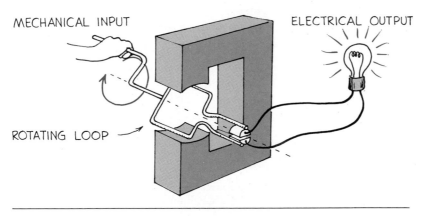

MECHANICAL INPUT

ELECTRICAL OUTPUT

ROTATING LOOP

Fig. 37-5 A simple generator. Voltage is induced in the loop when it is rotated in the magnetic field.

▶ **Answer**

Simply put, more work is required to induce the greater voltage induced by more loops. You can also look at it this way: When the magnetic fields of two magnets (electro or permanent) overlap, the two magnets are either forced together or forced apart. When one of the fields is induced by motion of the other, the polarity of the fields is always such as to force the magnets apart. This produces the resistive force you feel. Inducing more current in more coils simply increases the induced magnetic field and hence the resistive force.

When the loop of wire is rotated in the magnetic field, there is a change in the number of magnetic field lines within the loop (Figure 37-6). In sketch *a* the loop has the largest number of lines inside it. As the loop rotates (sketch *b*), it encircles fewer of the field lines until it lies along the field lines (sketch *c*) and encloses none at all. As rotation continues, it encloses more field lines (sketch *d*) and reaches a maximum when it has made a half revolution (sketch *e*). As rotation continues, the magnetic field inside the loop changes in cyclic fashion.

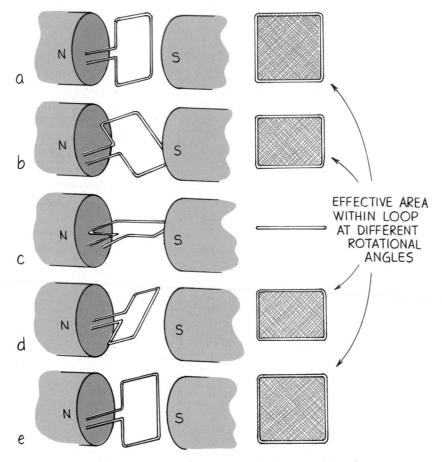

Fig. 37-6   As the loop rotates, there is a change in the number of magnetic field lines it encloses. It varies from a maximum (a) to a minimum (c) and back to a maximum again (e).

The voltage induced by the generator alternates, and the current produced is alternating current (ac). It changes magnitude and direction periodically (Figure 37-7). The standard alternating current in North America changes its magnitude and direction during 60 complete cycles per second—60 hertz.

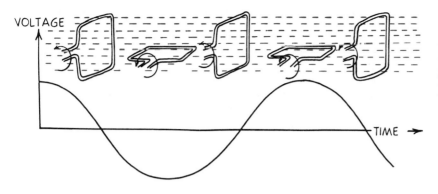

Fig. 37-7 As the loop rotates, the magnitude and direction of the induced voltage (and current) change. One complete rotation of the loop produces one complete cycle in voltage (and current).

The generators used in power plants are much more complex than the model discussed here. Huge coils made up of many loops of wire are wrapped in an iron core, to make an armature much like the armature of a motor. They rotate in the very strong magnetic fields of powerful electromagnets. The armature is connected externally to an assembly of paddle wheels called a *turbine*. Energy from wind or falling water can produce rotation of the turbine, but those of most commercial generators are driven by moving steam. The steam itself requires an energy source, which is usually fossil or nuclear fuel.

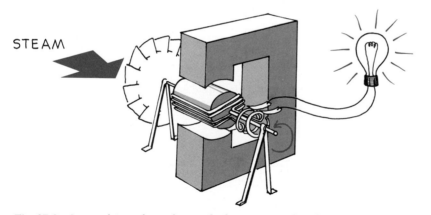

Fig. 37-8 Steam drives the turbine, which is connected to the armature of the generator.

It is important to emphasize that an energy source of some kind is required to operate a generator. Energy from the source, usually some type of fuel, is converted to mechanical energy to drive the turbine, and the generator converts most of this to electric energy. The electricity that is produced simply carries this energy to distant places. Some people think that electricity is a *source* of energy. It is not. It is a *form* of energy that must have a source.

37.4   Motor and Generator Comparison

Chapter 36 discussed how an electric current is deflected in a magnetic field, which underlies the operation of the motor. This discovery occurred ten years before Faraday and Henry discovered electromagnetic induction, which underlies the operation of a generator. Both of these discoveries, however, stem from the same single fact: Moving charges experience a force that is perpendicular to both their motion and the magnetic field they traverse (Figure 37-9). We will call the deflected wire the *motor effect* and the law of induction the *generator effect*. Each of these effects is summarized in the figure. Study them. Can you see that the two effects are related?

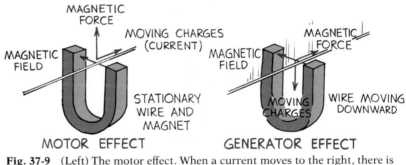

Fig. 37-9   (Left) The motor effect. When a current moves to the right, there is a perpendicular upward force on the electrons. Since there is no conducting path upward, the wire is tugged upward along with the electrons. (Right) The generator effect. When a wire with no initial current is moved downward, the electrons in the wire experience a deflecting force perpendicular to their motion. There *is* a conducting path in this direction, and the electrons follow it, thereby constituting a current.

37.5   Transformers

Consider a pair of coils, side by side (Figure 37-10). One is connected to a battery and the other is connected to a galvanometer. It is customary to refer to the coil connected to the power source as the *primary* (input), and the other as the *secondary* (output). As soon as the switch is closed in the primary and current passes through its coil, a current occurs in the secondary also—even though there is no material connection between the two coils. Only a brief surge of current occurs in the secondary, however. Then when the primary switch is opened, a surge of current again registers in the secondary but in the opposite direction.

The explanation is that the magnetic field that builds up around the primary extends to the secondary coil. Changes in the magnetic field of the primary are sensed by the nearby secondary. These changes of magnetic field intensity at the secondary induce voltage in the secondary, in accord with Faraday's law.

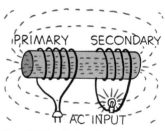

Fig. 37-10 Whenever the primary switch is opened or closed, voltage is induced in the secondary circuit.

> ▶ **Question**
>
> When the switch of the primary in Figure 37-10 is opened or closed, the galvanometer in the secondary registers a current. But when the switch remains closed, no current is registered on the galvanometer of the secondary. Why?

If we place an iron core inside the primary and secondary coils of the arrangement of Figure 37-10, the magnetic field about the primary is intensified by the alignment of magnetic domains in the iron. The magnetic field is also concentrated in the core which extends into the secondary, so the secondary intercepts more of the field change. The galvanometer will show greater surges of current when the switch of the primary is opened or closed.

Instead of opening and closing a switch to produce the change of magnetic field, suppose that alternating current is used to power the primary. Then the rate at which the magnetic field changes in the primary (and hence in the secondary) is equal to the frequency of the alternating current. Now we have a **transformer** (Figure 37-11).

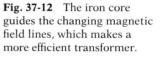

Fig. 37-11 A simple transformer arrangement.

A more efficient arrangement is shown in Figure 37-12, where the iron core forms a complete loop to guide all the magnetic field lines through the secondary. All the magnetic field lines in the primary are intercepted by the secondary.

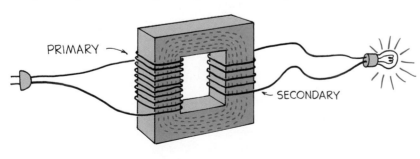

Fig. 37-12 The iron core guides the changing magnetic field lines, which makes a more efficient transformer.

▶ **Answer**

When the switch remains in the closed position, there is a steady current in the primary, and a steady magnetic field about the coil. This field extends to the secondary, but unless there is a *change* in the field, electromagnetic induction does not occur.

Voltages may be stepped up or stepped down with a transformer. To see how, consider the simple case shown in sketch *a* of Figure 37-13. Suppose the primary consists of one loop connected to a 1-V alternating source. Consider the symmetrical arrangement of a secondary of one loop that intercepts all the changing magnetic field lines of the primary. Then a voltage of 1 V is induced in the secondary.

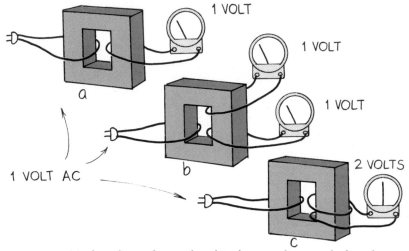

Fig. 37-13   (a) The voltage of 1 V induced in the secondary equals the voltage of the primary. (b) A voltage of 1 V is induced in the added secondary also because it intercepts the same magnetic field change from the primary. (c) The voltages of 1 V each induced in the two one-turn secondaries are equivalent to a voltage of 2 V induced in a single two-turn secondary.

If another loop is wrapped around the core so the transformer has two secondaries (sketch *b*), it intercepts the same magnetic field change. A voltage of 1 V is induced in it too. There is no need to keep both secondaries separate, for we could join them (sketch *c*) and still have a total induced voltage of 1 V + 1 V, or 2 V. This is equivalent to saying that a voltage of 2 V will be induced in a single secondary that has twice the number of loops as the primary. So twice as much voltage will be induced in a secondary that has twice as many loops as the primary.

If the secondary is wound with three times as many loops, or *turns* as they are called, then three times as much voltage will be induced. If the secondary has a hundred times as many turns as the primary, then a hundred times as much voltage will be induced, and so on. This arrangement of a greater number of turns on the secondary than on the primary makes up a *step-up transformer*. Stepped-up voltage may light a neon sign or operate the picture tube in a television receiver.

If the secondary has fewer turns than the primary, the alternating voltage produced in the secondary will be *lower* than that in the primary. The voltage is said to be stepped down. If the secondary has half as many turns as the primary, then only half as much voltage is induced in the secondary. Stepped-down voltage may safely operate a toy electric train.

So electric energy can be fed into the primary at a given alternating voltage and taken from the secondary at a greater or lower alternating voltage, depending on the relative number of turns in the primary and secondary coil windings.

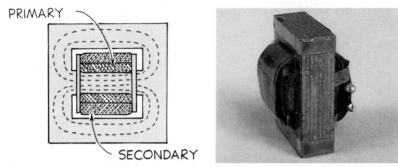

Fig. 37-14 A practical transformer.

The relationship between primary and secondary voltages with respect to the relative number of turns is:

$$\frac{\text{primary voltage}}{\text{number of primary turns}} = \frac{\text{secondary voltage}}{\text{number of secondary turns}}$$

It might seem that you get something for nothing with a transformer that steps up the voltage. But this is not without cost, which is increased current drawn by the primary. The transformer actually transfers energy from one coil to the other. The rate at which energy is transferred is the power. The power used in the secondary is supplied by the primary. The primary gives no more power than the secondary uses, in accord with the conservation of energy. If the slight power losses due to heating of the core are neglected, then:

power into primary = power out of secondary

Electric power is equal to the product of voltage and current, so we can say

$(\text{voltage} \times \text{current})_{\text{primary}} = (\text{voltage} \times \text{current})_{\text{secondary}}$

The ease with which voltages can be stepped up or down with a transformer is the principal reason that most electric power is ac rather than dc.

▶ **Questions**

The following questions refer to a transformer with 100 turns in the primary and 200 turns in the secondary.

1. If a voltage of 100 V is put across the primary, what will be the voltage output in the secondary?

2. The secondary is connected to a floodlamp with a resistance of 50 ohms. Assuming the answer to the last question is 200 V, what will be the current in the secondary circuit?

3. What is the power in the secondary coil?

4. What is the power in the primary coil?

5. What is the current drawn by the primary coil?

6. The voltage has been stepped up, and the current has been stepped down. Ohm's law says that increased voltage will produce increased current. Is there a contradiction here, or does Ohm's law not apply to transformers?

37.6 | Power Transmission

Almost all electric energy sold today is in the form of alternating current because of the ease with which it can be transformed from one voltage to another. Power is transmitted great distances at high voltages and correspondingly low currents, a process that otherwise would result in large energy losses owing to the heating of the wires. Power may be carried from power plants

---

▶ **Answers**

1. From (100 V)/(100 primary turns) = (? V)/(200 secondary turns), you can see that the secondary puts out 200 V.

2. From Ohm's law, (200 volts)/(50 ohms) = 4 A.

3. Power = (200 V) × (4 A) = 800 watts.

4. By the conservation of energy, the power in the primary is the same, 800 watts.

5. Power = 800 watts = (100 V) × (? A), so the primary must draw 8 A. (Note that the voltage is stepped up from primary to secondary and that the current is correspondingly stepped down.)

6. Ohm's law still holds and there is no contradiction. The voltage induced across the secondary circuit, divided by the load (resistance) of the secondary circuit, equals the current in the secondary circuit. The current is stepped down in comparison to the larger current that is drawn in the *primary* circuit.

to cities at about 120 000 volts or more, stepped down to about 2200 volts in the city, and finally stepped down again to provide the 120 volts used in household circuits.

Fig. 37-15 Power transmission.

Energy, then, is transformed from one system of conducting wires to another by electromagnetic induction. The same principles account for eliminating wires and sending energy from a radio-transmitter antenna to a radio receiver many kilometers away, and to the transformation of energy of vibrating electrons in the sun to life energy on earth. The effects of electromagnetic induction are very far-reaching.

37.7 | Induction of Electric and Magnetic Fields

Electromagnetic induction has thus far been discussed in terms of the production of voltages and currents. Actually, the more fundamental electric and magnetic fields underlie both voltages and currents. The modern view of electromagnetic induction holds that electric and magnetic *fields* are induced, which in turn give rise to the voltages we have considered. Induction takes place whether or not a conducting wire or any material medium is present. In this more general sense, Faraday's law states

> An electric field is induced in any region of space in which a magnetic field is changing with time. The magnitude of the induced electric field is proportional to the rate at which the magnetic field changes. The direction of the induced electric field is at right angles to the changing magnetic field.

There is a second effect, which is the counterpart to Faraday's law. It is the same as Faraday's law, only the roles of electric and magnetic fields are interchanged. It is one of the many symmetries in nature. This effect was advanced by the British physicist James Clerk Maxwell in about 1860. According to Maxwell:

A magnetic field is induced in any region of space in which an electric field is changing with time. The magnitude of the induced magnetic field is proportional to the rate at which the electric field changes. The direction of the induced magnetic field is at right angles to the changing electric field.

These statements are two of the most important statements in physics. They underlie an understanding of electromagnetic waves.

37.8   Electromagnetic Waves

Shake the end of a stick back and forth in still water and you will produce waves on the water surface. Similarly shake a charged rod to and fro in empty space and you will produce electromagnetic waves in space. This is because the shaking charge can be considered an electric current. What surrounds an electric current? The answer is a magnetic field. What surrounds a *changing* electric current? The answer is, a changing magnetic field. What do we know about a changing magnetic field? The answer is, it will induce a changing electric field, in accord with Faraday's law. What do we know about a changing electric field? The answer is, in accord with Maxwell's counterpart to Faraday's law, the changing electric field will induce a changing magnetic field.

An electromagnetic wave is composed of vibrating electric and magnetic fields that regenerate each other. No medium is required. The vibrating fields emanate (move outward) from the vibrating charge. At any point on the wave, the electric field is perpendicular to the magnetic field, and both are perpendicular to the direction of motion of the wave (Figure 37-17).

Fig. 37-16   Shake a charged object to and fro and you produce electromagnetic waves.

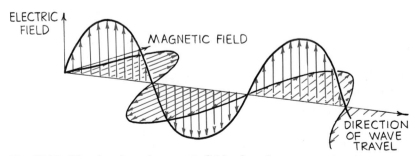

Fig. 37-17   The electric and magnetic fields of an electromagnetic wave are perpendicular to each other.

The intensity of the induced fields very much depends on this speed of emanation. Let's see why this is so.

The magnitude of each induced field depends not only on the vibrational rate, but on the *motion* of the other field—on the speed at which the other field emanates from the vibrating charge. The higher the speed, the greater the magnitude of the field that is induced. A high-speed changing magnetic field induces a stronger electric field than a low-speed changing magnetic field, and vice versa. If the speed of emanation is low, the strength of the induced field is weak. Too low a speed would mean the mutual induction would die out. But what of the energy of the fields in this case? The fields contain energy acquired from the vibrating charge. If the fields disappeared with no means of transferring energy to some other form, energy would be destroyed. So low-speed emanation of electric and magnetic fields is incompatible with the law of energy conservation.

At emanation speeds too high, on the other hand, the fields would be induced to greater and greater magnitudes, with a crescendo of ever increasing energies—again clearly a no-no with respect to energy conservation. At some critical speed, however, mutual induction would continue indefinitely, with neither a loss nor a gain in energy.

From his equations of electromagnetic induction, Maxwell calculated the value of this critical speed and found it to be 300 000 kilometers per second. But this is the speed of light! Maxwell quickly realized that he had discovered the solution to one of the greatest mysteries of the universe—the nature of light. If the electric charge is set into vibration within the incredible frequency range from 4.3×10^{14} to 7×10^{14} vibrations per second, the resulting electromagnetic wave will activate the "electrical antennae" in the retina of the eye. Light is simply electromagnetic waves in this range of frequencies! The lower frequency appears red, and the higher frequency appears violet.

On the evening of Maxwell's discovery, he had a date with a young lady he was later to marry. While walking in a garden, his date remarked about the beauty and wonder of the stars. Maxwell asked how she would feel to know that she was walking with the only person in the world who knew what the starlight really was. For it was true. At that time, James Clerk Maxwell was the only person in the world to know that light of any kind is energy carried in waves of electric and magnetic fields that continually regenerate each other.

37 | Chapter Review

Concept Summary

Electromagnetic induction is the inducing of voltage by changing the magnetic field in a conductor.

- According to Faraday's law, the induced voltage in a coil is proportional to the product of the number of loops and the rate at which the magnetic field within the loops changes.
- A generator uses electromagnetic induction to convert mechanical energy to electric energy.
- A transformer uses electromagnetic induction to induce a voltage in the secondary that is different from that in the primary coil.

Electromagnetic induction may be described in terms of fields.

- A changing magnetic field induces an electric field.
- A changing electric field induces a magnetic field.
- Electromagnetic waves are composed of vibrating electric and magnetic fields that regenerate each other.

Important Terms

electromagnetic induction (37.1)
Faraday's law (37.2)
generator (37.3)
transformer (37.5)

Review Questions

1. What did Michael Faraday and Joseph Henry discover? (37.1)

2. How can voltage be induced in a wire with the help of a magnet? (37.1)

3. A magnet moved into a coil of wire will induce voltage in the coil. What is the effect of moving a magnet into a coil with more loops? (37.1)

4. Why is it more difficult to move a magnet into a coil of more loops? (37.1)

5. Current, as well as voltage, is induced in a wire by electromagnetic induction. Why is Faraday's law expressed in terms of induced voltage, and not induced current? (37.2)

6. How does the frequency of the changing magnetic field compare to the frequency of the alternating voltage that is induced? (37.3)

7. What is a generator, and how does it differ from a motor? (37.3)

8. Why is alternating voltage induced in the rotating armature of a generator? (37.3)

9. The armature of a generator must rotate in order to induce voltage and current. What causes the rotation? (37.3)

10. A motor is characterized by three main ingredients: magnetic field, moving charges, and magnetic force. What are the three main ingredients that characterize a generator? (37.4)

11. How can a change in voltage in a coil of wire (the primary) be transferred to a neighboring coil of wire (the secondary) without physical contact? (37.5)

12. Why does an iron core that extends inside and connects the primary and secondary pair of coils intensify electromagnetic induction? (37.5)

13. What does a transformer actually transform—voltage, current, or energy? (37.5)

14. What does a step-up transformer step up—voltage, current, or energy? (37.5)

15. How does the relative number of turns on the primary and secondary coil in a transformer affect the step-up or step-down voltage factor? (37.5)

16. If the number of secondary turns is ten times the number of primary turns, and the input voltage to the primary is 6 volts, how many volts will be induced in the secondary coil? (37.5)

17. a. In a transformer, how does the power input to the primary coil compare to the power output of the secondary coil?
    b. How does the product of voltage and current in the primary compare to the product of voltage and current in the secondary? (37.5)

18. Why is it advantageous to transmit electrical power long distances at high voltages? (37.6)

19. What fundamental quantity underlies the concepts of voltages and currents? (37.7)

20. Distinguish between Faraday's law expressed in terms of fields and Maxwell's counterpart to Faraday's law. How are the two laws symmetrical? (37.7)

21. In what way is the speed of electromagnetic waves consistent with the law of energy conservation? (37.8)

22. What is light? (37.8)

Think and Explain

1. How is the amount of current induced in a loop of wire affected by the frequency of an oscillating magnetic field?

2. Why is a generator armature more difficult to rotate when it is connected to and supplying electric current to a circuit?

3. Some bicycles have electric generators that are made to turn when the bike wheel turns. These generators provide energy for the bike's lamp. Will a cyclist coast farther if the lamp connected to the generator is turned off? Explain.

4. Why does a transformer require alternating voltage?

5. Can an efficient transformer step up energy? Defend your answer.

6. A portable tape deck requires 12 volts to operate correctly. A transformer nicely allows the device to be powered from a 120-volt outlet. If the primary has 500 turns, how many turns should the secondary have?

7. A model electric train requires a low voltage to operate. If the primary coil of its transformer has 400 turns, and the secondary has 40 turns, how many volts will power the train when the primary is connected to a 120-volt household circuit?

8. An induction coil in an automobile is actually a form of transformer that boosts 12 volts to about 24 000 volts. What is the ratio of secondary to primary turns? (The 12 volts in a car is dc. In order to make the induction coil operate, the dc must be interrupted frequently. This is controlled by a switch in the distributor.)

9. If a car made of iron and steel moves over a wide closed loop of wire embedded in a road surface, will the magnetic field of the earth in the loop be altered? Will this produce a current pulse? (Can you think of a practical application of this?)

10. What is wrong with this scheme? To generate electricity without fuel, arrange a motor to run a generator that will produce electricity that is stepped up with a transformer so that the generator can run the motor while furnishing electricity for other uses.

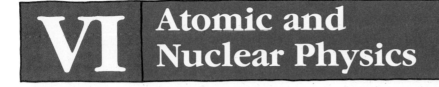

VI Atomic and Nuclear Physics

WHENEVER YOU MARVEL AT THE BEAUTY OF A NATURAL GEYSER SUCH AS OLD FAITHFUL, OR ENJOY THE WARMTH OF A NATURAL HOT SPRING, REMEMBER THAT THEIR HEAT COMES FROM NUCLEAR POWER--- THE RADIOACTIVITY OF MINERALS IN THE EARTH'S CORE. POWER FROM THE ATOMIC NUCLEUS HAS BEEN WITH US SINCE THE EARTH FORMED AND IS NOT RESTRICTED TO TODAY'S NUCLEAR REACTORS, OR "NUKES," AS THEY ARE CALLED. THIS LAST UNIT OF THE BOOK LAYS IMPORTANT GROUNDWORK IN THESE TIMES--- "KNOW NUKES!"

571

38 The Atom and the Quantum

The final unit of this book is about the realm of the unimaginably tiny atom. This chapter investigates atomic structure, which is revealed by analyzing light. Light has a dual nature, which in turn radically alters our understanding of the atomic world. The next chapter covers the structure of the atomic nucleus and radioactivity, and the concluding chapter is about the nuclear processes of fission and fusion.

38.1 Models

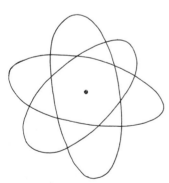

Fig. 38-1 The old planetary model of an atom with electrons orbiting like little planets around a tiny sun.

Nobody knows what an atom looks like, or even if it makes sense to suppose it has an appearance. To visualize the processes that occur at the subatomic realm, we construct models. In the planetary model—the one that most people think of when they picture an atom—the electrons orbit the nucleus like planets going around the sun. This was an early model of the atom suggested by the Danish physicist Niels Bohr in 1913. We still tend to think in terms of this simple picture, even though it has been replaced several times over by models that give progressively better results but become progressively more abstract. Old models sometimes steer us off track in our investigations of nature, and sometimes they provide a scaffolding that allows us to progress to more complicated models. Today's models of the atom, for example, have given way to mathematical theories that cannot be represented by pictures.

A useful model of the atom must be consistent with a model for light. All light has its source in the motion of electrons within the atom. Down through the centuries there have been two primary models of light: the particle model and the wave model. Isaac Newton believed in a particle model of light. He thought that light was composed of a hail of tiny particles. Christian Huygens stated that light was a wave phenomenon. About a century later this was reinforced when Thomas Young demonstrated interference. Later, Maxwell proposed and Hertz proved that light is a part of the electromagnetic spectrum. This seemed to verify the wave nature of light once and for all. But Maxwell's electromagnetic wave model was not the last word on the nature of light. In 1905 Albert Einstein resurrected the particle theory of light.

38.2   Light Quanta

Einstein viewed light as being made up of a hail of tiny particles—concentrated bundles of electromagnetic energy. Einstein extended the idea of a German physicist, Max Planck, who a few years earlier had proposed that energy in an atom is not continuous, but occurs in little chunks he called **quanta** (plural of *quantum*). Planck suggested that when an atom emits light, the energy of the atom changes by quantized amounts. Einstein carried this idea further and proposed that light is composed of quanta also. He called these quanta **photons**.

A quantum is an elemental unit—a smallest amount of something. The idea that certain quantities are quantized—that they come in discrete (separate) units—was known in Einstein's time. Matter is quantized. The mass of a gold ring, for example, is equal to some whole-number multiple of the mass of a single gold atom. Electricity is quantized, as all electric charge is some whole-number multiple of the charge of a single electron.

The newer physics tells us that other quantities are also quantized—quantities such as energy and angular momentum. The energy in a light beam is quantized and comes in packets, or quanta; only a whole number of quanta can exist. The quanta of light, or of electromagnetic radiation in general, are the photons.

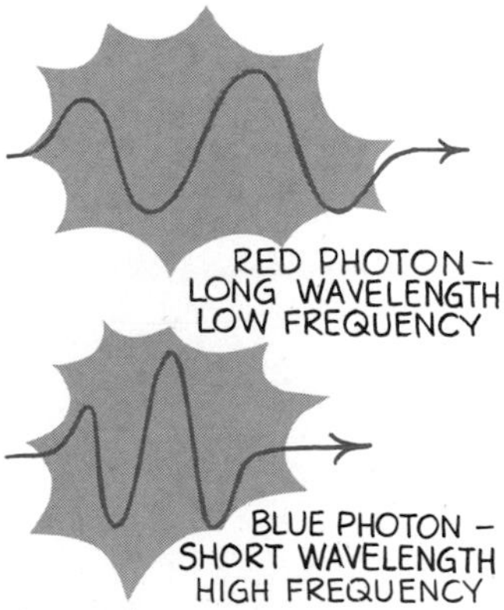

The energy of a photon is directly proportional to its frequency. When the energy E of a photon is divided by its frequency f, the quantity that results is always the same, no matter what the frequency. This quantity is a constant known as **Planck's constant**, h. The energy of every photon is therefore

$$E = hf$$

This equation gives the smallest amount of energy that can be converted to light of frequency f. Light is not emitted continuously, but is emitted as a stream of photons, each with an energy hf.

RED PHOTON—
LONG WAVELENGTH
LOW FREQUENCY

BLUE PHOTON —
SHORT WAVELENGTH
HIGH FREQUENCY

Fig. 38-2 The energy of a photon of light is proportional to its vibrational frequency.

38.3   The Photoelectric Effect

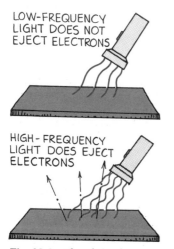

LOW-FREQUENCY
LIGHT DOES NOT
EJECT ELECTRONS

HIGH-FREQUENCY
LIGHT DOES EJECT
ELECTRONS

Fig. 38-3   The photoelectric effect depends on the frequency of light.

Einstein was led to the quantum theory of light by his study of the photoelectric effect. The **photoelectric effect** is the ejection of electrons from certain metals when light falls upon them. These metals are said to be *photosensitive* (that is, sensitive to light). This effect is used in electric eyes, in the photographer's light meter, and in the sound tracks of motion pictures.

Early investigators knew that high-frequency light, even from a dim source, was capable of ejecting electrons from a photosensitive metal surface; yet low-frequency light, even from a very bright source, could not dislodge electrons. They knew that bright light carried more energy than dim light. It was thought that very bright red light, for example, should dislodge electrons more easily than dim blue or violet light could. But this was not the case. Only light of high frequencies was capable of supplying sufficient energy to eject electrons from the metals.

Einstein explained the photoelectric effect by thinking of light in terms of photons. The absorption of a photon by an atom in the metal surface is an all-or-nothing process. One and only one photon is completely absorbed by each electron ejected from the metal. This means that the *number* of photons that hit the metal has nothing to do with whether a given electron will be ejected. If the energy per photon is too small, then the brightness or intensity of light does not matter. The critical factor is the frequency or color of the light. A few photons of blue or violet light can eject a few electrons, but hordes of red or orange photons cannot eject a single electron. Only high-freqency photons have the concentrated energy needed for ejection.

A wave has a broad front, and its energy is spread out along this front. For the energy of a light wave to be concentrated enough to eject a single electron from a metal surface is as unlikely as for an ocean wave to hit a beach and knock a single seashell far inland with an energy equal to the energy of the whole wave. The photoelectric effect suggests that we think of light as a succession of particle-like photons, rather than as a continous train of waves. The number of photons in a light beam controls the brightness of the beam, while the energy of each photon is proportional to the frequency of the light.

Experimental verification of Einstein's explanation of the photoelectric effect was made 11 years later by the American physicist Robert Millikan. Every aspect of Einstein's interpretation was confirmed. It was for this (and not for his theory of relativity) that Einstein received the Nobel prize.

▶ **Questions**
1. Will brighter light eject more electrons from a photosensitive surface than dimmer light of the same frequency?

2. Will high-frequency light eject a greater number of electrons than low-frequency light?

38.4 | Waves as Particles

Figure 38-4 is a striking example of the particle nature of light. The photograph was taken with exceedingly feeble light. Each frame shows the image progressing photon by photon. Note also that the photons seem to strike the film in an independent and random manner.

a   b   c

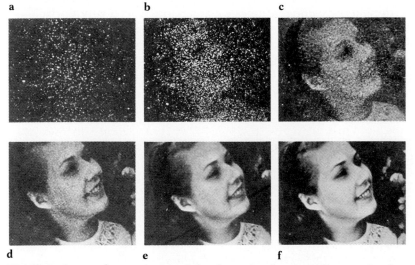

d   e   f

Fig. 38-4 Stages of exposure reveal the photon-by-photon production of a photograph. The approximate numbers of photons at each stage were (a) 3×10^3, (b) 1.2×10^4, (c) 9.3×10^4, (d) 7.6×10^5, (e) 3.6×10^6, (f) 2.8×10^7.

▶ **Answers**
1. Yes (if the frequency is great enough for any electrons to be ejected). The number of ejected electrons depends on the number of incident photons.

2. Not necessarily. The answer is yes if electrons are ejected by the high-frequency light but not by the low-frequency light, because its photons do not have enough energy. If the light of both frequencies can eject electrons, then the number of electrons ejected depends on the brightness of the light, not on its frequency.

38.5 | Particles as Waves

If light can have particle properties, cannot particles have wave properties? This question was posed by the French physicist Louis de Broglie in 1924, while he was still a student. His answer gave him a Ph.D. in physics, and later won him the Nobel prize in physics.

De Broglie suggested that all matter could be viewed as having wave properties. All particles—electrons, protons, atoms, bullets, and even humans—have a wavelength that is related to the momentum of the particles by

$$\text{wavelength} = \frac{h}{\text{momentum}}$$

where h is, lo and behold, Planck's constant again. The wavelength of a particle is called the *de Broglie wavelength*. A particle of large mass and ordinary speed has too small a wavelength to be detected by conventional means. However, a tiny particle—such as an electron—moving at high speed has a detectable wavelength.* It is smaller than the wavelength of visible light but large enough for noticeable diffraction. A beam of electrons, interestingly enough, behaves like a beam of light. It can be diffracted and undergoes wave interference under the same conditions that light does (Figure 38-5).

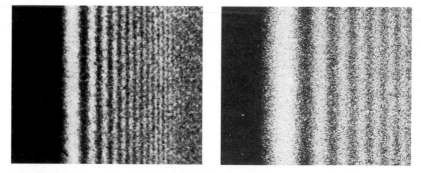

Fig. 38-5 Fringes produced by the diffraction (left) of an electron beam and (right) of a light.

---

* A bullet of mass 0.10 kg traveling at 330 m/s, for example, has a de Broglie wavelength of $(6.6 \times 10^{-34} \text{ J·s})/[(0.02 \text{ kg}) \times (330 \text{ m/s})] = 10^{-34}$ m, an incredibly small size that is a million million million millionth the diameter of a hydrogen atom. An electron traveling at 2 percent of the speed of light, on the other hand, has a wavelength of 10^{-10} m, which is equal to the diameter of the hydrogen atom. Diffraction effects for electrons are measurable, whereas diffraction effects for bullets are not.

An electron microscope makes practical use of the wave nature of electrons. The wavelength of electron beams is typically thousands of times shorter than the wavelength of visible light, so the electron microscope is able to distinguish detail not possible with optical microscopes (Figure 38-6).

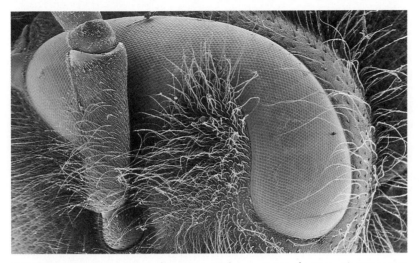

Fig. 38-6   Detail of a wasp eye as seen with a scanning electron microscope at a "low" magnification of 200 times.

38.6   Electron Waves

More far-reaching than the diffraction of electrons is de Broglie's model of matter waves in the atom. The planetary model of the atom developed by Niels Bohr was useful in explaining the atomic spectra of the elements. It explained why elements emitted only certain frequencies of light. An electron has different amounts of energy when it is in different orbits around the nucleus. From an energy point of view, an electron is said to be in different *energy levels* when it is in different orbits. The electrons in an atom normally occupy the lowest energy levels available.

An electron is boosted by various means to higher energy levels. As it returns to its stable level, it emits a photon. The energy of the photon is exactly equal to the difference in the energy levels in the atom. The characteristic pattern of lines in the spectrum of an element corresponds to electron transitions between the energy levels characteristic of the atoms of that element. By examining spectra, physicists were able to determine the various energy levels in the atom. This was a tremendous triumph for atomic physics.

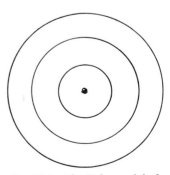

Fig. 38-7   The Bohr model of the atom.

One of the difficulties of this model of the atom, however, was reconciling why electrons occupied certain energy levels in the atom. Evidence was that they occupied one level or the other, but never a level in between. The orbital paths of electrons in the atom seemed to be discrete. That is, they were at definite distances from the atomic nucleus.

This seemed strange. There is nothing to suggest, as a comparative example, that a body couldn't orbit at any distance from the sun. There are no places in the solar system where orbiting is forbidden. Why, then, were planetary orbits in the atom so different? The planetary model of the atom was on shaky ground. This was because the electron was considered to be a particle, a tiny BB whirling around the nucleus like a planet whirling around the sun.

Why the electron occupies only discrete levels is understood by considering the electron to be not a particle but a *wave*. According to de Broglie's theory of matter waves, an orbit exists where an electron wave closes in on itself in phase. In this way it reinforces itself constructively in each cycle, just as the standing wave on a music string is constructively reinforced by its successive reflections. In this view, the electron is visualized not as a particle located at some point in the atom, but as though its mass and charge were spread out into a standing wave surrounding the nucleus. The wavelength of the electron wave must fit evenly into the circumferences of the orbits (Figure 38-8).

Fig. 38-8 (a) Orbital electrons form standing waves only when the circumference of the orbit is equal to a whole-number multiple of wavelengths. (b) When the wave does not close in on itself in phase, it undergoes destructive interference.

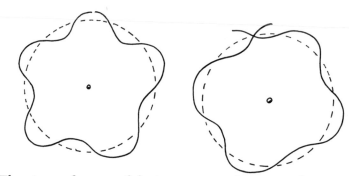

The circumference of the innermost orbit is equal to one wavelength of the electron wave. The second orbit has a circumference of two electron wavelengths, the third three, and so on (Figure 38-9). This is similar to a "chain necklace" made of paper clips. No matter what size necklace is made, its circumference is equal to some multiple of the length of a single paper clip.*

---

*   Electron wavelengths are successively longer for orbits of increasing radii; so for a more accurate analogy, the construction of longer necklaces requires using not only *more* paper clips, but *larger* paper clips as well.

Since the circumferences of electron orbits are discrete, it follows that the radii of these orbits, and hence the energy levels, are also discrete.

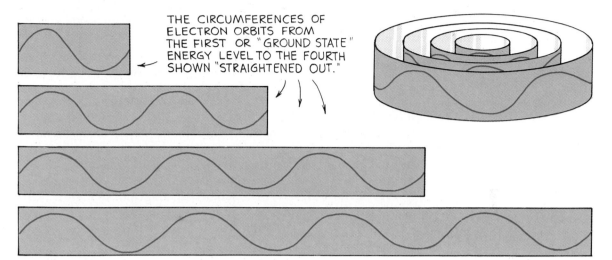

THE CIRCUMFERENCES OF ELECTRON ORBITS FROM THE FIRST OR "GROUND STATE" ENERGY LEVEL TO THE FOURTH SHOWN "STRAIGHTENED OUT."

Fig. 38-9   The electron orbits in an atom have discrete radii because the circumferences of the orbits are whole-number multiples of the electron wavelengths, which differ for the various elements (and also for different orbits within the elements). This results in discrete energy levels, which characterize each element. The figure is greatly oversimplified, as the standing waves make up spherical and ellipsoidal shells rather than flat, circular ones.

This view explains why electrons do not spiral closer and closer to the nucleus when photons are emitted. If each electron orbit is described by a standing wave, the circumference of the smallest orbit can be no smaller than one wavelength—no fraction of a wavelength is possible in a circular (or elliptical) standing wave.

38.7 Relative Sizes of Atoms

The radii of the electron orbits in the Bohr model of the atom are determined by the amount of electric charge in the nucleus. For example, the single positively charged proton in the hydrogen atom holds one negatively charged electron in an orbit at a particular radius. If we double the positive charge in the nucleus, the orbiting electron will be pulled into a tighter orbit with half its former radius since the electrical attraction is doubled. This doesn't quite happen, however, because the double charge in the nucleus normally holds a second orbital electron, which diminishes the effect of the positive nucleus. This added electron makes the atom electrically neutral. The atom is no longer hydrogen, but is helium. The two orbital electrons assume an orbit characteristic of helium. An additional proton in the nucleus pulls the electrons into an even closer orbit and, furthermore, holds a third electron in a second orbit. This is the lithium atom, atomic number 3. We can continue with this process, increasing the positive charge of the nucleus and adding successively more electrons and more orbits all the way up to atomic numbers above 100, to the synthetic radioactive elements.*

As the nuclear charge increases and additional electrons are added in outer orbits, the inner orbits shrink in size because of the stronger electrical attraction to the nucleus. This means that the heavier elements are not much larger in diameter than the lighter elements. The diameter of the uranium atom, for example, is only about three hydrogen diameters even though it is 238 times more massive. The schematic diagrams in Figure 38-10 are drawn approximately to the same scale.

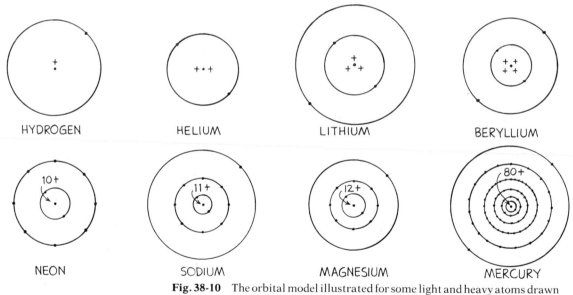

Fig. 38-10 The orbital model illustrated for some light and heavy atoms drawn to approximate scale. Note that the heavier atoms are not appreciably larger than the lighter atoms.

---

*   Each orbit will hold only so many electrons. A rule of quantum mechanics states that an orbit is filled when it contains a number of electrons given by $2n^2$, where n is 1 for the first orbit, 2 for the second orbit, 3 for the third orbit, and so on. For $n = 1$, there are 2 electrons; for $n = 2$, there are $2(2^2)$, or 8, electrons; for $n = 3$, there are a maximum of $2(3^2)$, or 18, electrons, etc. The number n is called the *principal quantum number*.

> ▶ **Question**
> What fundamental force dictates the size of an atom?

Each element has an arrangement of electron orbits unique to that element. For example, the radii of the orbits for the sodium atom are the same for all sodium atoms, but different from the radii of the orbits for other kinds of atoms. When we consider the 92 naturally occurring elements, we find that there are 92 distinct patterns or orbits. There is a different pattern for each element.

The Bohr model of the atom solved the mystery of the atomic spectra of the elements. It accounted for X rays that were emitted when electrons made transitions from outermost to innermost orbits. Bohr was able to predict X-ray frequencies that were later experimentally confirmed. He calculated the *ionization energy* of the hydrogen atom—the energy needed to knock the electron out of the atom completely. This also was verified by experiment. The Bohr model accounted for the general chemical properties of the elements and predicted properties of a missing element (hafnium), which led to its discovery.

The Bohr model was impressive. Nonetheless, Bohr was quick to point out that his model was to be interpreted as a crude beginning, and the picture of electrons whirling like planets about the sun was not to be taken literally (to which popularizers of science paid no heed). His discrete orbits were conceptual representations of an atom whose later description involved a purely mathematical rather than a visual model. Still, his planetary model of the atom with electrons occupying discrete energy levels underlies the more complex models of the atom today, which are built upon a completely different structure from that built by Newton and other physicists before the twentieth century. This is the structure called *quantum mechanics*.

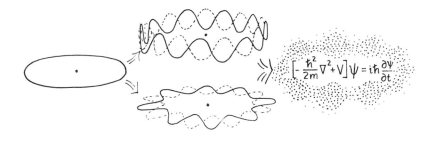

Fig. 38-11 The model of the atom has evolved from the Bohr planetary model (left) to the modified model with de Broglie waves (center) to the mathematical model (the Schroedinger wave equation, right).

▶ **Answer**
The electrical force.

38.8 Quantum Physics

The more that physicists studied the atom, the more convinced they became that the Newtonian laws that work so well for large objects such as baseballs and planets simply do not apply to events in the microworld of the atom. Whereas in the macroworld the study of motion is called *mechanics*, in the microworld the study of the motion of quanta is called **quantum mechanics**. The more general study of quanta in the microworld is simply called **quantum physics**.

While one can be quite certain about careful measurements in the macroscopic world, there are fundamental uncertainties in the measurements of the atomic domain. Macroscopic measurements, such as the temperature of materials, the vibrational frequencies of certain crystals, and the speeds of light and sound, can be made as accurate as the experimenter wishes. But subatomic measurements such as the momentum and position of electrons in an atom and the decay rate of individual radioactive atoms are entirely different. In this domain, the uncertainties in many measurements are comparable to the magnitudes of the quantities themselves. The structure of quantum mechanics is based on probabilities, a notion that is difficult for many people to accept. Even Einstein did not accept this, which prompted his often-quoted statement, "I cannot believe that God plays dice with the universe."

If you continue to expose yourself to physics, you will likely study quantum mechanics in the future. Then you'll find that subatomic interactions are unpredictable, and that the notion of certainty is replaced with probability. It's fascinating material.

38 | Chapter Review

Concept Summary

Early models attempted to explain atoms and light.
- In Bohr's planetary model of the atom, electrons orbit the nucleus.
- Newton proposed a particle model of light.
- Huygens' wave model of light was reinforced by Maxwell's electromagnetic wave model.

Einstein proposed that light is composed of discrete quanta of energy called photons.
- The energy of a photon is proportional to its frequency.
- Energy and frequency are related by Planck's constant, h.

The photoelectric effect, the ejection of electrons from certain metals that are struck by light, reinforced the particle theory of light.
- When the frequency of light is so low that the photons have insufficient energy to cause the effect, then increasing the intensity of the light does not matter.

De Broglie suggested that all matter has wave properties.
- Electrons have a detectable wavelength, can be diffracted, and undergo interference.

Bohr's planetary model explained the atomic spectra of the elements.
- Spectral lines are due to transitions of electrons between energy levels that correspond to different electron orbits.
- Bohr's model could not explain why only certain electron energy levels were possible, but de Broglie's concept of unique electron wavelengths for each element could.

The radius of the atoms of each element is unique to that element.
- As the nuclear charge increases, inner electron orbits shrink. As a result, heavier elements are not much larger in diameter than lighter elements.

Quantum mechanics is the study of motion of quanta.
- Newtonian mechanics does not apply to events on the atomic scale.
- Quantum measurements, such as the momentum and position of an electron, involve fundamental uncertainties.

Important Terms

photoelectric effect (38.3)
photon (38.2)
Planck's constant (38.2)
quantum (pl. quanta) (38.2)
quantum mechanics (38.8)
quantum physics (38.8)

Review Questions

1. What is a model? Give two examples for the nature of light. (38.1)

2. What exactly is a quantum? Give two examples. (38.2)

3. What is a quantum of light called? (38.2)

4. What is Planck's constant, and how does it relate to the frequency and energy of a quantum of light? (38.2)

5. Which has more energy per photon—red light or blue light? (38.2)

6. What is the photoelectric effect? (38.3)

7. Why does blue light eject electrons from a photosensitive surface, whereas red light has no effect? (38.3)

8. Will bright blue light eject more electrons than dim light of the same frequency? (38.3)

9. Does the photoelectric effect support the particle model or the wave model of light? (38.3)

10. a. Do particles have wave properties?
    b. Who was the first physicist to give a convincing answer to this question? (38.4)

11. As the speed of a particle increases, does its associated wavelength increase or decrease? (38.5)

12. Does the diffraction of an electron beam support the particle model or the wave model of electrons? (38.5)

13. How does the energy of a photon compare to the difference in energy levels of the atom from which it is emitted? (38.6)

14. What does it mean to say that an electron occupies discrete energy levels in an atom? (38.6)

15. Does the particle view of an electron or the wave view of an electron better explain the discreteness of electron energy levels? Why? (38.6)

16. What does wave interference have to do with the electron energy levels in an atom? (38.6)

17. Why is a helium atom smaller than a hydrogen atom? (38.7)

18. Why are the heaviest elements not appreciably larger than the lightest elements? (38.7)

19. What is quantum mechanics? (38.8)

20. Can the momentum and position of elec-trons in an atom be measured with certainty? (38.8)

Think and Explain

1. What does it mean to say that a certain quantity is quantized?

2. What evidence can you cite for the wave nature of light? For the particle nature of light?

3. A very bright source of red light has much more energy than a dim source of blue light, but the red light has no effect in ejecting electrons from a photosensitive surface. Why is this so?

4. Which photon has the most energy—one from infrared, visible, or ultraviolet light?

5. If a beam of red light and a beam of green light have exactly the same energy, which beam contains the greater number of photons?

6. Suntanning produces cell damage in the skin. Why is ultraviolet light capable of producing this damage while infrared radiation is not?

7. Electrons in one electron beam have a greater speed than those in another. Which electrons have the longer de Broglie wavelength?

8. We do not notice the wavelength of moving matter in our ordinary experience. Is this because the wavelength is extraordinarily large or extraordinarily small?

9. The equation $E = hf$ describes the energy of each photon in a beam of light. If Planck's constant, h, were larger, would photons of light of the same frequency be more energetic or less energetic?

10. Why will helium rather than hydrogen more readily leak through an inflated rubber balloon?

39 The Atomic Nucleus and Radioactivity

The configuration of electrons in an atom determines whether and how the atom bonds to form compounds. It also dictates the melting and freezing temperatures, the thermal and electrical conductivity, as well as the taste, texture, appearance, and color of substances. Small changes in electron energy levels produce visible light. Large changes produce X rays. This chapter goes even deeper into the atom—to the atomic nucleus.

39.1 The Atomic Nucleus

The atomic nucleus in an atom occupies only a few quadri-millionths the volume of an atom. It is incredibly tiny, yet much has been learned about its structure. The nucleus is composed of particles called **nucleons**, which when electrically charged are protons, and when electrically neutral are neutrons.* Neutrons and protons have close to the same mass, with the neutron's being slightly greater. Nucleons have nearly 2000 times the mass of electrons, so the mass of an atom is practically equal to the mass of its nucleus alone.

The positively charged protons in the nucleus hold the nega-tively charged electrons in their orbits. Each proton has the

---

* Protons and neutrons are themselves composed of subnuclear particles called *quarks*. Are quarks themselves made of still smaller particles? They may be, but so far there is no evidence or theoretical reason for believing so. Theoreti-cal physicists say today that quarks are the elementary particles of which all nucleons are made.

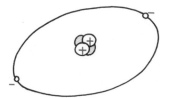

Fig. 39-1 The number of electrons that surround the atomic nucleus is matched by the number of protons in the nucleus.

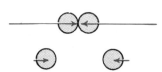

Fig. 39-2 The nuclear strong force is a very short-range force. For nucleons very close or in contact, it is very strong. But a few nucleon diameters away it is nearly zero.

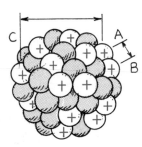

Fig. 39-3 Proton A both attracts and repels proton B but mainly repels proton C. The greater the distance between A and C, the more unstable the nucleus.

ability to hold one electron. So in an electrically neutral atom there are as many protons in the nucleus as there are electrons outside. Each proton has exactly the same magnitude of charge as the electron, but opposite in sign. The number of protons in the nucleus therefore determines the chemical properties of that atom, because the positive nuclear charge determines the possible structures of electron orbits that can occur.

The number of neutrons in the nucleus has no direct effect on the electron structure, and hence does not affect the chemistry of the atom. The principal role of the neutrons is to act as a sort of nuclear cement to hold the nucleus together. Nucleons are bound by an attractive nuclear force appropriately called the **strong force**, which holds the nucleus together.

The strong nuclear force of attraction is very short-ranged (Figure 39-2). Whereas the electrical force between charges decreases as the square of the distance, the nuclear strong force decreases approximately as the inverse fifth or sixth power of distance. Its strength is nearly zero for nucleons spaced a few nucleon diameters apart.

Although the strong force is very strong at close distances, it is not strong enough to hold two protons together by themselves. The electrical force of repulsion between a pair of bare protons in contact is even stronger (recall that like charges repel one another). So the "tug-of-war" between the attractive strong force and the repelling electrical force favors the electrical force. When neutrons are present, however, the strong force alone is increased (since neutrons have no charge). Thus, the presence of neutrons adds to the nuclear attraction and keeps protons from flying apart.

The more protons there are in a nucleus, the more neutrons are needed to hold them together. For light elements, it is sufficient to have about as many neutrons as protons. For heavy elements, extra neutrons are required. The most common form of lead, for example, has 82 protons and 126 neutrons, or about one and a half times as many neutrons as protons. For elements with more than 83 protons, even the addition of extra neutrons cannot stabilize the nucleus.

39.2 | Radioactive Decay

One of the factors that limits the size of the atomic nucleus is the fact that neutrons are not stable by themselves. A lone neutron will spontaneously decay into a proton plus an electron. Out of a bunch of lone neutrons, about half of them will decay in 12 minutes. A neutron seems to need protons around to keep this from happening. After the size of a nucleus reaches a certain point, the neutrons so outnumber the protons that there are not enough protons in the mix to prevent the neutrons from decaying. Neutrons without nearby protons decay into a proton by expelling an electron. Nuclei that decay in this or in similar ways are said to be **radioactive**.

All the elements heavier than bismuth decay in one way or another. Thus, these elements are radioactive. Their atoms emit three distinct types of rays. The rays are called by the first three letters of the Greek alphabet, α, β, and γ, named *alpha*, *beta*, and *gamma*, respectively. Alpha rays have a positive electric charge, beta rays are negative, and gamma rays are electrically neutral. The three rays can be separated by putting a magnetic field across their path (Figure 39-5).

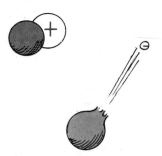

Fig. 39-4   A neutron with a proton is stable, but a neutron by itself is unstable and turns into a proton by emitting an electron.

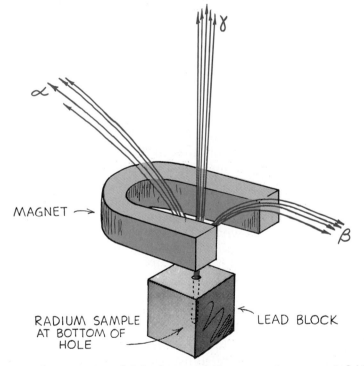

Fig. 39-5   The separation of alpha, beta, and gamma rays in a magnetic field. The rays come from a radioactive source placed at the bottom of a hole drilled in a lead block.

An alpha ray is a stream of particles that are made of two protons and two neutrons and that are identical to the nuclei of helium atoms. These are called *alpha particles* (Figure 39-6).

Fig. 39-6   An alpha particle contains two protons and two neutrons bound together and is identical to a helium nucleus.

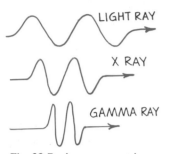

Fig. 39-7 A gamma ray is simply electromagnetic radiation, much higher in frequency and energy than light and X rays.

A beta ray is simply a stream of electrons. These are ejected from neutrons in the nucleus, which then become protons. It may seem that the electrons are "buried" inside the neutron, but this is not true. An electron does not exist in a neutron any more than a spark exists inside a rock about to be scraped across a rough surface. The electron that pops out of the neutron, like the spark that pops out of the scraped stone, is produced during an interaction.

A gamma ray is pure energy. Like visible light, gamma rays are simply photons of electromagnetic radiation, but of much higher frequency and energy. Visible light is emitted when electrons jump from one orbit to another of lower energy. Gamma rays are emitted when nucleons seem to do a similar sort of thing inside the nucleus. There are great differences in nuclear energy levels, so the photons emitted carry a large amount of energy.

39.3 | Radiation Penetrating Power

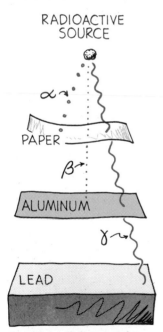

Fig. 39-8 Alpha particles penetrate least and can be stopped by a few sheets of paper; beta particles by a sheet of aluminum; gamma rays by a thick layer of lead.

There is a great difference in the penetrating power of the three types of rays. Alpha rays are the easiest to stop. They can be stopped by a reasonably heavy piece of paper or a few sheets of thin paper. Beta rays go right through paper but are stopped by several sheets of aluminum foil. Gamma rays are the most difficult to stop and require lead or other heavy shielding to block them.

An alpha particle is easy to stop because it is relatively big and its double positive charge interacts with molecules it encounters. It slows down as it shakes many of these molecules apart and leaves positive and negative ions in its wake. Even traveling through nothing but air, it comes to a stop after only a few centimeters. It soon grabs up a couple of stray electrons and becomes nothing more than a harmless helium atom.

A beta particle is normally faster than an alpha particle, carries only a single charge, and travels much farther through the air. Most beta particles lose their energy during the course of a large number of glancing collisions with atomic electrons. Except for rare direct hits, energy is lost in many small steps. Beta particles slow down until they are stopped altogether, becoming a part of the material they are in, like any other electron.

Gamma rays are the most penetrating of the three because they have no charge. With no electric attraction or deflection, a gamma ray photon interacts with the absorbing material only via a direct hit with an atomic electron or a nucleus. Unlike charged particles, it can be removed from its beam in a single encounter. Materials composed of heavy metals such as lead are good absorbers mainly because of their high electron density.

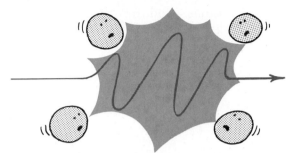

Fig. 39-9   A gamma ray photon is absorbed only when it makes a direct hit.

39.4 | Radioactive Isotopes

It has already been stated that the number of protons in an atomic nucleus determines the number of electrons surrounding the nucleus. If there is a difference in the number of electrons and protons, the atom is called an *ion*. An ionized atom is one that has a different number of electrons compared to nuclear protons.

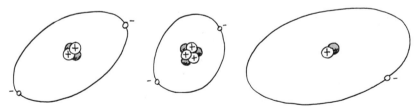

Fig 39-10   Which of these atoms is an ion?

The number of neutrons in the nucleus, however, has no bearing on the number of electrons the atom may have. The number of neutrons, then, has no direct bearing on the chemistry of an atom. Consider a hydrogen atom. The common form of hydrogen has a bare proton as its nucleus. Any nuclear configuration that has only one proton in its nucleus *is* hydrogen—by definition. There can be different kinds, or **isotopes** of hydrogen, however. In one isotope, the nucleus consists of only a single proton. In a second isotope, the proton is accompanied by a neutron. In a third isotope, there are two neutrons. All the isotopes of a particular element are chemically identical. The orbital electrons are affected only by the positive charge in the nucleus, not by its neutrons.

We distinguish between the different isotopes of hydrogen by $^1_1$H, $^2_1$H, and $^3_1$H, where the lower number is the **atomic number** —the number of protons—and the upper number is the **atomic mass number**—is the total number of nucleons.

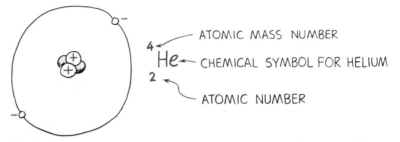

Fig. 39-11   The atomic number is equal to the number of protons in the nucleus, and the atomic mass number is equal to the number of nucleons in the nucleus (both protons and neutrons).

The common isotope of hydrogen, $^1_1$H, is a stable element. So is the isotope $^2_1$H, called *deuterium.* "Heavy water" is the name usually given to H_2O in which the H's are deuterium atoms. The triple weight hydrogen isotope $^3_1$H, called *tritium,* however, is unstable and undergoes decay. This is the radioactive isotope of hydrogen. All elements have isotopes. Some are radioactive and some are not. All the isotopes of elements above atomic number 83, however, are radioactive.

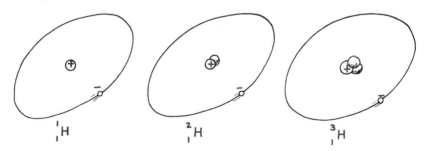

Fig. 39-12   Three isotopes of hydrogen. Each nucleus has a single proton, which holds a single orbital electron, which, in turn, determines the chemical properties of the atom. The different number of neutrons changes the mass of the atom, but not its chemical properties.

The common isotope of uranium is $^{238}_{92}$U, or U-238 for short. It is radioactive, but not nearly as radioactive as $^{235}_{92}$U, or U-235. Any nucleus with 92 protons is uranium, by definition. Nuclei with 92 protons but different numbers of neutrons are simply different isotopes of uranium.

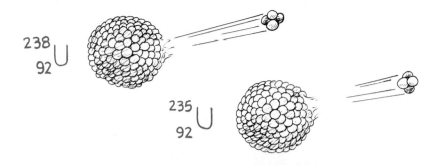

$^{238}_{92}U$

$^{235}_{92}U$

Fig. 39-13 All isotopes of uranium are unstable and undergo radioactive decay.

> ▶ **Questions**
>
> 1. Why is the helium nucleus $^4_2$He stable?
>
> 2. Why would you expect a hydrogen isotope with many neutrons to be radioactive?
>
> 3. Why are large nuclei, such as radium and uranium, radioactive?

39.5   Radioactive Half Life

Radioactive isotopes decay at different rates. The radioactive decay rate is measured in terms of a characteristic time, the **half life**. The half life of a radioactive material is the time needed for half of the active atoms to decay. Radium-226, for example, has a half life of 1620 years. This means that half of any given specimen of Ra-226 will be converted into some other element by the end of 1620 years. In the next 1620 years, half of the remaining radium will decay, leaving only one-fourth the original number of radium atoms. The other three-fourths are converted, by a

▶ **Answers**

1. The two neutrons provide the additional attractive nuclear force to counteract the repulsive electric force between the two protons. Interestingly enough, except for the bare proton of $^1_1$H, this configuration of nucleons is the most stable of all; it is an alpha particle.

2. Neutrons by themselves are unstable. Next to a proton, however, a neutron is stable. In a hydrogen isotope with more than 2 neutrons, one or more of those neutrons will at times be distant from the proton. The distant neutron will act as though it is alone; it will be unstable and likely decay by beta emission.

3. In a large nucleus, each proton feels electric repulsive forces from *all* the other protons, but the short-range nuclear strong forces from nucleons on the far side of the nucleus are negligible. Thus, the net attractive force on an outer proton is barely stronger than the net repulsive force on it.

succession of disintegrations, to lead. After 20 half lives, an initial quantity of radioactive atoms will be diminished to one millionth that quantity.

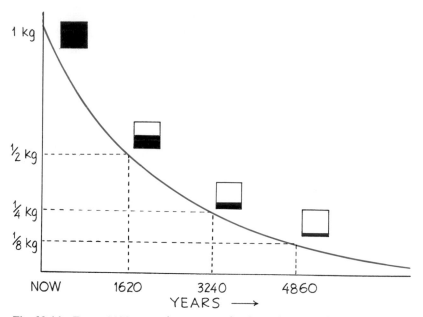

Fig. 39-14 Every 1620 years the amount of radium decreases by half.

The isotopes of some elements have a half life of less than a millionth of a second, while U-238, for example, has a half life of 4.5 billion years. The isotopes of each radioactive element have their own characteristic half lives.

Rates of radioactive decay are remarkably constant and are not affected by any external conditions, however drastic. High or low pressures, high or low temperatures, strong magnetic or electric fields, and even violent chemical reactions have no detectable effect on the rate of decay of an element. Any of these stresses, however severe by ordinary standards, is far too mild to affect the nucleus deep in the interior of the atom.

How do physicists measure radioactive half lives? They certainly do not do it by observing a specimen and waiting until the quantity reduces to half. This is often much longer than a human lifespan! One can measure, however, the rate at which a substance decays. There are various radiation detectors for doing this (Figure 39-15). The half life of an isotope is related to its rate of disintegration. In general, the shorter the half life of a substance, the faster it disintegrates, and the more active is the substance. The half life can be computed from the rate of disintegration, which can be measured in the laboratory.

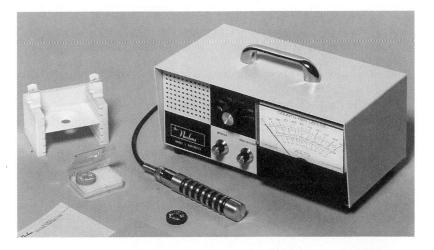

Fig. 39-15 Some radiation detectors. (Top) A Geiger counter detects incoming radiation by its ionizing effect on enclosed gas in the tube. (Bottom) A scintillation counter detects incoming radiation by flashes of light that are produced when charged particles or gamma rays pass through it.

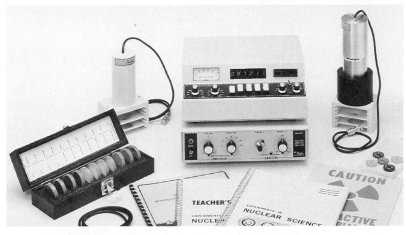

▶ **Questions**

1. If a sample of radioactive isotopes has a half life of 1 year, how much of the original sample will be left at the end of the second year?

2. Which will give a higher reading on a radiation detector—equal amounts of radioactive material that has a short half life, or radioactive material that has a long half life?

▶ **Answers**

1. One quarter of the original sample will be left. The three quarters that underwent decay is now a different element altogether.

2. The material with the shorter half life is more active and will give a higher reading on a radiation detector.

39.6   Natural Transmutation of Elements

When a nucleus emits an alpha or a beta particle, a different element is formed. The changing of one element to another is called **transmutation**. Consider common uranium, for example. Uranium has 92 protons. When an alpha particle is ejected, the nucleus is reduced by two protons and two neutrons. This is because they make up the alpha particle that leaves. The 90 protons and 144 neutrons left behind are then the nucleus of a new element. This element is *thorium*. This reaction is expressed as

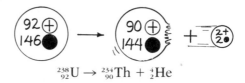

$$^{238}_{92}U \rightarrow \, ^{234}_{90}Th + \, ^{4}_{2}He$$

An arrow is used here to show that the $^{238}_{92}U$ changes into the other elements. When this happens, energy is released in three forms: gamma radiation, the kinetic energy of the alpha particle ($^{4}_{2}He$), and the kinetic energy of the thorium atom. Be sure to notice in the nuclear equation that the mass numbers at the top balance (238 = 234 + 4) and that the atomic numbers at the bottom also balance (92 = 90 + 2).

Thorium-234, the product of this reaction, is also radioactive. When it decays, it emits a beta particle. Recall that a beta particle is an electron—not an orbital electron, but one from the nucleus—from one of the neutrons. In a loose sense, we can think of a neutron as a combined proton and electron; so when the neutron emits an electron, it becomes a proton. In the case of thorium, which has 90 protons, beta emission leaves it with one fewer neutron and one more proton. The new nucleus then has 91 protons and is no longer thorium. It is the element *protactinium*. The reaction is

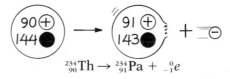

$$^{234}_{90}Th \rightarrow \, ^{234}_{91}Pa + \, ^{0}_{-1}e$$

Note that although the atomic number has increased by 1 in this process, the mass number (number of nucleons) remains the same. Also, note that the beta particle (electron) is written as $^{0}_{-1}e$. The -1 is the charge of the electron. The 0 indicates that its mass is insignificant when compared to that of the protons and neutrons that alone contribute to the mass number. Beta emission has hardly any effect on the mass of the nucleus; only the charge (atomic number) changes.

From the previous two examples, you can see that when an atom ejects an alpha particle from its nucleus, the mass number of the resulting atom decreases by 4, and its atomic number decreases by 2. The resulting atom belongs to an element two spaces back in the periodic table.* When an atom ejects a beta particle from its nucleus, it loses no nucleons, so there is no change in mass number, but its atomic number *increases* by 1.** The resulting atom belongs to an element one place forward in the periodic table. A radioactive nucleus may emit gamma radiation along with an alpha particle or beta particle. Gamma emission has no effect on the mass number or the atomic number. Thus, radioactive elements decay backward or forward in the periodic table.

The radioactive decay of $^{238}_{92}U$ to $^{206}_{82}Pb$, an isotope of lead, is shown on the next page in Figure 39-16. The steps in the decay process are shown in the diagram, where each nucleus that plays a part in the series is shown by a burst. The vertical column that contains the burst shows its atomic number, and the horizontal column shows its mass number. Each arrow that slants downward toward the left shows an alpha decay. Each arrow that points to the right shows a beta decay. Notice that some of the nuclei in the series can decay either way. This is one of several similar radioactive series that occur in nature.

▶ **Questions**
1. Complete the following nuclear reactions.
   a. $^{226}_{88}R \rightarrow ^{?}_{?}? + ^{0}_{-1}e$
   b. $^{209}_{84}Po \rightarrow ^{205}_{82}Pb + ^{?}_{?}?$

2. What finally becomes of all the uranium that undergoes radioactive decay?

▶ **Answers**
1. a. $^{226}_{88}Ra \rightarrow ^{226}_{89}Ac + ^{0}_{-1}e$
   b. $^{209}_{84}Po \rightarrow ^{205}_{82}Pb + ^{4}_{2}He$

2. All uranium will ultimately become lead. On the way to becoming lead, it will exist as a series of other elements, as indicated in Figure 39-16.

---

*   For a periodic table, see Figure 17-11 back on page 247. Look in the periodic table for the elements mentioned in this section.
**   Sometimes a nucleus emits a *positron*, which is the antiparticle of an electron. A positron has a charge of +1 and the same mass as the electron. In this case, a proton in the nucleus becomes a neutron, and the atomic number is decreased by 1 with no change in mass number.

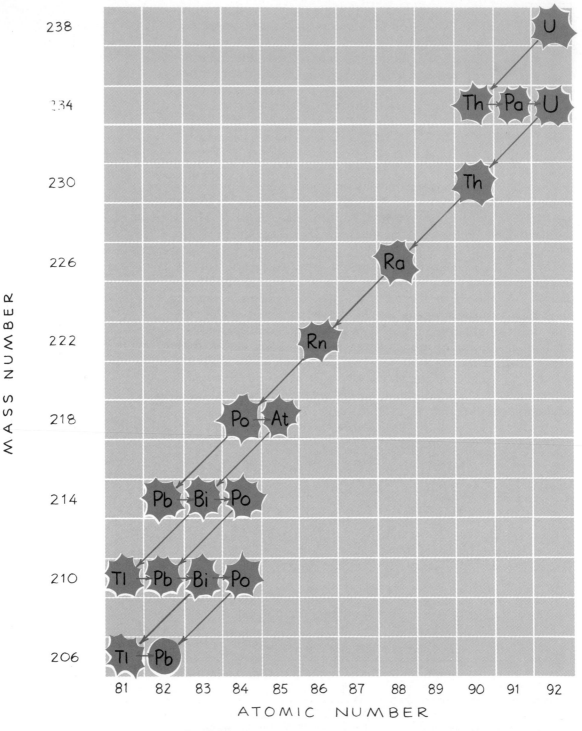

Fig. 39-16   U-238 decays to Pb-206 through a series of alpha and beta decays.

39.7 | Artificial Transmutation of Elements

The British physicist Ernest Rutherford, in 1919, was the first of many investigators to succeed in transmuting a chemical element. In a sealed container he bombarded nitrogen nuclei with alpha particles and then found traces of oxygen and hydrogen that were not there before. He accounted for the presence of the oxygen with the nuclear equation

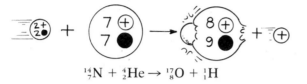

$$^{14}_{7}N + {}^{4}_{2}He \rightarrow {}^{17}_{8}O + {}^{1}_{1}H$$

After Rutherford's experiment there followed many such nuclear reactions—first with natural bombarding particles from radioactive ores, and then with more energetic particles (protons and electrons) hurled by giant atom-smashing accelerators. Artificial transmutation is an everyday fact of life to researchers today.

Fig. 39-17 Artificial transmutation can be accomplished by simple means or by elaborate means.

The elements beyond uranium in the periodic table—the *transuranic* or synthetic elements—are the result of artificial transmutation. All these artificially made elements have short half lives. If they ever existed naturally when the earth was formed, they have long since decayed.

39.8 | Carbon Dating

The earth's atmosphere is continuously bombarded from above by *cosmic rays*—high-energy particles and gamma rays from beyond the earth. This results in the transmutation of many atoms in the upper atmosphere. Protons and neutrons are scattered throughout the atmosphere. Most of the protons quickly capture stray electrons and become hydrogen atoms in the upper atmosphere, but the neutrons keep going for long distances because they have no charge and do not interact electrically with matter. Sooner or later many of them collide with the nuclei of atoms in the lower atmosphere. If they are captured by the nucleus of a nitrogen atom, the following reaction takes place:

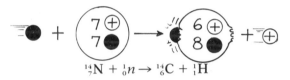

$$^{14}_{7}\text{N} + ^{1}_{0}n \rightarrow ^{14}_{6}\text{C} + ^{1}_{1}\text{H}$$

The reaction shows that the nitrogen captures a neutron $^{1}_{0}n$ and breaks up into an isotope of carbon by emitting a proton $^{1}_{1}\text{H}$.

Most of the carbon that exists is the stable $^{12}_{6}\text{C}$, carbon-12. Because of the cosmic bombardment, less than one-millionth of 1 percent of the carbon in the atmosphere is carbon-14. Both of these isotopes join with oxygen and become carbon dioxide, which is taken in by plants. This means that all plants have a tiny bit of radioactive carbon-14 in them. All animals eat plants (or eat plant-eating animals), and therefore have a little carbon-14 in them. All living things contain some carbon-14.

Carbon-14 is a beta emitter and decays back into nitrogen by the following reaction:

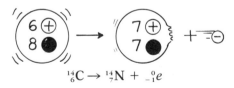

$$^{14}_{6}\text{C} \rightarrow ^{14}_{7}\text{N} + ^{0}_{-1}e$$

Because living plants continue to take in carbon dioxide, this decay is accompanied by a replenishment of carbon-14. As carbon-14 atoms decay, new ones take their place. In this way a radioactive equilibrium is reached where there is a fixed ratio of carbon-14 to carbon-12. When a plant or animal dies, replenishment stops. Then the percentage of carbon-14 decreases—at a known rate. The longer an organism is dead, the less carbon-14 remains.

The half life of carbon-14 is about 5730 years. This means that half of the carbon-14 atoms that are now present in a body, plant, or tree will decay in the next 5730 years. Half the remaining carbon-14 atoms will then decay in the following 5730 years, and so forth. The radioactivity of living things therefore gradually decreases at a steady rate after they die (Figure 39-18).

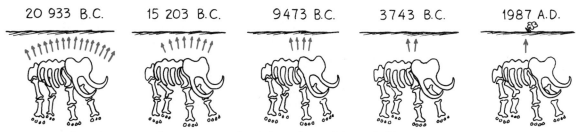

Fig. 39-18 The radioactive carbon isotopes in the skeleton diminish by one-half every 5730 years.

Archeologists use the carbon dating technique to establish the dates of wooden artifacts and skeletons. Because of fluctuations in the production of carbon-14 through the centuries, this technique gives an uncertainty of about 15%. This means, for example, that the straw in an old adobe brick that is dated to be 500 years old may really be only 425 years old on the low side, or 575 years old on the high side. For many purposes this is an acceptable level of uncertainty. If greater accuracy is desired, then other techniques must be employed.

Fig. 39-19 If an old sample of carbon is half as radioactive as a new sample of the same amount, then the age of the old sample is equal to the half life of carbon-14.

▶ **Questions**

1. An archeologist extracts a gram of carbon from an ancient ax handle and measures between 7 and 8 beta emissions per minute from the sample. A gram of carbon extracted from a fresh piece of wood gives off 15 betas per minute. Estimate the age of the ax handle.

2. Suppose the carbon sample from the ax handle were found to be only one-fourth as radioactive as a gram of carbon from new wood. Estimate the age of the ax handle.

▶ **Answers**

1. Since beta emission for the old sample is one half that of the fresh sample, about one half life has passed, 5730 years.

2. The ax handle is two half lives of carbon-14, about 11 460 years.

39.9 Uranium Dating

The dating of older, but nonliving, things is accomplished with radioactive minerals, such as uranium. The naturally occurring isotopes U-238 and U-235 decay very slowly and ultimately become isotopes of lead—but not the common lead isotope Pb-208. For example, U-238 decays through several stages to finally become Pb-206, whereas U-235 finally becomes the isotope Pb-207. The lead isotopes 206 and 207 that exist were at one time uranium. The older the uranium-bearing rock, the higher the percentage of these lead isotopes.

From the half lives of the uranium isotopes, and the percentage of lead isotopes in uranium-bearing rock, a calculation can be made of the date the uranium in the rock first started to decay. Rocks dated in this way have been found to be as much as 3.7 billion years. Samples from the moon, where there has been less obliteration of the early rocks than occurs on earth, have been dated at 4.2 billion years, which begins to impinge closely upon the well-established 4.6-billion-year age of the earth and solar system.

39.10 Radioactive Tracers

Radioactive isotopes of all the elements have been made by bombardment with neutrons and other particles. These isotopes are inexpensive, quite available, and useful in scientific research and industry.

A small amount of radioactive isotopes can be mixed with fertilizer before being applied to growing plants. Once the plants are growing, the amount of fertilizer taken up by the plant can be easily measured with radiation detectors. From such measurements, farmers know the proper amount of fertilizer to use. When used in this way, radioactive isotopes are called *tracers* (Figure 39-20).

Tracers are used in medicine to study the process of digestion and the way in which chemical substances move about in the body. Food that contains a small amount of radioactive isotopes is fed to a patient and traced through the body with a radiation detector. The same method is used to study the circulation of the blood.

Engineers can study how the parts of an automobile test engine wear away by making the cylinder walls in the engine radioactive. While the engine is running, the piston rings rub

Fig. 39-20 Radioactive isotopes are used to check the action of fertilizers in plants and the progress of food in digestion.

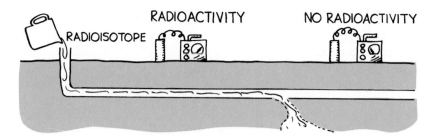

Fig. 39-21   Tracking pipe
leaks with radioactive
isotopcs.

against the cylinder walls. The tiny particles of radioactive metal that are worn away fall into the lubricating oil, where they can be measured with a radiation detector. This test is repeated with different oils. In this way the engineer can determine which oil gives the least wear and longest life to the engine.

There are hundreds more examples of the use of radioactive isotopes. The important thing is that this technique provides a way to detect and count atoms in samples of materials too small to be seen with a microscope.

39.11 | Radiation and You

Radioactivity has been around longer than the human race. It is as much a part of our environment as the sun and the rain. It is what warms the interior of the earth and makes it molten. In fact, radioactive decay inside the earth is what heats the water that spurts from a geyser or that wells up from a natural hot spring. Even the helium in a child's balloon is nothing more than the alpha particles that were produced by radioactive decay.

As Figure 39-22 shows, most of the radiation you encounter originates in the natural surroundings. It is in the ground you stand on, and in the bricks and stones of surrounding build-

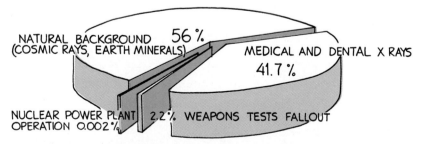

Fig. 39-22   Origins of radiation exposure for an average individual in the United States.

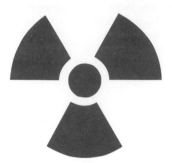

Fig. 39-23 This is the internationally used symbol to indicate an area where radioactive material is being handled or produced.

ings. This natural background radiation was present before humans emerged in the world. If our bodies couldn't tolerate it, we wouldn't be here.

Even the cleanest air we breathe is slightly radioactive. This is because of the cosmic rays that penetrate our atmosphere and shred the nuclei of the various molecules in the air. At sea level the atmosphere provides a protective blanket against cosmic rays, but at higher altitudes, radiation is more intense. In Denver, the "mile-high city," you receive more than twice the cosmic radiation you receive at sea level. A couple of round-trip flights between places as distant as New York and San Francisco exposes you to as much radiation as you receive in a normal chest X ray. The air time of airline personnel is limited because of this extra radiation.

Exposure to radiation greater than normal background should be avoided. This is because of the damage it can do. The cells of living tissue are composed of intricately structured molecules in a watery, ion-rich brine. When nuclear radiation encounters this highly ordered soup, it produces chaos on the atomic scale. A beta particle, for example, passing through living matter collides with a small percentage of the molecules and leaves a randomly dotted trail of altered or broken molecules in its wake. Similar damage is done by a variety of particles. These altered molecules are more often harmful than useful to life processes.

Cells are able to repair most kinds of molecular damage if the radiation is not too intense. This is how we are able to tolerate natural background radiation. On the other hand, people working for long periods in uranium mines or around high concentrations of radioactive materials can expect the likelihood of cancer and a shorter life.

Whenever possible, exposure to radiation should be avoided. Unavoidable, however, is the natural radiation that all living beings have always absorbed.

39 | Chapter Review

Concept Summary

The atomic nucleus is composed of nucleons that consist of positively charged protons and electrically neutral neutrons.
- The number of protons determines the number of electrons and the chemical properties of an atom.
- Nucleons are bound together by the strong force.
- As the number of protons increases, more neutrons are needed to stabilize the nucleus.

Radioactive elements have unstable nuclei and emit various nuclear particles.
- Alpha particles consist of two protons and two neutrons and can be stopped by heavy paper.
- Beta particles are electrons ejected from neutrons. They can be stopped by aluminum foil.
- Gamma rays are high-energy photons. Heavy shielding, such as lead, is required for protection.

Isotopes of an element are chemically identical but differ in their number of neutrons.
- They have the same atomic number but different atomic mass numbers.
- Some isotopes are radioactive.
- Half life is a measure of the decay rate of radioactive isotopes.

Transmutation is the changing of one element into another that occurs when a radioactive nucleus emits an alpha or beta particle.
- The transuranic elements were created by artificial transmutation.

Radioactive isotopes have several important uses.
- Carbon-14 is used to date wooden artifacts and remains of plants and animals.

- Radioactive minerals such as uranium-238 or uranium-235 are used to date older, non-living materials.
- Radioactive tracers are used in agriculture and medicine.

Natural environmental radiation constantly bombards us.
- Additional exposure to radiation should be avoided whenever possible because it is damaging to living molecules and cells.

Important Terms

atomic mass number (39.4)
atomic number (39.4)
half life (39.5)
isotope (39.4)
nucleon (39.1)
radioactive (39.2)
strong force (39.1)
transmutation (39.6)

Review Questions

1. Which of the following are nucleons—protons, neutrons, or electrons? (39.1)

2. Do electrical forces tend to hold a nucleus together or push it apart? (39.1)

3. Between what kinds of particles does the nuclear strong force act? (39.1)

4. Why does the presence of neutrons in a nucleus make it more stable? (39.1)

5. Why do too many neutrons in a nucleus make it unstable? (39.1)

6. Distinguish between alpha, beta, and gamma rays. (39.2)

7. How do the penetrating powers of the three types of radiation compare? (39.3)

8. Distinguish between an ion and an isotope. (39.4)

9. How does the number of electrons in a normal atom compare to the number of protons in its nucleus? (39.4)

10. Which isotope has the greater number of neutrons, U-235 or U-238? (39.4)

11. What is the meaning of radioactive half life? (39.5)

12. If the radioactive half life of a certain isotope is 1620 years, how much of that substance will be left at the end of 1620 years? After 3240 years? (39.5)

13. When an atom undergoes radioactive decay, does it turn into a completely different element? (39.6)

14. a. What happens to the atomic number of an atom when it ejects an alpha particle?
    b. What happens to its atomic mass number? (39.6)

15. a. What happens to the atomic number of an atom when it ejects a beta particle?
    b. What happens to its atomic mass number? (39.6)

16. a. What element does thorium become if it emits an alpha particle?
    b. What if it emits a beta particle instead? (39.6)

17. a. What is a transuranic element?
    b. Why are there no ore deposits of them in the earth? (39.7)

18. Which of the following is radioactive: C-12 or C-14? (39.8)

19. Why is there more C-14 in new bones compared to old bones of the same mass? (39.8)

20. Why would the carbon dating method be

useless in dating old coins but not old pieces of adobe bricks? (39.8)

21. Why are there deposits of lead in all deposits of uranium ores? (39.9)

22. What isotopes accumulate in old, uranium-bearing rock? (39.9)

23. What is a radioactive tracer? (39.10)

24. From where does most of the radiation you encounter originate? (39.11)

25. Why is radiation more intense at high altitudes and near the earth's poles? (39.11)

Think and Explain

1. Why are larger nuclei more unstable than smaller nuclei?

2. a. How does the atomic nucleus change if a proton is added? If a neutron is added?
   b. Which determines the chemical nature of the nucleus?

3. How is a gamma ray different from either an alpha or a beta ray?

4. Why would you expect alpha particles to penetrate materials less readily than beta particles?

5. Why is a sample of radioactive material always a little warmer than its surroundings? Can you think of a reason why the center of the earth is so hot?

6. One of the isotopes that is a product of nuclear power plants is cesium-137, which has a half life of 30 years. How many years will be required for this isotope to decay to one-sixteenth its amount?

7. Coal contains minute quantities of radioactive materials, yet there is more environmental radiation surrounding a coal-fired power plant than a fission power plant. What

does this tell you about the shielding that typically surrounds these power plants?

8. When the isotope bismuth-213 emits an alpha particle, it transforms into a new element.
   a. What is the atomic number and atomic mass number of the new element?
   b. What are they if bismuth-213 emits a beta particle instead?

9. a. State the numbers of neutrons and protons in each of the following nuclei: $^6_3$Li, $^{14}_6$C, $^{56}_{26}$Fe, $^{201}_{80}$Hg, and $^{239}_{94}$Pu.
   b. How many electrons will typically surround each of these nuclei?

10. Radiation from a point source follows an inverse-square law. If a Geiger counter that is 1 m away from a small source reads 100 counts per minute, what will be its reading 2 m from the source? 3 m from it?

11. How is it possible for an element to decay "forward in the periodic table"—that is, to decay to an element of higher atomic number?

12. People who work around radioactivity monitor the amount of radiation that reaches their bodies by wearing film badges. These badges consist of small pieces of photographic film that are enclosed in a light-proof wrapper. How can these devices determine the amount of radiation the people receive?

40 Nuclear Fission and Fusion

The discovery of radioactivity just before the beginning of this century sparked much interest among many kinds of people. Some people thought it was no more than a scientific curiosity, some thought it was a cure for medical ailments, and a few thought it would have its use as a source of plentiful energy to heat homes, power factories, and light up cities at night.

Radioactive reactions do release energy, but very little. Then in 1939, just before World War II, a nuclear reaction similar to radioactivity was discovered that released enormous amounts of energy. This was the splitting of the atom, or *nuclear fission*. Within a decade, a very different nuclear reaction was discovered, *nuclear fusion*, which also released huge amounts of energy. Both processes were found to produce more energy per kilogram of matter than had ever been imagined by atomic scientists. The awesome release of this energy in atomic and hydrogen bombs ushered in the present "nuclear age." Out of the ashes of despair brought about by these bombs, hope grew that atoms could be used for peaceful purposes—that the awesome energy of nuclear reactions could be used for domestic power instead of as arsenals of war.

What exactly are nuclear fission and nuclear fusion? How do they differ? What is the physics that underlies why so much energy is released by these reactions? The answers to these questions are what this chapter is about.

40.1 Nuclear Fission

Biology students know that living tissue grows by the division of cells. The splitting in half of living cells is called *fission*. In a similar way, the splitting of atomic nuclei is called **nuclear fission**.

Nuclear fission involves the delicate balance between the attraction of nuclear strong forces and the repulsion of electrical forces within the nucleus. In nearly all nuclei the nuclear strong forces dominate. In uranium, however, this domination is tenuous. If the uranium nucleus is stretched into an elongated shape (Figure 40-1), the electrical forces may push it into an even more elongated shape. If the elongation passes a critical point, electrical forces overwhelm nuclear strong forces, and the nucleus splits. This is nuclear fission.

The absorption of a neutron by a uranium nucleus is apparently enough to cause such an elongation. The resulting fission process may produce any of several combinations of smaller nuclei. A typical example is

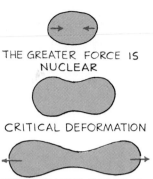

THE GREATER FORCE IS NUCLEAR

CRITICAL DEFORMATION

THE GREATER FORCE IS ELECTRICAL

Fig. 40-1 Nuclear deformation results when repelling electrical forces are stronger than attracting nuclear strong forces. The nucleus undergoes fission.

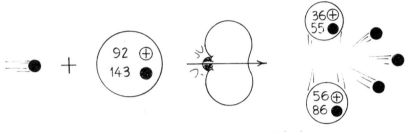

$$\,_{0}^{1}n + \,_{92}^{235}U \rightarrow \,_{36}^{91}Kr + \,_{56}^{142}Ba + 3(\,_{0}^{1}n)$$

The energy that is released by the fission of one U-235 atom is enormous—about seven million times the energy released by the explosion of one TNT molecule.

Note that one neutron starts the fission of the uranium atom, and three more neutrons are produced when the uranium fissions. Between two and three neutrons are produced in most nuclear fission reactions. These new neutrons can, in turn, cause the fissioning of two or three other atoms, releasing from four to nine more neutrons. If each of these succeeds in splitting just one atom, the next step in the reaction will produce between 8 and 27 neutrons, and so on. This makes a **chain reaction** (Figure 40-2).

Why do chain reactions not occur in naturally occurring uranium ore deposits? They would if all uranium atoms fissioned so easily. Fission occurs mainly for the rare isotope U-235, which makes up only 0.7 percent of the uranium in pure uranium metal. When the prevalent isotope U-238 absorbs neutrons from fission, it does not undergo fission. So any chain reaction is snuffed out by the neutron-absorbing U-238. Uranium deposits in nature do not spontaneously undergo a chain reaction.

If a chain reaction occurred in a chunk of pure U-235 the size of a baseball, an enormous explosion would likely result. If the chain reaction were started in a smaller chunk of pure U-235, however, no explosion would occur. This is because of geometry. A small piece has relatively more surface compared to

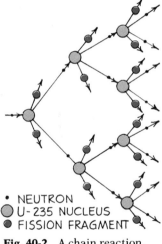

• NEUTRON
◯ U-235 NUCLEUS
● FISSION FRAGMENT

Fig. 40-2 A chain reaction.

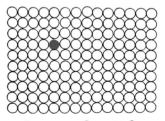

● U 235 ◯ U 238

Fig. 40-3 Only 1 part in 140 of naturally occurring uranium is U-235.

volume (or mass) than a large piece. (Similarly, there is more skin on a kilogram of small potatoes than on a single, large, one-kilogram potato.) In a small piece of uranium, neutrons would leak through the surface before an explosion could occur. In a bigger piece, the chain reaction builds up to enormous energies before the neutrons get to the surface and escape (Figure 40-4). Above a certain amount, called the **critical mass**, an explosion of enormous magnitude takes place.

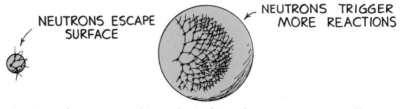

Fig. 40-4   The exaggerated view shows that a chain reaction in a small piece of pure U-235 runs its course, because neutrons leak from the surface too soon. The small piece has a lot of surface compared to mass. In a larger piece, more uranium and less surface is presented to the neutrons.

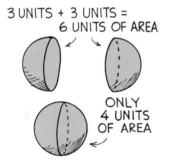

Fig. 40-5 Each piece is sub-critical. The amount of sur-face area is relatively large compared to the mass of ura-nium. When combined, the total surface area decreases, and neutron escape decreases.

In Figure 40-5 there are two pieces of pure U-235, each of them below the critical mass. They are subcritical. Neutrons readily reach a surface and escape before a sizeable chain reaction can build up. But if the pieces are suddenly driven together, the total surface area is decreased. If the timing is right and the com-bined mass is greater than critical, the device is a nuclear fis-sion bomb.

The construction of a fission bomb is not a formidable task. The difficulty is separating enough U-235 from the more abun-dant U-238. It took project scientists more than two years to ex-tract enough U-235 from uranium ore to make the bomb that was detonated on Hiroshima in 1945. Uranium isotope separa-tion is still a difficult process today.

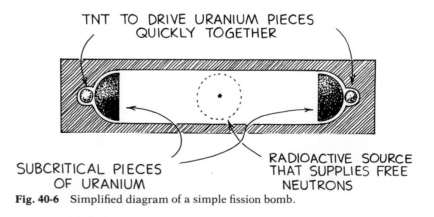

Fig. 40-6 Simplified diagram of a simple fission bomb.

> ► **Questions**
> 1. What is nuclear fission?
>
> 2. What is a chain reaction?
>
> 3. A kilogram of U-235 broken up into small chunks is not critical, but if the chunks are put together in a ball shape, it is critical. Why?

40.2   The Nuclear Fission Reactor

A better use for uranium is not bombs but power reactors. More than 20% of our electric energy is generated by nuclear fission reactors. These reactors are simply nuclear furnaces, which (like fossil fuel furnaces) do nothing more elegant than boil water to produce steam for a turbine (Figure 40-7). The greatest practical difference is the amount of fuel involved. One kilogram of uranium fuel, less than the size of a baseball, yields more energy than 30 freight-car loads of coal.

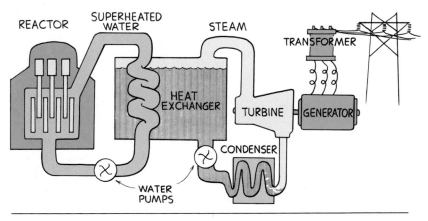

Fig. 40-7   Diagram of a nuclear fission power plant.

► **Answers**
1. Nuclear fission is the splitting of the atomic nucleus. When a heavy atom such as U-235 splits into two main parts, there is a release of much energy.

2. A chain reaction is a self-sustaining reaction that, once started, steadily provides the energy and matter necessary to continue the reaction.

3. A kilogram of U-235 in small chunks will not support a sustained reaction because of the relatively greater surface area of the chunks (like the greater combined surface area of gravel compared to the surface area of a boulder of the same mass). Neutrons escape from the surface before a sustained chain reaction can build up. But when the pieces are brought together, there is more uranium compared to surface, and a chain reaction occurs.

A reactor contains three components: the nuclear fuel, the control rods, and water used to transfer heat from the reactor to the generator. The nuclear fuel is primarily U-238 with about 3 percent U-235. Because the U-235 isotopes are so highly diluted with U-238, an explosion like that of a nuclear bomb is not possible. Control rods inserted into the reactor control the number of neutrons that initiate the fission of other U-235 nuclei. The control rods are made of a material, usually the metal cadmium or boron, which readily absorbs neutrons without undergoing a reaction itself. Heated water around the nuclear fuel transfers heat to a second water system, which operates the electric generator in a conventional fashion. There are two water systems because the water circulated near the nuclear fuel can become radioactive. For safety reasons, this water is kept away from the turbine and generator.

A major drawback to fission power is the waste products of fission. Recall that light atomic nuclei are most stable when composed of equal numbers of protons and neutrons, and that heavy nuclei need more neutrons than protons for stability. So there are more neutrons than protons in uranium—143 neutrons compared to 92 protons in U-235, for example. When uranium fissions into two medium-weight elements, the extra neutrons in their nuclei make them unstable. They are radioactive, with half lives that are typically thousands of years. Safely disposing of these waste products is a major problem.

> ▶ **Question**
>
> What is the function of the control rods in a nuclear reactor?

40.3 | Plutonium

When a neutron is absorbed by a U-238 nucleus, no fission results. The nucleus emits a beta particle instead and becomes the first synthetic element beyond uranium—the transuranic element called *neptunium* (named after the first planet discovered

---

▶ **Answer**

Control rods absorb neutrons and thereby control the amount of neutrons that participate in a chain reaction.

from the application of Newton's law of gravitation).* Neptunium, in turn, very soon emits a beta particle and becomes *plutonium* (named after Pluto, the second planet to be discovered via Newton's law). The isotope plutonium-239, like U-235, will undergo fission when it captures a neutron.

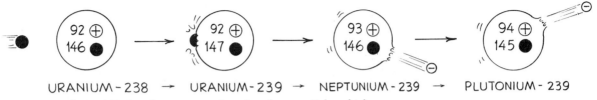

URANIUM - 238   →   URANIUM - 239   →   NEPTUNIUM - 239   →   PLUTONIUM - 239

Fig. 40-8 After U-238 absorbs a neutron, it emits a beta particle, which means that a neutron in the nucleus becomes a proton. The atom is no longer uranium, but neptunium. After the neptunium atom emits a beta particle it becomes plutonium.

The half life of neptunium is only 2.3 days, while the half life of plutonium is about 24 000 years. Since plutonium is an element distinct from uranium, it can be separated from uranium by ordinary chemical methods. Unlike the difficult process of separating U-235 from U-238, it is relatively easy to separate plutonium from uranium.

The element plutonium is chemically a poison in the same sense as are lead and arsenic. It attacks the nervous system and can cause paralysis. Death can follow if the dose is sufficiently large. Fortunately, plutonium does not remain in its elemental form for long because it rapidly combines with oxygen to form three compounds: PuO, PuO_2, and Pu_2O_3, all of which are chemically inert. They will not dissolve in water or in biological systems. These plutonium compounds do not attack the nervous system and have been found to be biologically harmless.

Plutonium in any form, however, is radioactively toxic. It is more toxic than uranium, although less toxic than radium. Plutonium emits high-energy alpha particles, which kill cells rather than simply disrupting them and leading to mutations. Interestingly enough, damaged cells rather than dead cells contribute to cancer. This is why plutonium ranks low as a cancer-producing substance. The greatest danger that plutonium presents to humans is its potential for use in nuclear fission bombs. Its usefulness is in breeder reactors.

---

* Transuranic elements at this writing extend to atomic number 109. See the periodic table of the elements on page 247.

40.4   The Breeder Reactor

When small amounts of Pu-239 are mixed with U-238 in a reactor, the fissioning of plutonium liberates neutrons that convert the relatively abundant, nonfissionable U-238 into more of the fissionable Pu-239. Not only is abundant energy produced, but fission fuel is bred from the relatively abundant U-238 in the process. A reactor with this fuel is a **breeder reactor**. Using a breeder reactor is like filling a gas tank in a car with water, adding some gasoline, then driving the car and having more gasoline after the trip than at the beginning, at the expense of common water! After the initial high costs of building such a device, this is a very economical method of producing vast amounts of energy. With a breeder reactor, a power utility can, after a few years of operation, breed twice as much fuel as it starts with.

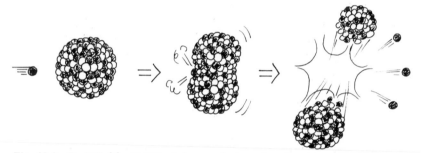

Fig. 40-9   Pu-239, like U-235, undergoes fission when it captures a neutron.

Fission power has several benefits. First, it supplies plentiful electricity. Second, it conserves the many billions of tons of coal, oil, and natural gas that every year are literally turned to heat and smoke, and which in the long run may be far more precious as sources of organic molecules than as sources of heat. Third, it eliminates the megatons of sulfur oxides and other poisons that are put into the air each year by the burning of these fuels.

The drawbacks include the problems of storing radioactive wastes, the production of plutonium and the danger of nuclear weapons proliferation, low-level release of radioactive materials into the air and ground water, and the risk of an accidental release of large amounts of radioactivity.

Reasoned judgement is not made by considering only the benefits or the drawbacks of fission power. You must also compare its benefits and its drawbacks to those of alternate power sources. All power sources have drawbacks of some kind. The benefits versus the drawbacks of fission power is a subject of much debate.

> ▶ **Question**
> In a breeder reactor, what is bred from what?

40.5 | **Mass-Energy Equivalence**

The key to understanding why a great deal of energy is released in nuclear reactions has to do with the equivalence of mass and energy. Recall from your study of special relativity in Chapter 16 that mass and energy are essentially the same—they are two sides of the same coin. Mass can be thought of as "squashed-up energy."

The more energy associated with a particle, the greater is the mass of the particle. Is the mass of a nucleon inside a nucleus the same as that of the same nucleon outside a nucleus? This question can be answered by considering the work that would be required to separate all the nucleons from a nucleus.

Recall that work, which is expended energy, is equal to the product of force and distance. Then think of the amount of force that would be required to pull nucleons apart through a sufficient distance to overcome the attractive nuclear strong force. Enormous work would be required. The work you would have to put into such a task would be manifest in the energy of the protons and neutrons that are pulled out. They would have more energy outside the nucleus—an amount equal to the energy or work required to separate them. This energy, in turn, would be manifest in mass. Therefore the mass of nucleons outside a nucleus is greater than the mass of the same nucleons when locked inside a nucleus. For example, in the units used to measure masses of atoms and atomic particles, a carbon-12 atom has a mass of exactly 12 units.* However, a proton has a mass of 1.00728 units (outside the nucleus), and a neutron has a mass of 1.00866 units. The combined mass of six free protons and six free neutrons is greater than the mass of one carbon-12 atom with a nucleus of six protons and six neutrons.

The masses of ions of the isotopes of various elements can be very accurately measured with a *mass spectrograph* (Fig-

Fig. 40-10 Work is required to pull a nucleon from an atomic nucleus. This work increases the energy and hence the mass of the nucleon outside the nucleus.

▶ **Answer**
Fissionable Pu-239 is bred from nonfissionable U-238.

---

* These units are called *atomic mass units* and are the units for atomic mass used in chemistry.

ure 40-11). This important device is immersed in a magnetic field, which deflects ions into circular arcs. The greater the inertia (mass) of the ion, the more it resists deflection, and the greater the radius of its curved path. The magnetic force sweeps heavier ions into larger arcs and lighter ions into shorter arcs.

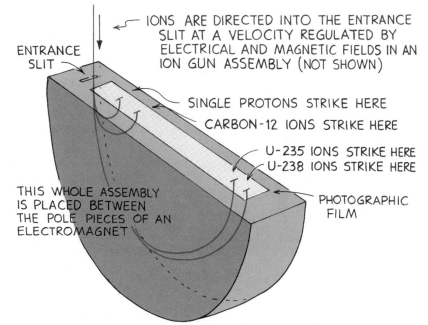

ENTRANCE SLIT

IONS ARE DIRECTED INTO THE ENTRANCE SLIT AT A VELOCITY REGULATED BY ELECTRICAL AND MAGNETIC FIELDS IN AN ION GUN ASSEMBLY (NOT SHOWN)

SINGLE PROTONS STRIKE HERE

CARBON-12 IONS STRIKE HERE

U-235 IONS STRIKE HERE
U-238 IONS STRIKE HERE

THIS WHOLE ASSEMBLY IS PLACED BETWEEN THE POLE PIECES OF AN ELECTROMAGNET

PHOTOGRAPHIC FILM

Fig. 40-11   The mass spectrograph. Ions are directed into the semicircular "drum," where they are swept into semicircular paths by a strong magnetic field. Because of inertia, heavier ions are swept into curves of large radii and lighter ions are swept into curves of smaller radii. The radius of the curve is directly proportional to the mass of the ion.

A graph of the nuclear masses for the elements from hydrogen through uranium is shown in Figure 40-12. The graph slopes upward with increasing atomic number as expected—elements are more massive as atomic number increases. The slope curves because there are proportionally more neutrons in the more massive atoms.

A more important graph results from the plot of nuclear mass *per nucleon* from hydrogen through uranium (Figure 40-13). To obtain the nuclear mass per nucleon, the nuclear mass is simply divided by the number of nucleons in the particular nucleus. (If you divided the mass of your whole class by the number of people in your class, you would get the average mass per person.) Note that the masses of the nucleons are different when combined in different nuclei. A proton has the greatest mass as the nucleus of a hydrogen atom, and has progressively less and less mass as it occurs in atoms of increasing atomic number. The proton has

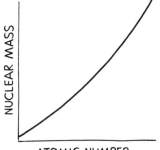

NUCLEAR MASS

ATOMIC NUMBER

Fig. 40-12   A graph that shows how nuclear mass increases with increasing atomic number.

the least mass when it is in the nucleus of the iron atom. Beyond iron, the process reverses itself as protons (and neutrons) have progressively more and more mass in atoms of increasing atomic number. This continues all the way to uranium and the transuranic elements.

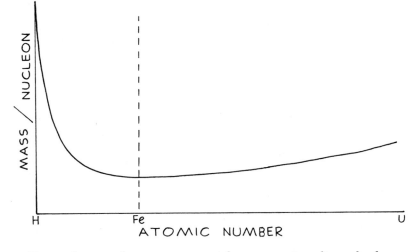

Fig. 40-13 The graph shows that the average mass of a nucleon depends on which nucleus it is in. Individual nucleons have the most mass in the lightest nuclei, the least mass in iron, and intermediate mass in the heaviest nuclei.

From the graph you can see why energy is released when a uranium nucleus is split into nuclei of lower atomic number. If a uranium nucleus splits in two, the masses of the fission fragments lie about halfway between uranium and hydrogen on the horizontal scale of the graph. Most importantly, note that the masses of these nucleons in the fission fragments are *less than* the masses of the same nucleons when combined in the uranium nucleus. The protons and neutrons decrease in mass. When this decrease in mass is multiplied by the speed of light squared, it is equal to the energy yielded by each uranium nucleus that undergoes fission.

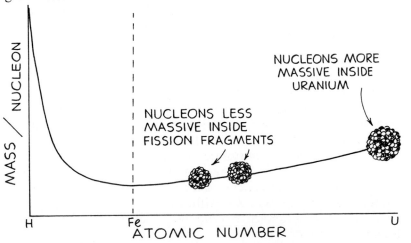
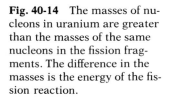

Fig. 40-14 The masses of nucleons in uranium are greater than the masses of the same nucleons in the fission fragments. The difference in the masses is the energy of the fission reaction.

You can think of the mass-per-nucleon graph as an energy hill which starts at hydrogen (the highest point) and drops steeply to the lowest point (iron), then rises gradually to uranium. Iron is at the bottom of the energy hill and is the most stable nucleus. It is also the strongest nucleus; more energy is required to pull each proton or neutron from the iron nucleus than from any other.

Fig. 40-15 The mass of a nucleus is *not* equal to the sum of the masses of its parts. Fission fragments of a heavy nucleus have less mass than the nucleus. Why is there a difference in mass?

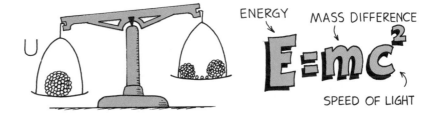

So the release of mass is detectable in the form of energy—much energy—when heavy nuclei undergo fission. A drawback to this process involves the fission fragments. They are radioactive isotopes because of the greater-than-normal number of neutrons for elements of these atomic numbers. A more promising source of energy is to be found on the left side of the energy hill.

40.6   Nuclear Fusion

Inspection of the graph of Figure 40-13 will show that the steepest part of the energy hill is from hydrogen to iron. Energy is gained as light nuclei *fuse*, or combine, rather than split apart. This process is **nuclear fusion**, the opposite of nuclear fission. Whereas energy is released when heavy nuclei split apart in the fission process, energy is released when light nuclei fuse together. After fusion, the nucleons of the light nuclei that have fused have less mass than they had before fusion (Figure 40-16).

Atomic nuclei are positively charged. For fusion to occur, they must collide at very high speeds in order to overcome electrical repulsion. The required speeds correspond to the extremely high temperatures found in the sun and stars. Fusion brought about by high temperatures is called **thermonuclear fusion**—that is, the welding together of atomic nuclei by high temperature. In the high temperatures of the sun, approximately 657 million tons of hydrogen are converted into 653 million tons of helium each second. The missing 4 million tons of mass is discharged as radiant energy. Such reactions are, quite literally, nuclear burning.

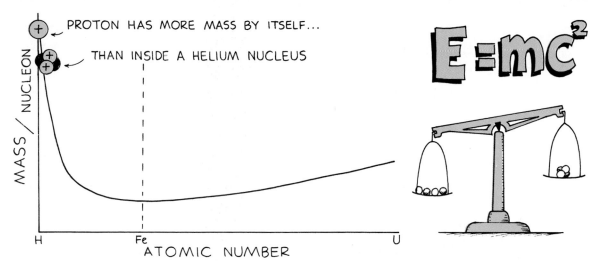

Fig. 40-16 (Left) A proton has more mass when it is the nucleus of the hydrogen atom than when it is part of a helium nucleus. The difference in mass is the energy of the fusion reaction. (Right) Two protons and two neutrons have more mass in their free states than when combined to form a helium nucleus.

Thermonuclear fusion is analogous to ordinary chemical combustion. In both chemical and nuclear burning, a high temperature starts the reaction; the release of energy by the reaction maintains a high enough temperature to spread the fire. The net result of the chemical reaction is a combination of atoms into more tightly bound molecules. In nuclear reactions, the net result is more tightly bound nuclei. The difference between chemical and nuclear burning is essentially one of scale.

▶ **Questions**

1. First it was stated that nuclear energy is released when atoms split apart. Now it is stated that nuclear energy is released when atoms combine. Is this a contradiction? How can energy be released by opposite processes?

2. To get energy from the element iron, should iron be fissioned or fused?

▶ **Answers**

1. Energy is released only in a nuclear reaction in which the mass of the nucleons decreases. Light nuclei, such as hydrogen, lose mass when they combine (fuse) to form heavier nuclei. They release energy in this reaction. Heavy nuclei, such as uranium, lose mass when they split to become lighter nuclei, and release energy in so doing.

2. Iron will release no energy at all, because it is at the very bottom of the energy hill. If fused, it climbs the right side of the hill and gains mass. If fissioned, it climbs the left side of the hill and gains mass. In gaining mass, it absorbs energy instead of releasing energy.

40.7 | Controlling Nuclear Fusion

To produce fusion reactions under controlled conditions requires temperatures of millions of degrees. Producing and sustaining such high temperatures is the goal of much current research. There are a variety of techniques for attaining high temperatures. No matter how the temperature is produced, a problem is that all materials melt and vaporize at the temperatures required for fusion. The solution to this problem is to confine the reaction in a nonmaterial container.

A magnetic field is nonmaterial, can exist at any temperature, and can exert powerful forces on charged particles in motion. "Magnetic walls" of sufficient strength provide a kind of magnetic straightjacket for hot gases called *plasmas*. Magnetic compression further heats the plasma to fusion temperatures.

Fig. 40-17 A magnetic bottle used for containing plasmas for fusion research.

At a temperature of about a million degrees, some nuclei are moving fast enough to overcome electrical repulsion and slam together, but the energy output is small compared to the energy used to heat the plasma. Even at 100 million degrees, more energy must be put into the plasma than will be given off by fusion. At about 350 million degrees, the fusion reactions will produce enough energy to be self-sustaining. At this *ignition temperature*, nuclear burning yields a sustained power output without further input of energy. A steady feeding of nuclei is all that is needed to produce continuous power.

Fusion has already been achieved in several devices, but instabilities in the plasma current have thus far prevented a sustained reaction. A big problem is devising a field system that will hold the plasma in a stable and sustained position while an ample number of nuclei fuse. A variety of magnetic confinement devices are the subject of much present-day research.

Another promising approach bypasses magnetic confinement altogether with high-energy lasers. One technique is to align an array of laser beams at a common point and drop solid pellets composed of hydrogen isotopes through the synchronous cross fire (Figure 40-18).

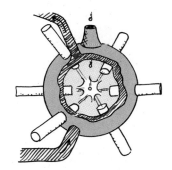

Fig. 40-18 Fusion with multiple laser beams. Pellets of deuterium are rhythmically dropped into synchronized laser cross fire. The resulting heat is carried off by molten lithium to produce steam.

Fig. 40-19   Pellet chamber at Lawrence Livermore Laboratory. The laser source is Nova, the most powerful laser in the world, which directs 10 beams into the target region.

Other fusion schemes involve the bombardment of fuel pellets not by laser light but by beams of electrons, light ions, and heavy ions. As this book goes to press we are still looking forward to the great day, *break-even day*, when one of the variety of fusion schemes will sustain a yield of at least as much energy as is required to initiate it.

Fusion power is near ideal. Fusion reactors cannot blow up like a bomb because fusion requires no critical mass. Furthermore, there is no air pollution because there is no combustion. Except for some radioactivity in the inner chamber of the fusion device because of high-energy neutrons, the byproducts of fusion are not radioactive. Fusion produces clean nonradioactive helium (good for children's balloons).

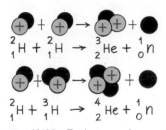

$$^2_1H + ^2_1H \rightarrow ^3_2He + ^1_0n$$

$$^2_1H + ^3_1H \rightarrow ^4_2He + ^1_0n$$

Fig. 40-20 Fusion reactions of hydrogen isotopes.

The fuel for nuclear fusion is hydrogen, the most plentiful element in the universe. The simplest reaction is the fusion of the hydrogen isotopes deuterium (H-2) and tritium (H-3), both of which are found in ordinary water. For example, 30 liters of sea water contains 1 gram of deuterium, which when fused releases as much energy as 10 000 liters of gasoline or 80 tons of TNT. Natural tritium is much scarcer, but given enough to get started, a controlled thermonuclear reactor will breed it from deuterium in ample quantities. Because of the abundance of fusion fuel, the amount of energy that can be released in a controlled manner is virtually unlimited.

Fusion power is expected not only to supply abundant energy, but will be a source of material as well. All the elements result from fusing more and more hydrogen nuclei together. For example, if you fuse 8 deuterium nuclei, you have oxygen; fuse 26 deuterium nuclei and you have iron, and so forth. Future humans will likely synthesize their own elements and produce energy in the process—just as the stars do.

Humans may one day travel to the stars in ships fueled by the same energy that makes the stars shine.

40 | Chapter Review

Concept Summary

Nuclear fission, the splitting of atomic nuclei, occurs when the repelling electrical forces in the nucleus exceed the attracting nuclear strong forces.

• Fission is initiated by the absorption of a neutron by a nucleus. It continues in a chain reaction in which more neutrons are ejected and triggers fission in other nuclei.

• A critical mass of a fissionable element is required for sustained fission to occur.

• In a breeder reactor, a small amount of fissionable plutonium-239 is mixed with non-fissionable uranium-238 and converts it into plutonium-239.

• Fission reactors are efficient energy producers but generate radioactive wastes.

Individual nucleons have most mass in nuclei of the lightest elements, least mass in iron, and intermediate mass in the heaviest elements.

• When fission occurs, the combined masses of the nucleons in the fission fragments are less than the masses of the same nucleons when in the nucleus that has undergone fission. The missing mass is equivalent to the great energy released.

In nuclear fusion, hydrogen nuclei fuse to form helium nuclei, releasing large amounts of energy.

• After fusion, the nucleons of the light nuclei that have fused have less mass than they had before fusion.

• Thermonuclear fusion occurs at the high temperatures found in the sun and stars.

• Sustained fusion under controlled conditions is not yet possible.

Important Terms

breeder reactor (40.4)
chain reaction (40.1)
critical mass (40.1)
nuclear fission (40.1)
nuclear fusion (40.6)
thermonuclear fusion (40.6)

Review Questions

1. What is the role of electrical forces in nuclear fission? (40.1)

2. What is the role of a neutron in nuclear fission? (40.1)

3. Of what use are the neutrons that are produced when a nucleus undergoes fission? (40.1)

4. Why does a chain reaction not occur in uranium mines? (40.1)

5. a. Which isotope of uranium is most common?
   b. Which isotope of uranium will fission? (40.1)

6. Which has more total surface area—an apple or an apple cut in half? (40.1)

7. Which has more total surface area—two separate pieces of uranium or the same pieces stuck together? (40.1)

8. Which will leak more neutrons—two separate pieces of uranium or the same pieces stuck together? (40.1)

9. Will an energetic chain reaction be more likely in two separate pieces of U-235 or in the same pieces stuck together? (40.1)

10. What controls the chain reaction in a nuclear reactor? (40.2)

11. Are the fission fragments from a nuclear reactor light, medium, or heavy elements? (40.2)

12. Why are the fission-fragment elements radioactive? (40.2)

13. What happens when U-238 absorbs a neutron? (40.3)

14. How can plutonium be created? (40.3)

15. Is plutonium an isotope of uranium or is it a completely different element? (40.3)

16. What is the effect of putting a little Pu-239 with a lot of U-238? (40.4)

17. Does a nucleon have more mass, or less mass, when outside of an atomic nucleus? (40.5)

18. What device can be used to measure the relative masses of ions of isotopes? (40.5)

19. What is the primary difference in the graphs shown in Figures 40-12 and 40-13? (40.5)

20. What becomes of the loss in mass of nucleons when heavy atoms split? (40.5)

21. Why does helium not yield energy if fissioned? (40.5)

22. Why does uranium not yield energy if fused? (40.6)

23. Why does iron not yield energy if fissioned or fused? (40.6)

24. What becomes of the loss in mass of nucleons when light atoms fuse to become heavier ones? (40.6)

25. Why are fusion reactors not yet a present day reality like fission reactors? (40.7)

Think and Explain

1. Why does a neutron make a better nuclear bullet than a proton?

2. Suppose that fission or fusion were the only two choices for future generation of electric energy. Which of the two would you prefer? Why?

3. How is chemical burning similar to a nuclear chain reaction?

4. If a piece of uranium is flattened into a pancake shape, will this make an energetic chain reaction more likely or make it less likely? Why?

5. Does the total surface area increase or decrease when pieces of fissionable material are assembled into one piece?

6. Why are there no appreciable deposits of plutonium in the earth's crust?

7. Is the mass of an atomic nucleus greater or less than the masses of the nucleons that compose it?

8. The energy release of nuclear fission is tied to the fact that nucleons in the heaviest nuclei weigh about 0.1 percent more than when in medium-weight nuclei. What would be the effect on energy release if the 0.1 percent figure were 1 percent?

9. To predict the approximate energy release of either a fission or a fusion reaction, explain how a physicist makes use of the curve of Figure 40-13, or a table of nuclear masses, and the equation $E = mc^2$.

10. Which process would release energy from lead—fission or fusion? From aluminum? From iron?

11. If a uranium nucleus were to split into three pieces of approximately the same size instead of two, would more energy or less energy be released? Defend your answer in terms of Figure 40-13.

12. If three hydrogen nuclei were fused instead of two, would more energy or less energy be released? Defend your answer in terms of Figure 40-13.

Appendix A: Units of Measurement

The units of measurement primarily used in this book are those used by scientists throughout the world—the International System of Units, or SI (after the French name, Système International). SI units are the outgrowth of the metric system of units. While familiar to scientists, many SI units are not generally familiar to students in high school. The SI units used in this book are the following.

Meter

The meter (m) is the SI unit of length. The standard of length for the metric system originally was defined in terms of the distance from the North Pole to the equator. This distance is close to (or was thought to be at the time) 10 million meters. So one meter equals one ten-millionth of the distance from the North Pole to the equator. A more exact definition is that one meter equals the length of the path traveled by light in a vacuum during a time interval of $\frac{1}{299\,792\,458}$ of a second.

Common SI length units based on the meter are the *centimeter*, *millimeter*, and *kilometer*.

$$1 \text{ centimeter (cm)} = \frac{1}{100} \text{ meter}$$

$$1 \text{ millimeter (mm)} = \frac{1}{1000} \text{ meter}$$

$$1 \text{ kilometer (km)} = 1000 \text{ meters}$$

Kilogram

The kilogram (kg), the SI unit of mass, is defined as the mass of a block of platinum preserved at the International Bureau of Weights and Measures in France. The kilogram originally was defined as the mass of one liter (1000 cubic centimeters) of water at the temperature at which it is most dense (now known to be 4° Celsius). (The mass of a one-pound object is equal to 0.4536 kilogram.)

Other common mass units are the *gram* and *milligram*.

$$1 \text{ gram (g)} = \frac{1}{1000} \text{ kilogram}$$

$$1 \text{ milligram (mg)} = \frac{1}{1000} \text{ gram}$$
$$\frac{1}{1\,000\,000} \text{ kilogram}$$

Second

The second (s) is the SI unit of time. Until 1956 the second was defined in terms of the mean solar day, which was divided into 24 hours. Each hour was divided into 60 minutes and each minute into 60 seconds. Thus there were 86 400 seconds per day, and the second was defined as $\frac{1}{86\,400}$ of the mean solar day. This was found to be unsatisfactory because the rate of rotation of the earth is gradually becoming slower. In 1956 the mean solar day of the year 1900 was chosen as the standard on which to base the second. Since 1964, the second has been officially defined as the time taken by a cesium-133 atom to make 9 192 631 770 vibrations.

Newton

The newton (N), the SI unit of force, is named after Sir Isaac Newton. One newton is the force required to give an object with a mass of one kilogram an acceleration of one meter per second squared. (One newton is approximately equal to 0.2 pound.)

Joule

The joule (J), the SI unit of energy, is named after James Joule. One joule is equal to the amount of work done by a force of one newton acting over a distance of one meter. For heat energy, the joule replaces the calorie. (One calorie is equal to 4.187 joules.)

The unit for power is derived from the unit for energy. Power is the rate at which energy is expended. Work done at the rate of one joule per second is equal to a power of one *watt* (W). The *kilowatt* (kW) equals 1000 watts. From the definition of power, it follows that energy can be expressed as the product of power and time. Electric energy is often expressed in units of *kilowatt-hours* (kWh), where

$$1 \text{ kilowatt-hour} = 3.60 \times 10^6 \text{ joules}$$

Ampere

The ampere (A), the SI unit of electric current, is named after André Marie Ampère. In this text the ampere is defined as the rate of flow of one coulomb of charge per second, where one coulomb is the charge of 6.25×10^{18} electrons. The official definition of the ampere is the intensity of constant electric current maintained in two parallel conductors of infinite length and negligible cross section that when placed one meter apart in a vacuum would produce between them a force of 2×10^{-7} newton per meter of length.

Kelvin

The kelvin (K), the SI unit of temperature, is named after the scientist Lord Kelvin. The kelvin is defined as $1/273.16$ of the temperature change between absolute zero (the coldest possible temperature) and the triple point of water (the fixed temperature at which ice, liquid water, and water vapor coexist in equilibrium). Temperatures are expressed in kelvins, and not in "degrees kelvin." On the Kelvin scale, absolute zero is 0 K. The temperature of melting ice at atmospheric pressure is 273.15 K, the triple point of water is 273.16 K, and the temperature of pure boiling water at atmospheric pressure is 373.15 K. There are 100 kelvins between the melting and boiling points of water, just as there are 100 Celsius degrees between these points. Kelvins and Celsius degrees have the same spacings on the temperature scale.

Measurements of Area and Volume

Area   Area refers to the amount of surface. The unit of area is the surface of a square that has a standard unit of length as a side. In the SI system it is a square with sides one meter in length, which makes a unit of area of one square meter ($1 \ m^2$). A smaller unit area is represented by a square with sides one centimeter in length, which makes a unit of area of one square centimeter ($1 \ cm^2$).

The area of a rectangle equals the base times the height. The area of a circle is equal to πr^2, where $\pi = 3.14$ and r is the radius of the circle. Formulas for the surfaces of other shapes can be found in geometry textbooks.

Volume   The volume of an object refers to the space it occupies. The unit volume is the space taken up by a cube that has a standard unit of length for its edge. In the SI system, it is the space occupied by a cube whose sides are one meter. This volume is one cubic meter ($1 \ m^3$) and is a relatively large volume by everyday standards. A smaller unit volume is the space occupied by a cube whose sides are one centimeter. Its volume is one cubic centimeter ($1 \ cm^3$), the space taken up by one gram of water at 4°C.

A liter (L) is equal to 1000 cm^3, and is a common measure of volume for liquids.

Appendix B: Working with Units in Physics

A quantity in science is expressed by a number and a unit of measurement. Quantities may be actual measurements, or they may be obtained by performing calculations on measurements. Quantities may be added, subtracted, multiplied, or divided. There are rules for handling both the numbers and the units of measurement during these mathematical operations.

Addition

When you add quantities, all must have the *same* units. Add up the numbers. The sum has the same unit as well.

Example:
(4 m) + (8 m) + (3 m) = 15 m

Subtraction

When you subtract one quantity from another, both must have the *same* units. Subtract the numbers. The difference has the same unit.

Example:
(5.2 s) − (3.8 s) = 1.4 s

Multiplication

Quantities that are multiplied together need *not* have the same units. Multiply the numbers. Multiply the units just as if they are algebraic variables.

When full names of units are used, use a hyphen between the units that are multiplied together.

Example:
(3 newtons) × (2 meters) = 6 newton-meters

When symbols are used, use a raised dot between the unit symbols that are multiplied together.

Example:
(3 N) × (2 m) = 6 N·m

When the units being multiplied are the same, the product is called the square (or cubic) unit.

In symbols, a raised 2 after the unit symbol is used for the square. A raised 3 after the unit symbol is used for the cubic unit. These raised numerals are known as *exponents*.

Examples:
(3 meters) × (2 meters) = 6 meter-meters
$\qquad\qquad\qquad\qquad$ = 6 square meters

(3 m) × (2 m) = 6 m·m = 6 m^2

(3 meters) × (2 meters) × (4 meters)
$\qquad\qquad\qquad$ = 24 meter-meter-meters
$\qquad\qquad\qquad$ = 24 cubic meters

(3 m) × (2 m) × (4 m) = 24 m·m·m = 24 m^3

Division

Quantities that are divided by each other need *not* have the same units. Divide the numbers. Divide the units as though they are algebraic variables.

When the units are full names, use the word *per* after the unit that is being divided.

Example:
(100 kilometers) ÷ (2 hours)

$$= \frac{100 \text{ kilometers}}{2 \text{ hours}}$$

$$= 50 \text{ kilometers per hour}$$

When the units are symbols, use a slash after the unit symbol that is being divided.

Example:

$$(100 \text{ km}) \div (2 \text{ h}) = \frac{100 \text{ km}}{2 \text{ h}}$$

$$= 50 \text{ km/h}$$

When both units are the same, they "cancel" out and do not appear in the quotient.

Example:

$$(6 \text{ m}) \div (3 \text{ m}) = \frac{6 \; \cancel{m}}{3 \; \cancel{m}}$$

$$= 2$$

Complicated Multiplication and Division

In multiplication, when the quantities have units which are quotients of units, treat them as algebraic variables. Identical units in the numerator and denominator may be "canceled" out.

Example:
(25 meters per second) × (6 seconds)

$$= \left(25 \ \frac{\text{meters}}{\text{second}} \right) \times (6 \text{ seconds})$$

$$= 25 \times 6 \ \frac{\text{meters-seconds}}{\text{second}}$$

$$= 150 \text{ meters}$$

$$(25 \text{ m/s}) \times (6 \text{ s}) = \left(25 \ \frac{\text{m}}{\text{s}} \right) \times (6 \text{ s})$$

$$= 25 \times 6 \ \frac{\text{m·s}}{\text{s}}$$

$$= 150 \text{ m}$$

In division, when the quantities have units which are quotients of units, it is easiest to express the division in numerator and denominator form. That is, the number to be divided is the numerator (top value) and the divisor is the denominator (bottom value). Divide the numbers. Treat units as algebraic variables.

Examples:
(8.2 meters per second) ÷ (2.0 seconds)

$$= \frac{8.2 \text{ meters per second}}{2.0 \text{ seconds}}$$

$$= \frac{8.2}{2.0} \ \frac{\text{meters}}{\text{second-second}}$$

$$= 4.1 \text{ meters per second squared}$$

$$(8.2 \text{ m/s}) \div (2.0 \text{ s}) = \frac{8.2 \text{ m/s}}{2.0 \text{ s}}$$

$$= \frac{8.2}{2.0} \ \frac{\text{m}}{\text{s·s}}$$

$$= 4.1 \text{ m/s}^2$$

Note that when *second* is multiplied by itself in the denominator, it is changed to *per second squared* (and not to *per square second*). Similarly, the symbols "m/s²" are read as "meters per second squared."

Scientific Notation

It is convenient to use a mathematical abbreviation for large and small numbers. The number 40 000 000 can be obtained by multiplying 4 by 10, and again by 10, and again by 10, and so on until 10 has been used as a multiplier seven times. The short way of showing this is to write the number 40 000 000 as 4×10^7.

The number 0.0004 can be obtained from 4 by using 10 as a divisor four times. The short way of showing this is to write the number 0.0004 as 4×10^{-4}. Thus,

$$2 \times 10^5 = 2 \times 10 \times 10 \times 10 \times 10 \times 10 = 200\,000$$
$$5 \times 10^{-3} = 5/(10 \times 10 \times 10) = 0.005$$

Numbers expressed in this shorthand manner are said to be in *scientific notation*.

$$
\begin{aligned}
1\,000\,000 &= 10 \times 10 \times 10 \times 10 \times 10 \times 10 = 10^6 \\
100\,000 &= 10 \times 10 \times 10 \times 10 \times 10 = 10^5 \\
10\,000 &= 10 \times 10 \times 10 \times 10 = 10^4 \\
1\,000 &= 10 \times 10 \times 10 = 10^3 \\
100 &= 10 \times 10 = 10^2 \\
10 &= 10 = 10^1 \\
1 &= 1 = 10^0 \\
0.1 &= 1/10 = 10^{-1} \\
0.01 &= 1/100 = 10^{-2} \\
0.001 &= 1/1\,000 = 10^{-3} \\
0.0001 &= 1/10\,000 = 10^{-4} \\
0.00001 &= 1/100\,000 = 10^{-5} \\
0.000001 &= 1/1\,000\,000 = 10^{-6}
\end{aligned}
$$

We can use scientific notation to express some of the physical data often used in physics.

Table B-1   Some Important Values in Physics

Speed of light in a vacuum $= 2.9979 \times 10^8$ m/s
Average earth-sun distance (1 astronomical unit (A.U.)) $= 1.50 \times 10^{11}$ m
Average earth-moon distance $= 3.84 \times 10^8$ m
Average radius of the sun $= 6.96 \times 10^8$ m
Average radius of Jupiter $= 7.14 \times 10^7$ m
Average radius of the earth $= 6.37 \times 10^6$ m
Average radius of the moon $= 1.74 \times 10^6$ m
Average radius of the hydrogen atom $= 5 \times 10^{-11}$ m
Mass of the sun $= 1.99 \times 10^{30}$ kg
Mass of Jupiter $= 1.90 \times 10^{27}$ kg
Mass of the earth $= 5.98 \times 10^{24}$ kg
Mass of the moon $= 7.36 \times 10^{22}$ kg
Proton mass $= 1.6726 \times 10^{-27}$ kg
Neutron mass $= 1.6749 \times 10^{-27}$ kg
Electron mass $= 9.1 \times 10^{-31}$ kg
Electron charge $= 1.602 \times 10^{-19}$ C

Appendix C: Vector Applications

Appendix C is a continuation of the material in Chapter 6, "Vectors," and Chapter 27, "Light." One of the most fascinating illustrations of the vector approach to looking at things is the sailboat—how it sails in directions other than with the wind, and even into the wind. Understanding the role of vectors in sailing is a bit more complicated than in other examples treated in Chapter 6, so it is set aside here for more careful study.

Another striking illustration of vectors involves the vector nature of light and its behavior in passing through polarization filters. The vector treatment of light here is a continuation of Chapter 27. Vector explanations for both the sailboat and the transmission of light through polarizing filters involve a blend of geometry and physics.

The Sailboat

Sailors have always known that a sailboat can sail downwind (in the same direction as the wind). The ships of Columbus were designed to sail only downwind. Not until modern times did sailors learn that a sailboat can sail upwind (against the wind). It turns out that many types of sailboats can sail faster "cutting" upwind than when sailing directly with the wind. The oldtimers didn't know this, probably because they didn't understand vectors and vector components. Luckily, we do, and today's sailboats are far more maneuverable than the sailboats of the past.

To understand all this, first consider the relatively simple case of sailing downwind. Figure C-1 shows a force vector F due to the impact of the wind against the sail. This force tends to increase the speed of the boat. If it were not for resistive forces, mainly water drag, the speed of the boat would build up to nearly the speed of the wind. (It could be pushed no faster than wind speed because the wind would no longer make impact with the sails. They would sag and the force F would shrink to zero.) It is important to note that the faster the boat goes, the smaller will be the magnitude of F. So we see that a sailboat sailing directly with the wind can sail no faster than the wind.

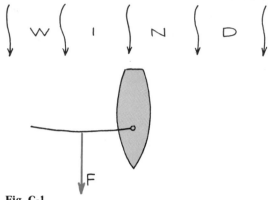

Fig. C-1

If the sail is oriented as shown in Figure C-2 left, the boat will move forward but with less increase in speed for two reasons. First, the force F on the sail is less because the sail does not intercept as much wind at this angle. Second, the force on the sail is not in the direction of the boat's motion. It is instead perpendicular to the sail's surface. Generally speaking, whenever any fluid (liquid or gas) interacts with a smooth surface, the force of interaction is perpendicular to the smooth surface. In this case the boat will not move in the direction of F because of its deep finlike keel, which knifes through the water and resists motion in sideways directions.

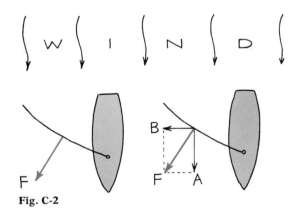

Fig. C-2

We can understand the motion of the boat by resolving F into perpendicular components, as shown in Figure C-2 right. The important com-

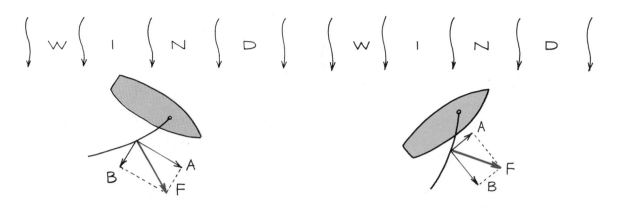

Fig. C-3

ponent is the one parallel to the keel and is labeled *A*. Component *A* propels the boat forward. The other component (*B*) is useless and tends to tip the boat over and move it sideways. The tendency to tip is offset by the heavy deep keel. Again, maximum speed can only approach wind speed.

When a sailboat's keel points in a direction other than exactly downwind and its sails are properly oriented, it can exceed wind speed. In the case of cutting across at an angular direction to the wind (Figure C-3 left), the wind continues to make impact with the sail even after the boat achieves wind speed. A surfer, in a similar way, exceeds the speed of the propelling wave by angling her surfboard across the wave. Greater angles to the propelling medium (wind for the boat, water wave for the surfboard) result in greater speeds. Can you see why a sailcraft can sail faster cutting across the wind than it can sailing downwind?

As strange as it may seem to people who do not understand vectors, maximum speed is attained by cutting into (against) the wind—that is, by angling the sailcraft in a direction upwind (Figure C-3 right)! Although a sailboat cannot sail *directly* upwind, it can reach a destination upwind by angling back and forth in zigzag fashion. This is called *tacking*. As the speed increases, the wind impact, rather than decreasing, actually *increases*. (If you run outdoors in a slanting rain, the drops will hit you harder if you run into the rain than away from the rain!) The faster the

boat moves as it tacks upwind, the greater the magnitude of *F*. Thus, component *A* will continue pushing the boat along in the forward direction. The boat reaches its terminal speed when opposing forces, mainly water drag, balance the force of wind impact.

Icecraft which are equipped with runners for sliding on ice encounter no water drag. They can travel at several times wind speed when they tack upwind. Terminal speed is reached not so much because of resistive forces but because the wind direction shifts relative to the moving craft. When this happens, the wind finally moves parallel to the sail rather than against it. This appendix will not go into detail about this complication; nor will it discuss the curvature of the sail, which also plays an important role.

The central idea underlying sailcraft is the concept of vectors. It was this concept that ushered in the era of clipper ships and revolutionized the sailing industry. Sailing, like most things, is more enjoyable if you understand what is happening.

The Vector Nature of Light

Recall from Chapter 27 that light is electromagnetic energy that travels as a transverse, electromagnetic wave. The wave is made up of an oscillating electric field vector and an oscillating magnetic field vector that is at right angles to the electric vector (Figure C-4 on the next page). It is the orientation of the electric vector that defines the direction of polarization of light.

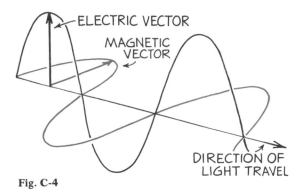

Fig. C-4

The electric vectors in waves of light from the sun or from a lamp vibrate in all conceivable directions as they move. Such light is non-polarized. When the electric vectors of the waves are aligned parallel to each other, the light is considered to be polarized. Light can be polarized when it passes through polarizing filters. The most familiar are Polaroid sunglasses. Regular light incident upon a polarizing filter emerges as polarized light.

Think of a beam of nonpolarized light coming straight toward you. Consider the electric vectors in that beam. Some of the possible directions of the vibrations are as shown in Figure C-5 left. There are as many vectors in the horizontal direction as there are in the vertical direction, since the light is nonpolarized. The center sketch shows the light falling on a polarizing filter with its polarization axis vertically oriented. Only vertical components of light pass through the filter, and the light that emerges is vertically polarized, as shown on the right.

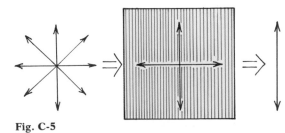

Fig. C-5

Figure C-6 shows that no light can pass through a pair of Polaroid sunglasses when their axes are at a right angle to one another, but some light does pass through when their axes are at a non-

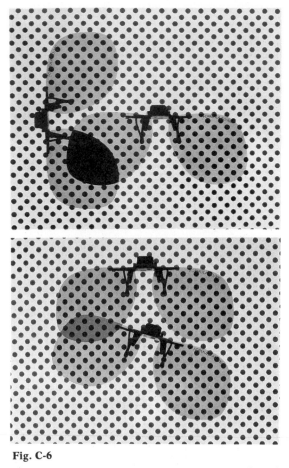

Fig. C-6

right angle. This fact can be understood with vectors and vector components.

Recall from Chapter 6 that any vector can be thought of as the sum of two components at right angles to each other. The two components are often chosen to be in the horizontal and vertical directions, but they can be in *any* two perpendicular directions. In fact, the number of sets of perpendicular components possible for any

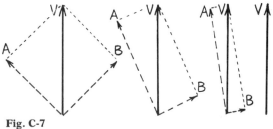

Fig. C-7

vector is infinite. A few of them are shown for the vector V in Figure C-7. In every case components A and B make up the sides of a rectangle that has V as its diagonal.

You can see this somewhat differently by thinking of component A as always being vertical and B as being horizontal, and picturing vector V as rotating instead (Figure C-8). This time the different orientations of V are superimposed on a polarizing filter with its polarization axis vertical. In the first sketch on the left,

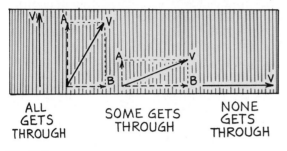

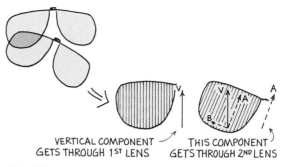

VERTICAL COMPONENT GETS THROUGH 1ST LENS   THIS COMPONENT GETS THROUGH 2ND LENS

Fig. C-9

ALL GETS THROUGH   SOME GETS THROUGH   NONE GETS THROUGH

Fig. C-8

all of V gets through. As V rotates, only the vertical component A passes through, and it gets shorter and shorter until it is zero when V is completely horizontal.

Can you now understand how light gets through the second pair of Polaroid sunglasses in Figure C-6? Look at Figure C-9, where for clarity the two crossed lenses of Figure C-6 that are one atop the other are instead shown side by side. The vector V that emerges from the first lens is vertical. However, it has a component A in the direction of the polarization axis of the second lens. Component A passes through the second lens, while component B is absorbed.

To really appreciate this, you must toy around with a couple of polarizing filters, which you can do in a lab exercise. Rotate one above the other and see how you can regulate the amount of light that gets through. Can you think of practical uses for such a system?

▶ **Question**

Consider a pair of polarizing filters crossed so that no light gets through. If you place a third filter *in front of* the pair, still no light gets through. The same is true if you place a third filter *in back of* the pair. But if you sandwich a third filter *between* the two with its polarization axis in a different direction from the other two, light *does* get through! (See Figure 27-20 on page 394.) Magic? No, just physics. Can you explain why this happens?

▶ **Answer**

Work on this one and doodle with vectors. If you solve it, help your classmates if they've tried without success and want help. As a last resort, ask your teacher for help.

Appendix D: Exponential Growth and Doubling Time*

You can't fold a piece of paper in half, then fold it again upon itself successively for 9 times. It gets too thick to keep folding. And if you could fold a fine piece of tissue paper upon itself 50 times, it would be more than 20 million kilometers thick! The continual doubling of a quantity builds up astronomically. Double one penny 30 times, so that you begin with one penny, then have two pennies, then four, and so on, and you'll accumulate a total of $10 737 418.23! One of the most important things we have trouble perceiving is the process of exponential growth, and why it proliferates out of control.

When a quantity such as money in the bank, population, or the rate of consumption of a resource steadily grows at a fixed percent per year, the growth is said to be *exponential*. Money in the bank may grow at 8 percent per year; world population is presently growing at about 2 percent per year; the electric power generating capacity in the United States grew at about 7 percent per year for the first three quarters of the century. The important thing about exponential growth is that the time required for the growing quantity to double in size (increase by 100 percent) is constant. For example, if the population of a growing city takes 10 years to double from 10 000 to 20 000 people and its growth remains steady, in the next 10 years the population will double to 40 000, and in the next 10 years to 80 000, and so on.

---

* This appendix is adapted from material written by University of Colorado physics professor Albert A. Bartlett, who strongly asserts, "The greatest shortcoming of the human race is man's inability to understand the exponential function." Look up Professor Bartlett's still timely and provocative article, "Forgotten Fundamentals in the Energy Crisis," in the September 1978 issue of the *American Journal of Physics*, or his revised version in the January 1980 issue of the *Journal of Geological Education*.

** For exponential decay we speak about *half life*, the time for a quantity to reduce to half its value. An example of this case is radioactive decay, treated in Chapter 39.

There is an important relationship between the percent growth rate and its *doubling time*, the time it takes to double a quantity:**

$$\text{doubling time} = \frac{69.2\%}{\text{percent growth per unit time}}$$

$$\approx \frac{70\%}{\text{percent growth rate}}$$

This means that to estimate the doubling time for a steadily growing quantity, we simply divide 70% by the percentage growth rate. For example, the 7-percent-per-year growth rate of electric power generating capacity in the United States means that in the past the capacity has doubled every 10 years (since [70%]/[7%/year] = 10 years). A 2-percent-per-year growth rate for world population means that the population of the world doubles every 35 years (since [70%]/[2%/year] = 35 years). A city planning commission that accepts what seems like a modest 3.5-percent-per-year growth rate may not realize that this means that doubling will occur in 20 years (since [70%]/[3.5%/year] = 20 years). That means double capacity for such things as water supply, sewage-treatment plants, and other municipal services every 20 years.

Steady growth in a steadily expanding environment is one thing, but what happens when steady growth occurs in a finite environment? Consider the growth of bacteria that grow by division, so that one bacterium becomes two, the two divide to become four, the four divide to become eight, and so on. Suppose the division time for a certain kind of bacteria is one minute. This is then steady growth—the number of bacteria grows exponentially with a doubling time of one minute. Further, suppose that one bacterium is put in a bottle at 11:00 a.m. and that growth

▶ **Question**
When was the bottle half full?

▶ **Answer**
At 11:59 a.m., since the bacteria will double in number every minute!

continues steadily until the bottle becomes full of bacteria at 12 noon. Consider the question at the bottom of the previous page.

It is startling to note that at 2 minutes before noon the bottle was only ¼ full, and at 3 minutes before noon only ⅛ full. Table D-1 summarizes the amount of space left in the bottle in the last few minutes before noon. If bacteria could think, and if they were concerned about their future, at which time do you think they would sense they were running out of space? Do you think a serious problem would have been evident at, say, 11:55 a.m., when the bottle was only 3-percent full (1/32) and had 97 percent open space (just yearning for development)? The point here is that there isn't much time between the moment the effects of growth become noticeable and the time when they become overwhelming.

Table D-1	The last minutes in the bottle	
Time	Portion full	Portion empty
11:54 a.m.	$\frac{1}{64}$ (1.5%)	$\frac{63}{64}$ (98.5%)
11:55 a.m.	$\frac{1}{32}$ (3 %)	$\frac{31}{32}$ (97 %)
11:56 a.m.	$\frac{1}{16}$ (6 %)	$\frac{15}{16}$ (94 %)
11:57 a.m.	$\frac{1}{8}$ (12 %)	$\frac{7}{8}$ (88 %)
11:58 a.m.	$\frac{1}{4}$ (25 %)	$\frac{3}{4}$ (75 %)
11:59 a.m.	$\frac{1}{2}$ (50 %)	$\frac{1}{2}$ (50 %)
12:00 noon	Full (100 %)	None (0 %)

Suppose that at 11:58 a.m. some farsighted bacteria see that they are running out of space and launch a full-scale search for new bottles. And further suppose they consider themselves lucky, for they find three new empty bottles. This is three times as much space as they have ever known. It may seem to the bacteria that their problems are solved—and just in time.

▶ **Question**
If the bacteria are able to migrate to the new bottles and their growth continues at the same rate, what time will it be when the three new bottles are filled to capacity?

▶ **Answer**
All four bottles will be filled to capacity at 12:02 p.m.!

Table D-2 illustrates that the discovery of the new bottles extends the resource by only two doubling times. In this example the resource is

Table D-2	Effects of the discovery of three new bottles
Time	Effect
11:58 a.m.	Bottle 1 is ¼ full; bacteria divide into four bottles, each $\frac{1}{16}$ full
11:59 a.m.	Bottles 1, 2, 3, and 4 are each ⅛ full
12:00 noon	Bottles 1, 2, 3, and 4 are each ¼ full
12:01 p.m.	Bottles 1, 2, 3, and 4 are each ½ full
12:02 p.m.	Bottles 1, 2, 3, and 4 are each all full

space—such as land area for a growing population. But it could be coal, oil, uranium, or any nonrenewable resource.

Continued growth and continued doubling lead to enormous numbers. In two doubling times, a quantity will double twice ($2^2 = 4$), or quadruple in size; in three doubling times, its size will increase eightfold ($2^3 = 8$); in four doubling times, it will increase sixteenfold ($2^4 = 16$); and so on.

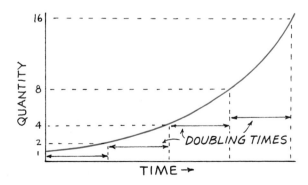

Fig. D-1   Graph of a quantity that grows at an exponential rate. Notice that the quantity doubles during each of the successive equal time intervals marked on the horizontal scale. Each of these time intervals represents the doubling time.

> **Question**
> According to a French riddle, a lily pond starts with a single leaf. Each day the number of leaves doubles, until the pond is completely full on the thirtieth day. On what day was the pond half covered? One-quarter covered?

This is best illustrated by the story of the court mathematician in India who years ago invented the game of chess for his king. The king was so pleased with the game that he offered to repay the mathematician, whose request seemed modest enough. The mathematician requested a single grain of wheat on the first square of the chessboard, two grains on the second square, four on the third square, and so on, doubling the number of grains on each succeeding square until all squares had been used. At this rate there would be 2^{63} grains of wheat on the sixty-fourth square alone. The king soon saw that he could not fill this "modest" request, which amounted to more wheat than had been harvested in the entire history of the earth!

As Table D-3 shows, the number of grains on any square is one grain more than the total of all grains on the preceding squares. This is true anywhere on the board. For example, when four grains are placed on the third square, that number of grains is one more than the total of three grains already on the board. The number of grains (eight) on the fourth square is one more than the total of seven grains already on the board. The same pattern occurs everywhere on the board. In any case of exponential growth, a greater quantity is represented in one doubling time than in all the preceding growth. This is important enough to be repeated in different words: Whenever steady growth occurs, the numerical count of a quantity that exists after a single doubling time is one greater than the total count of that quantity in the entire history of growth.

Fig. D-2 A single grain of wheat placed on the first square of the chess board is doubled on the second square, and this number is doubled on the third square, and so on. There is not enough wheat in the world for this process to continue to the 64th square!

Table D-3 Filling the squares on the chessboard		
Square number	Grains on square	Total grains thus far
1	1	1
2	2	3
3	$4 = 2^2$	7
4	$8 = 2^3$	15
5	$16 = 2^4$	31
6	$32 = 2^5$	63
7	$64 = 2^6$	127
•	•	•
•	•	•
•	•	•
64	2^{63}	$2^{64} - 1$

The consequences of unchecked exponential growth are staggering. It is very important to ask: Is growth really good? Is bigger really better? Is it true that if we don't continue growing we will stagnate?

> **Answer**
> The pond was half covered on the 29th day, and was one-quarter covered on the 28th day!

Glossary

This glossary gives meanings for all terms printed in boldface within the text, and explains commonly used symbols as well. The section reference at the end of each meaning is that of the section where the term is introduced.

A simple, phonetic spelling is given for the terms that may be unfamiliar or hard to pronounce. CAPITAL LETTERS indicate the syllable that receives the heaviest stress. Accent marks are used when two syllables in a word are stressed; a lower-case syllable followed by an accent mark receives the secondary stress. The phonetic spellings are simple enough so that most can be interpreted without referring to the following key, which gives examples for the vowel sounds and for consonants that are commonly used for more than one sound.

Pronunciation Key

a	cat	ew	new	or	for
ah	father	g	grass	ow	now
ar	car	i, ih	him	oy	boy
aw	walk	ī	kite	s	so
ay	say	j	jam	sh	shine
ayr	air	ng	sing	th	thick
e, eh	hen	o	hot	u, uh	sun, forces
ee	meet	ō	hole	z	zebra
eer	deer	oo	moon	zh	pleasure
er	her	o͝o	pull		

A The symbol for *ampere*. (34.2) Also, when in lower-case italic, the symbol for *acceleration*. (2.4)

aberration (ab-er-RAY-shun) The unavoidable distortion in an image produced by a lens. (30.8)

absolute zero The temperature at which a substance has no kinetic energy to give up. This temperature corresponds to 0 K, or to -273°C. (21.1)

acceleration (ak-sel'-er-RAY-shun) The rate at which velocity is changing. The change may be in magnitude, direction, or both. (2.4)

action force One of the pair of forces described in Newton's third law. (5.2)

air resistance The friction that acts on something moving through air. (4.5)

alternating current (ac) Electric current that rapidly reverses in direction, usually at the rate of 60 hertz (in North America) or 50 hertz (in most other places). (34.7)

ampere (AM-peer) The SI unit of electric current. One ampere (symbol A) is equal to a flow of one coulomb of charge per second. (34.2)

amplitude (AMP-lih-tewd) The distance from the midpoint to the crest of a wave or, equivalently, from the midpoint to the trough. (25.2)

aneroid barometer (AN-er-oyd buh-ROM-uh-ter) An instrument used to measure atmospheric pressure; based on the movement of the lid of a metal box, rather than the movement of a liquid. (20.4)

angle of incidence (IN-sih-dens) The angle between an incident ray and the normal to a surface (see Figure 29–3). (29.2)

angle of reflection The angle between a reflected ray and the normal to a surface (see Figure 29–3). (29.2)

angular momentum (mo-MEN-tum) The "inertia of rotation" of a rotating object, equal to the product of rotational inertia and rotational velocity. (14.6)

apogee (AP-uh-jee) The point in an elliptical orbit where an object is farthest away from the object about which it orbits. (12.4)

apparent weightlessness The feeling of weightlessness that one has when falling toward or around the earth (as in an orbiting spacecraft). True weightlessness, however, requires that an object be far out in space, where gravitational forces are negligible. (11.3)

Archimedes' principle (ark-uh-MEE-deez) The relationship between buoyancy and displaced fluid: An immersed object is buoyed up by a force equal to the weight of the fluid it displaces. (19.3)

astigmatism (uh-STIG-muh-tizm) A defect of the eye caused when the cornea is curved more in one direction than in another. (30.7)

atom The smallest particle of an element that can be identified with that element. It consists of protons and neutrons in a nucleus surrounded by electrons. (17.1)

atomic mass number The total number of nucleons (neutrons and protons) in the nucleus of an atom. (39.4)

atomic number The number of protons in the nucleus of an atom. (17.7, 39.4)

average speed The total distance covered divided by the time interval. (2.2)

axis (AK-sis) The straight line about which rotation takes place. (13.1)

barometer An instrument used for measuring the pressure of the atmosphere. (20.3)

beats A throbbing variation in the loudness of sound caused by interference when two tones of slightly different frequencies are sounded together. (26.9)

Bernoulli's principle (ber-NOO-leez) The statement that the pressure in a fluid decreases as the speed of the fluid increases. (20.7)

bimetallic strip (bī'-meh-TAL'-ik) Two strips of different metals, such as one of brass and one of iron, that are welded or riveted together; used in thermostats. Because the two substances expand at different rates, when heated or cooled they bend in different directions. (22.1)

black hole A massive star that has collapsed to so great a density that its enormous local gravitational field prevents light from escaping and it thus appears black. (11.6)

blue shift An increase in the measured frequency of light from an approaching source; called the *blue shift* because the increase is toward the high-frequency, or blue, end of the spectrum. (25.9)

boiling The change of state from liquid to gas that occurs beneath the surface of the liquid. The gas that forms beneath the surface occurs as bubbles, which rise to the surface and escape. (24.4)

bow wave The V-shaped wave produced by an object moving across a liquid surface at a speed greater than the wave speed. (25.10)

Boyle's law The statement that the product of pressure and volume for a given mass of gas is a constant as long as the temperature does not change. (20.5)

breeder reactor A nuclear fission reactor that not only produces power but produces more nuclear fuel than it consumes by converting a nonfissionable uranium isotope into a fissionable plutonium isotope. (40.4)

Brownian motion The random movement observed among microscopic particles suspended in a fluid medium. (17.4)

buoyancy (BOY-un-see) The apparent loss of weight of an object submerged in a fluid. (19.2)

buoyant force (BOY-unt) The net upward force exerted by a fluid on a submerged object. (19.2)

C The symbol for *coulomb*. (32.3) Also, when preceded by the degree symbol °, the symbol for *Celsius*. (21.1)

cal The symbol for *calorie*. (21.5)

calorie (KAL-er-ee) A unit of heat. One calorie (symbol cal) is the amount of heat required to raise the temperature of one gram of water by 1°C. One Calorie (with a capital *C*) is equal to one thousand calories and is the unit used in describing the energy available from food. (21.5)

Celsius scale (SEL-see-us) A temperature scale in which the number 0 is assigned to the temperature at which water freezes, and the number 100 is assigned to the temperature at which water boils (at standard pressure). (21.1)

center of gravity The point at the center of an object's weight distribution, where the force of gravity can be considered to act. (9.1)

center of mass The point at the center of an object's mass distribution, where all its mass can be considered to be concentrated. For everyday conditions, it is the same as the center of gravity. (9.2)

centrifugal force (sen-TRIH-fuh-gul) An apparent outward force experienced by a rotating body. It is ficticious in the sense that it is not part of an interaction but is due to the tendency of a moving body to follow a straight-line path. (13.4)

centripetal force (sen-TRIH-peh-tul) A center-seeking force that causes an object to follow a circular path. (13.3)

chain reaction A self-sustaining reaction that, once started, steadily provides the energy and matter necessary to continue the reaction. (40.1)

charge The property to which is attributed the mutual repulsion of two electrons or two protons, and the mutual attraction of an electron and a proton. Also, the sum of all the electron and proton charges on an object (allowing for the cancelling effect of equal numbers of like and unlike charges). (32.1)

chemical formula A description, in terms of numbers and of symbols for elements, of the proportions of each kind of atom in a compound. (17.6)

circuit (SER-kit) Any complete path along which charge can flow. (35.1)

coherent (kō-HEER-ent) Type of light beam in which the waves all have the same frequency, phase, and direction. Lasers produce coherent light. (31.7)

complementary colors (kom'-pluh-MENT'-uh-ree) Two colors of light beams which when added together appear white. (28.6)

component (kom-PŌ-nent) One of the vectors in different directions whose vector sum is equal to a given vector. Any single vector may be regarded as the vector sum of two components, each of which acts in a different direction. (6.6)

compound A chemical substance made of atoms of two or more different elements combined in a fixed proportion. (17.6)

condensation (kon'-den-SAY'-shun) (a) The change of state of a gas into a liquid; the opposite of evaporation. (24.2) (b) In sound, a pulse of compressed air (or other matter). (26.2)

conduction A means of heat transfer within certain materials and from one material to another when the two are in direct contact. It involves the transfer of energy from atom to atom. (23.1)

conductor (a) A material through which heat can flow. (23.1) (b) A material, usually a metal, through which electric charge can flow. Good conductors of heat are generally good conductors of charge. (32.4)

conservation of charge The principle that net electric charge is neither created nor destroyed but simply transferred from one material to another. (32.2)

conserved Term applied to a physical quantity, such as momentum, energy, or electric charge, that remains unchanged after some interaction. (7.4)

constructive interference Addition of two waves when the crest of one wave overlaps the crest of another, so that their individual effects add together. The result is a wave of increased amplitude. (25.7)

convection A means of heat transfer by movement of the heated substance itself, such as by currents in a fluid. (23.2)

converging lens A lens that is thickest in the middle and that causes parallel rays of light to converge to a focus. (30.1)

cornea (KOR-nee-uh)   The transparent covering over the eye. (30.6)

correspondence principle   The principle that for a new theory to be valid, it must account for the verified results of the old theory in the region where both theories are applicable. (16.4)

coulomb (KOO-lōm)   The SI unit of charge. One coulomb (symbol C) is equal to the total charge of 6.25×10^{18} electrons. (32.3)

Coulomb's law   The relationship among electrical force, charges, and distance: The electrical force between two charges varies directly as the product of the charges and inversely as the square of the distance between them. (32.3)

crest   One of the places in a wave where the wave is highest or the disturbance is greatest. (25.2)

critical angle   The minimum angle of incidence at which a light ray is totally reflected within a medium. (29.12)

critical mass   The minimum mass of fissionable material in a nuclear reactor or nuclear bomb that will sustain a chain reaction. (40.1)

crystal (KRIS-tul)   A regular geometric shape found in a solid in which the component particles are arranged in an orderly, three-dimensional, repeating pattern. (18.1)

current   See *electric current.*

density (DEN-sih-tee)   A property of a substance, equal to the mass divided by the volume; commonly thought of as the "lightness" or "heaviness" of a substance. (18.2)

destructive interference   Addition of two waves when the crest of one wave overlaps the trough of another, so that their individual effects cancel each other. The result is a wave of decreased amplitude. (25.7)

diffraction (dih-FRAK-shun)   The bending of a wave around a barrier, such as an obstacle or the edges of an opening. (31.2)

diffraction grating   A series of closely-spaced parallel slits which are used to separate colors of light by interference. (31.4)

diffuse reflection (dih-FYOOS)   The reflection of waves in many directions from a rough surface (see Figure 29–7). (29.4)

direct current (dc)   Electric current whose flow of charge is always in one direction only. (34.7)

dispersion (dih-SPER-zhun)   The separation of light into colors arranged according to their frequency, for example by interaction with a prism or diffraction grating. (29.10)

displaced   Term applied to the fluid that is moved out of the way when an object is placed in the fluid. A completely submerged object always displaces a volume of fluid equal to its own volume. (19.2)

diverging lens   A lens that is thinnest in the middle and that causes parallel rays of light to diverge. (30.1)

Doppler effect (DOP-ler)   The apparent change in frequency of a wave due to the motion of the source or the receiver. (25.9)

eddy   Changing, curling paths in turbulent flow of a fluid. (20.7)

efficiency   The ratio of useful work output to total work input, or the percentage of the work put into a machine that is converted to useful work output. (8.8)

elapsed time   The time that has passed, or elapsed, since the beginning of the time measurement. (2.5)

elastic   Term applied to a material that returns to its original shape after it has been stretched or compressed. (18.3)

elastic collision   A collision in which colliding objects rebound without lasting deformation or the generation of heat. (7.5)

elasticity (ih-las-TIH-sih-tee)   The property of a body or material by which it experiences a change in shape when a deforming force acts on it and by which it returns to its original shape when the deforming force is removed. (18.3)

elastic limit   The distance of stretching and compressing beyond which an elastic material will not return to its original state. (18.3)

electrical force   A force one electric charge exerts on another. When the charges are both positive or both negative, the force is repulsive; when the charges are unlike, the force is attractive. (32.1)

electrical resistance   The resistance of a material to the flow of an electric current through it; measured in ohms. (34.4)

electrically polarized   Term applied to an atom or molecule in which the charges are aligned so that one side is slightly more positive or negative than the opposite side. (32.7)

electric charge   See *charge.*

electric current   The flow of electric charge; measured in amperes. (34.2)

electric field   A force field that fills the space around every electric charge or group of charges. Another electric charge introduced into this region will experience an electric force acting on itself. (33.1)

electric potential The electric potential energy per charge at a location in an electric field; measured in volts and often called *voltage*. (33.5)

electric potential energy The energy a charge possesses by virtue of its location in an electric field. (33.4)

electric power The rate at which electric energy is converted into another form, such as light, heat, or mechanical energy (or converted *from* another form). (34.10)

electromagnet (ih-lek′-trō-MAG′-net) A magnet whose field is produced by an electric current. Usually in the form of a wire coil with a piece of iron inside the coil. (36.5)

electromagnetic induction (ih-lek′-trō-mag-NET′-ik in-DUK-shun) The phenomenon of inducing a voltage in a conductor by changing the magnetic field around the conductor. (37.1)

electromagnetic spectrum The range of electromagnetic waves extending from radio waves to gamma rays. (27.3)

electromagnetic wave A wave that is partly electric and partly magnetic and that carries energy emitted by vibrating electric charges in atoms. (27.3)

electrostatics (ih-lek′-trō-STAT′-iks) The study of electric charges at rest. (32.1)

element A substance made of only one kind of atom. Examples of elements are carbon, hydrogen, oxygen, and nitrogen. (17.1)

ellipse (ih-LIPS) An oval-shaped curve that is the path taken by a point that moves such that the sum of its distances from two fixed points (foci) is constant (see Figure 12–7). (12.3)

energy That property of an object or a system which enables it to do work; measured in joules. (8.3)

equilibrium (ee-kwih-LIH-bree-um) In general, a state of balance. In particular: (a) The state of a body on which no net force acts. (6.5) (b) The state of a body on which no net torque acts. (14.2) (c) The state of a liquid in which the processes of evaporation and condensation are taking place at equal rates. (24.3)

escape speed The minimum speed necessary for an object to escape permanently from a gravitational field which holds it. (12.5)

evaporation (ih-vap′-or-AY′-shun) The change of state from liquid to gas that takes place at the surface of a liquid. (24.1)

eyepiece The lens of a telescope that is closer to the eye; enlarges the real image formed by the first lens. (30.5)

fact A close agreement by competent observers of a series of observations of the same phenomena. (1.4)

Fahrenheit scale (FA-ren-hīt) The temperature scale in common use in the United States. The number 32 is assigned to the freezing point of water, and the number 212 to the boiling point of water (at standard atmospheric pressure). (21.1)

family A group of elements in the same column of the periodic table. Elements within a family have similar chemical properties and have the same number of electrons in the outer shell. (17.8)

Faraday's law (FA-ruh-dayz) The statement that the induced voltage in a coil is proportional to the product of the number of loops and the rate at which the magnetic field changes within those loops. (37.2) In general, the statement that an electric field is induced in any region of space in which a magnetic field is changing with time. The magnitude of the induced electric field is proportional to the rate at which the magnetic field changes. (37.7)

farsighted Term applied to a person who has trouble focusing on nearby objects because the eyeball is so short that images form behind the retina. (30.7)

field See *force field*.

first postulate of special relativity The statement that all the laws of nature are the same in all uniformly moving frames of reference. (15.4)

fission See *nuclear fission*.

fluid Anything that flows; in particular, any liquid or gas. (4.5)

focal length The distance between the center of a lens and either focal point. (30.1)

focal plane A plane that passes through either focal point of a lens and is perpendicular to the principal axis. For a converging lens, any incident parallel beam of light converges to a point somewhere on a focal plane. For a diverging lens, such a beam appears to come from a point on a focal plane. (30.1)

focal point For a converging lens, the point at which a beam of light parallel to the principal axis converges. For a diverging lens, the point from which such a beam appears to come. (30.1)

focus (FŌ-kus); pl. foci (FŌ-sī) For an ellipse, one of the two points for which the sum of the distances to any point on the ellipse is a constant. A satellite orbiting the earth moves in an ellipse which has the earth at one focus. (12.3)

force Any influence that tends to accelerate an object; commonly, a push or pull; measured in newtons. (3.3)

forced vibration   The vibration of an object that is made to vibrate by another vibrating object that is nearby. The sounding board in a musical instrument amplifies the sound through forced vibration. (26.5)

force field   What fills the space around a mass, electric charge, or magnet, so that another mass, electric charge, or magnet introduced to this region will experience a force. Examples of force fields are gravitational fields, electric fields, and magnetic fields. (11.2)

free fall   Motion under the influence of the gravitational force only. (2.5)

freezing   The change in state from liquid to solid. (24.5)

frequency (FREE-kwen-see)   The number of vibrations per unit of time; measured in hertz. (25.2)

friction   The force that acts to resist the relative motion (or attempted motion) of objects or materials that are in contact. (3.3)

fulcrum (FOOL-krum)   The pivot point of a lever. (8.7)

fusion   See *nuclear fusion*.

g   The symbol for *gram*.   Also, when in lower-case italic, the symbol for *the acceleration due to gravity at the earth's surface*, that is, 9.8 m/s^2. (2.5) When in upper-case italic, the symbol for *the universal constant of gravitation*, that is, 6.67 x 10^{-11} N·m^2/kg^2. (10.4)

generator   A machine that produces electric current by rotating a coil within a stationary magnetic field. (37.3)

gravitational field (grav'-ih-TAY'-shun-ul)   A force field that fills the space around every mass. Another mass in this region will experience a gravitational force. (11.2)

greenhouse effect   The warming effect whose cause is that short-wavelength radiant energy from the sun can enter the atmosphere and be absorbed by the earth more easily than long-wavelength energy from the earth can leave. (23.7)

grounding   Allowing charges to move freely along a connection between a conductor and the ground. (32.6)

h   The symbol for *hour*. (2.2)   Also, when in italic, the symbol for *Planck's constant*. (38.2)

half life   The time required for half the atoms of a radioactive isotope of an element to decay. (39.5)

heat   The energy that is transferred from one material to another because of a temperature difference

between the materials. Once the energy is absorbed by matter, it is internal energy. (21.2)

hertz (HERTS)   The SI unit of frequency. One hertz (Hz) equals one vibration per second. (25.2)

hologram (HOL-uh-gram)   A three-dimensional version of a photograph produced by interference patterns of laser beams. (31.8)

Hooke's law   The statement that the amount of stretch or compression of an elastic material is directly proportional to the applied force. (18.3)

Huygens' principle (HĪ-gunz)   The statement that every point on any wave front can be regarded as a new point source of secondary waves. (31.1)

hypothesis (hī-POTH-uh-sis)   An educated guess; a reasonable explanation of an observation or experimental result that is not fully accepted as factual until tested over and over again by experiment. (1.3)

Hz   The symbol for *hertz*. (25.2)

impulse (IM-puls)   The product of force multiplied by the time interval during which the force acts. The impulse is equal to the change in momentum. (7.2)

incoherent (in'-kō-HEER'-ent)   Type of light beam in which the waves are out of phase with each other. (31.7)

induced (in-DEWSD)   Term applied to electric charge that has been redistributed on an object because of the presence of a charged object nearby. (32.6)   Also, term applied to a voltage, electric field, or magnetic field that is created due to a change in or motion through a magnetic field or electric field. (37.1, 37.7)

induction (in-DUK-shun)   The charging of an object without direct contact. (32.6) See also *electromagnetic induction*.

inelastic   Term applied to a material that does not return to its original shape after it has been stretched or compressed. (18.3)

inelastic collision   A collision in which the colliding objects become distorted and generate heat during the collision. (7.5)

inertia (ih-NER-shuh)   The resistance of any material object to change in its state of motion. (3.3)

infrared   Electromagnetic waves of frequencies lower than the red of visible light. (27.3)

infrasonic (in'-fruh-SON'-ik)   Term applied to sound of pitch too low to be heard by the human ear, that is, of pitch below 20 hertz. (26.1)

in parallel   Term applied to portions of an electric circuit that are connected at two points and provide alternative paths to the current between those two points. (35.2)

in phase (FAYZ)   Term applied to two or more water waves whose crests (and troughs) arrive at a place at the same time, so that their effects reinforce each other. (25.7)

in series   Term applied to portions of an electric circuit that are connected in a row so that the current that goes through one must go through all of them. (35.2)

instantaneous speed (in-stan-TAY-nee-us)   The speed at any instant of time. (2.2)

insulator (IN-suh-lay-ter)   A material that is a poor conductor of heat and that delays the transfer of heat. (23.1)   Also, a material that is a poor conductor of electricity. (32.4)

interference pattern (in'-ter-FEER'-ens)   A pattern formed by the overlapping of two or more waves that arrive in a region at the same time. (25.7)

internal energy   The total energy inside a substance. (21.4)

inversely   When two values change in opposite directions, so that if one is doubled, the other is reduced to one half, they are said to be inversely proportional to each other. (4.2)

ion (Ī-un)   An atom (or group of atoms bound together) with a net electric charge, due to the loss or gain of electrons. (17.8)

iridescence (ih-rih-DES-ens)   The phenomenon whereby the interference of light waves of mixed frequencies reflected from the top and bottom of thin films causes a spectrum of colors. (31.6)

iris (Ī-ris)   The colored part of the eye, which surrounds the black opening through which light passes and which regulates the amount of light entering the eye. (30.6)

isotope (Ī-suh-tōp)   A form of an element having a particular number of neutrons in the nuclei of its atoms. Different isotopes of a particular element have the same atomic number but different atomic mass numbers. (17.7, 39.4)

J   The symbol for *joule*. (8.1)

joule (JOOL)   The SI unit of work and of all other forms of energy as well. One joule (symbol J) of work is done when a force of one newton is exerted on an object that moves a distance of one meter in the direction of the force. (8.1)

K   The symbol for *kelvin*. (21.1)   Also, when in lower case, the symbol for the prefix *kilo-*.

kcal   The symbol for *kilocalorie*. (21.5)

kelvin (KEL-vin)   The SI unit of temperature. A temperature measured in kelvins (symbol K) indicates the number of units above absolute zero. Since the divisions on the Kelvin scale and the Celsius scale are the same size, a change in temperature of one kelvin is equal to a change in temperature of 1°C. (21.1)

Kelvin scale   A temperature scale calibrated in terms of energy itself as well as in terms of the freezing and boiling points of water. Absolute zero (-273°C) is taken as 0 K. There are no negative temperatures on the Kelvin scale. (21.1)

kg   The symbol for *kilogram*. (3.5)

kilocalorie (KIL'-uh-kal'-er-ee)   A unit of heat. One kilocalorie equals 1000 calories, or the amount of heat required to raise the temperature of one kilogram of water by 1°C. (21.5)

kilogram (KIL-uh-gram)   The fundamental SI unit of mass. One kilogram (symbol kg) is the amount of mass in one liter of water at 4°C. (3.5)

kinetic energy (kih-NET-ik)   The energy of motion. It is equal to half the mass multiplied by the square of the speed. (8.5)

km   The symbol for *kilometer*. (2.2)

L   The symbol for *liter*. (19.3)

laser (LAY-zer)   An optical instrument that produces a beam of coherent light, that is, a beam in which the waves all have the same frequency, phase, and direction. (31.7)

law   A general hypothesis or statement about the relationship of natural quantities that has been tested over and over again and has not been contradicted. Also known as a *principle*. (1.4)

law of conservation of angular momentum   The statement that an object or system of objects will maintain a constant angular momentum unless acted upon by an unbalanced external torque. (14.7)

law of conservation of energy   The statement that energy cannot be created or destroyed; it may be transformed from one form into another, but the total amount of energy never changes. (8.6)

law of conservation of momentum   The statement that in the absence of a net external force, the momentum of an object or system of objects remains unchanged. (7.4)

law of inertia   The statement that every body continues in its state of rest, or of motion in a straight line at constant speed, unless it is compelled to change that state by a net force exerted upon it. Also known as *Newton's first law*. (3.4)

law of reflection   The statement that when a wave strikes a surface, the angle of incidence is equal to the angle of reflection. It holds true for both partially and totally reflected waves. (29.2)

law of universal gravitation   The statement that for any pair of objects, each object attracts the other object with a force that is directly proportional to the mass of each object, and inversely proportional to the square of the distance between their centers of mass. (10.4)

lens (LENZ)   A piece of glass (or other transparent material) that can bend parallel rays of light so that they cross, or appear to cross, at a single point. (30.1)

lever (LEH-ver, LEE-ver)   A simple machine, made of a bar that turns about a fixed point. (8.7)

lever arm   For a force that tends to cause rotation about an axis and that is perpendicular to the line between the point of contact and the axis, the length of that line. (14.1)

lift   In the application of Bernoulli's principle, the net upward force produced by the difference between upward and downward pressures. When the lift equals the weight, horizontal flight is possible. (20.8)

light year   The distance traveled by light in one year. (27.2)

line spectrum   The pattern of distinct lines of color, corresponding to particular wavelengths, that are seen in the spectroscope when a hot gas is viewed. (28.11)

linear momentum   The product of the mass and the velocity of an object. Also called simply *momentum*. (14.6)

linear speed   The distance moved per unit of time. Also called simply *speed*. (13.2)

longitudinal wave (lon-jih-TEWD-ih-nul)   A wave in which the vibration is in the same direction as that in which the wave is traveling, rather than at right angles to it. (25.6)

lunar eclipse   The cutoff of light from the full moon when the earth is directly between the sun and the moon, so that the earth's shadow is cast on the moon. (11.4)

m   The symbol for *meter*. (2.2)   Also, when in italic, the symbol for *mass*. (3.5)

machine   A device for multiplying (or decreasing) forces or simply changing the direction of forces. (8.7)

magnetic domain   A microscopic cluster of atoms with their magnetic fields aligned. (36.4)

magnetic field   A force field that fills the space around every magnet or current-carrying wire. Another magnet or current-carrying wire introduced into this region will experience a magnetic force acting on itself. (36.2)

magnetic pole   One of the regions on a magnet that produce magnetic forces. (36.1)

mass   A measure of the quantity of matter a body contains; may also be considered a measure of the inertia of an object. (3.5)

mechanical advantage   The ratio of output force to input force for a machine. (8.7)

mechanical energy   The energy due to the position or the movement of something; potential or kinetic energy (or a combination of both). (8.3)

mirage (mih-RAHZH)   A floating image that appears in the distance and is due to the refraction of light in the earth's atmosphere. (29.9)

molecule (MOL-uh-kyool)   Two or more atoms of the same or different elements joined to form a larger particle. (17.5)

momentum   The product of the mass and the velocity of an object. Has direction as well as size. Also called *linear momentum*. (7.1)

monochromatic (mon'-ō-krō-MAT'-ik)   Having a single color or frequency. (31.4)

N   The symbol for *newton*.

natural frequency   A frequency at which an elastic object naturally tends to vibrate, so that minimum energy is required to produce a forced vibration or to continue vibration at that frequency. (26.6)

neap tide   A tide that occurs when the moon is halfway between a new and full moon, in either direction. The tides due to the sun and the moon partly cancel, so that the high tides are lower than average and the low tides are not as low as average. (11.4)

nearsighted   Term applied to a person who can see nearby objects clearly but not distant objects. (30.7)

net force   The combination of all the forces that act on an object. (4.1)

neutral equilibrium   The state of an object balanced so that any small rotation neither raises nor lowers its center of gravity. (9.5)

neutron   An electrically neutral particle that is one of the two kinds of particles found in the nucleus of an atom. (17.7)

newton   The SI unit of force. One newton (symbol N)

is the force that will give an object of mass one kilogram an acceleration of one meter per second squared. (3.5)

Newton's first law See *law of inertia*. (3.4)

Newton's law of cooling The statement that the rate of cooling of an object—whether by conduction, convection or radiation—is approximately proportional to the temperature difference between the object and its surroundings. (23.6)

Newton's second law The statement that the acceleration produced by a net force on a body is directly proportional to the magnitude of the net force, in the same direction as the net force, and inversely proportional to the mass of the body. (4.3)

Newton's third law The statement that whenever one body exerts a force on a second body, the second body exerts an equal and opposite force on the first. (5.2)

node Any part of a standing wave that remains stationary. (25.8)

normal A line that is perpendicular to a surface. (29.2)

normal force For an object resting on a horizontal surface, the upward force that balances the weight of the object. Also called the *support force*. (4.4)

nuclear fission (FIH-shun) The splitting of an atomic nucleus, particularly that of a heavy element such as uranium-235, into two main parts, accompanied by the release of much energy. (40.1)

nuclear fusion (FYOO-zhun) The combining of nuclei of light atoms, such as hydrogen, into heavier nuclei, accompanied by the release of much energy. (40.6)

nucleon (NEW-klee-on) The principal building block of the nucleus; a neutron or a proton. (17.7, 39.1)

nucleus The positively charged center of an atom, which contains protons and neutrons and has almost all the mass of the entire atom but only a tiny fraction of the volume. (17.7)

objective lens In an optical device using compound lenses, the lens closest to the object observed. (30.5)

ohm (ŌM) The SI unit of electrical resistance. One ohm (symbol Ω) is the resistance of a device that draws a current of one ampere when a voltage of one volt is impressed across it. (34.4)

Ohm's law The statement that the current in a circuit is directly proportional to the voltage impressed across the circuit and inversely proportional to the resistance of the circuit. (34.5)

opaque Term applied to materials that absorb light without re-emission and thus do not allow light through them. (27.5)

optical fiber A transparent fiber, usually of glass or plastic, that can transmit light down its length by means of total internal reflection. (29.12)

out of phase Term applied to two waves for which the crest of one wave arrives at a point at the same time as a trough of the second wave arrives. Their effects cancel each other. (25.7)

parallel circuit An electric circuit in which devices are connected to the same two points of the circuit, so that any single device completes the circuit independently of the others. (35.4)

pascal (pas-KAL) The SI unit of pressure. One pascal (symbol Pa) of pressure exerts a force of one newton per square meter of surface. (4.6)

Pascal's principle The statement that changes in pressure at any point in an enclosed fluid at rest are transmitted undiminished to all points in the fluid and act in all directions. (19.6)

penumbra A partial shadow which appears where some of the light is blocked and other light fills it in. (27.6)

perigee (PEH-rih-jee) The point in an elliptical orbit where an object is nearest the object about which it orbits. (12.4)

period The time required for a complete orbit. (12.2) Also, the time required for a pendulum to make one to-and-fro swing. In general, the time required to complete one cycle. (25.1)

periodic table A chart that lists elements by their atomic number and by their electron arrangements, so that elements with similar chemical properties are in the same column (see Figure 17–11). (17.8)

perturbation The deviation of an orbiting object from its normal path, caused by an additional gravitational force. (10.6)

photoelectric effect The ejection of electrons from certain metals when exposed to light. (38.3)

photon (FŌ-ton) In the particle model of electromagnetic radiation, a particle that travels at the speed of light and whose energy is related to the frequency of the radiation in the wave model. (27.1, 38.2)

pigment A material that selectively absorbs colored light. (28.3)

pitch Term that refers to how high or low a sound appears to be. (26.1)

Planck's constant   The quantity that results when the energy of a photon is divided by its frequency. (38.2)

plasma (PLAZ-muh)   A fourth state of matter, in addition to a solid, liquid, and gas. In the plasma state, which exists only at high temperatures, matter consists of bare atomic nuclei and free electrons. (17.9)

polarization (pō′-ler-ih-ZAY′-shun)   The filtering out of all vibrations in a transverse wave, such as a light wave, that are not in a given direction. (27.7)

postulate (POS-tyoo-lit)   A fundamental assumption. (15.3)

potential   See *electric potential*.

potential difference   The difference in electric potential, or voltage, between two points. Charge will flow when there is a difference, and will continue until both points reach a common potential. (34.1)

potential energy   Energy that is stored and held in readiness by an object by virtue of its position. In this stored state it has the potential for doing work. (8.4)

power   The rate at which work is done, equal to the amount of work done divided by the amount of time during which the work is done; measured in watts. (8.2)

pressure   The force per unit of surface area, where the force is perpendicular to the surface; measured in pascals. (4.6)

principal axis   The line joining the centers of curvature of the surfaces of a lens. (30.1)

principle   A general hypothesis or statement about the relationship of natural quantities that has been tested over and over again and has not been contradicted. Also known as a *law*. (1.4)

principle of flotation   The statement that a floating object displaces a quantity of fluid of weight equal to its own weight. (19.5)

projectile   Any object that is projected by some force and continues in motion by its own inertia. (6.8)

proton   A positively charged particle that is one of the two kinds of particles found in the nucleus of an atom. (17.7)

pulley   A type of lever that is a wheel with a groove in its rim and that is used to change the direction of a force. A pulley or system of pulleys can also multiply forces. (8.7)

pupil   The opening in the eyeball through which light passes. (30.6)

quantum (pl. **quanta**) (KWONT-um)   An elemental unit; the smallest amount of something. One quantum of light energy is called a *photon*. (38.2)

quantum mechanics   The branch of physics that is the study of the motion of quanta in the microworld of the atom. (38.8)

quantum physics   The branch of physics that is the general study of quanta in the microworld of the atom. (38.8)

quark (KWORK, KWARK)   One of the elementary particles of which all nucleons (protons and neutrons) are made. (39.1)

radiant energy   Any energy, including heat, light, and X rays, that is transmitted by radiation. It occurs in the form of electromagnetic waves. (23.3)

radiation   (a) The transmission of energy by electromagnetic waves. (23.3) (b) The particles given off by radioactive atoms such as uranium and radium. (39.2)

radioactive   Term applied to an atom with a nucleus that is unstable and that can spontaneously emit a particle and become the nucleus of another element. (39.2)

rarefaction (rayr-uh-FAK-shun)   A disturbance in air (or other matter) in which the pressure is lowered. (26.2)

rate   How fast something happens, or how much something changes per unit of time; a change in a quantity divided by the time it takes for the change to occur. (2.1)

ray   A thin beam of light. (27.6)

ray diagram   A diagram showing the principal rays that can be drawn to determine the size and location of an image formed by a mirror or lens. (30.3)

reaction force   The force that is equal in strength and opposite in direction to the action force and that acts on whatever is exerting the action force. (5.2)

real image   An image that is formed by converging light rays and that can be displayed on a screen. (30.2)

red shift   A decrease in the measured frequency of light (or other radiation) from a receding source; called the *red shift* because the decrease is toward the low-frequency, or red, end of the color spectrum. (25.9)

reflection   The bouncing back of a particle or wave that strikes the boundary between two media. (29.1)

refraction The change in direction of a wave as it crosses the boundary between two media in which it travels at different speeds. (29.6)

regelation The phenomenon of ice melting under pressure and freezing again when the pressure is reduced. (24.7)

relative Regarded in relation to something else. (2.1)

relative humidity A ratio between how much water vapor is in the air, and the limit for the same air temperature. (24.2)

relativistic mass The total mass of a moving object, taking into account any increase in mass due to its kinetic energy. (16.2)

resolution (rez-uh-LOO-shun) The process of breaking up a vector into components. (6.6)

resonance (REZ-uh-nuns) A phenomenon that occurs when the frequency of forced vibrations on an object matches the object's natural frequency, and a dramatic increase in amplitude results. (26.7)

rest mass The mass of an object at rest. (16.2)

resultant (rih-ZUL-tunt) The geometric sum of two vectors. (6.2)

retina (RET-ih-nuh) The layer of light-sensitive tissue at the back of the eye. (30.6)

reverberation (rih-verb-er-AY-shun) The persistance of a sound, as in an echo, due to multiple reflections. (29.5)

revolution Motion in which an object turns about an axis outside the object. (13.1)

rotation The spinning motion that takes place when an object moves about an axis that is located within the object. (13.1)

rotational inertia The resistance of an object to changes in its state of rotation, determined by the distribution of the mass of the object and the location of the axis of rotation or revolution. (14.4)

rotational speed The number of rotations or revolutions per unit of time; often measured in rotations or revolutions per second or per minute. (13.2)

rotational velocity Rotational speed, together with a direction of rotation or revolution. (14.6)

s The symbol for *second*. (2.2)

saturated Term applied to a substance, such as air, that contains the maximum amount of another substance, such as water vapor, at a given temperature. (24.2)

scalar quantity A quantity in physics, such as mass, volume, and time, that can be completely specified by its magnitude, without regard to direction. (6.1)

scaling The study of how size affects the relationship between weight, strength, and surface area. (18.5)

scatter To absorb sound or light and re-emit it in all directions. (28.8)

schematic diagram Diagram that describes an electric circuit, using special symbols to represent different devices in the circuit. (35.5)

scientific method An orderly method for gaining, organizing, and applying new knowledge. (1.3)

second postulate of special relativity The statement that the speed of light in empty space will always have the same value regardless of the motion of the source or the motion of the observer. (15.5)

semiconductor Material that can be made to behave as either a conductor or an insulator of electricity. (32.4)

series circuit An electric circuit in which devices are arranged so that charge flows through each in turn. If one part of the circuit should stop the current, it will stop throughout the circuit. (35.3)

shadow A shaded region that results when light falls on an object and thus cannot reach into the region on the far side of the object. (27.6)

shell model of the atom A model in which the electrons of an atom are pictured as grouped in concentric, spherical shells around the nucleus. (17.8)

shock wave A cone-shaped wave produced by an object moving at supersonic speed through a fluid. (25.11)

sine curve A curve whose shape represents the crests and troughs of a wave traced out by a swinging pendulum that drops a trail of sand over a moving conveyor belt. (25.2)

solar eclipse The cutoff of light from the sun to an observer on the earth when the moon is directly between the sun and the earth. (11.4)

sonic boom The sharp crack heard when the shock wave that sweeps behind a supersonic aircraft reaches the listener. (25.11)

spacetime A combination of space and time, which are viewed in special relativity as two parts of one whole. (15.1)

special theory of relativity The theory, introduced in 1905 by Albert Einstein, that describes how time is affected by motion in space at a constant velocity, and how mass and energy are related. (15.1)

specific heat The quantity of heat required to raise the temperature of a unit mass of a substance by one degree. (21.6)

spectroscope   An instrument used to separate the light from a hot gas or other light source into its constituent frequencies. (28.11)

spectrum   For sunlight and other white light, the spread of colors seen when the light is passed through a prism. In general, the spread of radiation by frequency, so that each frequency appears at a different position. (28.1)

speed   How fast something is moving; the distance moved per unit of time. (2.2)

spring tide   A high or low tide that occurs when the sun, earth, and moon are all lined up, so that the tides due to the sun and moon coincide, making the high tides higher than average and the low tides lower than average. (11.4)

stable equilibrium   The state of an object balanced so that any small rotation raises its center of gravity. (9.5)

standing wave   Wave in which parts of the wave remain stationary and the wave appears not to be traveling. The result of interference between an incident (original) wave and a reflected wave. (25.8)

state   One of the four possible forms of matter: solid, liquid, gas, and plasma. (24.1)

streamline   The smooth path of a small region of fluid in steady flow. (20.7)

strong force   The force that attracts nucleons to each other within the nucleus, and that is very strong at close distances but decreases rapidly as the distance increases. (39.1)

superconductor   Material that has near infinite conductivity at very low temperatures, so that charge flows through it without resistance. (32.4)

support force   Force that completely balances the weight of an object at rest. (4.4)

tangential velocity   For an object orbiting around another object, the sideways component of velocity; that is, the component of velocity parallel to the second object's surface and thus perpendicular to the line joining the centers of the two objects. (10.2)

telescope   Optical instrument that forms enlarged images of very distant objects. (30.5)

temperature   The property of a material that tells how warm or cold it is with respect to some standard. (21.1)

terminal speed   The speed at which the acceleration of a falling object terminates because friction balances the weight. (4.8)

terminal velocity   Terminal speed together with the direction of motion (down for falling objects). (4.8)

terrestrial radiation   Radiant energy emitted from the earth after having been absorbed from the sun. (23.7)

theory   A synthesis of a large body of information that encompasses well-tested and verified hypotheses about certain aspects of the natural world. (1.4)

thermal contact   The state of two or more objects or substances in contact such that it is possible for heat to flow from one object or substance to another. (21.2)

thermal equilibrium   The state of two or more objects or substances in thermal contact when they have reached a common temperature. (21.3)

thermonuclear fusion   A nuclear fusion reaction brought about by extremely high temperatures. (40.6)

thermostat   A type of valve or switch that responds to changes in temperature and that is used to control the temperature of something. (22.1)

time dilation   An observable stretching, or slowing, of time in a frame of reference moving past the observer at a speed approaching the speed of light. (15.1)

torque (TORK)   The tendency of a force to cause rotation about an axis; the product of the force and the lever arm; measured in newton-meters. (14.1)

total internal reflection   The 100% reflection (with no transmission) of light that strikes the boundary between two media at an angle greater than the critical angle. (29.12)

transformer   A device for increasing or decreasing voltage by means of electromagnetic induction. (37.5)

transmutation   The conversion of an atomic nucleus of one element into an atomic nucleus of another element through a loss or gain in the number of protons. (39.6)

transparent   Term applied to materials that allow light to pass through them in straight lines. (27.4)

transverse wave   A wave in which the vibration is at right angles to the direction in which the wave is traveling. (25.5)

trough (TRAWF)   One of the places in a wave where the wave is lowest or the disturbance is greatest in the opposite direction from a crest. (25.2)

ultrasonic   Term applied to sound frequencies above 20 000 hertz, the normal upper limit of human hearing. (26.1)

ultraviolet   Electromagnetic waves of frequencies higher than those of violet light. (27.3)

umbra   The darker part of a shadow where all the light is blocked. (27.6)

unstable equilibrium   The state of an object balanced so that any small rotation lowers its center of gravity. (9.5)

universal constant of gravitation   The constant G in the equation for Newton's law of universal gravitation; changes the units of mass and distance on the right side of the equation to the units of force on the left side. (10.4)

V   The symbol for *volt*. (33.5)   Also, when in lower-case italic, the symbol for *speed* or *velocity*. (2.2, 2.3)   When in upper-case italic, the symbol for *voltage*. (33.5)

vector   An arrow whose length represents the magnitude of a quantity and whose direction represents the direction of the quantity. (6.1)

vector quantity   A quantity in physics, such as force or velocity, that has both magnitude and direction. (6.1)

velocity   Speed together with the direction of motion. (2.3)

vibration   A "wiggle in time"; a repeating, to-and-fro motion of something (such as a pendulum or the particles of an elastic body or a fluid) when displaced from the position of equilibrium. (25.1)

virtual image   An image formed through reflection or refraction that can be seen by an observer but that cannot be projected on a screen because light from the object does not actually come to a focus. (29.3)

volt   The SI unit of electric potential. One volt (symbol V) is the electric potential at which one coulomb of charge would have one joule of potential energy. (33.5)

voltage (VŌL-tij)   (a) Electric potential; measured in volts. (33.5)   (b) Potential difference; measured in volts. (34.1)

voltage source   A device, such as a dry cell, battery, or generator, that provides a potential difference. (34.3)

W   The symbol for *watt*. (8.2)   Also, when in italic, the symbol for *work*. (8.1)

watt (WAWT)   The SI unit of power. One watt of power is expended when one joule of work is done in one second. (8.2)

wave   A "wiggle in space and time"; a disturbance that repeats regularly in space and time and that is transmitted progressively from one particle or region in a medium to the next with no actual transport of matter. (25.1)

wave front   The crest, trough, or any continuous portion of a two-dimensional or three-dimensional wave in which the vibrations are all the same way at the same time (see Figure 29–14). (29.6)

wavelength   The distance from the top of crest of a wave to the top of the following crest, or equivalently, the distance between successive identical parts of the wave. (25.2)

weight   The force on a body of matter due to the gravitational attraction of another body (commonly the earth). (3.5)

weight density   The weight of a substance divided by its volume. (18.2)

white light   Light, such as sunlight, that is a combination of all the colors. Under white light, white objects appear white and colored objects appear in their individual colors. (28.1)

work   The product of the force on an object and the distance through which the object is moved (when the force is constant and the motion takes place in a straight line in the direction of the force); energy expended when the speed of something is increased or when something is moved against the influence of an opposing force; measured in joules. (8.1)

Index

15